Earth´s Materials: Minerals and Rocks

Gautam Sen

Florida International University

Upper Saddle River, New Jersey 07458

Library of Congress Cataloging-in-Publication Data

Sen, Gautam
    Earth's materials: minerals and rocks / Gautam Sen.
    p. cm.
    Includes bibliographical references and index.
    ISBN 0-13-081295-1 (alk. paper)
    1. Minerals. 2. Rocks. I. Title.

  QE363.2 .S46 2001
  549—dc21   00-050214

Senior Editor: **Patrick Lynch**
Assistant Managing Editor: **Beth Sturla**
Manufacturing Manager: **Trudy Pisciotti**
Manufacturing Buyer: **Michael Bell**
Cover Designer: **Jayne Conte**
Production supervision/composition: **Prepare, Inc.**

 © 2001 by Prentice-Hall, Inc.
Upper Saddle River, New Jersey 07458

Printed in the United States of America
10 9 8 7 6 5 4 3 2 1

ISBN 0-13-081295-1

Prentice-Hall International (UK) Limited, *London*
Prentice-Hall of Australia Pty. Limited, *Sydney*
Prentice-Hall Canada Inc., *Toronto*
Prentice-Hall Hispanoamericana, S.A., *Mexico*
Prentice-Hall of India Private Limited, *New Delhi*
Prentice-Hall of Japan, Inc., *Tokyo*
Pearson Education Asia Pte. Ltd.
Editora Prentice-Hall do Brasil, Ltda., *Rio de Janeiro*

To my wife, Sanjukta, and children, Gaurav and Smita, for their
love, enthusiasm, and patience.

Brief Contents

Contents

Chapter 6   Quantitative Analysis of Minerals   159

Chapter 7   Chemical Mineralogy: Mineral Phase Relations   178

Chapter 15   Metamorphism and Metamorphic Rocks   423

Chapter 16   Metamorphic Facies, Reactions, and P–T–t Paths   444

Preface

I had three goals in mind when I began writing this book: (1) to fill a void created by the recent curricular changes in undergraduate education in earth sciences at most colleges and universities, (2) to leave the student with a deep sense of how quantitative this field really is, and (3) to expose the student to exciting new developments in the study of earth materials, particularly in the field of super-high pressure experiments. These amazing experiments have led to paradigm shifts in our understanding of the constitution and origin of the deep earth layers over the last decade; and yet the undergraduate student in an average college or university does not get to feel this sense of excitement even after taking courses in Physical Geology, Mineralogy, and Petrology.

The trend of the 1990's earth science curriculum has been one of a dramatic shift in emphasis from traditional "hard rock geology" to more "environmentally oriented" (near-surface, for most part) geology, primarily in response to a shift in "market demand" motivated by employment. Many universities and colleges have since added courses in Hydrogeology, Remote Sensing, Geographic Information Systems, Environmental Geochemistry, and related fields. This curricular "re-orientation" put enormous pressure on the Earth Science faculty who had to re-design the curriculum so that a student can get a well-rounded education and still graduate within a desirable time frame. Using my department as an example, I feel that the "passive target" of these curricular changes is the field of mineralogy and petrology. Courses on Mineralogy, Optical Mineralogy, Igneous Pterology, Metamorphic Petrology, and Sedimentary Petrology have since been merged into a straight two-semester (or even two-quarter) sequence of courses in many colleges and universities. This forced merger has put the faculty in great difficulty on two fronts: First, a choice must now be made as to the content of the two-semester sequence of courses (and also, in what logical sequence such materials must be taught). Second, what text book(s) should be adopted for the courses. The second issue has been particularly difficult because although there are excellent individual books on petrology, mineralogy, and optical mineralogy, not a single book exists that *adequately* covers all these aspects. Following in the footsteps of the classic 1969 (out of print) reference book on "Earth Materials" by W. Gary Ernst, two such books have since been published, and one of these has been out of print as well. At any rate, in my opinion, these books are essentially descriptive and/or conceptually out-of-date.

The present book is meant to be a up-to-date textbook for a two-semester undergraduate course on "Earth Materials" (or, Min-Pet I and Min-Pet II in some colleges). The fields of mineralogy and petrology are too vast for any textbook to cover all areas. The knowledge base has exploded particularly over the past decade with easy access to high technology and computational facilities. Instead of attempting to cover every possible aspect, I have focused on the theoretical foundation as much as possible with an emphasis on modern research. This book covers important scientific aspects of the study of earth's materials (minerals and rocks) by blending basic descriptive aspects with theory and quantitative analysis. Electron probe analysis of minerals have become a standard practice in this field, and therefore, I have included a chapter on this topic. This will spare the average student from buying another text book in trying to learn about probe analysis.

The reader may feel that the treatment is somewhat "uneven" throughout the text. This was intentional. While much of the text is written for the average student, I had to introduce more quantitative/difficult concepts at places for the students who may have a quest for more advanced knowledge. It is up to the instructor to pick and choose materials from the various chapters for class discussions as he/she deems necessary. I would greatly appreciate comments from students and faculty (no matter how brutal they may be) on the delivery of individual subject matters.

Each chapter starts with a summary of topics discussed within that chapter, and ends with a summary of the major concepts covered. Some worked out examples will help the student in developing quantitative knowledge of the subject. "Information boxes" have been added to illustrate how a simple theory or observation can lead to global scale ideas on planetary composition and evolution. My hope is that the student will recognize that simple observations can lead to elegant and high-impact scientific models provided that one understands the theory well and makes that ever-important connection between laboratory-based theory with observations on natural rocks. "Advanced Study Boxes" have also been added for the advanced undergraduate student. These provide an expanded knowledge of some of the more advanced topics.

The perceptive faculty will easily recognize the emphasis on igneous petrology. This is intentional because of my own orientation in the subject and for the reason that igneous processes provide fundamental clues as to the nature and evolution of planet earth. The text of this book gradually developed over my 15 year academic career at Florida International University and at the National Science Foundation. Its final format greatly benefitted from the comments made by a very perceptive undergraduate student—Mr. Mitchell Gold. Many other students, particularly, Amitava Gangopadhyay, Zachary Atlas, and Shantanu Keshav, and Dr. Huai-Jen Yang (post-doctoral fellow) have read and made comments that helped me revise the text. I would like to thank the following undergraduate students—Tammy Eisner, Darr Halstead, Deanna Hart, David Lopez, and Karim Onsi, a fine motivated group.

I hope that students and faculty at other universities will use this book and will find the time to provide me feedback via e-mail (seng@fiu.edu) at their convenience. I am in the process of developing a website that will have calculation software, current science topics, photo gallery, and other important links that will be of interest to the student/faculty. Based on student concerns, I have made an effort to produce black-and-white diagrams (most of which I drew) in order to make this textbook affordable.

I am particularly grateful to Professors Florentin Maurrasse and Dr. Claudia Owen for going through some of the chapters with a fine comb, and to Professors Grenville Draper and Rosemary Hickey-Vargas for their support. It was Professor Draper who encouraged me to contact the publishers before I started writing this book. My many thanks to Professor Simon Peacock (Arizona State University) for almost totally rewriting the section on "Seismometamorphism". I also thank Diane Pirie for her help with many of the photographs, and Bob Merkel for providing me with a sample of a graphic granite. I thank Iberta Patiño-Douce for permission to reproduce a figure used in this book. A special thanks to the reviewers of this text:

Andrew Wulff,	*Whittier College*
Stewart Farrar,	*Eastern Kentucky University*
Bradley Hacker,	*University of California -Santa Barbara*

Kathleen Johnson,	*University of New Orleans*
John Butler,	*University of Houston*
James Brophy,	*Indiana University.*

I was taught by some of the best teachers in this field, most notably Professors Aniruddha De (Calcutta University), Dean Presnall (University of Texas at Dallas), and W. Gary Ernst (then at University of California, Los Angeles), and their characteristic imprints may be found scattered across this text.

Finally, this book would not have been possible without the enormous support and understanding of my wife, Sanjukta, two terrific children—Gaurav and Smita Kakoli, and a moody cat—Kinko. In spite of the fact that they missed me for much of the last year, they kept their hopes up that someday this project would finish so that we can all go back to leading normal life again!

About the Author

Gautam Sen received his Ph.D from the University of Texas at Dallas and after a Post-doctoral fellowship with Professor Gary Ernst (then at UCLA) joined the faculty of Florida International University. He served as a Program Officer with the Ocean Sciences Division of National Science Foundation and is currently Chair of Earth Sciences at Florida International University.

CHAPTER 1

Introduction

The following topics are covered in this chapter:

Earth and a brief account of its origin
    Physical and chemical properties of the earth
    Origin and early history
Plate tectonics
    Definition of plate
    Plate boundaries and present distribution of plates
    Hot spots and giant mantle plumes
    Wilson cycle
Earth materials: rocks and minerals

ABOUT THIS CHAPTER

This chapter provides a brief review of the concepts about the solid earth that are normally developed in a freshman-level earth science or physical geology course. This review provides the background that is particularly necessary for the "rocks" part of the book. Perhaps the most important among all the concepts is the grand unification idea of *plate tectonics*, which proposes that the earth's outer shell behaves like a composite system of plates that are constantly shifting positions in response to internal forces. Most or all observable phenomena on earth, such as earthquakes and volcanism, are somehow tied to the movement of plates. Some basic and pertinent data on planet earth are first presented, then the concept of plate tectonics is briefly discussed.

SOME BASIC FACTS ABOUT THE EARTH

Earth is one of the four *inner planets* (so called because they are closer to the sun) in our solar system; the others are Mercury, Venus, and Mars. The *outer planets* consist of Jupiter, Saturn, Uranus, Neptune, and Pluto. The outer planets are very cold and largely composed of gases (mostly hydrogen and helium). Pluto, however, is an exception in that it is more like a satellite (such as the moon) in terms of its size and it is composed of a frozen mixture of ice, rock, and gas. An *asteroid belt*, composed of asteroids and gases, separates the inner planets from the outer planets. Most meteorites that fall on earth from outer space are believed to originate from the asteroid belt. They provide important clues to how our solar system formed.

    Earth has the shape of an oblate spheroid (i.e., it is not a perfect sphere but a sphere that is somewhat flattened at the poles), which accounts for its smaller polar radius relative to the equatorial radius (Table 1.1). Studies of earthquake-generated waves, or *seismic waves*, indicate that the earth has a concentrically layered structure. When a major earthquake occurs, some seismic waves generated from it can travel through the entire interior of the earth. Along their paths, such waves bend (refract) or reflect as they encounter layers having different physical properties (e.g., differing densities). A boundary between any two such layers is referred to as a *seismic discontinuity*.

TABLE 1.1   *Some basic parameters of the earth.*

Mean density	5.515 g cm^{-3}
Surface gravity	9.8202 m s^{-2}
Mass	5.9736 $\times$ 10^{24} kg
Mean radius	6371.01 $\pm$ 0.02 km
Equatorial radius	6378.136 km
Polar radius	6356.752 km
Thicknesses of layers	
Inner core	1215 km
Outer core	2159 km
D″ layer	200–400 km
Lower mantle	2331 km
Transition zone	260 km
Upper mantle	392 km beneath mature ocean basins
	360 km beneath stable continental shields
Oceanic crust	6–8 km
Continental crust	35–45 km
Densities of common crystal rocks	
Sedimentary	2–2.8
Metamorphic	2.6–3.0
Igneous	2.5–3.1
Density of upper mantle (Lherzolite)	3.3–3.6 g cm^{-3}

Based on the existence of seismic discontinuities, the following major layers within the earth are recognized: a solid *inner core* at the center of the earth, surrounded by a molten *outer core*, a rocky *mantle*, and a thin outer layer, called the *crust* (Figure 1.1). A diffuse layer of uneven thickness (200–400 km) occurs between the mantle and the core. This layer is called the *D″ layer*. Laboratory studies of seismic wave velocities through different planet-forming materials, along with other types of studies, allows the scientist to evaluate the composition of the earth's internal layers. Based on such evaluations, it appears that the inner core is made of metallic iron, whereas the outer core is likely composed of a mixture of iron and iron sulfide, iron oxide, or other compounds of iron. The mantle and crust are dominantly made of silicate minerals (minerals that are dominantly composed of silicon and oxygen).

The mantle itself has several distinct layers within it. Of particular significance is the *transition zone* that occurs between 400 km and 670 km discontinuities. The transition zone divides the earth's mantle into a shallower *upper mantle* and a deeper *lower mantle*. It is generally believed that the upper and lower mantles have different compositions: the mineral *olivine* is the dominant constituent of the upper mantle, and another silicate mineral, *perovskite*, is the principal component of the lower mantle. Until recently most scientists accepted that the upper and lower mantles were convecting independently of each other, and the transition zone served as a boundary layer through which material and heat are episodically exchanged by the two mantle layers. However, this two-layered mantle convection model has recently been challenged on the basis of new tomographic images of the mantle. The newest models, proposed by a number of groups (Kellogg et al. 1999; Kaneshima and Helffrich 1999; van der Hilst and Karason 1999), suggest that the boundary between the two convecting layers is probably determined by a chemical change (i.e., neither

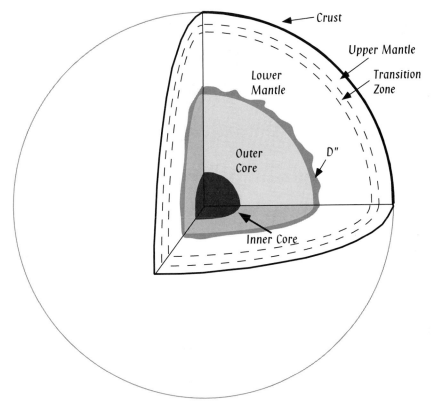

FIGURE 1.1   *Earth's internal layered structure.* (See Table 1.1 for thickness and the text for composition of the individual layers.)

associated with a phase change nor a thermal effect), and it lies somewhere around 1400–1600 km (mid-lower mantle).

The earth's *crust* is significantly richer in silica and lighter than the mantle. The crust that forms continents is considerably thicker, lighter, and richer in silica, soda, and potash than the oceanic crust. Rocks or minerals older than approximately 270 million years of age do not occur in the oceans, whereas much older rocks occur in all continents. The oldest continental rocks are about 3.6–3.8 billion years of age.

EARTH'S BEGINNING

The sun, the earth, and the other planets of the *solar system* are all believed to have formed from the *solar nebula*, a rotating, disk-shaped cloud of dust and gas, some 5 billion years ago. Much of the mass of the nebula quickly concentrated in the center due to gravitational attraction of the particles to each other, leading to the birth of the "baby" sun, or *protosun* (Figure 1.2). As the protosun grew, it began to compress under its own weight and became hot, which eventually led to the start of *nuclear fusion* at around 1,000,000 °C. The nuclear fusion that continues to occur today in the sun to provide its energy is the same reaction that occurs in a hydrogen bomb explosion; it involves the "fusion" of two atoms of hydrogen to form an atom of helium. The heat and solar wind during the formation of the protosun drove away the bulk of the gases to the cooler, outer reaches of the solar system where they eventually formed gaseous giant outer planets, such as Jupiter and Saturn, and ice giants, such as Uranus and Neptune. Pluto's origin is somewhat complicated. Many scientists

1. Cloud rotates
   more rapidly
   as it contracts

2. Cloud flattens
   to pancake-like
   form

3. Rings form

4. Planets form
   at their present
   distances from sun

FIGURE 1.2   *Formation of the solar system from a nebula.* (From Davidson et al., 1997, *Exploring Earth: An Introduction to Physical Geology*, Prentice Hall)

believe that Pluto is a captured comet or a "leftover planetesimal" (defined in the next paragraph) that did not grow to sufficiently large size.

As the inner parts of the solar nebula cooled, refractory elements (those that melt at very high temperatures) formed by the condensation of vapor at about 1800 K and 10^{-4} bar pressure. Refractory metals, such as iridium and osmium, condensed first at approximately 1800 K. Refractory oxides of Ca, Al, Mg, and Ti were next to form as condensates at around 1700 K (Table 1.2). Metallic iron (Fe) and nickel (Ni) then condensed from the cooling nebula, followed later by silicates. Particles formed of these elements eventually clumped (or *accreted*) together to form meter-sized bodies, which further accreted to form asteroid-sized to Mars-sized bodies, called *planetesimals*. The larger planetesimals "swept up" the smaller ones by their gravitational pull and became planets. Thus, some 4.6 billion years ago the earliest, rocky inner planets (also called *terrestrial planets*), including *protoearth*, were formed.

TABLE 1.2   *Approximate condensation temperatures of materials forming in the solar nebula at 10^{-4} bars pressure.* (Grossman and Larimer 1974 & Brown & Mussett 1993).

Mineral	Chemistry	Temp. (K)	Nature
Corundum	Al_2O_3	1680	More Refractory
Perovskite	$CaTiO_3$	1560	
Melilite	$Ca_2Al_2SiO_7$–$CaMgSi_2O_7$	1470	
Diopside	$CaMgSi_2O_6$	1410	↑
Spinel	$MgAl_2O_4$	1390	
Iron metal	Fe(Ni)	1380	↓
Forsterite	Mg_2SiO_4	1370	
Enstatite	$MgSiO_3$	1360	Less Refractory
Anorthite	$CaAl_2Si_2O_8$	1230	
Alkali feldspars	$(Na,K)AlSi_3O_8$	1060	
Troilite	FeS	650	More Volatile
Magnetite	$FeO.Fe_2O_3$	410	
Hydrous silicates		300	↑
Water ice	H_2O	240	
Ammonia ice	$NH_3.H_2O$	130	↓
Methane ice	$CH_4.6H_2O$	90	
Nitrogen ice	$N_2.6H_2O$	90	Less Volatile

ORIGIN OF THE EARTH'S INTERNAL LAYERING

How and when the earth developed its internal layering are topics that continue to intrigue the student of earth sciences. It is commonly accepted that the earth's internal layering developed very early—perhaps within the first 500 million years or less. However, the record of such early history is not preserved in rocks and must be inferred from meteorites, because many meteorites contain the same materials that composed the protoearth. Broadly speaking, two classes of hypotheses, commonly referred to as the homogeneous accretion and heterogeneous accretion hypotheses, have been put forward about the origin of the earth's internal layering. The majority of scientists appear to believe that the protoearth became very hot as it grew in size through continuous bombardment (and absorption) of asteroids and meteorites. The heating resulted mainly from the conversion of mass as the planet became dense due to compression under its own weight and from the decay of short-lived radioactive isotopes. In the *homogeneous accretion model*, it is believed that the earth started out as a homogeneous body composed of variously sized planetesimals and some trapped gases, and the intense internal heating subsequently led to the melting of iron metal, which segregated to the earth's center to form the core (Figure 1.3). The term *"iron catastrophe"* has been used to describe the iron melting event. The process through which layers developed from an initially homogeneous body is called *planetary differentiation.*

The *heterogeneous accretion model* calls for the accretion of already differentiated planetesimals. The fact that individual meteorites consist of many different phase types (e.g., minerals, chondrules, and metal) has led some scientists to propose that the core formed by accretion of planetesimals in which metallic iron had already separated from silicates. The presence of strong isotopic compositional variations in meteorites has been used by many scientists to argue for the existence of strong heterogeneities,

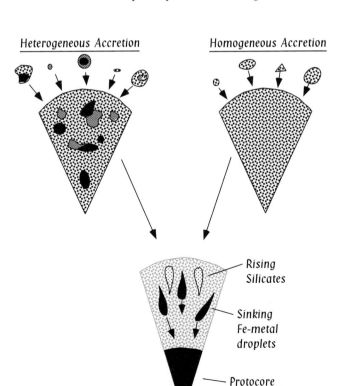

Heterogeneous Accretion   Homogeneous Accretion

Rising Silicates

Sinking Fe-metal droplets

Protocore

FIGURE 1.3 *Origin of the earth by accretion of planet-forming materials.* The heterogeneous accretion model calls for the accretion of already differentiated asteroid-like bodies, whereas the homogeneous accretion model suggests the formation of protoearth from undifferentiated planetesimals. The newly accreted protoearth was much smaller in size than the earth we know today. Early in its history, the earth underwent a major heating event in which metallic iron melted and sank to form the core, and the lighter silicates "floated up to" or remained at higher levels.

even in the solar nebula (Taylor 1992). The existence of distinct reservoirs within the earth, as required by a number of different radioactive isotopic systems, also supports the maintenance of strong vertical and lateral heterogeneities within the earth since its beginning some 4.6 billion years ago. Such heterogeneities may have had their origin in the accretion history. The heterogeneous accretion model does not rule out the former occurrence of an iron catastrophe; it only proposes that the materials which accreted were already differentiated and thus, the earth had a heterogeneous beginning.

Many scientists believe that at some point (perhaps 4.4 billion years ago) the earth's surface became largely molten and an approximately 450 km deep *magma ocean* formed (Figure 1.4). Some authors have suggested that the mineral olivine floated in this magma ocean to ultimately form an olivine-rich upper mantle, whereas the silicate mineral perovskite sank and formed the lower mantle. Some authors have also sought a link (direct as well as indirect) between this "surficial" magma ocean and the sinking of molten droplets of iron (Fe) and iron sulfide (FeS) from the bottom portion of this magma ocean to the core by percolation through the partially molten zone beneath the magma ocean. The magma ocean hypothesis has not gained universal acceptance, and therefore, many questions about it, which are beyond the scope of this text, remain unresolved.

Eventually the magma ocean cooled enough so that a solid, rigid "skin" (called the *lithosphere*) of igneous rocks formed on the surface. This early lithosphere may have had a thin crust composed of solidified ultramafic lavas, called *komatiite*. This idea is based on the fact that komatiite is almost exclusively found in the oldest terranes on continents. The lower part of the lithosphere was made of *peridotite*, an olivine-rich coarse-grained rock that is the dominant component of the earth's upper mantle today. A highly uneven lithosphere, with a crust that is very thick (35 km) at some places but thin in others, may have developed around 4.3 billion years ago. Even today the world's oceanic crust is somewhat uneven. On the average it is about 8 km thick, but in the so-called oceanic plateaus (for example, the Ontong–Java plateau in the Pacific Ocean) and in some seas (e.g., the Caribbean Sea) it is significantly thicker (about 12–30 km).

The history of continents is truly an interesting problem. The discovery of a 4.2 billion year old mineral zircon in some very old sedimentary rocks from Australia offers an interesting insight into early continent formation. This mineral is a common but minor constituent of granitoid rocks; and granitoids in turn are a fundamental component of continental crust. Therefore, the finding of such an old zircon suggests that at least some continental crust had already formed by 4.2 billion years ago! In general, continents are much older than oceans. Whereas the oldest continental rocks are generally 3.4–3.8 billion years old, the oldest oceanic crustal rocks are only about 270 million years old. The theory of plate tectonics attributes this age difference to the constant recycling of the older oceanic crust back into the mantle through a process known as "subduction," whereas the continental crust, being less dense, is unsubductible and collects on the surface of this dynamic planet like froth on milk. In summary, it is evident that continents were well in existence within the first billion years of earth's formation from the solar nebula.

The mineral diamond has provided an additional critical clue to the puzzle of continental lithosphere thickness and temperature (Haggerty 1999). Diamond is found in an igneous rock, called *kimberlite*, in the oldest parts of continents where the rocks are generally older than 2.7 billion years. Silicate mineral inclusions in some diamonds date back to 3.5 billion years and appear to have been brought up from a depth of around 200 km where the temperature was close to 950°C. This temperature is not much higher than the temperatures estimated to exist within the present-day continental lithosphere. Thus, one is forced to conclude that cool, thick, continental lithosphere

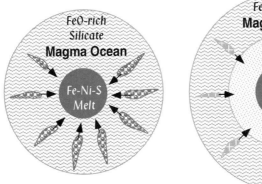

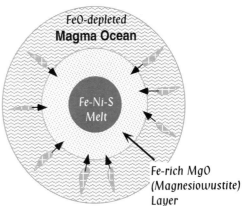

a. *Protocore formation with a silicate magma ocean envelope. Planet size smaller than present.*

b. *Formation and segregation of a Fe-rich magnesiowustite (Mg,Fe)O. Planet size is more-or-less the same as today.*

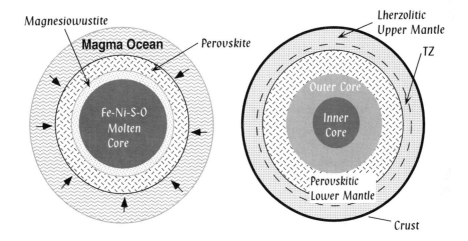

c. *Silicate perovskite [(Mg,Fe)SiO]₃ fractionation and the formation of the lower mantle*

d. *Flotation of olivine and eventual crystallization of magma ocean to form the upper mantle and transition zone (TZ; top of TZ is dashed)*

FIGURE 1.4 *A recently proposed hypothesis by Agee (1990) that calls for a complicated evolutionary history of the earth.* (The diagrams here are somewhat simplified from the original author's work.) (a) Early formation of a FeO-rich silicate magma ocean and of the molten protocore. (b) At some point the mineral magnesiowustite (MgO with a considerable amount of Fe in solution) formed and segregated to form a shell around the core. Separation of this mineral from the magma ocean caused a depletion in FeO in the magma ocean. (c) Then, the silicate mineral perovskite formed [composition-(Mg,Fe)SiO₃], which is the dominant constituent of the lower mantle today. The magnesiowustite layer became smaller because it lost iron to the core. The core became enlarged. Note that the stages (b) and (c) may have happened sequentially or simultaneously. (d) The magma ocean solidified to form the mantle and crust. The inner core solidified as well. Note that for brevity, subduction and other plate tectonic processes are not shown in these diagrams. (From Agee, 1990, *Nature* 346, Mcmillan and Co.)

existed 1 billion years after the birth of the planet. The inclusions in diamonds further tell us that such primitive continental lithosphere was already 200 km thick.

Water molecules probably did not form around this time because temperatures at the surface of the earth were still much above the boiling point of water. However, it is likely that steam from the primitive mantle escaped and formed a thin envelope of atmosphere around the planet. Many authors also believe that comets crashing into earth during its early history may have provided some of the water.

PLATE TECTONICS

In the Beginning

It is likely that the early lithosphere, however unevenly thick, did not remain static: hot mantle jets or plumes beneath the lithosphere kept poking it until it cracked. Through these cracks magma gushed out and solidified in the crack as the cracked lithospheric "plates" were carried away from each other by the convection currents in the hot mantle below. At the other ends, these plates collided with each other, one being pushed up over the other. Thus began what we call *plate tectonics*.

It is likely that the earliest lithospheric plates were constantly breaking up and colliding with each other. Plate movements were quite vigorous and were driven by strong convection currents and hot spots in the mantle. Although true water-filled ocean basins probably had not yet developed, the subduction of "proto-oceanic" lithosphere, similar to the present day situation, was occurring where the lithosphere was relatively thin. By virtue of having an overall relatively lower density, the early continental lithosphere resisted subduction much like the way it has throughout earth's history. About 4 billion years ago, the conditions became right, such that water was in sufficiently high atmospheric concentration to precipitate as rain and fill up the "basins" formed by the thinner lithospheric basins surrounded by the high continents. Thus, we finally had the beginning of ocean formation. At that point, earth looked vaguely familiar: tiny but thick continental plates; giant, thin oceanic plates; and abundant hot-spot–generated volcanoes probably dotted this planet.

Modern Scenario: A Brief Presentation

Plates and Plate Boundaries

Before proceeding further on the theory of plate tectonics, it is pertinent to examine the concept of the lithosphere, which plays a fundamental role in the theory. The term "lithosphere," as commonly used, has a mechanical connotation that stemmed from Barrell's (1914) idea of a "strong outer layer (about 100 times stronger than the asthenosphere) overlying a weak asthenosphere that could flow to maintain isostatic compensation" (cited in Anderson 1995, p. 125). Although there are thermal (as defined by the contrasting thermal behaviors of the lithosphere and the asthenosphere) and chemical definitions of the lithosphere, in this chapter we follow the mechanical definition of the lithosphere in the most general sense. Because the mechanical behavior of rocks, specifically their relative brittle and ductile behavior, is related to their mineralogy, temperature, stress, and strain rate, the thickness of the lithosphere (as defined in a mechanical sense) must also vary throughout the globe. The average thickness of the modern oceanic lithosphere is about 100 km and that of continental lithosphere is about 150 km.

According to the plate tectonics theory, the lithosphere is not laterally continuous but is broken into eight large pieces or "plates" and many small plates (Figure 1.5). Size

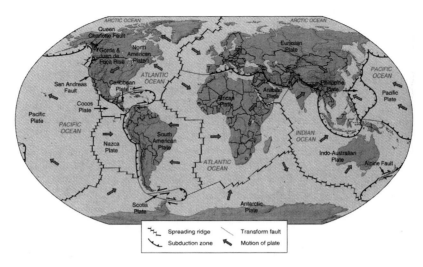

FIGURE 1.5   *Earth's surface showing the major plates.* (From Davidson et al., 1997, *Exploring Earth: An Introduction to Physical Geology,* Prentice Hall)

and shape of the plates change with geological time. Boundaries of each of these plates are based on earthquakes and volcanism. Three types of plate boundaries are recognized:

1. *Divergent plate boundary.*   Along such a boundary, two plates move apart from each other as new lithospheric material is added by magmas generated at depth (Figure 1.6a).

2. *Convergent plate boundary.*   Along such a boundary, two plates converge, one of which gets "subducted" or consumed back into the mantle as the other overrides it. In oceanic plate–continental plate convergence zones, it is the oceanic plate that always subducts into the mantle because the continental lithosphere is lighter than the oceanic plate (e.g., subduction of the Nazca plate beneath the South American plate) (Figure 1.6b). In case of continent–continent plate convergence, one of the plates slides under the other but because both are buoyant, enormously thickened continental mass results. The Himalayan mountains formed this way during the collision of the Indian plate with the Eurasian plate.

3. *Transform boundaries.*   Along such boundaries, the adjacent plates do not converge or diverge but simply move past each other along large faults (Figure 1.6c).

The best example of plate divergence is a mid-oceanic ridge, such as the Mid-Atlantic Ridge. (See Figure 1.5.) The earth's internal forces cause the seafloor to passively "spread" in opposite directions along the ridge axis (or *spreading center*). New magma batches generated down below well up at the ridge axis and solidify to hard rock, creating new lithospheric material. The seafloor spreading rate can vary considerably from one ocean to another and even along the same ridge axis! A fast spreading rate is 11 cm per year, whereas 1 cm per year is considered very slow.

Plate convergence zones are typically associated with explosive volcanic activity. The most prominent zone is the "Pacific Ring of Fire," which marks the volcanic boundary along the entire boundary of the Pacific plate. Though relatively rare, deepest focus earthquakes (around 650 km) are found only along such boundaries, whereas they are absent in plate divergent boundaries.

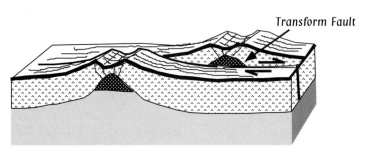

(a) Divergent Plate Boundary

(b) Convergent Plate Boundary

(c) Transform Fault Boundary

FIGURE 1.6   *Types of plate boundary.* (a) Divergent plate boundary. The schematic drawing shows the separation of two plates (plates A and B) at a mid-oceanic ridge axis, where magma formed in the asthenosphere ascends and makes new crust at the ridge axis. (b) Convergent plate boundary. This example shows the convergence of two oceanic plates, C and D. C is subducting underneath D. A trench zone marks the plate boundary. Magma production above the subduction zone generates a volcanic front on the overriding plate. (c) Transform Fault. Along some plate boundaries the plates may not separate from or converge toward each other, but simply slide past one another. A transform fault is a special type of fault that offsets mid-oceanic ridge axis segments. The famous San Andreas fault, which runs through the state of California, is such a fault boundary along which the North American plate is sliding past the Pacific plate. San Francisco is located on the North American plate, whereas Los Angeles is located on the Pacific plate and is moving northward.

Transform fault boundaries are typically found in ocean basins where a mid-ocean ridge axis is not continuous but is offset by a series of transform faults (Figure 1.6c). These faults develop due to the curvature of the earth's surface and differential spreading rates along the ridge axis. An important transform fault on land is the San Andreas fault in California, along which the Pacific plate is sliding past the North American plate along a northwesterly direction.

Plate boundaries can change over geologic time: some 200 million years ago, all the continents formed a single, giant landmass that was given the name *Pangaea*. This single plate eventually broke into many continental plates, each of which subsequently underwent modifications due to plate tectonic processes. Oceans can "close and open" as well, several times! It is clear that the Atlantic Ocean closed at least once before, during which the North American, Eurasian, and African plates collided and created the Appalachian mountains.

Hot Spots and Giant Mantle Plumes

There are many areas on earth where active volcanism is occurring in the middle of a plate rather than at a plate boundary. The best example of this is the "Big Island" of Hawaii in the middle of the Pacific plate, where two active volcanoes, Kilauea and Mauna Loa, have been pouring out large quantities of basaltic lavas. It is generally accepted that there exists an anomalously hot area (or *hot spot*) in the mantle beneath Hawaii. This hot spot has remained essentially stationary for at least the last 90 million years and has been supplying magmas to the Hawaiian–Emperor island-seamount chain. The volcanic chain formed due to the passage of the oceanic lithosphere over a relatively fixed hot spot or plume (Figure 1.7). There are other hot-spot tracks in the Pacific that parallel the Hawaiian–Emperor track, and all are believed to have resulted from hot-spot volcanism. A prominent bend occurs in the Hawaiian–Emperor track, which corresponds to an abrupt change in the direction of movement of the Pacific plate that occurred 40 million years ago.

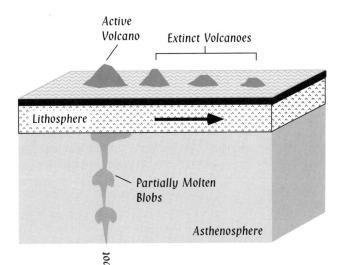

FIGURE 1.7 *Formation of intraplate volcanic chains, such as the Hawaiian–Emperor volcanic chain in the middle of the Pacific plate, is best explained by the "hot spot" theory.* It is believed that a hot spot has been residing beneath Hawaii for at least 90 million years and supplying magmas to form the volcanic islands. As the Pacific plate migrates away, a new volcano begins to form atop the hot spot, and the old one becomes extinct and migrates away as a "piggyback" rider. Such volcanic eruptions over millions of years of geologic time give rise to a chain of volcanic islands or seamounts.

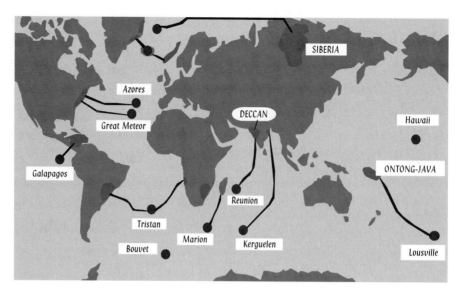

FIGURE 1.8 *Some well known large igneous provinces and their allegedly related hot spot tracks are shown.*

Throughout the earth's geologic past, there have been episodes of massive outpourings of basaltic lavas on the ocean floor and on continents. The large geographic plateaus that were formed by such volcanism have been variously called *flood-basalt provinces* or *large igneous provinces*. In some ways, they are anomalous because their estimated eruption rates were generally very high relative to the rate at which basalt magma erupts at mid-oceanic ridges today. They occur in the middle of a plate, and, in most cases of continental eruptions, they appear to occur on the edge of the continent. Some notable examples of continental flood-basalt provinces are the Deccan Traps (India), Parana basalts (South America), Karoo basalts (Southern Africa), and Columbia River basalts (western United States). The Ontong–Java and Manihiki plateaus (Pacific Ocean) as well as the Caribbean Sea floor are thought to be oceanic examples of flood-basalt provinces. These large igneous provinces are generally accepted to result from partial melting of giant plume "heads" that rise from the deep mantle (Richards et al. 1989). Some scientists believe that the narrow "tails" of such giant plumes continue to supply magmas for millions of years after the plume head "melts" away, producing hot spot tracks (Figure 1.8; discussed further in Chapter 11). While some authors believe that such plume eruptions are responsible for breaking up single continental plates and for mass extinctions, there are many examples where such hypotheses cannot work. Nonetheless, eruption of flood basalts remains a topic of considerable interest.

Wilson Cycle

The late J. Tuzo Wilson, a Canadian geologist, provided a unifying hypothesis that relates plume activity, seafloor spreading, and subduction to continental breakup, ocean formation, and growth of continents. This is known as the *Wilson cycle*. The cycle begins with a plume or hot spot impinging upon the base of a continental plate. This causes the overlying lithosphere to "dome up" (Figure 1.9a). Such uplift and heating of the lithosphere initiates thermal cracking and rift valley formation in the shape of a three-armed star (like the Mercedes–Benz logo: Figure 1.9b). Volcanic activity and continued rifting cause two of the rift systems to spread away from each other, thus resulting in the formation of a new ocean basin (Figure 1.9c,d). The third arm fails to

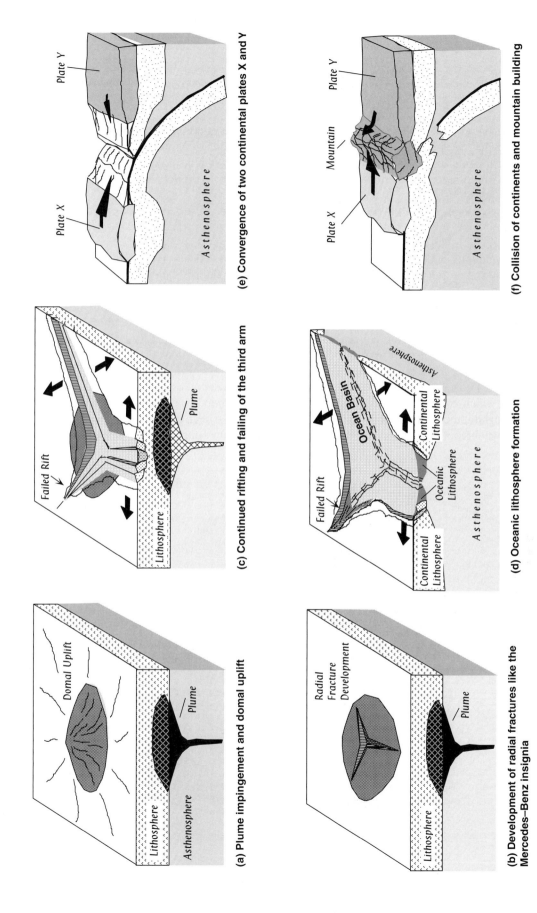

(a) Plume impingement and domal uplift

(b) Development of radial fractures like the Mercedes–Benz insignia

(c) Continued rifting and failing of the third arm

(d) Oceanic lithosphere formation

(e) Convergence of two continental plates X and Y

(f) Collision of continents and mountain building

FIGURE 1.9 *The Wilson cycle.* (See the text for a detailed discussion of these diagrams.)

rift further. The formation of the Red Sea, the Gulf of Aden, and the Ethiopian rift system in the Middle East is commonly cited as the best example of the early stage of the Wilson cycle. As the oceanic plates in the newly formed oceanic-ridge system spread away from each other, they cool, thicken, and become heavy. After about 200 million years or so, the oceanic lithosphere becomes heavy enough to break and begin to subduct. Such subduction may ultimately result in collision of two continents, and the sediments accumulating at the leading edge of the two continents would deform and form a large folded mountain belt between the two continents (Figure 1.9e,f). Such mountain building "bonds" the two previously separate continents into a single large continent (Figure 1.9f). This continent may undergo the entire Wilson cycle again at some point in its future.

As an aside, it should be noted that Anderson suggested a different mechanism for the initiation of the Wilson cycle (Anderson 1992). Specifically, his model does not require the presence of a plume or a hot spot to trigger the rift valley formation and continental splitting. Anderson pointed out that the continental crust is a good insulator. Therefore, large continental masses (called *supercontinents*), such as Pangaea (Figure 1.10), can effectively trap heat in the underlying mantle over geologic time. Such heat buildup in the mantle eventually triggers melting, which leads to immense flood-basalt volcanism, rifting, and the formation of new divergent plate boundaries (Figure 1.11). Anderson's model is based on the theoretical models developed by Michael Gurnis who showed that single large continental masses can effectively trap heat, which eventually leads to the breakup and dispersal of continental plates. Throughout the earth's geologic past, there appear to have been only three times when large supercontinents existed—Pangaea (which broke up approximately 200 million

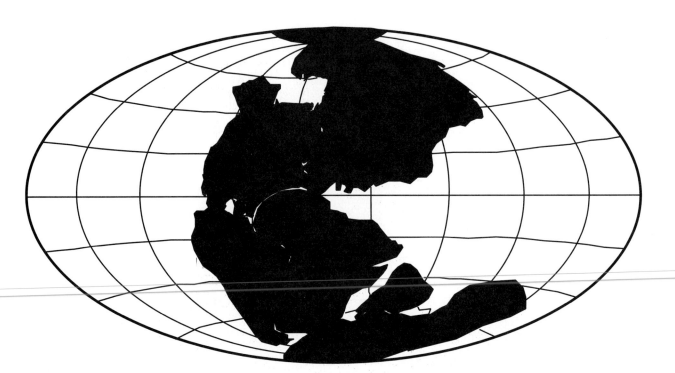

FIGURE 1.10 *Pangaea*—a supercontinent that is believed to have split up around 200 million years ago.

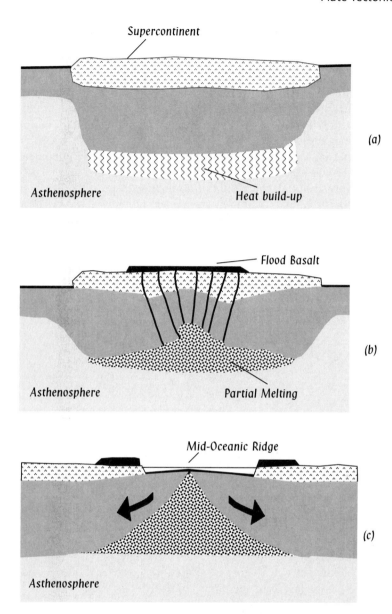

FIGURE 1.11 *Anderson's model of flood-basalt volcanism, splitting of continents, and formation of ocean basins.* Heat builds up in the mantle beneath large continental landmasses (like Pangaea), which leads to doming of the lithosphere and flood-basalt volcanism. Finally, rifting and seafloor spreading take over, and the ocean basin forms.

years ago [Ma]), Rodinia (split up approximately 800 Ma), and Mid-Proterozoic Supercontinent (split up around 1.4–1.5 billion years ago) (Condie and Sloan 1997; Larson 1995). Roger Larson invoked a model that involves large "superplumes" that rise from the deep mantle and break up such supercontinents (imagine a plume that is the size of a continent—4000 km wide and 2000 km in height; Larson 1995). The Larson superplume model is clearly very different from Anderson's shallow melting model.

STUDY OF EARTH MATERIALS

The solid earth has been studied by geologists in a number of different ways. For example, the subject of *seismology* deals with earthquakes and allows us an unparalleled insight into the deeper earth structure and processes. *Paleontology* (the study of fossils) is concerned with ancient life on earth. Similarly, the study of minerals and rocks that constitute the solid earth gives us fundamental information on the constitution of the earth and on such diverse processes as volcanism, sedimentation, and the earth's internal forces that are responsible for uplift and denudation of mountains. *Mineralogy* is a subject that deals with minerals; and the study of rocks is referred to as *petrology*. Although earth's materials include everything that comprise the earth, the focus of this book is on minerals and rocks. *Crystallography*, the study of crystals, is an essential subfield within mineralogy because all minerals are composed of crystals. In "real life," however, crystallography as a subject not only includes crystal structures of minerals but also the structures of complex organic molecules. An exhaustive coverage of mineralogy, crystallography, and petrology is well beyond the scope of any book. This book is structured to provide a useful amount of coverage of all three fields and of modern analytical tools used to study minerals and rocks, with additional (but brief) focus on frontiers in mineralogy and petrology.

SUMMARY

[1] Earth is one of the rocky inner planets, which formed about 4.6 billion years ago.

[2] The velocity of a seismic wave changes abruptly as it passes through layers of contrasting physical properties. Such boundary layers are called seismic discontinuities.

[3] Seismic discontinuities require that the earth is composed of several concentric layers or shells that are very different in density.

[4] The principal layers of the earth are the inner core, the outer core, the lower mantle, the transition zone, the upper mantle, and the crust. The lower mantle is (probably) chemically stratified into two fairly distinct layers; and the boundary between the upper and lower layers occurs at approximately 1600 km. The earth's core is dominantly composed of metallic iron, whereas the mantle and crust are almost entirely composed of silicate minerals.

[5] The core differentiated from the mantle and crust by an initial melting episode that occurred early in the earth's history.

[6] Further differentiation within the mantle and crust occurred during the formation of a global magma ocean.

[7] The oldest continental rocks and minerals are significantly older than oceanic rocks. Such an observation is compatible with the idea that oceanic rocks are destroyed due to their recycling back into the mantle via plate tectonics, whereas unsubductible continental rocks collect on the surface.

[8] Very old mineral inclusions in diamond indicate that some parts of the continental lithosphere had become approximately 200 km thick some 3.5 billion years ago.

[9] Earth's seismic and volcanic activities are best explained in terms of the unifying hypothesis of plate tectonics. Plates include continental and oceanic masses. They are created at divergent plate boundaries and are destroyed via subduction at sites of plate convergence.

[10] Continental plates grow by accretion and are broken up by plume activity. The Wilson cycle explains the evolution of continents and oceans in terms of plate tectonics. The cycle begins with the impingement of a hot spot or plume onto the base of the continental lithosphere. Eventually the lithosphere breaks into three radial rift zones, two of which eventually become ocean basins. The collision of continents related to subduction is the end of the cycle. In Anderson's model, continental splitting results in flood-basalt volcanism, which, according to him, is due to long-term entrapment of heat in the subcontinental lithosphere rather than plume impingement.

BIBLIOGRAPHY

Agee, C.B. (1990) "A new look at differentiation of the earth from melting experiments on the Allende meteorite." *Nature* **346**, 834–837.

Anderson, D.L. (1995) "Lithosphere, asthenosphere, and perisphere." *Rev. Geophys.* **33**, 125–149.

Anderson, D.L. (1996) "Enriched asthenosphere and depleted plumes." *Internat. Geol. Rev.* **38**, 1–21.

Brown, G.E. and Mussett, A.E. (1993) *The Inaccessible Earth: An Integrated View of its Structure and Composition* (2nd ed.). Chapman & Hall (New York).

de Wit, M.J. and 8 others (1992) "Formation of Archaean continent." *Nature* **357**, 553–562.

Condie, K.C. and Sloan, R.E. (1998) *Origin and Evolution of Earth: Principles of Historical Geology.* Prentice Hall (New Jersey).

Davidson, J. Reed, W.E., and Davis, P.M. (1997) *Exploring Earth.* Prentice Hall (New York).

Duncan, R.A. and Richards, M.A. (1991) "Hot spots, mantle plumes, flood basalts, and true polar wander." *Rev. Geophys.* **29**, 31–50.

Frankel, H. (1988) "From continental drift to plate tectonics." *Nature* **335**, 127–130.

Grossman, L. and Larimer, J.W. (1973) "Early chemical history of the solar system." *Rev. Geophys. Space Phys.* **12**, 71–101.

Gurnis, M. (1988) "Large-scale mantle convection and the aggregation and dispersal of continents." *Nature* **332**, 695–699.

Haggerty, S.E. (1999) "A diamond trilogy: superplumes, supercontinents, and supernovae." *Science* **285**, 851–860.

Kaneshima, S. and Helfrich, G. (1999) "Dipping low-velocity layer in the mid-lower mantle: evidence for geochemical heterogeneity." *Science* **283**, 1888–1891.

Kellogg, L.H., Hager, B.H., and van der Hilst, R.D. (1999) "Compositional stratification in the deep mantle." *Science* **283**, 1881–1884.

Larson, R.L. (1991) "Latest pulse of the earth: evidence for a mid-Cretaceous superplume." *Geology* **19**, 547–550.

Larson, R.L. and Kincaid, C. (1996) "Onset of mid-Cretaceous volcanism by elevation of the 670 km thermal boundary layer." *Geology* **24**, 551–554.

Lay, T. (1994) "The fate of descending slabs." *Ann. Rev. Earth Planet. Sci.* **22**, 33–61.

Polet, J. and Anderson, D.L. (1995) "Depth extent of cratons as inferred from tomographic studies." *Geology* **23**, 205–208.

Sen, G. (1993) "Earth's mantle, plumes, and volcanism in Hawaii." *Proc. Nat. Acad. Sci. India* **63(A)**, I, 15–32.

Stoffler, D. (1997) "Minerals in the deep earth: a message from the asteroid belt." *Science* **278**, 1576–1577.

Taylor, S.R. (1992) "The origin of the earth: In *Understanding the Earth.*" (Brown, G.C. et al., eds.), 25-42.

van der Hilst, R.D. and Karason, H. (1999) "Compositional heterogeneity in the bottom 1000 kilometers of earth's mantle: toward a hybrid convection model." *Science* **283**, 1885–1888.

USEFUL WEBSITES

http://fermat.geol.uconn.edu/Wk1ES.html—For a nice summary of the origin of the solar system.

http://fermat.geol.uconn.edu/Wk2ES.html—For an introduction to seismology, earth structure, and plate tectonics.

Mineralogy: An Introduction

Minerals are basic units of which rocks are composed. **Mineralogy** is the study of minerals, and the scientists who specialize in such studies are called **mineralogists**. Minerals are made of **crystals. Crystallography** is the study of crystals. Mineralogy is an essential subject of earth and materials science disciplines. The first few chapters of this book are devoted to the study of minerals and crystals. They include a cursory look at the organization of atoms in minerals, physical properties of minerals, chemical composition of minerals, and their behavior under the petrographic microscope. Chapter 6 is dedicated to modern instrumentation that is commonly used for mineralogical studies. The last chapter of this section looks at how minerals behave, individually and with other minerals, when subjected to pressure and temperature conditions that are appropriate within the earth where magmas (molten rock) are produced.

CHAPTER 2

Atoms and Crystals

The following topics are covered in this chapter:

Atoms, bonding, and coordination
Crystal symmetry
    Symmetry elements and operations; classes
Crystal systems and crystallographic axes
Unit cell and bravais lattices
Axial ratios, Miller indices, and interfacial angles
Common crystal forms
Imperfections in crystals
Twinning

ABOUT THIS CHAPTER

The earth is made of rocks, minerals, and other matter. Rocks are aggregates of minerals, and minerals are composed of a lattice—a three-dimensional array of atoms. Atoms are fundamental units of which all matter is composed, much like bricks are to buildings. This chapter broadly covers some fundamental aspects of the structure of atoms, of chemical forces or "bonds" that hold atoms in place, and of atomic construction of crystalline solids. An important chapter objective is to bring out the internal regularities (or *symmetry*) and irregularities (*imperfections* or *defects*) in crystal structure because an understanding of the fundamental concepts of crystalline structure is critical to understanding why some minerals behave similarly and others differently: for example, why some minerals are harder than others, why some are more soluble in water than others, and why some are better conductors of heat and electricity than others. After all, such properties of a mineral determine its utility—whether as refractory bricks on a space shuttle or as an electrical wire.

ATOMS AND BONDING

Structure of an Atom

An atom is an extremely small particle whose size is expressed in Angstroms $(\text{Å}; 1\text{Å} = 1/100,000,000 \text{ cm})$. Although there is much new knowledge about the structure of atoms, a relatively simple model describing the structure of an atom, the simplified *Rutherford–Bohr model*, is sufficient for our purpose (Figure 2.1). In this model, an atom is composed of a *nucleus* that holds more than 99.9% of the mass; and the nucleus is surrounded by a cloud of virtually weightless *electrons*. The nucleus is composed of *protons* and *neutrons*: each proton carries one positive charge, whereas neutrons are not charged particles.

Each electron carries a negative charge, weighs about 1/1837 of a proton, and moves at the speed of light. Electrons revolve around the nucleus along concentric shells, or *orbitals*, much like the planets of our solar system revolve around the sun. The positively charged nucleus tries to pull the negatively charged electrons toward it, but its

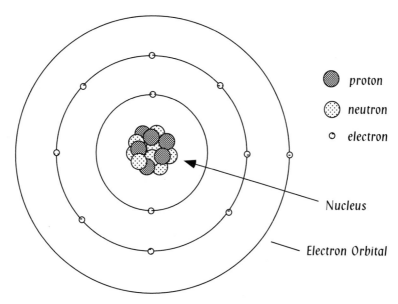

FIGURE 2.1   *A simplified Rutherford–Bohr model of an atom.* The nucleus is composed of protons and neutrons, and electrons revolve around the nucleus in orbitals.

attractive force is countered by the centrifugal force of the revolving electron. Each electron orbital carries an energy. The energy level of an electron orbital shell differs from that of another shell by a discrete amount called a *quanta*. The innermost shell, designated the *K shell* (principal quantum number, $n = 1$), can contain a maximum of 2 electrons. The subsequent outer shells, designated $L(n = 2)$, $M(n = 3)$, $N(n = 4)$, ... Q ($n = 7$), are characterized by increasingly higher energy. L, M, and N shells can contain a maximum of 8, 18, and 32 electrons, respectively. A number of additional complications exist in the atomic structure: for example, L and the other outer shells are composed of a number of subshells, s, p, d, and f, in increasing levels of energy. Table 2.1 shows the

TABLE 2.1   *Electron occupancy in the atoms of some elements.*

Element subshell	K	L		M			N			
	1s	2s	2p	3s	3p	3d	4s	4p	4d	4f
H	1									
He	2									
C	2	2	2							
N	2	2	3							
Na	2	2	6	1						
Mg	2	2	6	2						
Al	2	2	6	2	1					
Si	2	2	6	2	2					
P	2	2	6	2	3					
S	2	2	6	2	4					
Cl	2	2	6	2	5					
K	2	2	6	2	6		1			
Ca	2	2	6	2	6		2			
Ti	2	2	6	2	6	2	2			
Fe	2	2	6	2	6	6	2			

occupancy of electrons in the atoms of some elements that are common in earth materials. From this table it is apparent that s subshell can hold a maximum of 2 electrons, and p subshell can hold 6. Although not apparent in Table 2.1, d and f can hold a maximum of 10 and 14 electrons, respectively.

An atom contains equal number of protons and electrons, and therefore its positive and negative charges cancel each other out, and the atom is said to be electrically neutral. An *ion* is a charged atom in which the number of electrons and number of protons are not equal. When an ion has more electrons than protons, it is negatively charged and is called an *anion* (pronounced "an' ion"). Positively charged ions have more protons than electrons and are called *cations* (pronounced "cat' ion"). The *atomic number*, Z, of an element is its number of protons, and elements are distinguished from one another on the basis of their atomic numbers. An element can have a number of different *isotopes* that have identical number of protons but different number of neutrons. For example, common oxygen (^{16}O) has 8 protons and 8 neutrons; but a heavier isotope of oxygen, known as ^{18}O, has 8 protons and 10 neutrons. While some isotopes are "stable," there are many unstable isotopes in nature that break down into isotopes of different elements emitting radioactive energy. Both stable and unstable isotopes provide very important clues to the earth's geologic past and geologic processes.

CHEMICAL BONDS

The forces that hold atoms together in the atomic structure of a crystal are called bonds. There are four principal types of bonds: metallic, covalent, ionic, and van der Waal's bonds (Figure 2.2). A fifth type of bond, hydrogen bond, is a type of covalent bond specific to water (H_2O) molecules.

Metallic Bond

The bonds that hold the atoms in a metal together are called *metallic bonds*. In a metal, the atoms are so densely spaced that their outer electrons move freely from atom to atom (Figure 2.2). This free mobility of electrons makes a metal an excellent conductor of electricity and heat because their mobile electrons are quick to transport heat and electricity from one point to another through the metal.

Ionic Bond

An *ionic bond* is an electrostatic attraction between a cation and an anion. Such bonding forms when one atom gives up its outer electron(s) to another atom creating the ions needed to form a stable, electrically neutral, molecule. This exchange occurs because of a natural tendency of atoms to fill their outermost shells with the maximum number of electrons they can hold. If they succeed in "satisfying" their need, they will not react further. Consider the classic example of ionic bonds in the mineral halite (or common table salt, NaCl): the sodium (Na) atom has only 1 electron in its outermost M-shell, and if it can somehow loose this electron, it would be "satisfied" because its L-shell already has 8 electrons. On the other hand, a chlorine (Cl) atom has 7 electrons in its outermost M-shell, and it would be very "happy" to have sodium's electron so that it would then have 8 electrons in its M-shell, which is the condition that would make its outer shell "satisfied." Thus, the sodium atom gives up the single electron, the chlorine atom accepts it, and the two resulting ions are held together by an ionic bond in the NaCl structure (Figure 2.2). In this process the Na atom becomes a cation (Na^+) because it lost an electron; and the Cl atom becomes an anion (Cl^-) because it gained an electron.

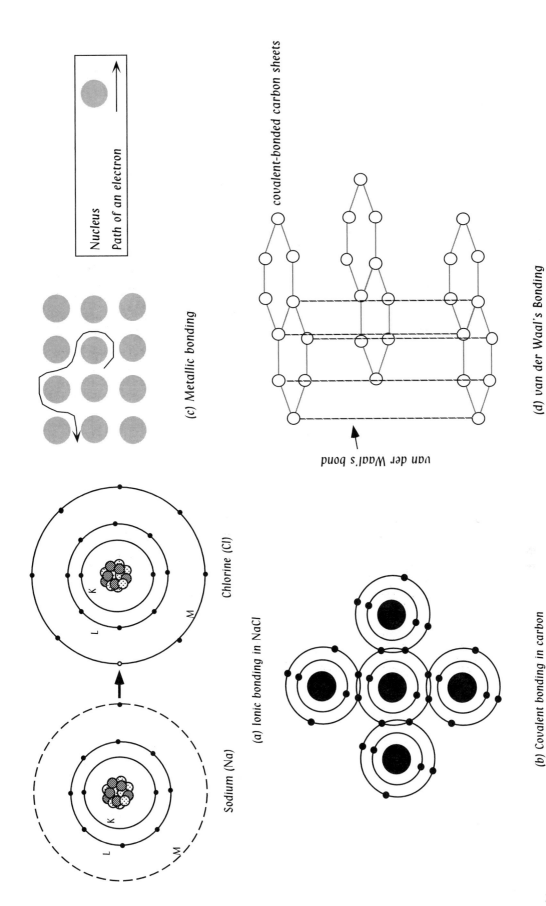

FIGURE 2.2 *Nature of chemical bonds.* (a) Ionic bonding in halite (NaCl). (b) Covalent bonding in carbon formed by the sharing of outer electrons between five adjacent C atoms. (c) Metallic bonding. The nuclei of the atoms are so densely packed that their outer electrons freely move from nucleus to nucleus. (d) van der Waal's bonding in graphite.

23

Covalent Bond

Some chemical compounds contain atoms that fill their outer shells with electrons by sharing electrons rather than exchanging them (as in an ionic bond). Such sharing of electrons is known as *covalent bonding*. In the case where two adjacent atoms share electrons and form a covalent bond, the shared electrons orbit the nuclei of both atoms (Figure 2.2). A familiar example of covalent bonding is the mineral diamond, in which adjacent carbon atoms share electrons. Covalent bonded minerals are generally hard and dense and are generally insoluble in water. The most abundant group of minerals, the silicate minerals, are characterized by covalent bonding between silicon and oxygen atoms (further discussed later).

Van der Waal's Bond

A particularly weak type of electrostatic attraction, known as a *van der Waal's bond*, may occur between atoms or ions in a mineral's atomic structure. They are important in minerals like mica, in which the individual sheets are formed of covalent or ionic bonded atoms and adjacent sheets are held together by weak van der Waal's bonds (Figure 2.2). The bonds offer little resistance when the mica sheets are pried open with a fingernail! Minerals with van der Waal's bonds tend to be lighter and softer than ionically and covalently bonded minerals. Clay minerals and graphite are important additional mineral examples that have van der Waal's bonds between the sheets.

One or more types of bonds may characterize an individual mineral. A crystal in which only one type of bond is present is called *homodesmic*. However, most minerals are *heterodesmic*—characterized by more than one type of bond. The mineral diamond (composed of carbon) is characterized by covalent bonds, whereas graphite (also carbon) has covalent as well as van der Waal's bonds. Thus, diamond and graphite are examples of homodesmic and heterodesmic minerals, respectively.

PAULING'S RULES, COORDINATION NUMBER AND RADIUS RATIO

How atoms or ions are organized in a crystal is fundamentally dictated by two things: the relative sizes of the neighboring atoms/ions and the local balance of electrical charges such that the total atomic structure is electrically neutral (i.e., has no residual electric charges). In 1929, Linus Pauling, winner of two Nobel prizes (the first one was in chemistry and the second one for his contributions to world peace), presented a set of five rules that describe how anion–cation relations affect stability of atomic structures. The rules are as follows:

Rule 1.   *A polyhedron (a geometrical figure with multiple faces, Figure 2.3) of anions forms around a cation; and the distance between a cation and an anion is equal to the sum of their radii.* This polyhedron is generally known as a *coordination polyhedron.* Its outlines may be imagined to form by connecting the centers of the anions. The number of anions that can surround a cation is referred to as the cation's *coordination number* (CN). In the example shown in Figure 2.3, four anions surround a cation; therefore, its coordination number is 4. The geometrical figure obtained by drawing imaginary lines through the centers of the anions is the coordination polyhedron, which in this case is a tetrahedron. A tetrahedron is characterized by four equally sized faces, each of which is an equilateral triangle.

Rule 2.   *In stable ionic structures, the total strength of valence bonds reaching an anion from all neighboring cations in a coordination polyhedron is equal to*

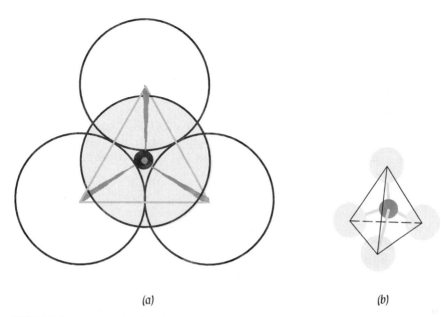

(a)   (b)

FIGURE 2.3   *An example of a coordination polyhedron formed by connecting the centers of 4 anions surrounding a small cation.* In this case it is a tetrahedron. (a) Shows the appropriate relative sizes (i.e., "scale") of the ions. (b) A perspective drawing ("stick-and-ball model") of this tetrahedron is drawn to show its geometric form and the bonds ("light gray") that extend between the anions and the cation.

the total charge of the anion. This value is known as the *electrostatic valence principle.* The *electrostatic valence* (e.v.) of an anion is defined as the ratio of the valence charge to its coordination number. This rule requires that local electrical neutrality be maintained around individual anions in a coordination polyhedron. Halite's (NaCl) crystal structure may be used to explain this rule: in halite each Na^+ is surrounded by six Cl^- ions, and therefore, the e.v. of Na^+ is 1/6. Each Cl^- is also surrounded by six Na^+ ions, and its e.v. is 1/6 as well. Thus, the e.v. of Na^+ and the e.v. of Cl^- are equal; that is, all bonds in the NaCl structure are of equal strength.

Rule 3.   *The sharing of edges, and especially of faces, by two coordination polyhedra decreases the stability of the crystal structure.* Two cations cannot share two anions (the equivalent of sharing an edge by two neighboring coordination polyhedra) or three (or more) anions (the equivalent of sharing a face by two adjacent polyhedra). Such sharing would bring the cations too close to each other; and since cations are both positively charged, repulsive forces between them would render the crystal structure unstable.

Rule 4.   *In a crystal containing different cations, those of high valence and small coordination number tend not to share polyhedral elements with each other.* This rule is, in a way, a corollary of rule 3. Cations having a high valence and a small coordination number (for example, Si^{4+} and Al^{3+}) are less likely to be expected to share their polyhedral elements (such as edges), simply because the repulsive forces between them would be stronger than those between two cations with small valence and higher coordination number.

Rule 5.   *The numbers of essentially different kinds of cations and anions in a crystal structure tend to be small.* This rule, known as the *principle of parsimony,* basically indicates that crystal structures are relatively simple: each has a limited number of cation and anion sites where only a limited number of cations and anions can fit without disrupting its structural stability or electrical neutrality.

Ionic Size and Radius Ratio

Ions of a single atom can be considered approximately spherical objects; however, the size of an ion is difficult to define or measure. In an ionically bonded structure, the electrostatic force (*Coulomb's force*) between the cation and anion is proportional to the product of the charges of the two ions and inversely proportional to the square of the distance between their centers:

$$\mathbf{F} \propto \frac{I_1 \cdot I_2}{d^2}$$

where $\mathbf{F}$ is Coulomb's force, I_1 and I_2 are the charges on the ions, and d is the interatomic (or interionic) distance. As the oppositely charged ions are attracted toward each other, repulsive forces build up between the two nuclei that do not allow the ions to collapse into each other. Within a crystal structure, the sum of the radii of two atoms/ions is equal to the center-to-center distance (called *bond length*) between two neighboring ions or atoms. The bond length between any two particles is a measurable quantity using X-ray methods; however, it is difficult to measure the radius of a single atom or ion because of variations in the bonding mechanism (i.e., covalent versus ionic or van der Waal's bond) for the same atom or ion in different crystals and the difficulty of determining how much of the bond length is due to the size of the particles.

The ratio of the cation's radius to the anion's radius is known as the *radius ratio*. The radius ratio dictates the shape of the coordination polyhedron (Figure 2.4): for example, when the radius ratio is between 0.155 and 0.225, the coordination polyhedron is a triangle whose apices are formed by three anions surrounding a cation at its center. Figure 2.4 shows other examples of coordination numbers and polyhedra resulting from different radius ratios.

CRYSTAL

A *crystal* is defined as a solid with an orderly arrangement of atoms or ions. Such orderly atomic structure is often reflected in the development of smooth natural planar surfaces, called *crystal faces* on a crystal (Figure 2.5). A naturally occurring crystalline substance may or may not possess well defined faces; however, that does not mean that it lacks an orderly atomic structure. How well such crystal faces will grow depends on the conditions under which the substance crystallizes. Extremely slow crystallization and, particularly, crystallization from vapor allow the growth of well-formed crystals. On the other hand, rapid crystallization from a magma (rock melt) may inhibit the growth of well-formed crystals. A crystal with well-formed faces is called an *euhedral* crystal. An *anhedral* crystal is one in which none of the faces is developed. A *subhedral* crystal has imperfectly developed faces. Note that in a strict sense, the terms euhedral, subhedral, and anhedral are used only when the crystals are igneous in origin (i.e, formed from magma). When they occur in metamorphic rocks, the equivalent terms used are idioblastic or idiomorphic (euhedral); subidioblastic, hypidioblastic, or hypidiomorphic (subhedral); and xenomorphic or xenoblastic (anhedral). It is also important to note that the cooling rate is not the only important factor that controls the form of a crystal (discussed in Chapter 15).

R^+/R^-	CN	Configuration		
<0.155	2	Linear		
0.155–0.225	3	Trigonal planar		
0.225–0.414	4	Tetrahedral		
0.414–0.732	6	Octahedral		
0.732–1.000	8	Cubic		

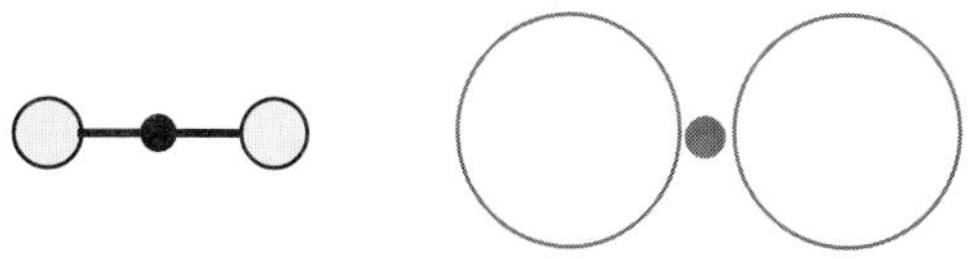

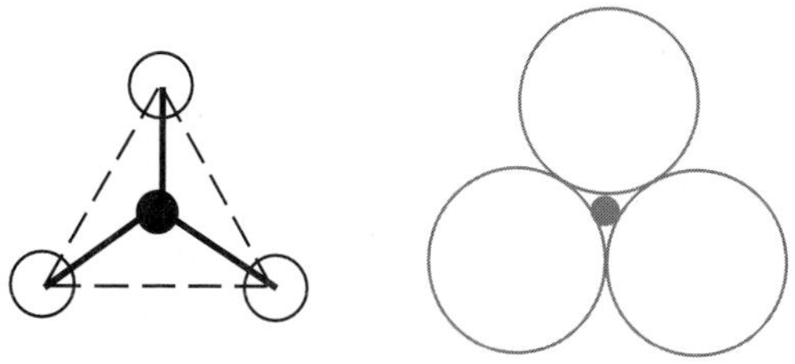

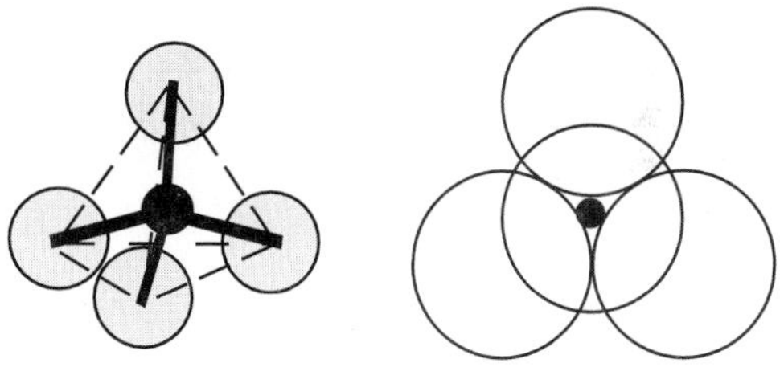

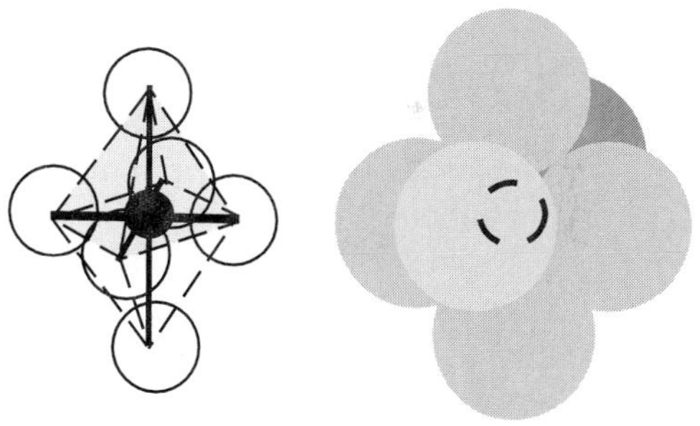

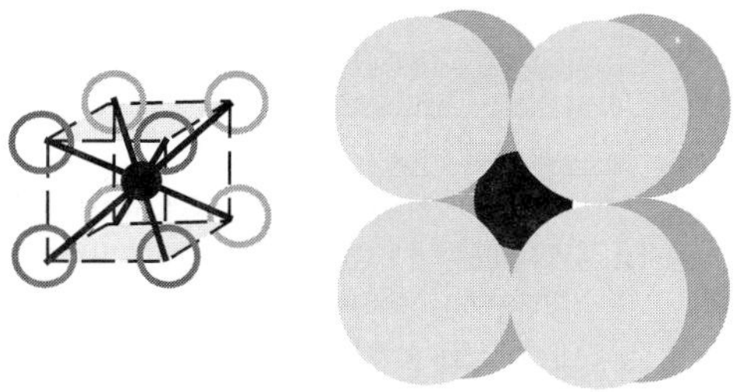

FIGURE 2.4   *Relationships between the cation/anion radius ratio* (R^+/R^-), *the coordination number (CN), and the geometry of the coordination polyhedron.* The middle column shows the geometry of the polyhedra corresponding to appropriate CN (cation—solid circles, anion—unfilled circles, bonds—dark lines, edges of polyhedra—dashed lines, cation and anion sizes are not to scale). The diagrams on the extreme right column show properly scaled relative sizes of cations and anions.

FIGURE 2.5   *Photograph of a crystal.*
(Courtesy of Gautam Sen)

Crystal Symmetry

Thousands of different sizes and shapes of crystals occur in nature. If we are to compare them and understand why one "looks" different from another, we must be able to use some property common to all crystals. *Symmetry* is such a property, whose principal feature is repetition, and may be best understood in terms of motif and operations. *Motif* is the smallest representative unit of a structure, which, when repeated in three dimensions, describes the entire structure. In the case of the atomic structure of any mineral, the smallest structural unit (or motif) that upon repetition of itself in three dimensions generates its crystal form is referred to as the *unit cell*. *Operation* means moving the motif in some way. As a simple example, let us consider the movement of a car in order to understand what operation is. When the car moves forward on a straight road, such motion along a line is an operation called *translation*. A second major type of operation is *rotation*, in which the motif is rotated around some axis to produce a geometric figure. An example of rotation is driving a car (motif) around a circle (Figure 2.6a,b). Figure 2.6a and b also show the translation and rotation of a square motif. The motif's repetition in the two cases has produced two different geometric figures: in the first case it is a type of box, and in the second it is a circle. The interesting thing is that although the basic structural unit (motif) is a simple square, very different geometric forms were generated by different types of symmetry operations!

A third fundamental type of operation is *inversion*, in which the motif is simply inverted. In nature, the symmetry of all objects can be described by using the three operations, individually or in combination, on a motif. Figure 2.6c shows an example of a combination of rotation and translation operations (*screw motion*). In this case the geometric figure generated is a cylinder.

In a *congruent pattern*, a motif can be exactly superimposed on another. When two motifs are mirror images of each other, they are said to be related by a *mirror plane* (also called a mirror, the symbol conventionally used is *m*). Figure 2.7a shows that a mirror (*m*) can be placed horizontally across the middle of the letter "K" to reflect that the upper

Symmetry Operations

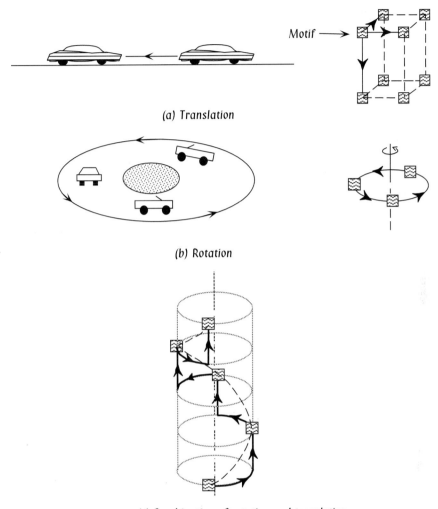

(a) Translation

(b) Rotation

(c) Combination of rotation and translation

FIGURE 2.6   *Motif (square) and various operations on the motif:* translation, rotation, and a combined rotation and translation.

half of this letter is a mirror image of its lower half. A *center of symmetry* (or center of inversion, with symbol *i*) is one in which the original motif and the derived motif are related by an inversion (Figure 2.7b). In Figure 2.7b, the triangular motif on the front face of the crystal and its inversion-derived motif on the back face indicate that a center of symmetry is present in this crystal. A combination of translation and mirror produces a *glide plane* (not shown here), a discussion of which is beyond the scope of this book. An *axis of rotation* (or *axis of symmetry*) is one around which when the motif is rotated 360°, it may reproduce itself once or more (Figure 2.7c,d,e,f). Mirror, axis of rotation, and inversion (center of symmetry) are generally referred to as *symmetry elements*.

There can only be five types of "proper" axes of symmetry: 1-fold, 2-fold, 3-fold, 4-fold, and 6-fold, depending upon how many times the motif repeats itself during a 360° rotation. In a 1-fold axis of symmetry, the motif repeats itself only once (i.e., the original position itself). It is obviously a minimum level of symmetry, and every object must have at least this much of symmetry. In the alphabet, the letter "Q" has only

**Plane of Symmetry
or Mirror**

**Center of Symmetry
or Inversion Center**

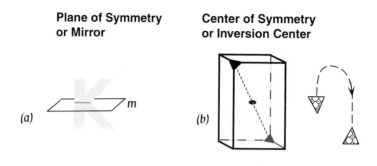

(a)

(b)

Axis of Symmetry

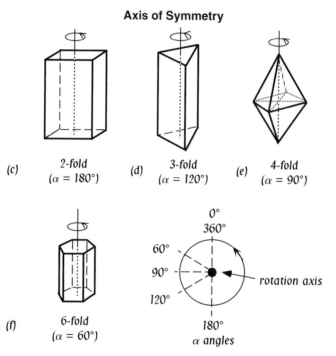

(c)
2-fold
$(\alpha = 180°)$

(d)
3-fold
$(\alpha = 120°)$

(e)
4-fold
$(\alpha = 90°)$

(f)
6-fold
$(\alpha = 60°)$

α angles

FIGURE 2.7 *Principal symmetry elements:* plane of symmetry or mirror (m), center of symmetry (i, also inversion center), and axes of symmetry. α is the angle of rotation of the axis needed to reproduce the same motif (face or edge).

a 1-fold symmetry. In a 2-fold axis of rotation, the motif appears twice during a 360° rotation, and so on (Figure 2.7). *Rotation angle* (α) is the minimum angle of repetition of a motif (Figure 2.7). The angle α is related to the *x*-fold axis of symmetry in the following way: $\alpha = 360°/x$. Therefore, in a 1-fold axis of rotation $\alpha = 360°$, in a 2-fold axis $\alpha = 180°$, and in a 3-fold axis $\alpha = 120$.

ROTATION AND INVERSION

Repetition of a motif by rotation combined with an inversion is called a *rotoinversion* operation. In a 1-fold rotation axis, a motif is turned 360° to reproduce itself (it is the same one). In a 1-fold rotoinversion (conventionally called "bar 1" or $\bar{1}$), the motif is turned 360° and inverted. It is the same thing as a center of symmetry (Figure 2.8). 2-fold rotoinversion ($\bar{2}$) occurs when a motif in the upper hemisphere is turned 180°

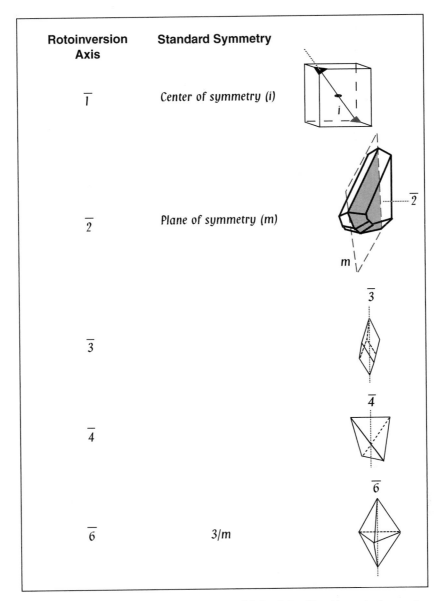

Rotoinversion Axis	Standard Symmetry
$\bar{1}$	Center of symmetry (i)
$\bar{2}$	Plane of symmetry (m)
$\bar{3}$	
$\bar{4}$	
$\bar{6}$	3/m

FIGURE 2.8   *Types of rotoinversion symmetry and their relationship with standard symmetry.*

and inverted to obtain the same motif. As shown in Figure 2.8, it is the same as a mirror plane. A 3-fold rotoinversion $\left(\bar{3}\right)$ axis is a combination of a 3-fold rotation axis plus a center of symmetry. A 4-fold rotoinversion $\left(\bar{4}\right)$ occurs when a motif is turned in 90° increments and inverted 4 times to repeat itself. A 6-fold rotoinversion $\left(\bar{6}\right)$ axis is a 3-fold rotation axis with a mirror perpendicular to the axis (Figure 2.8).

REPRESENTATION OF SYMMETRY ELEMENTS

Figure 2.9a shows a set of internationally accepted notations, commonly known as *Hermann–Mauguin notations*, used by crystallographers to represent various symmetry elements. Drawing the crystal faces and all the symmetry elements becomes a cumbersome process when dealing with complex crystals having a large

(a)

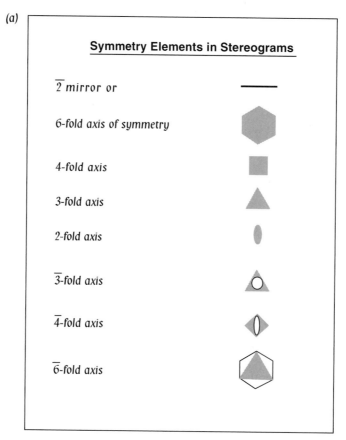

Symmetry Elements in Stereograms

$\overline{2}$ *mirror or*

6-fold axis of symmetry

4-fold axis

3-fold axis

2-fold axis

$\overline{3}$*-fold axis*

$\overline{4}$*-fold axis*

$\overline{6}$*-fold axis*

FIGURE 2.9   (a) The conventional representation of symmetry elements on stereograms. (b) Planes of symmetry (mirrors) through a crystal model: *h*—horizontal plane, *w, x, y, z*—vertical planes. (c) Mirrors and axes of symmetry are shown: 4 horizontal 2-fold axes—1-1′, 2-2′, 3-3′, 4-4′; 1 vertical 4-fold axis. (d) The concept of projection of symmetry elements on a stereogram is illustrated in terms of a sphere. A stereogram represents projections of all symmetry elements onto the horizontal plane (the deeply shaded slice through the middle of the sphere). Therefore, a horizontal plane would be the same as the perimeter of the circle, the vertical planes would project as straight lines, the vertical axis would plot at the center as a point, horizontal axes would fall as points on the perimeter. Obliquely inclined planes plot as curved lines ("great circles" or lines of longitude) inside the stereogram, and oblique axes also plot inside the stereogram as points (not shown). How far from the center the oblique plane (or axis) would plot depends upon the angle it makes with the horizontal plane—the shallower the angle, the closer it would plot toward the perimeter. In the example shown, the 4-fold vertical axis plots at the center of the stereogram; the horizontal mirror (*h*) is represented by the perimeter of the circle. The remaining vertical mirrors coincide with the horizontal 2-fold axes and plot as straight lines, whose orientation is dependent upon the angle they make with the N–S axis. (e) The stereogram of the crystal.

number of symmetry elements. (Figure 2.9b and c represent a somewhat simple example of a crystal with fewer symmetry elements.) Therefore, it is generally convenient to show their symmetry elements by projecting them onto a circular plane, and the center of the circle is the projection point of the vertical axis passing through the center of the crystal (Figure 2.9d). This circular plane is referred to as a *stereogram* (Figure 2.9e). The perimeter of the stereogram is the horizontal plane. On a stereogram, continuous lines represent mirrors and dashed lines indicate the absence of mirrors. Thus, when the horizontal plane is a mirror, the perimeter of the circle is drawn as a continuous line. (See "*h*" in Figure 2.9e.) When a horizontal mirror does not occur in the crystal, the perimeter is dashed. (See Figure 2.10b.) Vertical mirrors

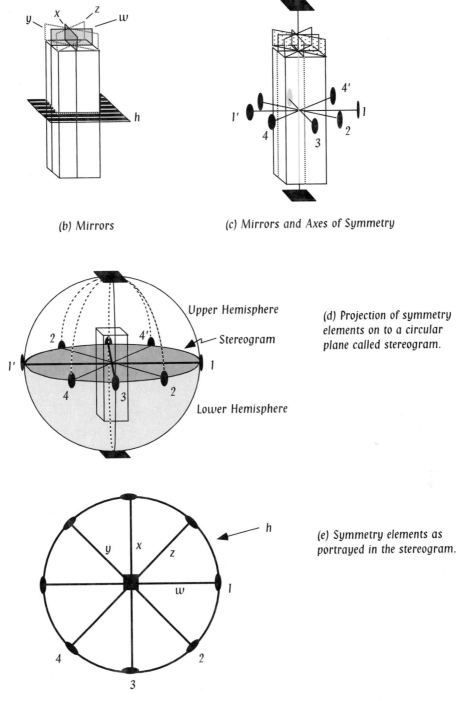

(b) Mirrors

(c) Mirrors and Axes of Symmetry

(d) Projection of symmetry elements on to a circular plane called stereogram.

(e) Symmetry elements as portrayed in the stereogram.

FIGURE 2.9   *(continued)*

plot as straight lines going through the center of the circle (e.g., w, x, y, and z in Figure 2.9e). Any plane or line, such as a crystal face or an axis of symmetry, that is not horizontal or vertical must plot on the inside of the circle, the exact location of which will depend upon the angle and direction of its tilt from the horizontal. (A detailed discussion of these concepts may be found in Battey 1981.)

COMBINATIONS OF ROTATIONS, MIRRORS, AND ROTOINVERSIONS

In crystals, crystal faces (as for motifs) can often be related by a combination of a number of symmetry operations. In these cases, symbolism such as *422* and *4/m2/m2/m* are used to depict spatial relationships between various symmetry elements, where the numbers represent the axes of symmetry and *m* stands for mirror. The highest symmetry axis is written first. When an axis of symmetry, for example a 4-fold axis, is perpendicular to a mirror, it is written as a fraction, *4/m*. If the 4-fold axis is parallel to a mirror, it would be written on the same line, *4m*. It is important to note that even though a larger number of symmetry elements may be present in a crystal, one needs to mention only a minimum of the elements because the others are automatically generated by their combined operation. For example, note that in Figure 2.9 one needs to mention only the 4-fold axis and two of the 2-fold axes; the others are automatically generated by a 4-fold rotation. Thus, the symmetry of this crystal may be expressed as *4/m2/m2/m*.

The above example and two additional examples of combined symmetry operations are compared in Figure 2.10. The spatial relations of the crystal faces and mirrors and axes of rotation and rotoinversion are shown, along with their corresponding stereograms. As indicated in the above discussion, *4/m2/m2/m* has four 2-fold rotation axes that are perpendicular to the one 4-fold axis (Figure 2.10a). Also, each axis is perpendicular to a mirror. The 4-fold axis is vertical, and the 2-fold axes are all horizontal. On the other hand, in a *422*, there are no mirrors (hence the dashed lines in its stereogram). The 4-fold axis is vertical, and the four 2-fold axes are horizontal (Figure 2.10b). Notice that each face of the crystal shown in Figure 2.10b has the shape of a trapezoid; therefore, this crystalline form (defined later) is referred to as *trapezohedron* (hedron means face). The third example, $\overline{4}2m$ has a vertical 4-fold rotoinversion axis, which is perpendicular to the two 2-fold axes. It also contains two vertical mirrors, but notice that none of the mirrors is perpendicular to any of the axes of rotation.

It turns out that the ten basic elements—the five rotation axes (6, 4, 3, 2, 1) and five rotoinversion axes ($\overline{6}(=3/m)$, $\overline{4}$, $\overline{3}$, $\overline{2}(=m)$, $\overline{1}(=i)$) or their combinations—are sufficient to describe the external symmetry of any crystal in the universe! Figure 2.11 shows several examples of combining various symmetry operations in reproducing crystal faces. Surprisingly, it turns out that only 32 symmetry operations (individually or in combination) are possible. These correspond to the 32 *crystal classes* that can be distinguished based on their external morphology (Figure 2.11). The classes can be further grouped into six *crystal systems* based on their common features.

CRYSTALLOGRAPHIC AXES AND CRYSTAL SYSTEMS

Crystallographic axes assigned to a crystal serve as a set of reference axes, which help in an accurate description of the morphology of the crystal (Figure 2.12). In general, the following "requirements" should be remembered when choosing crystallographic axes, however, these "rules" are not set in concrete:

1. There are usually three axes (and four at most in the hexagonal system).
2. The axes are mutually perpendicular (except in monoclinic and triclinic systems).

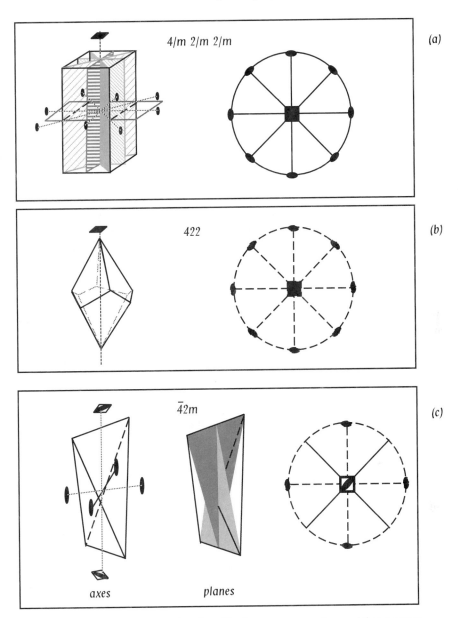

FIGURE 2.10 *Three simple examples of combined symmetry operations and the symmetry elements.* Stereograms on the right-hand side show the nature and orientation of the symmetry elements in each case.

3. The axes coincide with the crystal's symmetry axes, or lacking these, with the normals to mirrors.

4. In crystals that lack sufficient symmetry elements, the axes are selected to coincide with the lines of intersection between the faces of greatest area—either with actual intersections or with possible intersections obtained by "extending" these faces until they meet.

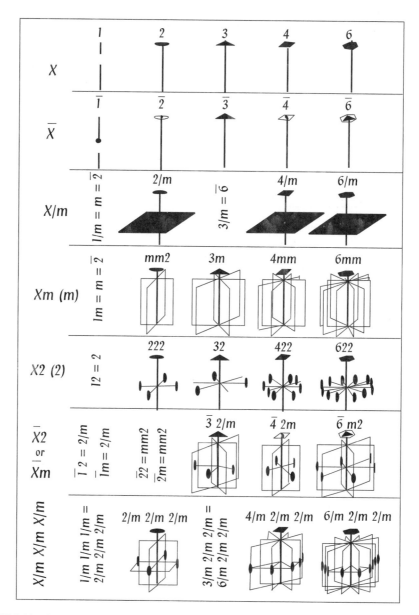

FIGURE 2.11   *Symmetry elements in 32 crystal classes.* (After Bloss, 1971, *Crystallography and Crystal Chemistry,* Holt, Rinehart and Winston)

With the exception of the hexagonal system, which has four crystallographic axes, all crystal systems are best described with reference to three crystallographic axes, conventionally called the $a, b,$ and c axes. By convention the orientation of the a and b axes are horizontal (note a is not horizontal in monoclinic and triclinic systems), and c is vertical (Figure 2.12). The a-axis is oriented front to back, and the b-axis runs right to left. The two ends of each of these axes are given the $+$ or $-$ notation by convention: the top of the c-axis is $c+$ and the bottom is $c-$; the front portion of the a-axis is $a+$, and the back portion of the a-axis is $a-$; and the right side of the b-axis is $b+$ and the left side is $b-$.

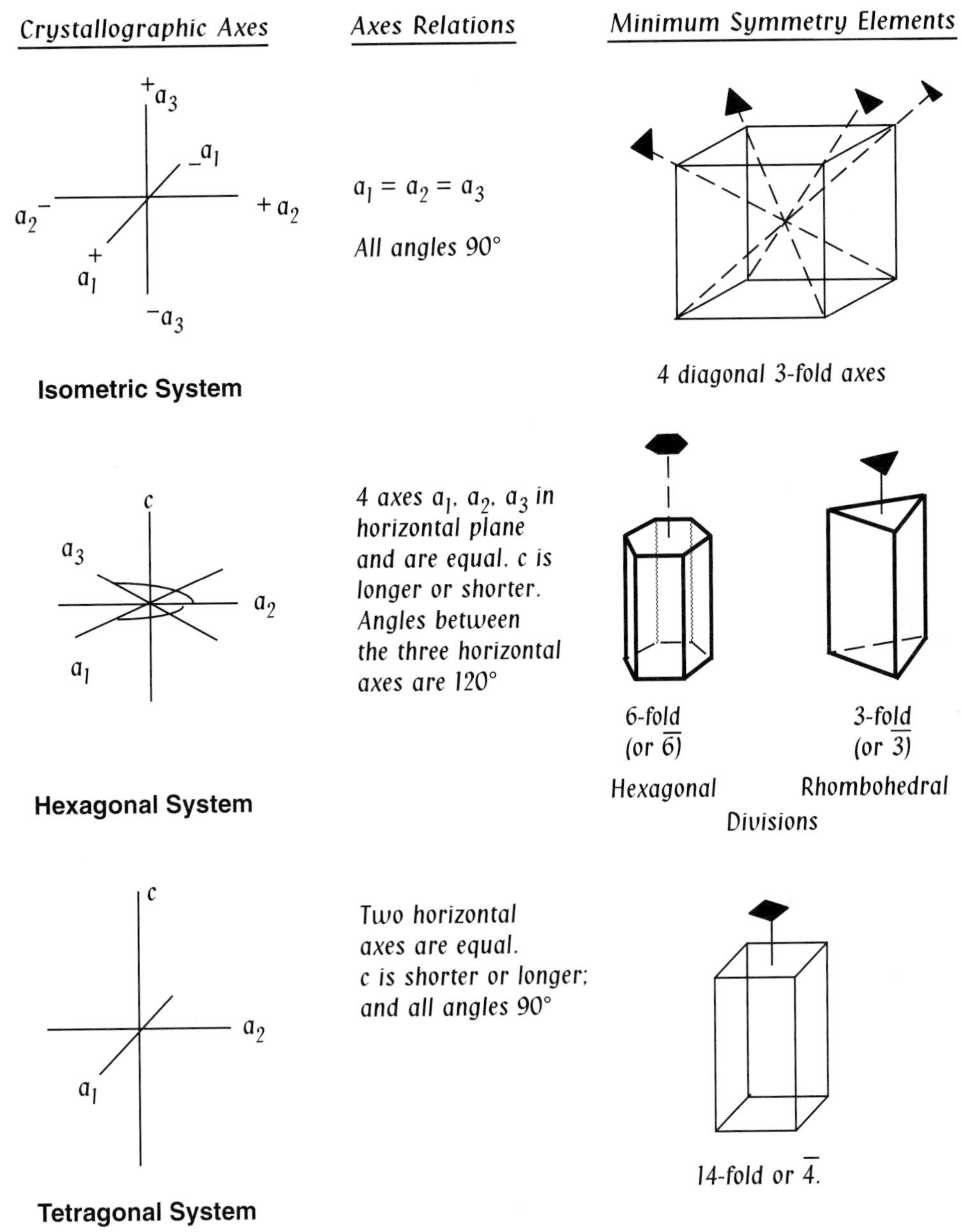

Crystallographic Axes   Axes Relations   Minimum Symmetry Elements

$a_1 = a_2 = a_3$

All angles 90°

4 diagonal 3-fold axes

Isometric System

4 axes a_1, a_2, a_3 in
horizontal plane
and are equal. c is
longer or shorter.
Angles between
the three horizontal
axes are 120°

6-fold   3-fold
(or $\overline{6}$)   (or $\overline{3}$)

Hexagonal   Rhombohedral

Divisions

Hexagonal System

Two horizontal
axes are equal.
c is shorter or longer:
and all angles 90°

14-fold or $\overline{4}$.

Tetragonal System

FIGURE 2.12   *Crystallographic axes in the six crystal systems, their interrelations, and some crystal
examples are shown.* The minimum symmetry elements in each system are displayed on the right.

(continues on next page)

Based on crystal symmetry and the crystallographic axes, all crystals may be grouped into six crystal systems: isometric, hexagonal, orthorhombic, tetragonal, monoclinic, and triclinic.

In the *isometric system* the three crystallographic axes are equal in length and are perpendicular to each other (Figure 2.12). In this system the three axes are referred to as the a_1, a_2, and a_3 axes. The crystal classes and their symmetry elements are listed in

Crystallographic Axes	Axes Relations	Minimum Symmetry Elements
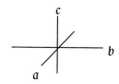	$b > a > c$ All angles 90°	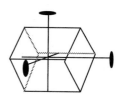
Orthorhombic System		3 perpendicular 2-fold axes
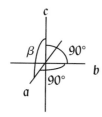	$a \neq b \neq c$ Two angles 90° $\beta > 90°$	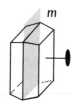 2/m 2-fold or a mirror
Monoclinic System		
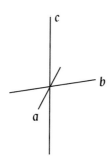	$a \neq b \neq c$ None of the angles are 90°	
Triclinic System		

FIGURE 2.12 *(continued)*

Table 2.2. In the *hexagonal system* there are four axes, three $(a_1, a_2,$ and $a_3)$ of which are horizontal and equal in length, and one axis (*c*-axis), which is vertical. The angle between any two adjacent horizontal axes is 120°. The hexagonal system is subdivided into *hexagonal division* and *rhombohedral division* based on the presence of the 6, 6 versus 3, or 3 symmetry axis, respectively. In the *tetragonal system* the two horizontal axes (a_1, a_2) are equal in length but are unequal to the *c*-axis; and they are all mutually perpendicular. In the *orthorhombic system* the three axes, a, b, and c, are mutually perpendicular but are unequal in length. In the *monoclinic system* all the axes are unequal in length, and the angles between b and c (i.e., $b \wedge c$) and b and a ($b \wedge a$) are 90°; however, the angle between $c+$ and $a+$, which is conventionally called β is >90°. In the *triclinic system* the three axes are unequal, and none of them are perpendicular to each other.

TABLE 2.2 *Symmetry elements of the thirty-two crystal classes.*

System	Class	Symmetry Elements					
		Axes				Planes	Center
		2-fold	3-fold	4-fold	6-fold		
Isometric	$4/m\,\bar{3}\,2/m$	6	4	3	0	9	present
	432	6	4	3	0	0	absent
	$\bar{4}\,3\,m$	3	4	0	0	6	absent
	$2/m\,\bar{3}$	3	4	0	0	3	present
	23	3	4	0	0	0	absent
Tetragonal	$4/m\,2/m\,3/m$	4	0	1	0	5	present
	422	4	0	1	0	0	absent
	$4\,mm$	0	0	0	0	4	absent
	$\bar{4}\,2m$	3	0	0	0	2	absent
	$4/m$	0	0	1	0	1	present
	4	0	0	1	0	0	absent
	$\bar{4}$	1	0	0	0	0	absent
Hexagonal	$6/m\,2/m\,2/m$	6	0	0	1	7	present
	622	6	0	0	1	0	absent
	$6mm$	0	0	0	1	6	absent
	$\bar{6}\,m2$	3	1	0	0	4	absent
	$6/m$	0	0	0	1	1	present
	6	0	0	0	1	0	absent
	$\bar{6}$	0	1	0	0	1	absent
Rhombo-hedral Division	$\bar{3}\,2/m$	3	1	0	0	3	present
	32	3	1	0	0	0	absent
	$3m$	0	1	0	0	3	absent
	$\bar{3}$	0	1	0	0	0	present
	3	0	1	0	0	0	absent
Orthorhombic	$2/m\,2/m\,2/m$	3	0	0	0	3	present
	222	3	0	0	0	3	present
	$mm2$	1	0	0	0	2	absent
Monoclinic	$2/m$	1	0	0	0	1	present
	2	1	0	0	0	0	absent
	m	0	0	0	0	1	absent
Triclinic	$\bar{1}$	0	0	0	0	0	present
	1	0	0	0	0	0	absent

It is interesting to note that more than 50% of minerals crystallize in monoclinic and orthorhombic systems (Figure 2.13), and 16.6% of minerals crystallize in the hexagonal system. Only 12.2% of minerals crystallize in the isometric system. Two factors give a reasonable explanation of these percentages: elemental abundances in the earth's crust and coordination of the atoms and ions of such elements (which in itself is controlled by the radius ratio and electrical neutrality requirements). As discussed in a later chapter, Si, O, Mg, Fe, Al, and Ca are the most abundant elements in the crust. A more detailed explanation is beyond the scope of this discussion.

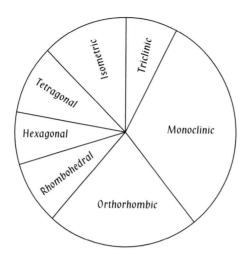

FIGURE 2.13   *The relative abundance of minerals crystallizing in each crystal system.*

DEFINING CRYSTAL FACES: UNIT CELL, PARAMETERS, AND MILLER INDICES

Long before the internal symmetry of crystals at the atomic scale was recognized, early crystallographers realized that the crystals of different minerals are different, and the number, shape, and relative sizes of faces are similar in crystals of same form and of the same mineral, even though the crystals themselves may be very different in size. In 1784, a French crystallographer, Hauy, proposed that all crystals are ultimately built of small polyhedral units, now called *unit cells* (Figure 2.14). When these units are repeated in three dimensions conforming to the crystal's symmetry, the crystal structure will result.

Much later in the nineteenth century, X-ray crystallography of crystals led to the recognition of the unit cell, which is essentially analogous to each of Hauy's polyhedral units defined at the atomic scale. Combining unit cells with symmetry operations, it has been shown by another French crystallographer, Auguste Bravais, that only 14 three-

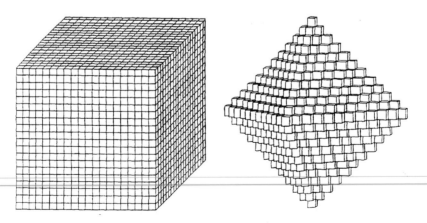

FIGURE 2.14   *The unit cell.* Two distinct crystal forms made of the same basic type of unit cell (a cube) illustrate the fact that the unit cells of one kind can be grouped in a number of different ways to result in distinct crystal forms. (Reprinted with permission, Ernst, 1969, *Earth Materials*, Mcmillan and Co.)

dimensional structures (called *space lattices*) are possible in building all crystalline structures. These 14 space lattices are called *Bravais lattices* (Figure 2.15). The Bravais lattices are of three types: *primitive* or *P* lattices, *body-centered* or *I* lattices, and *face-centered* or *F* lattices. In a primitive lattice the unit cell has a lattice point (or motif) at

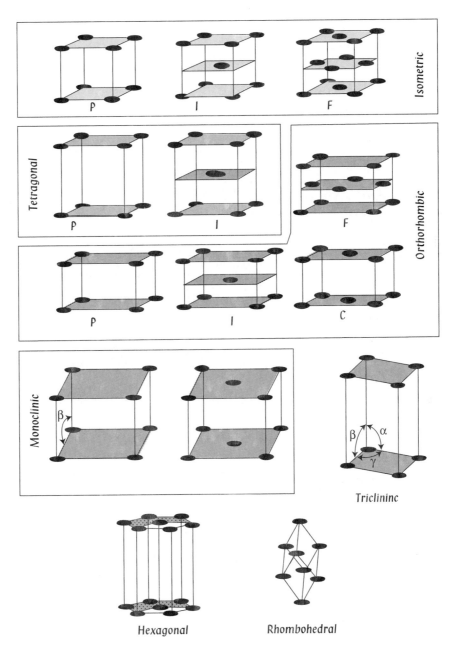

FIGURE 2.15   *Bravais lattices.* Primitive (P), body-centered (I), and face-centered (F, C,) lattices are shown for each system. When a lattice has a motif at the centers of all faces, the unit cell is called an F-type. When the motifs are at the center of only some faces, it may be an A, B, or C-type unit cell. (From Bloss, 1971)

each corner. In a body-centered lattice, the unit cell has a lattice point at the center in addition to the corner lattice points. In a face-centered lattice, a lattice point must be present at the centers of a pair of faces or of all faces (Figure 2.15).

Modern crystallographers can measure unit cell dimensions, which are presented in angstroms (Å). For example, the unit cell dimensions of sillimanite are as follows: $a = 7.47, b = 7.66$, and $c = 5.76$Å, where the numbers represent the dimensions of the cell along the $a, b,$ and c crystallographic directions. Following conventional practice, the three numbers are all normalized to the reference b (i.e., $a/b : b/b : c/b$), and the ratio is called the *axial ratio*. For sillimanite, the axial ratio is $7.47/7.66 : 7.66/7.66 : 5.76/7.66$, or $0.975 : 1 : 0.752$.

Long before the birth of X-ray crystallography, the need to define the geometry of the faces of crystals with respect to some common reference axes was recognized. This led to the subject of *morphological crystallography*, in which the crystal's geometry is studied at a megascopic or hand-specimen scale. In a manner similar to how points, lines, and planes are defined with respect to x- and y-coordinates in a Cartesian coordinate graph, early crystallographers felt that a method must be developed to define each face of a crystal in numerical terms. They developed numbers, called *parameters*, for each face of any crystal such that the crystal face (or its extensions) can be identified with reference to the intercepts it (or its extensions) makes on the crystallographic axes (Figure 2.16). The crystallographer first defines a reference face (called the *unit face*) of a crystal and then defines the other faces of the crystal with reference to that unit face. The unit face is chosen to be the one that cuts all three axes at arbitrarily set unit distances; the unit distances along the a and c axes must be proportional to their axial ratio. For example, a mineral that has the axial ratio $0.9 : 1 : 0.8$ should have its unit distance along the a-axis equal to 0.9 times the unit distance along the b-axis. Similarly, its c-axis unit distance should be 0.8 times the unit distance along b-axis. It is important to note that a set of parallel faces, no matter how large or small, will always have the same axial ratio, and can thus be given a single set of parameters for identification purposes.

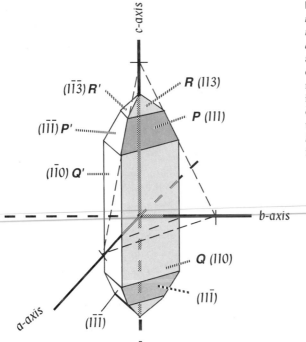

FIGURE 2.16   *Orientation and intercepts of planes P, Q, and R with reference to the crystallographic axes a, b, and c.* The unit face P intersects all three axes on their + side at equal distances; therefore, its axial ratio is 1a : 1b : 1c, and the Miller indices are (111). The Miller indices of the other faces are discussed in text. The prism face (110) cuts a and b at unit distances, respectively, but is parallel to c axis.

When more than one face of a crystal cuts all three crystallographic axes, the unit face is generally the largest face that intersects the +sides of all three axes at arbitrarily assigned distances. For example, in Figure 2.16 the face 'P' intercepts the *a, b*, and *c* axes at unit distances. Therefore, the axial ratios of this face may be given as *1a:1b:1c*. If, on the other hand, another face 'Q' intercepts the *a* and *b* axes at unit distances and is parallel to the *c* axis, then its axial ratios are *1a:1b:∞c*. It is obvious that our example belongs to a tetragonal system in which the unit distance on the *c*-axis is longer than on the *a* and *b* axes.

Consider another example in which a face intercepts the three axes at *(1/2)a:(1/2)b:3c*. Multiplying these by 2 would result in *1a:1b:6c*. This example illustrates the fact that these parameters of the face are not absolute values but relative values. Therefore, they have no relevance to the actual size of the face. That is, all parallel faces, whether large or small, will cut the axes in exactly the same *proportions*. In this context, note that the axial ratios (0.813:1:1.903) of sulfur determined in the nineteenth century (i.e., pre–X-ray days) were found to be almost identical to those (0.811:1:1.900) calculated from a 1912 X-ray determination of the unit cell dimensions of sulfur! This further verifies that the geometry of a sulfur crystal is constant regardless of whether it is at a megascopic scale or at an atomic scale. Thus, the absolute sizes of crystals are of no relevance in describing the geometry of crystals.

For identification and comparison purposes, defining each crystal face with a notation using fractions or decimals is rather cumbersome, and early crystallographers were searching for a more useful and simpler notation. The most commonly used notation is the system of *Miller indices*. The Miller indices of a face are a set of whole numbers that are obtained by inverting the axial ratios of the face and then clearing fractions by multiplying the numbers with a common value. Also, the commas or colons separating the numbers are taken out. The following example shows how the Miller indices of a face may be derived. Consider a face with an axial ratio of *1a:2b:(2/3)c*. In deriving its Miller indices we ignore the *a, b, c* and the colons and consider only the numbers. Thus, we obtain 1, 2, and 2/3; and inverting these, we obtain 1, 1/2, and 3/2. In order to clear the fractions, we multiply all three by the greatest common denominator, in this case 2, and obtain 2, 1, 3. The Miller indices of this face are written as (213), read as "*two–one–three*."

Now consider another face that cuts the *a* and *b* axes at fractional distances and is parallel to the *c* axis, such that its axial ratios are *(1/2)a:(3/2)b:∞c*, which through inversion becomes 2, 2/3, 0. Clearing the fractions gives 6, 2, 0; and dividing these numbers by 2, which is common to all, results in the Miller indices (310). The important thing to recognize here is the fact that the Miller indices of any crystal face are either small whole numbers or zero.

Note that the above examples considered faces that intersect on the positive side of the crystallographic axes. If a face intersects on the negative side of an axis, its Miller indices are given differently. For example, if the face makes unit intercepts on the + sides of *a* and *b* and unit intercept on the − side of *c*, its Miller indices are $(11\bar{1})$, which is read as "*one–one–bar one*." Note that crystals belonging to the hexagonal system have a fourth crystallographic axis, and therefore, the Miller indices of any face of a hexagonal crystal will have a four-digit Miller index, for example, $(11\bar{2}1)$ means a face that cuts the a_1, a_2, and *c*-axes at unit distances and $-a_3$ at 1/2 the unit distance.

Figure 2.16 shows a crystal with three different types of faces (called forms). Face Q, as we noted earlier, has an axial ratio of 1a:1b:∞c. Inverting these we obtain Miller indices of (110) for this face. The face Q' is similar in shape to Q, except that it has a different orientation as reflected in its Miller indices (110). Miller indices of the face P, derived from its axial ratios (1*a*:1*b*:1*c*), are (111). Parameters for the face R are 6*a*:6*b*:3*c*. Dividing by 6 (= *b* intercept) and then inverting, we obtain the Miller indices (113) for this face.

Sometimes when the axial ratios of a face are unknown, the general notations (hkl), or ($hkil$) for hexagonal crystals, may be used for their Miller indices. If a face of a non-hexagonal crystal intersects a and b but is parallel to the c axis, and the axial ratios are unknown, then its Miller index should be ($hk0$).

Consideration of angular relationships between crystal faces in a crystal requires measurement of *interfacial angles*. An interfacial angle is the angle between any two faces. The *law of constancy of interfacial angles*, put forth by Nicolas Steno in 1669, states that the interfacial angles measured between similar sets of faces are constant (Figure 2.17a). For example, in Figure 2.17, the angles between similar pairs of faces a–b and c–d are identical (i.e, angles $\alpha = \delta$ and $\beta = \gamma$. A *contact goniometer* are generally used to mea-

(a)

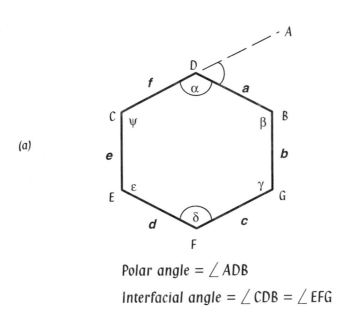

$$\text{Polar angle} = \angle ADB$$

$$\text{Interfacial angle} = \angle CDB = \angle EFG$$

(b)

Contact
Goniometer

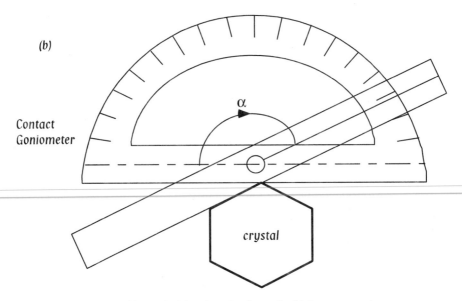

FIGURE 2.17   (a) Interfacial angle and polar angle. (b) A contact goniometer.

sure interfacial angles. It consists of a protractor and a movable plastic bar (Figure 2.17b). The goniometer is held perpendicularly on the first face, and the plastic bar is rotated until it sits against the second face. Two angles that add up to 180° may be read off the protractor: one of these is the internal solid angle (true interfacial angle), and the other is the polar angle, made between the two perpendiculars (or poles) to the faces. It is the second angle that is commonly referred to as the interfacial angle.

COMMON FORMS

Similar faces on a crystal compose a *form*. If all the faces belonging to a form enclose a space, it is referred to as a *closed form*, as opposed to *open form* when the faces do not (Figure 2.18). Miller indices are also used to decribe a form, but braces are used instead of parentheses to define a crystal face. For example, in Figure 2.16, the forms are {111}, {113}, and {110}. The form {111}, for example, contains the following eight faces:

$$(111), (\bar{1}11), (\bar{1}\bar{1}1), (1\bar{1}1) \quad \text{upper half of the crystal}$$

$$(11\bar{1}), (\bar{1}1\bar{1}), (\bar{1}\bar{1}\bar{1}), (1\bar{1}\bar{1}) \quad \text{lower half of the crystal}$$

The naming of forms is based on the shapes, the number of faces involved, and their mutual relationships. In one case, the name is based on the mineral that commonly occurs in that form—*pyritohedron* (the mineral concerned is obviously pyrite). Table 2.3 shows the names of some common types of forms, and Figures 2.18 and 2.19 show what the forms look like.

TABLE 2.3   *Some crystal forms.*

Form	Description
Open Forms	
Dome	Two nonparallel faces separated by a mirror
Pedion	A single face
Pinacoid	Two parallel faces belonging to a form
Prism	Three or more faces belonging to a form, which are all parallel to a common axis
Pyramid	Three or more nonparallel faces belonging to a form in which all intersect at a common point and are equally inclined to a common axis at the point of intersection
Sphenoid	2 nonparallel faces with a 2-fold rotation axis
Closed Forms	
Tetrahedron	Four equal faces, each is an equilateral triangle
Cube	Six equal square faces
Octahedron	Eight faces, each is an equilateral triangle
Rhombohedron	Six rhombus-shaped faces
Scalenohedron	Eight or twelve faces, each has the shape of a scalene triangle
Trapezohedron	Trapezoid-shaped faces
Disphenoid	Four faces, each is a nonequilateral triangle
Dipyramid	Two pyramids related by a mirror

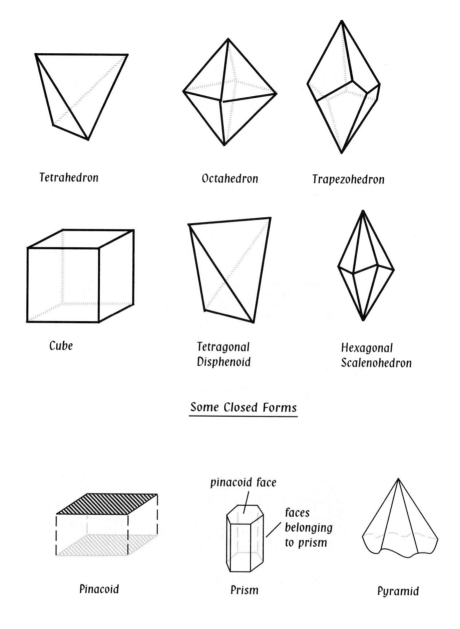

Tetrahedron   Octahedron   Trapezohedron

Cube   Tetragonal Disphenoid   Hexagonal Scalenohedron

Some Closed Forms

Pinacoid   Prism   Pyramid

pinacoid face

faces belonging to prism

Some Open Forms

FIGURE 2.18 *Examples of closed and open forms.*

IMPERFECTIONS IN CRYSTALS

Under "normal" conditions, it is very hard to grow perfect crystals without any imperfections or defects. Therefore, crystals are often far from being perfect, and such defects can be characteristic of a crystalline phase and thus be useful for identification of the mineral concerned. Such external imperfections of crystals are simply a megascopic manifestation of imperfections in their lattices. Commonly, such imperfections in crystal

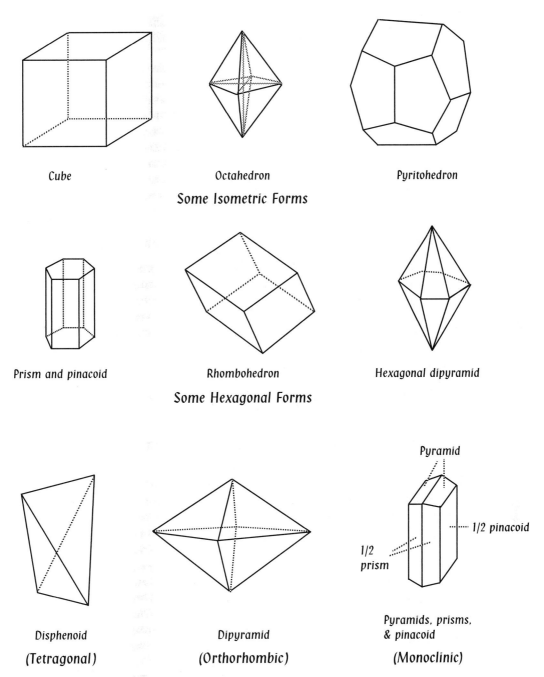

FIGURE 2.19   *Examples of some forms in the six crystal systems.* The 1/2 pinacoid shown for the monoclinic example refers to one of the two parallel and equivalent faces that compose the pinacoid. The same is true of the 1/2 pyramid.

lattices are limited to an atom, an ion, or a unit cell. Other imperfections may be more complex, more extensive, and involve larger groups of ions and unit cells. In general three types of defects are identified: *point defects, line defects, and plane defects*. Point defects are further subdivided into three types: the *Frenkel defect, Schottky defect*, and the *interstitial defect* (Figure 2.20). In a Frenkel defect, an atom or ion may be displaced

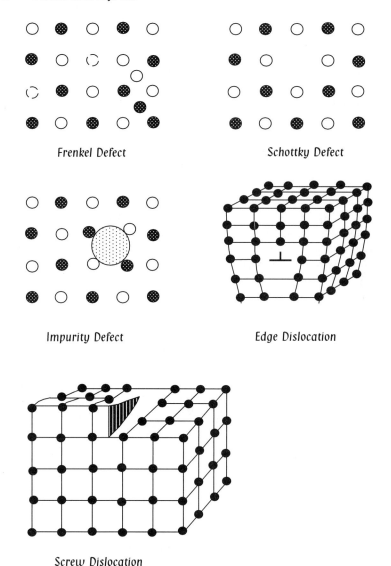

Frenkel Defect

Schottky Defect

Impurity Defect

Edge Dislocation

Screw Dislocation

FIGURE 2.20 *Imperfections in the atomic structures of crystals.*

from its normal site to a different site in the atomic structure. In a Schottky defect, an atom or ion may be missing in the structure. In an interstitial (impurity) defect, a foreign ion or atom may occupy interstitial sites between normal atomic/ionic sites in the mineral atomic structure. Point defects may lead to small chemical variations. Impurities are particularly significant in controlling a mineral's color: for example, emerald looks green because of a very tiny amount of Cr^{3+} present as an impurity.

Line defects result from the dislocation of an entire row or entire rows of atoms or ions. There are two kinds of linear defects: *edge dislocations* and *screw dislocations* (Figure 2.20). In an edge dislocation, a plane of atoms or ions stops at a line instead of continuing as it would have in the nondefective structure. In screw dislocation the dislocated atoms spiral around an axis like a corkscrew. Plane defects refer to the displacement of a two-dimensional array of atoms or ions.

TWINNING

Twinned crystals are composed of sectors or domains ("twinned halves") that are oriented differently with respect to crystallographic axes but are attached to each other nonrandomly along crystallographically controlled surfaces. The common plane along which two twin halves are joined is generally referred to as the *composition plane*. *Simple twins* are composed of only two halves of a crystal that share a common atomic/ionic plane. *Complex twins* are composed of many different sectors. The *twin law* describes the relationships between twinned parts of a crystal. Figure 2.21 shows some types of twinned crystals.

From the point of view of their genesis, twins can be classified as *growth twins*, which form during the formation and growth of the crystal, or *secondary twins*, which form subsequently in the crystal through polymorphic transformation or deformation in response to external forces. Growth twins are a characteristic feature of plagioclase crystals where repeated twinned domains lead to the development of the so-called lamellar twinning in plagioclase. (See the chapter on optical mineralogy.) In detail, the origin of growth twinning is quite complicated (and beyond the scope of this book), except to say that it developed to minimize the lattice energy of the crystal.

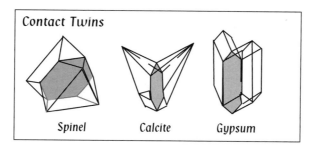

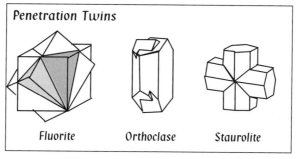

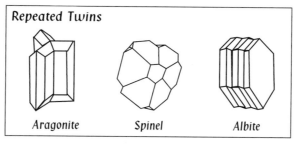

FIGURE 2.21   *Twinned crystals.*

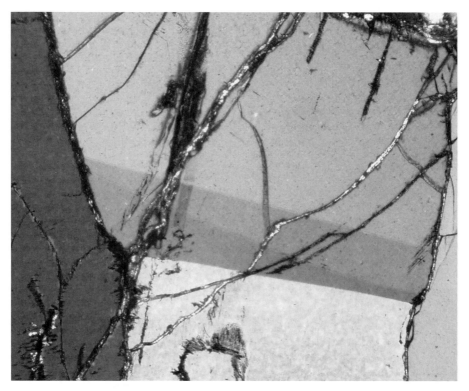

FIGURE 2.22   *Photomicrograph showing the deformation lamellae in olivine.* (Courtesy of Gautam Sen)

Polymorphism is the phenomenon in which an element or a compound can occur in two or more different solid forms distinguished by their internal atomic structures (further discussed in the next chapter). In the *displacive type* of polymorphic transformation, a crystalline phase transforms into another by small displacements of atoms in the structure in response to external pressure or temperature conditions. This often results in twinning, and is well exemplified by the Dauphine twins formed during the inversion of high quartz to low quartz. Deformation twinning is quite common in olivines of metamorphosed ultramafic rocks (Figure 2.22). These may form due to the development of slip planes in the atomic structure.

SUMMARY

[1]   An atom is composed of a nucleus surrounded by an electron cloud. The nucleus is composed of protons and neutrons. Electron orbitals carry specific energies.

[2]   Chemical bonds are like glue that hold the atomic structure together. There are four principal types of chemical bonds: ionic, covalent, metallic, and van der Waal's bonds. Ionic bonded minerals are easily soluble in water. Covalently bonded minerals are hard. Metallic minerals are excellent conductors of heat and electricity. Minerals with van der Waal's bonds are easily cleaved.

[3]   The cation/anion radius ratio, the coordination number, the nature of the polyhedron, and the stability of crystal structures are related. These relationships are described by Pauling's rules.

[4]   Crystal symmetry may be precisely described with the help of motifs and symmetry operations. "Standard" symmetry elements include the plane of symmetry (mirror), the axis of rotation, and the center of symmetry (inversion center). In addition, a rotoinversion operation can produce symmetry elements that cannot be entirely described by standard operations.

[5]   The representation of symmetry elements can be done most efficiently on a stereogram using conventional symbols. In a stereogram planes and axes of symmetry are all projected onto a horizontal circular plane which is essentially a slice through the middle of a sphere. A horizontal plane plots as the perimeter of the stereogram. Vertical planes plot as straight lines cutting across the stereogram. A vertical axis plots in the center, and horizontal axes plot on the perimeter of the stereogram.

[6]   The combination of all standard and rotoinversion operations leads to 32 possible symmetry operations, which define the 32 crystal classes.

[7]   Crystallographic axes are imaginary reference axes used to illustrate the geometry of a crystal and its symmetry elements. There are six crystal systems: isometric, tetragonal, orthorhombic, hexagonal (rhombohedral and hexagonal divisions), monoclinic, and triclinic systems.

[8]   The unit cell is the smallest unit composed of atoms that, when repeated in three dimensions, can define the larger crystal's structure. There are 14 Bravais lattices that are fundamental to all crystal structures.

[9]   Facial intercepts on crystallographic axes are converted to Miller indices in three steps: the division of all axial intercepts by a common denominator (the b-axis intercept or a_2 for isometric, tetragonal, and hexagonal systems), inverting these axial ratios, then clearing fractions.

[10] Crystallographic measurements involve the determination of interfacial angles with a contact goniometer.

[11] A form includes similar faces in a crystal. There are many different types of forms that are named mainly on the basis of the number and shapes of the faces.

[12] Imperfections in crystals can arise from the dislocations of atoms at points, lines, or planes. Twinned crystals result from growth, polymorphic transformation, or deformation.

BIBLIOGRAPHY

Battey, M.H. (1981) *Mineralogy for Students* (2nd ed.). Longman (New York).

Blackburn, W.H. and Dennen, W.H. (1994) *Principles of Mineralogy* (2nd ed.). W.C. Brown (Dubuque).

Bloss, F.D. (1971) *Crystallography and Crystal Chemistry.* Holt, Rinehart and Winston (New York).

Cepeda, J.C. (1994) *Introduction to Minerals and Rocks.* Prentice-Hall (Englewood Cliffs, New Jersey).

Ernst, W.G. (1969) *Earth Materials.* Mcmillan (New York).

Perkins, D. (1998) *Mineralogy.* Prentice Hall (Upper Saddle River, New Jersey).

USEFUL WEBSITES

www.geologylink.com
www.webmineral.com

CHAPTER 3

Minerals

The following topics are covered in this chapter:

Definition of mineral
Polymorphism and solid solution
The world of minerals—dominant minerals in the crust and mantle
Chemical classification of minerals
Physical properties of minerals

ABOUT THIS CHAPTER

This chapter is about minerals: what they are and how they may be studied, described, and identified in the field or the laboratory on the scale of a hand-sized specimen. Physical properties, such as color and hardness, can be diagnostic in some cases where the mineral grains are sufficiently large; however, when the grain sizes are small, more sophisticated techniques (described in later chapters) must be used for identification.

MINERALS

When we think of minerals, we think of diamond, emerald, garnet, ruby, jade, peridot, topaz, and other gemstones. The reason for our familiarity with these minerals is simply that they are used in jewelry. In nature, however, with the exception of peridot, garnet, and quartz, all of these minerals are rare. Minerals are important because virtually the entire earth and other terrestrial planets and their moons are made of them; our bones are made of minerals, and our existence depends on them. For example, the mineral quartz is the major component of glass, quartz watches, oscillators, and silicon chips in computers. Also, quartz is commonly used in the industry as an abrasive mineral, as fluxes, and as a building material. Without quartz, we would not be able to have a home with concrete walls, have a "glass" of water, keep time, or turn on a MacIntosh! Quartz played an important role in World War II because of its special property, called *piezoelectricity*, which made it useful in detecting submarines.

Exactly what is a *mineral*? In this book a mineral is flexibly defined as a naturally occurring crystalline solid with a definite chemical composition. A crystalline solid is made up of crystals, which are composed of atoms (or ions) arranged in specific geometrical frameworks. Such internal organization of atoms puts certain limits on the chemical composition (i.e, elements and their proportions) of a mineral. Although the chemical compositions of an individual mineral may vary, for example, due to solid solution (discussed in a later section) or the "impurities" of foreign ions/atoms, all such variations cannot exceed certain limits imposed by the mineral's crystal structure. Therefore, a mineral's chemical composition may not be absolutely fixed but may vary within certain limits. However, a distinction can still be made between two minerals on the basis of their chemical compositions, for example, between diamond (whose composition is elemental carbon) and quartz (SiO_2).

The definition of a mineral used in this book indicates that all minerals must form naturally. In modern times, however, many "minerals" are synthetically made in the laboratory from appropriate chemical ingredients for industrial use and for experimentation. Synthetic minerals are studied to understand how minerals form in nature and to learn about the minerals within the deep regions of the earth that cannot be drilled. Although the "natural" requirement in the definition excludes such synthetic minerals, a fair number of modern practitioners of the subject generally ignore the "naturally occurring" requirement and call the synthetics minerals anyway! The "natural" requirement is centuries old and does not contribute anything to modern mineralogy, where much of the exciting research is being done on minerals made in the laboratory at super high pressures and temperatures. Therefore, deleting the "natural" requirement is appropriate.

The requirement that a mineral must be a solid makes ice a mineral, but not water or water vapor—the two most important ingredients for life on earth! It also excludes petroleum, an important natural resource. Despite these "shortcomings," the definition given earlier is quite useful and appropriate and is therefore followed by most mineralogists.

Naming a mineral does not follow any specific set of rules, and may be based on (a) the place where it is commonly found or was first described, (b) a famous scientist, (c) a specific chemical character, or (d) a particular physical property, such as color. For example, the mineral magnetite is so named because of its special magnetic property. The mineral olivine has an olive-green color. The element calcium is an important component of the mineral calcite. The mineral stishovite is named after Stishov, the scientist who first discovered this mineral as an extremely high pressure equivalent (or polymorph, discussed in the next section) of the common mineral quartz. Stishovite is found around meteorite craters because only a large meteorite impact can generate the sort of pressure needed to produce stishovite from any SiO_2-rich mineral or material.

Polymorphism and Solid Solution

Polymorphism is the phenomenon by which minerals, called *polymorphs*, with distinct internal atomic structures but identical chemical composition can form. Note that, although polymorphs have identical chemical composition, they must be given separate names because of their difference in atomic structure. Many examples of polymorphism occur in rocks: for example, there are numerous SiO_2 polymorphs—low quartz, high quartz, tridymite, cristobalite, coesite, and stishovite. The pressure–temperature conditions of stability of each mineral are discussed in Chapter 7. Perhaps the most familiar examples of polymorphism are diamond and graphite; they are both composed of carbon, but diamond has a dense covalent-bonded structure, whereas graphite has a fairly open sheet structure with adjacent sheets being held together by weak chemical bonds. (See Figure 2.2b in Chapter 2.) These structural differences are also the reason diamond is the hardest and graphite is the softest mineral.

Solid solution is a phenomenon in which one or more ions can substitute for other ions in a mineral's atomic structure without seriously distorting the crystal structure or rendering it unstable by introducing chemical inequalities. Minerals that form a solid solution series are referred to as isomorphs and the series is called isomorphous series because they all have similar atomic structure and crystal morphology. In general, these substitutions are possible only when the substituting and substituted ions are within 15% of each other's ionic size, and when their positive and negative charges cancel out such that the resultant atomic structure does not have any remaining electrical charge. An example of this type of substitution is the mineral olivine, whose composition is generally written as $(Mg,Fe)_2SiO_4$ because Mg^{2+} and Fe^{2+} can substitute for one another in olivine's structure. Because of such substitution, a whole range of olivines may be found whose compositions vary between the "end members"—pure Mg_2SiO_4 (called *forsterite*, abbreviated as Fo) and Fe_2SiO_4 (*fayalite*, abbreviated as Fa)—as illustrated in Figure 3.1a. Thus, one olivine crystal may have a 60 mole% forsterite component and a 40 mole% fayalite component. (The concept of mole% is discussed elsewhere.) The composition of such a crystal may be written as $(Mg_{60}Fe_{40})_2SiO_4$ or as $Fo_{60}Fa_{40}$. Another olivine crystal in a different rock may have 80% forsterite and 20% fayalite components in solid solution; and its composition would be given as $(Mg_{80},Fe_{20})_2SiO_4$ or $Fo_{80}Fa_{20}$. This type of complete substitution extending from one end member to the other is called *diadochy*. In the case of olivines, note that Mg^{2+} and Fe^{2+} ions are very similar in size (Mg^{2+} and Fe^{2+} ions have radii of 0.78Å and 0.83Å, respectively) and both carry a 2+ charge. Thus, their mutual substitution does not affect the atomic structure in a serious way, nor is there any leftover charges. This makes the olivine structure stable and electrically neutral.

There is a second type of solid solution, which results from coupled substitution of two (or more) ions by two (or more) other ions in the atomic structure (rather than one ion by another as in the olivine example). A common rock-forming mineral group—the plagioclase feldspars, exhibits this type of solid solution. Plagioclases exhibit extensive solid solution between albite ($NaAlSi_3O_8$) and anorthite ($CaAl_2Si_2O_8$) end members (Figure 3.1b). In terms of cation size, Na is the largest, followed by Ca. Al and Si, both being much smaller, are similar to each other in size. The

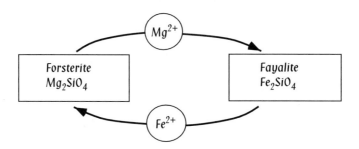

(a) Diadochy in olivine solid solution

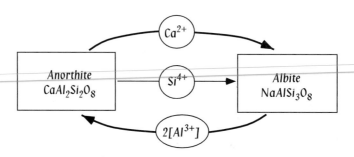

(b) Coupled substitution in plagioclase solid solution

FIGURE 3.1   *The nature of cation exchange in (a) olivine solid solution and (b) plagioclase solid solution.*

Na$^+$ ion has one positive charge, whereas Ca^{2+} has two positive charges. Therefore, when a Na$^+$ ion is replaced by a Ca^{2+} ion, an excess positive charge is produced. This extra + charge must be canceled by another ionic substitution in the structure in order to make the atomic structure of the mineral electrically neutral. This is achieved by a simultaneous substitution of a Si^{4+} ion by an Al^{3+} ion in the atomic structure. Solid solutions of this type are known as solid solution by coupled substitution.

THE WORLD OF MINERALS

Minerals are the building blocks of the earth much like bricks are the building blocks of houses. Although there are over 3000 minerals that occur in nature, only about 25 of these compose the bulk of the earth. The most abundant minerals have one thing in common: they are all silicates, that is, the two dominant elements in them are silicon and oxygen. This is perhaps not surprising when one considers the fact that silicon and oxygen are also the two most abundant constituents of the earth's crust and mantle (Figure 3.2). Together they make up almost three-fourths of the earth's crust and mantle.

CLASSIFICATION OF MINERALS

The Berzelian classification of minerals, in variously modified forms, is generally accepted by most scientists. The basis of this classification is the main anion or anion complex present in the mineral's atomic structure. Table 3.1 shows various classes into which minerals may be grouped on the basis of their chemical composition. Among all the groups, silicates comprise 95% of the crust and approximately 97% of the mantle. Carbonates are the major component of a group of sedimentary rocks called limestones and dolostones. Oxide minerals are a common constituent in many different types of rocks. Other mineral groups are not globally abundant.

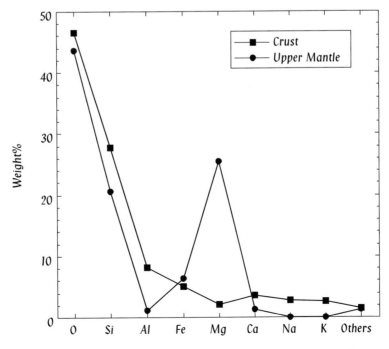

FIGURE 3.2   *Elemental abundances in the earth's crust and upper mantle.*

TABLE 3.1   *A simplified berzelian classification scheme.*

Group	Main Anion/Anion Complex	Notable Examples
I. Native elements	Metals and nonmetals	Diamond, graphite (C) Gold (Au) Copper (Cu) Sulfur (S)
II. Sulfides	S	Galena (PbS) Pyrite (FeS_2) Chalcopyrite ($CuFeS_2$) Sphalerite (Zn,Fe)S
Sulfosalts	S, As,Sb	Niccolite (NiAsS)
III. Oxides	O	Magnetite (Fe_3O_4) Hematite (Fe_2O_3) Ilmenite ($FeTiO_3$) Rutile (TiO_2) Uraninite (UO_2) Spinel ($MgAl_2O_4$) Chromite ($FeCr_2O_4$) Corundum (Al_2O_3)
Hydroxides	OH	Manganite (MnO(OH)) Gibbsite ($Al(OH)_3$)
IV. Halides	Cl, F, Br, I	Fluorite (CaF_2) Halite (NaCl) Sylvite (KCl)
V. Carbonates	CO_3	Calcite, aragonite ($CaCO_3$) Magnesite ($MgCO_3$) Siderite ($FeCO_3$) Dolomite ($CaMg(CO_3)_2$)
VI. Sulfates	SO_4	Barite ($BaSO_4$) Celestite ($SrSO_4$) Gypsum ($CaSO_4, 2H_2O$) Anhydrite ($CaSO_4$)
VII. Phosphates	PO_4	Apatite ($Ca_5(PO_4)_3(F,Cl,OH)$)
VIII. Silicates	SiO_4	Olivine (($Mg,Fe)_2SiO_4$) Diopside ($CaMgSi_2O_6$) Enstatite ($MgSiO_3$) Anorthite ($CaAl_2Si_2O_8$) and many more
IX. All others	Includes nitrates, borates, titanates, and others	

Silicates

The basic structural unit of all silicate minerals is the $[SiO_4]^{4-}$ tetrahedron. This tetrahedron results from a 4-fold coordination between a small Si^{4+} cation at the center and four O^{2-} anions occupying the apices of the tetrahedron (Figure 3.3). The bonds

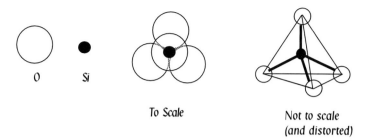

To Scale

Not to scale (and distorted)

Silicate Tetrahedron

FIGURE 3.3 *The nature of the silicate tetrahedron.* The sketch in the middle shows more or less the exact size of the Si and O atoms, and the sketch on the right shows the shape of the tetrahedron formed by the Si and O atoms and the bonds (heavy lines).

between the oxygen anions and the silicon cation are covalent; therefore, it is hard to break down the silicate chains even through melting.

Silicates are classified into several subclasses based on how the silicate tetrahedra are connected to one another (Figure 3.4). The simplest subclass is called *nesosilicate* (or *island silicate*) in which the tetrahedra are independent of each other—they do not share any corners, edges, or faces. Instead, they are linked within the atomic structure of the mineral by other cations, such as Mg^{2+}, Fe^{2+}, and Ca^{2+}.

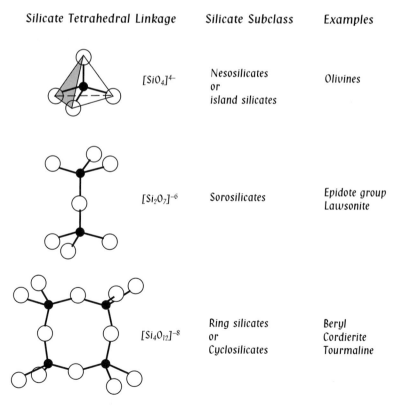

Silicate Tetrahedral Linkage	*Silicate Subclass*	*Examples*
$[SiO_4]^{4-}$	Nesosilicates or island silicates	Olivines
$[Si_2O_7]^{-6}$	Sorosilicates	Epidote group Lawsonite
$[Si_4O_{12}]^{-8}$	Ring silicates or Cyclosilicates	Beryl Cordierite Tourmaline

FIGURE 3.4 *Classification of silicate minerals into various subclasses based on the organization of silicate tetrahedra.* (From Ernst, 1969, *Earth Materials*, Mcmillan and Co.)

(continues on next page)

Silicate Tetrahedral Linkage	Silicate Subclass	Examples

Inosilicates
or
chain silicates

Single chain   Pyroxenes

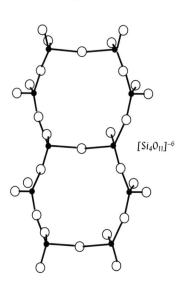

$[SiO_3]^{-2}$

Double chain   Amphiboles

$[Si_4O_{11}]^{-6}$

Tektosilicate

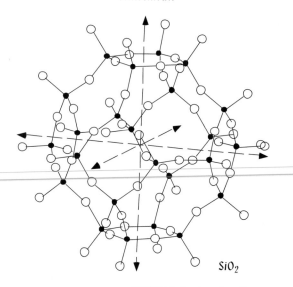

0   3
Angstrom

SiO_2

FIGURE 3.4   *(continued)*

A good example of the nesosilicates is forsterite, in which $[SiO_4]^{-4}$ tetrahedra are linked to one another by Mg^{2+} cations such that the composition becomes Mg_2SiO_4. The next subclass is *sorosilicate*, in which two adjacent $[SiO_4]^{-4}$ tetrahedra share a common apex such that for every two Si ions there are now seven O anions in the structure (instead of the eight that are present in nesosilicate structure). Thus, the anion formula of sorosilicates is $[Si_2O_7]^{-6}$. Two important minerals found in low to moderately low grade metamorphic rocks, the *epidote* group minerals and *lawsonite*, are examples of sorosilicates. In the third group, referred to as *cyclosilicates* (also called *ring silicates*), each silicate tetrahedron shares two oxygens with two other tetrahedra, and the resultant structure takes the form of a ring. Greater complications in tetrahedral arrangement lead to the formation of the other silicate groups—*inosilicates* (also known as *chain silicates*), *phyllosilicates* (also called *sheet silicates* and *layer-lattice silicates*), and *tektosilicates* (also known as *framework silicates*).

There are two types of inosilicates: *single chain* and *double chain* (Figure 3.4). The chains are oriented parallel to the *c*-crystallographic axis. Two important mineral groups are inosilicates—*pyroxenes* (single chain) and *amphiboles* (double chain). The sharing of oxygens by adjacent tetrahedra in a single chain leads to a basic compositional unit of $[SiO_3]^{-2}$. In enstatite, a type of pyroxene, Mg^{2+} cations laterally link the single chains in the atomic structure, such that its chemical formula becomes $MgSiO_3$. In another important type of pyroxene, called diopside, Ca^{2+} and Mg^{2+} cations together form the link between the chains leading to its chemical formula—$CaMgSi_2O_6$. In a double chain the fundamental unit is $[Si_4O_{11}]^{-6}$. Examples of this subgroup are the amphiboles *anthophyllite* $[Mg_7(Si_4O_{11})_2(OH)_2]$ and *tremolite* $[Ca_2Mg_5(Si_4O_{11})_2(OH)_2]$.

In phyllosilicates the tetrahedra form sheets composed of six-sided double rings that extend along the *a* and *b* crystallographic axes. These sheets are linked vertically by other cations and hydroxyl ions. The fundamental unit is $[Si_4O_{10}]^{-4}$. Examples of this group are micas and clay minerals. Note that Si is not the only cation to form tetrahedral units in many sheet silicates: for example, in micas Al forms one-fourth of the tetrahedral units, and thus the fundamental unit in micas is $[Si_3AlO_{10}]^{-5}$. The chemical formula for *muscovite* (white mica) is $KAl_2Si_3AlO_{10}(OH)_2$.

In tektosilicates each tetrahedron shares all four oxygens with neighboring tetrahedra forming a three-dimensional framework such that the basic unit of this group is two oxygens per tetrahedrally coordinated cation. Two abundant minerals of the earth, the feldspars and quartz, belong to this group. While Si is the only tetrahedrally coordinated cation in quartz (SiO_2), in feldspars Al also occurs in one-fourth of the tetrahedral sites: for example, orthoclase ($KAlSi_3O_8$).

Nonsilicates

Nonsilicate minerals include all other mineral classes listed in Table 3.1. Although nonsilicate minerals are much less abundant than silicate minerals, many of them are very important for economic and industrial reasons.

Native Elements.   Many minerals occur in their native state in nature, including sulfur, diamond (carbon), graphite (carbon), copper, gold, and silver. These minerals do not have much in common other than occurring as native elements in nature. As we know, copper, gold, and silver are metals and therefore conduct electricity and heat. Graphite is a semi-metal, whereas diamond is an insulator.

Sulfides and Sulfosalts.   In this group, S, As, Sb, Bi, Te, and Se combine with Fe, Pb, Cu, and many other elements. The more commonly known minerals belonging to this group include chalcopyrite ($CuFeS_2$), pyrite (FeS_2), pyrrhotite ($Fe_{1-x}S$), pentlandite ($(Fe,Ni)_9S_8$), sphalerite (ZnS), galena (PbS), and the two arsenic minerals—realgar (AsS) and orpiment (As_2S_3).

Halides.   In this group, Cl, F, Br, or I (the halogens) form the major anion. The two familar examples are fluorite (CaF_2) and halite ($NaCl$). In passing, note that some silicate minerals, such as apatite, amphibole, and micas, are also known to contain small amounts of Cl and F in their structure.

Oxides and Hydroxides.   Oxygen and hydroxyls form the main anion (anion complex) in this group. Several examples are listed in Table 3.1. Among them the iron-titanium oxide minerals (principally magnetite, hematite, ilmenite, and rutile) are relatively more abundant. A descriptive account of some of the oxide minerals are given in Chapter 5.

Carbonates. $(CO_3)^{2-}$ is the principal anion complex in these minerals. The $CaCO_3$-$MgCO_3$-$FeCO_3$ minerals are the most familiar members of this group. These include the two $CaCO_3$ polymorphs—calcite and aragonite; and dolomite, magnesite, siderite, and ankerite (Table 3.1). Because carbonates minerals are the most abundant constituents of limestones and dolostones, we will learn more about them in Chapters 5 and 14.

Sulfates.   In this group $(SO_4)^{2-}$ is the principal anion complex. Familar examples are gypsum and anhydrite. We discuss them in some detail in Chapter 14 because they are important constituents of sedimentary deposits in arid environments.

Phosphates and the Other Mineral Groups.   An important member of the phosphate group is the mineral apatite (Table 3.1). $(PO_4)^{3-}$ is the main anion complex in all phosphates. As Table 3.1 shows, apatite has structurally bound H_2O, and can also contain Cl or F. Similar to the phosphates, borates, tungstates, vanadates, and several other mineral groups have been recognized which are named as such on the basis of their dominant anion complex.

PHYSICAL PROPERTIES OF MINERALS

Minerals are often easily distinguishable by their physical properties, such as color, crystal shapes, and hardness. For the identification of minerals in the field, a few simple tests using a pocket lens, a knife, a magnet, or a dull surface of broken porcelain (called a "streak plate") can go a long way. Below is a list of physical properties that are examined in routine hand-specimen identification of minerals.

Crystal Form and Habit

Because the physical appearance of a crystal is guided by its internal arrangement of atoms, the larger crystals of individual minerals may have distinctive forms. However, a crystal can only grow to large sizes when its growth is not hindered by other crystals, which is generally not possible in nature. A very good example is quartz (Figure 3.5) whose hexagonal crystal form is best developed in the large crystals typically found in *geodes*. A geode is a cavity, generally in a basaltic lava, lined with large crystals that grow from a vapor-rich fluid phase. Another example is mica, which typically

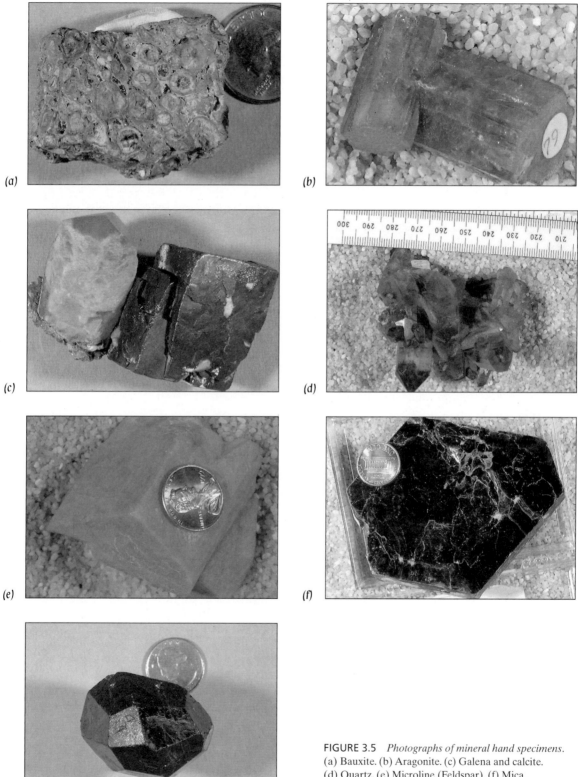

FIGURE 3.5 *Photographs of mineral hand specimens.*
(a) Bauxite. (b) Aragonite. (c) Galena and calcite.
(d) Quartz. (e) Microline (Feldspar). (f) Mica.
(g) Garnet.
(Courtesy of Gautam Sen)

has a flaky habit: that is, it peels off very easily along inherently weak planes called *cleavage planes*. Pyrite, or "fool's gold," is sometimes mistaken as gold because of its metallic gold appearance. Pyrite crystals typically have the form of a cube.

Even a group of finer size crystals of a mineral may develop characteristic structures that are different from other minerals. For example, the mineral bauxite, an important source for alumina, typically forms a *pisolitic* structure, which is an agglomerate of pea-like spheres (Figure 3.5). Psilomelane, a manganese ore mineral, has a characteristic *botroyoidal* structure—an agglomeration of smooth rounded surfaces.

Color

The color of a mineral is generally not a very reliable property in mineral identification. For example, the mineral quartz can be colorless, white, rose red, dark gray, or even black. In a few cases, however, color can be helpful as with the mineral olivine, which has a distinctive olive-green color. The mineral epidote is also green, but its green color is distinctly deeper in hue—more like grass. A mineral of much tectonic significance, glaucophane, has a distinctive blue color.

The color of a mineral results from its chemical composition. Some elements, even when present in minor amounts, give rise to specific colors: Cr yields green or red and Fe^{2+} yields yellow. For example, the mineral corundum (Al_2O_3) is commonly grayish, but with small amounts of Cr in solid solution, it becomes a red gem variety known as *ruby*. However, if minor amounts of Fe and Ti are incorporated in solid solution, corundum assumes a blue color, which is the gem *sapphire*. Garnets form an extensive solid solution. Almandine, an iron-bearing common garnet, is of deep brown color. On the other hand, Uvarovite, a rare chromium-garnet, is of deep green color.

Streak

Streak is the color of the finely powdered mineral. Although the color of a mineral may vary considerably from specimen to specimen, its streak is generally the same and, therefore, is quite useful in identifying minerals. This streak of a mineral is determined by grinding a sharp point or edge of the mineral on a dull porcelain plate, called a *streak plate*. Some typical streaks are as follows: magnetite—black, hematite—brownish red, pyrite—black, galena—grayish black. Silicate minerals, which dominate the earth's crust, generally have a colorless (hard minerals) or white (soft minerals) streak.

Luster

Luster refers to the way a mineral reflects light. Fundamentally, minerals exhibit *metallic* or *nonmetallic* lusters. Metals are generally opaque and reflect most of the light that strikes them. As a result they appear shiny, which indicates a metallic luster. Pyrite, magnetite, and galena are examples of minerals that show metallic luster. Minerals with nonmetallic luster are generally light colored and transmit light in a variety of ways. Appropriate names have been given to the different nonmetallic luster types. Some minerals, such as quartz, show the luster of a glass, which is called *vitreous* luster. *Dull* luster is one in which the mineral does not reflect light well at all, and as a result the mineral appears dull. Chalk (mineral name *kaolin*) is an example of dull luster. *Pearly* luster refers to the luster of pearl. *Resinous* luster is that of resin, an example of which is the mineral sulfur. Diamond shows an especially brilliant luster due to its ability to reflect light extremely well from all its faces. This luster is known as *adamantine* luster. When the luster is half-way developed, the prefix "sub-" is used to describe the texture as in *subvitreous* and *submetallic*.

Cleavage

Cleavage is a distinctive property of minerals. It is the natural tendency of minerals to break along regularly spaced, inherently weak planes. The weak planes form along directions where the bond strength is less relative to that in the other directions within a crystal's internal atomic structure. For example, micas and graphite have a very distinct *basal* set of cleavage planes. The reason for the development of such cleavage planes is simply that the van der Waal's bonds, which hold the covalent sheets together, are too weak.

The mineral calcite has three sets of cleavage planes, each having the shape of a rhombohedron; therefore, the calcite cleavage is referred to as rhombohedral cleavage (Figure 3.6). The mineral fluorite has four sets of octahedral cleavage. Two important minerals of the earth's crust, amphibole and pyroxenes, can be distinguished on the basis of their cleavage (Figure 3.6). Both minerals have two sets of cleavage planes; however, the angle subtended between the two sets is very different.

Parting looks much like cleavage but the planes are weakly developed and unevenly spaced. They form when cleavage is not well developed.

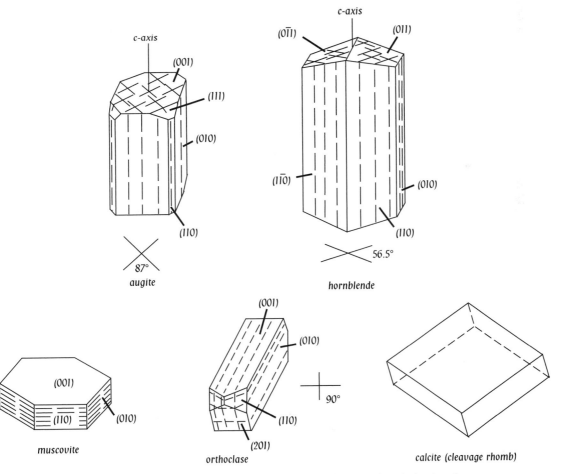

FIGURE 3.6 *Perspective sketches of the idealized crystals of certain minerals showing cleavage.*

FIGURE 3.7 *Photograph of conchoidal fracture.* (Courtesy of Gautam Sen)

Fracture and Tenacity

Minerals that lack distinct cleavage planes break in any possible direction, and the nature of the fracture surface can be a distinctive property for some minerals. For example, the mineral quartz lacks cleavage; when fractured, it develops smooth, curved fracture surfaces. This type of fracture is called *conchoidal* fracture (Figure 3.7). Other minerals have *uneven* or *irregular* or *hackly* fracture, and still others may be *splintery*.

Tenacity is a property that refers to how a mineral deforms plastically under external stress conditions. Minerals such as graphite and micas deform easily under stress whereas many other minerals (e.g., quartz) do not.

Hardness

A mineral's hardness refers to the degree of difficulty with which its surface may be scratched by another mineral or object of known hardness. In general, covalent-bonded minerals are harder than ionic-bonded minerals. Minerals possessing van der Waal's bonds, such as micas and graphite, are the softest.

In 1824, an Austrian scientist named Friedrich Mohs developed a scale for the relative comparison of mineral hardness. (See the following table.) Mohs' scale lists only 10 minerals, with diamond being the hardest having a hardness of 10. Talc, with a hardness of 1, is the softest. This scale is logarithmic—the numbers along the scale increase by an order of magnitude. For example, the *absolute* difference in hardness between talc and gypsum is 1, but the difference between gypsum and calcite is 10, and the difference between calcite and fluorite is 100, and so on.

1. Talc	*Softest*
2. Gypsum	
3. Calcite	
4. Fluorite	
5. Apatite	
6. Orthoclase	
7. Quartz	
8. Topaz	
9. Corundum	
10. Diamond	*Hardest*

This scale shows that the mineral apatite is harder than all the minerals from fluorite to talc; that is, apatite can scratch all of them but cannot be scratched by them. Mohs' scale shows that apatite cannot scratch any mineral with hardness of 6 or greater. In the field, it is more convenient to determine the hardness of a mineral with a fingernail (hardness of about 2), copper penny (~ 3), pocket knife (5), or glass (5.5).

The hardness of a mineral can vary with crystallographic directions. A notable example of this is the mineral kyanite (Al_2SiO_5), which has a hardness of 5 along the length of the crystal but 7 across it. In general, oxide minerals tend to be hard. Anhydrous silicates (e.g., olivine and pyroxene) are generally hard (H = 5.5–8), while hydrous silicate minerals and carbonate minerals are generally soft (micas and calcite; an exception is the amphiboles group).

Specific Gravity

The specific gravity of any substance is defined by the ratio of its weight to the weight of an equal volume of water at 4°C. In early days of mineralogy, when sophisticated analytical techniques did not exist, determining the specific gravity of a mineral was an important way of characterizing a mineral. Today, however, hardly anybody measures the specific gravity of a mineral, except to note that a few minerals, such as barite, are relatively heavy, which helps in their identification.

Special Properties

Some minerals possess special properties such as magnetism, luminescence, piezoelectricity, and radioactivity. As mentioned earlier, the mineral magnetite was named for its natural magnetic property. Radioactivity is a particularly important property of certain minerals and has seen many applications in modern times. From a geologist's point of view, two of these applications are the use of "radioactive clocks" to date various events in earth history, and the use of radioactive reactions to decipher the origin of rocks and global-scale processes deep within the earth.

Luminescence is a phenomenon in which light is emitted from a mineral in response to exposure to ultraviolet rays, X-rays, or cathode rays or heating (thermoluminescence). *Phosphorescence* is a type of luminescence in which light emission may continue even after the source of excitation, such as the ultraviolet rays, are cut off. On the other hand, if the mineral stops glowing as soon as the external source is removed, it is called *fluorescence*. In this phenomenon, electrons of certain atoms are excited by the external energy source and are raised to higher levels. Some of the energy is lost as light, which makes the mineral fluorescent. When the external energy source is cut off, the electrons may take a while to fall back to their "ground" levels (i.e., the original energy levels prior to excitation); thus, light may continue to emit from the mineral (phosphorescence). Such minerals are called phosphorescent because phosphorus is known to exhibit such behavior. Currently, synthetic phosphors are in heavy use in the making of fluorescent lamps, paints, and tapes.

3.1 METHOD OF LABORATORY/FIELD DESCRIPTION OF MINERALS

In describing a hand sample of a mineral in the field or in the laboratory, the following physical properties should be observed:

Crystal Form or Habit.   If the crystalline form (i.e., octahedron, rhombohedron, etc.) is clearly visible, it should be mentioned. If it is twinned, then the nature of the twins may be indicated. Such terms as bladed, columnar, acicular (needle-like), pisolitic etc. should be used to describe mineral structures.

Color/Opacity.   How transparent the mineral is to light should be described using such terms as transparent, translucent (semitransparent), and opaque. The mineral's color may be described as appropriately as possible. For example "olive" green is the color of olivine, whereas "grass" green is the color of the mineral epidote.

Luster.   Descriptive terms such as metallic, submetallic, vitreous, and dull should be used.

Streak.   A sharp point on the mineral grain is scratched against a streak plate, and the obtained streak is described using appropriate terms and adjectives, where necessary. For example, a "cherry"-red streak is typical of a mercury mineral called cinnabar, and reddish brown is more typical of the iron-oxide mineral hematite.

Cleavage.   First, one should note whether cleavage is well developed, moderately developed, or poorly developed. Then a description of how many cleavage sets are present and their interrelationships may be made. (See earlier discussion.)

Hardness.   Using fingernail, penny, glass plate, or other method, the hardness of the mineral should be determined.

Fracture.   Obtaining a fracture surface on a mineral does not generally require one to break it, since the specimens would soon disappear! Fracture characteristics should be easily visible if the mineral is hard and has poorly developed cleavage. In describing fracture, such terms as conchoidal (or subconchoidal if not as well developed), hackly, and splintery are used.

Specific Gravity.   As stated before, routine field or laboratory work does not require the precise measurement of specific gravity as in the "old times". It is generally sufficient to recognize how heavy the mineral is.

Special Properties.   Under this category one may describe whether the mineral is magnetic, fluorescent, or effervescent in dilute HCl (obviously a carbonate mineral).

While hand-specimen identification of minerals is a necessary first step, absolute identification generally requires optical examination of the specimen beneath a petrographic microscope (described later), and in specific cases, X-ray or electron microprobe techniques may need to be used.

SUMMARY

[1]   A mineral is a naturally occurring crystalline solid with a definite chemical composition and internal atomic structure.

[2]   Two or more different minerals, called polymorphs, can have identical chemical composition but very different atomic structures.

[3]   In minerals forming an isomorphous or solid solution series, two or more cations can substitute for each other.

[4]   Many different physical properties of minerals are described in routine field or hand specimen description of minerals, including color, luster, streak, cleavage, fracture, tenacity, specific gravity, and other special properties.

[5]   Special properties include magnetism, luminescence, piezoelectricity, and radioactivity.

CHAPTER 4

Optical Mineralogy

The following topics are covered in this chapter:

ABOUT THIS CHAPTER

Nature often does not permit minerals to grow sufficiently large that they can be identified with a pocket lens. Furthermore, we noted earlier that some of these physical properties, particularly color of hand specimens, can be quite misleading at times. A more definitive and most commonly used way of identifying minerals is to make a very thin slice (called a "thin section") of the mineral and study it with a petrographic microscope. *Optical mineralogy* is a subject that specifically deals with the optical behavior of minerals under a petrographic microscope. This chapter considers some of the fundamental principles that govern the subject of optical mineralogy.

SECTION 1

LIGHT AND PETROGRAPHIC MICROSCOPE

LIGHT

Light is a form of radiation that is detectable with the eye. For the purpose of studying minerals, light is best described as an electromagnetic wave that travels in a straight line, while its component electrical and magnetic vectors vibrate at right angles to the direction of propagation (Figure 4.1a). The direction of propagation of this wave

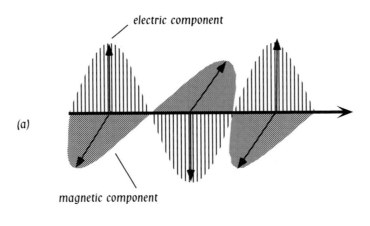

(a)

electric component

magnetic component

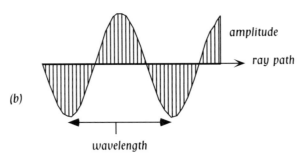

(b)

amplitude

ray path

wavelength

FIGURE 4.1 (a) Electric and magnetic components of a light wave. The two components vibrate perpendicularly to each other and to the direction of propagation of the ray. (b) Elements of a light wave.

is the light *ray path*. In optical mineralogy, only the electric vector is considered, and it is referred to as the *vibration direction* of the ray. A light wave is described in terms of its wavelength, amplitude, and frequency (Figure 4.1b). *Wavelength*, commonly denoted as λ, is the distance between two adjacent crests (or troughs) of a wave. The height of a crest is referred to as the *amplitude* of the wave. *Frequency* (f) is the number of wave crests that pass through a fixed reference point per second. In optical mineralogy, f is constant regardless of the material through which it travels. *Velocity* (V), frequency, and wavelength are related by the following equation:

$$V = f\lambda. \qquad \text{[Eq. 4.1]}$$

Light composed of a single wavelength is called *monochromatic light*. An example of this is the yellow light that is emitted by the sodium vapor lamps found in many mineralogy laboratories. Visible "*white light*" is composed of a spectrum of colored light components with wavelengths ranging from red ($\lambda = 700$ nanometers, or nm; 1 nm $= 10^{-7}$ cm) to violet ($\lambda = 400$ nm); blue, green, yellow, and orange form the intermediate components (Figure 4.2). Equation 4.1 tells us that in order to maintain a constant f, V must proportionally increase with λ. Therefore, during the propagation of white light, the red component must travel faster than the violet component.

It is important to point out that light does not consist of a single wave but of an infinite number of waves. A *wave front* is defined as an imaginary surface that connects equivalent points on adjacent waves propagating from a light source (Figure 4.3). A *wave normal* is an imaginary line perpendicular to the wave front. As Figure 4.3 shows, the wave normal and the ray path, which is the direction of propagation of a

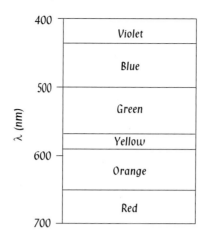

FIGURE 4.2   *Component wavelengths of "white light" expressed in nanometers (nm).*

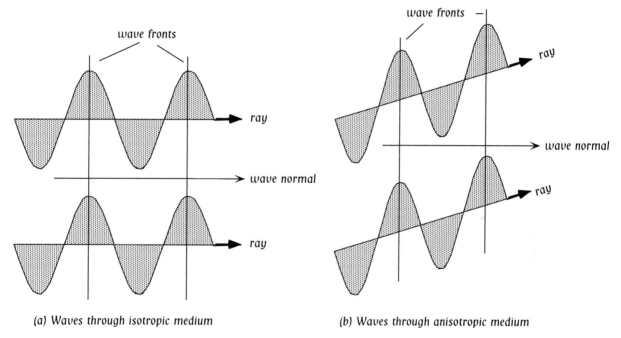

(a) Waves through isotropic medium

(b) Waves through anisotropic medium

FIGURE 4.3   *Wave fronts, ray path, and wave normal.* The wave normal coincides with the ray path in isotropic substances and follows a separate direction in an anisotropic medium.

wave, may or may not coincide depending upon whether the mineral is isotropic or anisotropic. In *isotropic* minerals, light travels in all directions with the same velocity. In *anisotropic* crystals, light is slowed down along specific crystallographic directions. The isotropic/anisotropic behavior of minerals is directly related to their atomic structures. Minerals that crystallize in the isometric system have uniform atomic structure in all dimensions; therefore, light travels with the same velocity in all directions through such material. In all other crystal systems, the atomic structure (spacing of atoms, strength of chemical bonds) varies along different crystallographic directions; and as a result, light traveling in some directions may meet greater resistance and may thus be preferentially slowed down.

REFLECTION, REFRACTION AND ABSORPTION

When light travels through two different media (say, air and water), it may be reflected or refracted (bent) at the interface between the two media (Figure 4.4). Other phenomena, such as total absorption, partial absorption, or absorption of only selective wavelengths may also occur (Figure 4.5).

In reflection, a light ray incident from a source is reflected (turned back) from the interface at which the two media come together. The *law of reflection* states that the angle of incidence (the angle subtended by the incident ray with the interface) must be equal to the angle of reflection (the angle between the reflected ray and the interface).

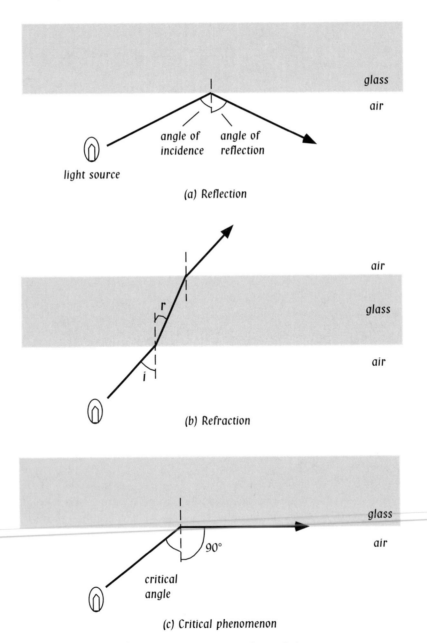

FIGURE 4.4   *Reflection, refraction, and critical phenomenon.*

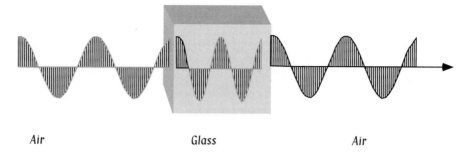

(a) Shortening of wavelength

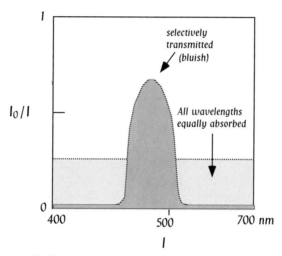

(b) Selective versus nonselective absorption

FIGURE 4.5 (a) Shortening of the λ of a light wave as it passes through glass. (b) Selective versus nonselective absorption of white light. The vertical axis shows transmissivity expressed in terms of the ratio I_0/I, where I is the intensity of the incident light and I_0 is the intensity of the transmitted light. Two examples are shown here: (1) Selective absorption. The medium absorbs all wavelengths except the blue wavelength of the incident white light. Thus, this medium would appear blue. (2) The second example in light gray is that of a medium that absorbs all wavelengths equally.

The phenomenon of *refraction* refers to the bending of light as it travels across the interface between two media. Exactly how much bending would occur is controlled by a property known as the refractive index. The *refractive index* of any material (conventionally expressed as *n*) is defined as the ratio of the velocity of light in vacuum (V_v) to the velocity of the same light in that material (V):

$$n = \frac{V_v}{V} \qquad \text{[Eq. 4.2]}$$

The refractive index of vacuum is therefore equal to 1.0. Most minerals have refractive indices between 1.4 and 2.0 (Nesse 1991). It is important to note from equation 4.2 that the refractive index of any material is inversely proportional to the velocity that travels through it (i.e., $n \propto (1/V)$). When a light wave travels from the air through glass, its velocity is reduced because glass has a higher refractive index than air. Recall that $V = f\lambda$ and *f* is constant; therefore, this lowering of the velocity of light through the glass results in a shortening of the wavelength (Figure 4.5a).

Note that an isotropic mineral is characterized by a unique refractive index value for a particular wavelength of light. This is because the velocity of such light will not vary with the crystallographic direction of such a mineral. However, we noted earlier that the velocity of such light will vary with crystallographic direction in an anisotropic mineral. This means that the refractive index value of an anisotropic mineral corresponding to such light will also vary with crystallographic direction.

Snell's law is an important law that relates the bending of a light ray to the refractive indices. It is best understood through the following expression:

$$n_i \sin i = n_r \sin r \qquad \text{[Eq. 4.3]}$$

where i is the angle of incidence, r is the angle of refraction, and n_i and n_r are refractive indices of the incident medium and the refractive medium, respectively (Figure 4.4b). Consider a case where a ray of light is traveling through an air (the refractive index is approximately 1) and glass (the refractive index is approximately 1.54) interface. Assuming that the angle of incidence (i) is 60°, the amount of "bending" (i.e., angle of refraction) by the ray may be calculated from Snell's law as follows:

$$1 \cdot \sin 60° = 1.54 \cdot \sin r$$

or

$$r = \sin^{-1}(0.649 \times 0.866) = 34.2°$$

When $i = 90°$ (the incident beam of light is perpendicular to the interface), the light does not bend but simply travels through the glass with a slightly lower velocity. When $r = 90°$, the refracted ray will neither refract nor reflect but pass along the interface between the two media (Figure 4.4c). This happens only for a specific value of i, which is called the *critical angle*, and the phenomenon is known as the *critical phenomenon*. For any angle of incidence greater than the critical angle, the ray is totally reflected back (a situation called *total internal reflection*). The critical angle depends upon the refractive indices of the two media that form the interface and can be calculated from Snell's law.

Total absorption of all components of incident white light by a mineral makes it appear opaque. However, if some white light with significantly reduced intensity passes through the mineral, the mineral appears translucent. Very little absorption makes a mineral look transparent. Some minerals may selectively absorb some wavelengths of white light while allowing others to go through and, as a result, may appear colored (Figure 4.5b). For example, a blue mineral appears blue because it absorbs all wavelengths of white light except blue. Such preferential absorption may vary with crystallographic directions of the individual crystals of a mineral—a property known as *pleochroism*—and the mineral is said to be *pleochroic*. This property is particularly useful in identifying minerals such as amphibole, biotite, staurolite, and tourmaline (to name just a few). For example, hornblende is strongly pleochroic and goes from light green to deep green as it is rotated through 360° on a microscope stage under plane polarized light (discussed in the next section).

When white light travels from air through a thick glass prism, it breaks down into its color components of different wavelengths due to a phenomenon called *dispersion*. Figure 4.6 shows how this happens. As noted before, various color components of the white light are characterized by different wavelengths with red ($\lambda = 700$ nm) and violet ($\lambda = 400$ nm) being the two extremes. Naturally, the red component would travel faster than the violet through the prism because $V = f\lambda$. Recall that velocity is inversely proportional to the refractive index, and therefore, the glass prism will have two different refractive indices for red versus violet components. Because of

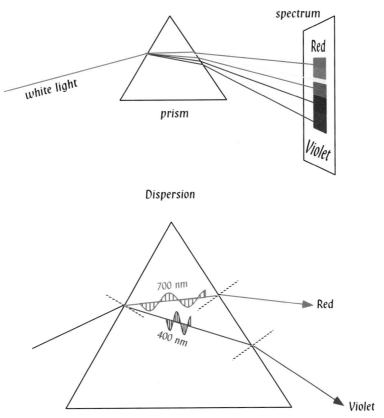

FIGURE 4.6 *Dispersion of white light into its color components as it passes through a glass prism.* The lower diagram shows that the two extreme wavelengths of white light (red and violet) refract differently at each of the two interfaces where glass meets air. Such differential bending of the component waves of various wavelengths is called dispersion.

this difference in refractive index, the red and violet component rays will bend by different amounts (i.e., their angles of refraction will be different) obeying Snell's law at each interface: first when the light enters the glass prism and a second time when the component waves exit the glass prism. The differential bending of the component waves causes them to separate or disperse from each other. As Figure 4.6 shows, the thicker the glass prism, the greater is the separation (dispersion).

POLARIZATION

Unpolarized light, like the light that is produced from a light bulb, is light that vibrates equally in all directions perpendicular to the direction of propagation (Figure 4.7). *Polarized light*, on the other hand, vibrates only in specific directions. Three types of polarization are possible depending upon whether the vibration directions of the polarized wave are confined to a plane (*plane polarization*), or to a helical surface whose cross section is a circle (*circular polarization*) or an ellipse (*elliptical polarization*). In optical mineralogy we will be mostly concerned with *plane-polarized light* (Figure 4.7). Polarization of light may result from a number of phenomena such as pleochroism, reflection, refraction, and scattering (Nesse 1991). The polarizing films used in sunglasses and petrographic microscopes are pleochroic materials.

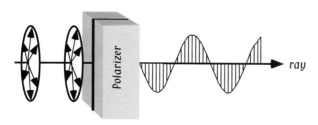

FIGURE 4.7 *Conversion of an unpolarized ray (vibrating in all directions perpendicular to the direction of propagation) to a plane-polarized ray (vibrating in the plane of the polarizing plate).*

Light vibrating
in all directions

Light vibrating
only in the plane
of the polarizing plate

THE PETROGRAPHIC MICROSCOPE

The petrographic microscope is one tool that every mineralogist (the specialist who studies minerals) and petrologist (a specialist on rocks) must use for routine identification and characterization of minerals and rocks. The primary function of any microscope, petrographic or not, is to magnify an object significantly so that detailed observations of the object may be made. All modern microscopes must have an illuminator to provide a light source and a number of magnifying lenses to magnify the object.

The standard petrographic microscope (Figure 4.8) comes with a substage assembly that houses a polarizer (or *lower polar*), which turns the incoming ordinary light into a plane-polarized light. Two *condenser lenses* are mounted in the same assembly—one fixed and another, which may be swung in and out of the light path. The fixed condenser lens provides *orthoscopic illumination*; that is, it illuminates the sample on the stage equally without significantly converging the light. The free-swinging condenser lens provides *conoscopic illumination*; that is, it strongly converges the light beam from the source to the sample such that the light emerging from it takes the geometric form of an inverted cone. Conoscopic illumination is required for obtaining interference figures on anisotropic minerals (discussed later). There is also an *iris diaphragm* in the substage assembly that has the sole purpose of adjusting the size of the light beam that may reach the mineral or rock placed on the *stage*. The stage is circular and rotates freely. The perimeter of the stage is marked in degrees so that the angles of rotation of the stage may be measured.

Immediately above the stage is a movable nosepiece that generally houses three *objective lenses* capable of providing different magnifications (most microscopes have $4.5\times$, $10\times$, and $40\times$ objectives). Each of these may be used at different times by simply rotating the nosepiece. The *numerical aperture (NA)* of an objective lens is a measure of the size of the cone of light that comes through the lens and is given by the equation $NA = n \sin(AA/2)$, where AA is the *angular aperture* (Figure 4.9). The *free working distance (FWD)* is the distance between the end of the lens and the top of the sample in focus (Figure 4.9). Different objective lenses have different FWD's: the higher the magnification the smaller the FWD. One needs to be particularly careful when using the $40\times$ objective because the FWD is so small (less than a millimeter) that one may accidentally run it right into the glass slide or thin section.

Above the objective lens housing assembly (the nosepiece) there is a slot through which an *accessory plate* may be inserted in the path of the light, which has gone through the polarizer and the mineral thin section. Three types of accessory plates are used: the *gypsum plate* (also called the first order ($1°$)-red plate), the *mica plate* (also called a quarter wavelength or $1/4$-λ plate), and the *quartz wedge*. These

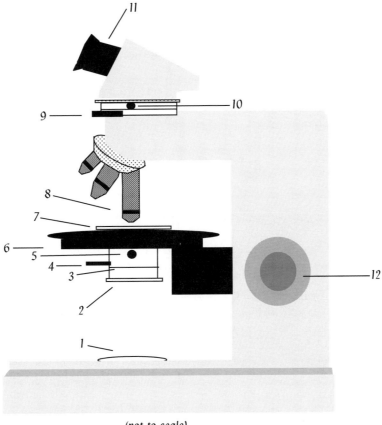

(not to scale)

1.	Light source	7.	Thin section
2.	Substage assembly	8.	Objective
3.	Polarizer	9.	Accessory plate
4.	Iris diaphragm	10.	Analyzer
5.	Condenser lens	11.	Ocular
6.	Stage	12.	Coarse and fine adjustments for vertical movement of stage

FIGURE 4.8   *The petrographic microscope and its components (side view).*

accessory plates are extremely useful in determining the optic sign and other properties of anisotropic minerals.

The *upper polar*, or *analyzer*, is a polarizing film located above the accessory slot that may be inserted into the path of the oncoming light from the objective lens via a sliding or rotating mechanism. The analyzer is oriented in such a way that its preferred polarization direction is exactly perpendicular to that of the lower polar or the polarizer. When the analyzer is inserted, the two polars (upper and lower) are said to be *crossed*, and the plane-polarized light coming through the polarizer is completely absorbed by the analyzer. Under crossed polars, isotropic materials always stay dark or *extinct*. This is because the plane-polarized light emerging through the lower polar will vibrate in the preferred planar direction of the polar and will continue to vibrate in the same plane as it goes through the isotropic material (although its wavelength will shorten). With the upper polar's vibration direction being exactly

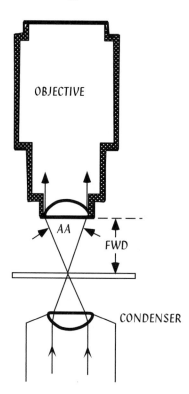

FIGURE 4.9 *The condenser and objective lenses.* AA = angular aperture. FWD = full working distance between the mineral and the objective lens. (See the text for further explanation.)

perpendicular to that of the lower polar, the upper polar will not allow any light to go through it (Figure 4.10).

Above the analyzer is the *Bertrand lens*, which is a small lens that sits on a pivot. It can be inserted into the path of light when needed. This lens is generally not needed in the routine description of rocks and minerals, but it needs to be used while determining the optic sign of an anisotropic mineral.

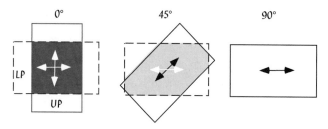

FIGURE 4.10 *The crossing of two polars by rotating one over the other through various angles.* The phenomenon of extinction and crossing vibration directions of the two light waves is illustrated here.

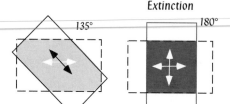

A cylindrical housing containing lenses called *oculars* or *eyepieces* slides into the microscope tube that forms the uppermost part of the petrographic microscope. These lenses magnify and focus the light to the eye of the observer. The magnification it provides is usually between 5 and 12×. The total magnification of the microscope is thus equal to the magnification of the objective lens multiplied by the magnification of the ocular.

In binocular petrographic microscopes, the right ocular is fitted with *crosshairs*, which are two thin black lines that form a cross. These crosshairs are generally oriented in north–south and east–west directions. In the routine use of a microscope, the crosshairs must be brought into focus first by rotating the upper part of the ocular. The image may then be focused with the same ocular by turning the coarse and fine focusing knobs that are located on the arm of the microscope to move the stage up or down. If using a binocular scope, the focusing mechanism of the second ocular may be rotated so that the image is focused on both eyes. Eyeglasses are not needed during the use of a microscope, as the same sort of focusing adjustments may be made.

Routine use of a microscope often results in distortion of the center of the field of view. That is, the axis of rotation of the stage no longer coincides with the center of the objective lens. Objective lenses come with a "centering device" that needs to be adjusted while rotating the stage for each of the objective lenses in order to make them "centered." Centering is more problematic for the high magnification objectives than the lower power objectives.

SECTION 2
OPTICAL PROPERTIES OF MINERALS

RELIEF

When a mineral grain mounted with epoxy on a glass slide is viewed with plane-polarized light through a petrographic microscope, the grain will seem to "stand out" of the glass slide provided that it has a significantly higher refractive index than the mounting medium (in this case it is epoxy, but could be immersion oil or something else). How high a mineral stands out in the surrounding mounting medium is referred to as its *relief*. The greater the difference in refractive index between the mineral and the mounting medium, the higher is its relief. Relief, which is really an approximate indicator of a mineral's relative refractive index, is an important property of a mineral. Minerals immersed in oil with a refractive index of 1.54 can be broadly classified into *low relief* (when the mineral does not stand out very well), *moderate relief* (the mineral "moderately" stands out), and *high relief* (the mineral stands out quite sharply). The same terms may be used even when a mineral grain is being viewed as part of a rock thin section. Among the common rock-forming silicate minerals, some that have conspicuously high relief are zircon, garnet, kyanite, sillimanite, andalusite, staurolite, and olivine. Micas, feldspars, and quartz are characteristically of low relief. Amphiboles and pyroxenes show moderately high relief but not nearly as high as garnet and other minerals that exhibit very high relief.

REFRACTIVE INDEX DETERMINATION: OIL IMMERSION TECHNIQUE

Quantitative determination of the refractive index of a mineral is done by the *oil immersion technique*, in which grains of the mineral are immersed in oils of known refractive indices. If the mineral and oil have identical refractive indices, light rays passing through the mineral and surrounding oil will not refract at the contact between the mineral and oil; therefore, the mineral grain boundary will not be visible (i.e., the mineral grain will "disappear"). If the refractive index of the oil and mineral are different, then refraction would occur at the contact between the two, which will make the grain visible. By carrying out such comparison with a number of different oils of known refractive indices, the mineral's refractive index is determined. This is carried out in several steps in the laboratory.

Grain Mount Preparation: Isotropic Mineral

In the laboratory, determination of refractive index generally requires a maximum of 1 gram of powdered mineral sample. Passing the powder through a 140-mesh sieve results in a collection of grains that are roughly 0.1 mm^3. In a grain mount, a few of the collected grains are picked up by a small spatula and sprinkled at the center of a glass slide. The grains are then covered by a piece of a cover slip. A single cover slip is usually cut into four pieces with a diamond pencil, and each piece is used to make an individual grain mount. After putting the cover slip on the grains, a few drops of oil are dropped next to the cover slip with a dropper. (Immersion oil vials come with droppers.) These oil drops quickly migrate underneath the cover slip by capillary action and immerse the grains.

Observations in Plane-polarized Light: The Becke Line

The grain mount is then placed on the microscope stage for refractive index determination. Plane-polarized light emerging through the lower polar is used for this purpose, and the upper polar stays out of the path of light. If there is sufficient difference in refractive index between the mineral and the surrounding oil, a bright rim of light, called the *Becke line (BL)*, forms along the grain boundary. The Becke line results from a focusing of light rays, which is caused by a lens effect or a total internal reflection effect along the grain boundary (Figure 4.11). If it is a little blurry, the Becke line can be sharpened by closing the iris diaphragm somewhat. Note also that the Becke line is best seen using a 10× objective. As the stage is lowered, the Becke line moves toward the medium of higher index. For example, if the mineral has a higher refractive index than oil, the Becke line will move into the mineral grain. Figure 4.11a shows how the Becke line migrates from mineral to oil or vice versa depending upon which has the greater refractive index.

It is unlikely that the experimenter will obtain a perfect match of refractive indices with the very first oil, so chances are that several grain mounts with oils of different indices have to be made and the Becke line test performed. The Becke line becomes gradually weaker as oils of progressively closer refractive index to that of the mineral are used. If a monochromatic light source such as the yellow light ($\lambda = 589$ nm) given off by a sodium vapor lamp is used, the grain will completely disappear when the oil and the mineral have a perfectly matching refractive index.

Ordinary white light, instead of a monochromatic light source, is routinely used in performing the Becke line test. In this case, the Becke line undergoes dispersion and forms bluish and reddish Becke lines when close to a match. As pointed out ear-

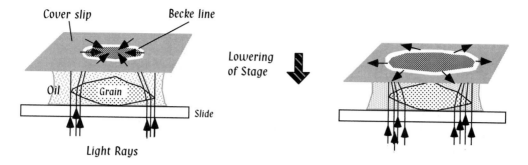

(a) Formation and movement of the Becke Line

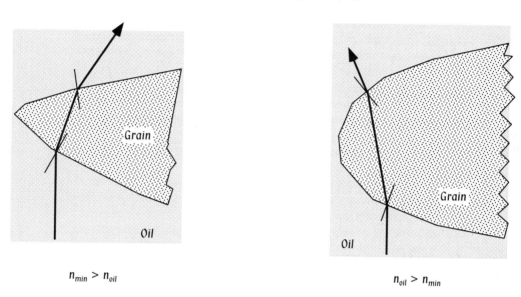

$n_{min} > n_{oil}$

$n_{oil} > n_{min}$

(b) Bending of a ray at the grain margin

FIGURE 4.11   (a) Formation and movement of the Becke line (shown as a light ring in the top plane). When the stage is lowered, the Becke line will always move toward the medium that has the higher refractive index (n): the example on the left-hand side is a case where the mineral has a greater n than the surrounding oil; therefore, the Becke line will move toward the mineral grain (shown by short radial arrows in the top plane). In the example on the right, the mineral has a lower refractive index than the oil; therefore, the Becke line will move toward the oil. A schematic drawing of the light rays in each example in the top diagram shows how the rays will bend as they travel through the two media in each case. (b) An expanded view of the grain margin area showing how a ray bends at the grain boundaries in each case.

lier, the refractive index of any material varies as a function of the wavelength of the light source used. Therefore, even an isotropic mineral will show different refractive index values for light sources of different wavelengths (Figure 4.12). As expected from our previous discussion, the refractive index values of both the oil and the mineral decrease with increasing wavelength. Note that the slopes of the two lines in the figure indicate that the oil's refractive index varies more strongly than that of the mineral as a function of wavelength. This difference in slopes of the refractive index lines in Figure 4.12 is responsible for the dispersion of the Becke line. In Figure 4.12c the two lines (mineral's and oil's refractive index variation lines) cross at 589 nm (yellow light); hence, the mineral has an identical refractive index to that of the oil at 589 nm. However, for blue light ($\lambda = 400$ nm) the oil has a greater refractive index

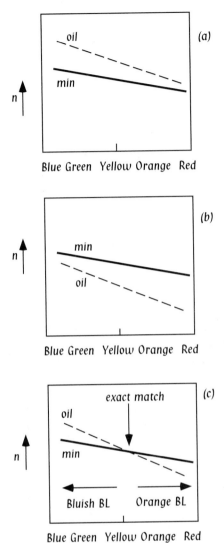

FIGURE 4.12 *The refractive indices (vertical axis) of three different oils (dashed lines in the three diagrams) and a mineral (solid line in all three diagrams) are compared for blue-to-red components of white light.* In case (a), in which the oil has a greater refractive index than the mineral for all wavelengths $(n_{Oil} > n_{Mineral})$, the white Becke line will move toward the oil when the stage is lowered. In case (b), where $n_{min} > n_{oil}$ for all wavelengths, the white Becke line will move toward the mineral. In case (c), $n_{min} = n_{oil}$ for yellow wavelength but $n_{min} < n_{oil}$ for blue wavelength, and for orange and red wavelengths, $n_{min} > n_{oil}$. Therefore, two Becke lines, bluish and orangish in color, will form in this case. And as the stage is lowered, the bluish Becke line will move toward the oil and the orangish Becke line will move toward the mineral.

than the mineral, and a bluish Becke line would form, which would move toward the oil as the stage is lowered. At the other end, the mineral has a higher refractive index than the oil for red wavelength ($\lambda = 700$ nm); therefore, the red Becke line would move toward the mineral with the lowering of the stage.

The oil immersion technique allows an accuracy of ± 0.003 in the determination of the refractive index. This may be compromised somewhat if the oil gets contaminated by other oil due to sloppy laboratory work. A second potential source of error is temperature because heating reduces the refractive index of an oil; however, this effect is insignificant as a 20°C rise would decrease the refractive index by only 0.008. It is generally impossible for the temperature in the laboratory to increase by that much!

ISOTROPIC INDICATRIX

The *indicatrix* is a three-dimensional geometric figure that defines the variation in the refractive index of a mineral as a function of all possible vibration directions of light passing through the mineral. The distance from the center to the surface at any point

Isotropic Indicatrices

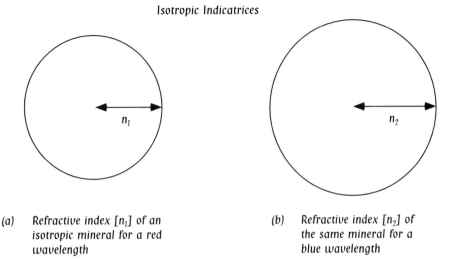

(a) *Refractive index [n_1] of an*
     isotropic mineral for a red
     wavelength

(b) *Refractive index [n_2] of*
     the same mineral for a
     blue wavelength

FIGURE 4.13   *Isotropic indicatrices of a mineral for two different wavelengths.*

of an indicatrix represents the refractive index along that vibration direction. In the case of an isotropic mineral, the indicatrix is a sphere because its refractive index does not vary with vibration direction (Figure 4.13). The radius of this sphere is the refractive index of the isotropic mineral. As pointed out earlier, the refractive index of any material, mineral or other, depends on the wavelength of light that passes through it. Therefore, the indicatrix of a given isotropic mineral is large or small for blue versus red wavelengths of light (Figure 4.13a,b). As we will see later, the concept of an indicatrix provides a useful reference frame within which the optical behavior of minerals may be understood.

OPTICAL BEHAVIOR OF ANISOTROPIC MINERALS

Optic Axis, Double Refraction, and Extinction

We noted earlier that the refractive index varies with vibration direction in an anisotropic mineral. However, there can be one or two specific directions, called *optic axes*, within an anisotropic mineral, along which the mineral behaves like an isotropic mineral. If an oriented section of a mineral is prepared so that the optic axis is perpendicular to it, that section (called a *circular section*) will give only one refractive index for a particular wavelength and will remain extinct between crossed polars, just like any isotropic mineral. All other sections of the anisotropic mineral will show interference colors (discussed later).

Anisotropic minerals with a single optic axis are called *uniaxial* minerals and those that have two such axes are known as *biaxial* minerals. There is a direct correspondence between crystallography and the optical property of minerals. As we have seen before, isotropic crystals also have isometric symmetry. Uniaxial minerals crystallize in the hexagonal and tetragonal systems; whereas biaxial minerals have orthorhombic, monoclinic, and triclinic symmetry. The optic axis in a uniaxial mineral coincides with the *c*-crystallographic axis. In all other directions within the mineral, any oncoming ray of light will split into two rays: the *ordinary ray (O-ray)* and the *extraordinary ray (E-ray)*. The two rays travel with different velocities and vibrate in

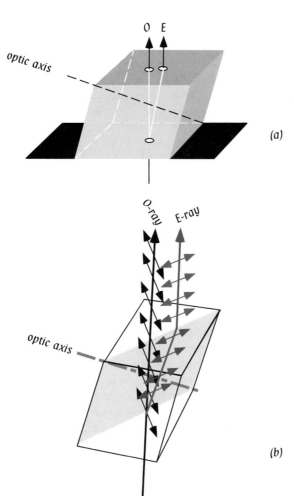

FIGURE 4.14 *The Iceland spar experiment showing double refraction.* (a) A transparent calcite rhomb is placed above a metal plate (black) with a hole in the center through which light is passed into the calcite. Two images of the hole will be seen at the top due to the phenomenon of double refraction. One of these images is caused by the ordinary ray (or O-ray) and the other is caused by extraordinary ray (or E-ray). (b) The mutually perpendicular vibration directions of the E- and O-rays are shown with reference to the optic-axial plane (shaded) of calcite.

planes that are perpendicular to each other (Figure 4.14a,b). This phenomenon where one incident ray splits into two rays that then refract differently from one another (due to different velocities) is called *double refraction*.

Double refraction is best demonstrated in the laboratory with the *Iceland spar experiment*. Iceland spar is a transparent variety of the mineral calcite. In the Iceland spar experiment, a thick rhomb-shaped crystal of calcite is placed on a metal plate with a hole beneath the calcite. In this position the optic axis of the calcite crystal lies diagonally (Figure 4.14a,b). When the hole in the metal plate is viewed through the Iceland spar from the top, two images of it appear. This is only possible if the incident ray (technically, a bunch of rays) splits into two rays while traveling through the calcite rhomb (i.e., double refraction takes place). Viewed under the microscope, one of the two images of the hole remains fixed at the center, whereas the other image revolves around it. The ray that generates the fixed image is the O-ray, and the other is the E-ray. One may also observe that the two images do not focus on the same plane: if the fixed image is focused, the moving image would be slightly blurry and the stage would have to be moved (Up or down? Experiment for yourself.) in order to get it in focus. This is additional evidence that the extraordinary and ordinary rays travel with different velocities.

If a polaroid sheet is placed over the calcite and gently rotated through 360°, the two images will go dark (*extinct*) alternately at 90° intervals. When one of the images goes extinct, the other will be sharply visible. This can only happen because ordinary and extraordinary rays are both plane-polarized and vibrate at right angles to each other (Figure 4.14b). Thus, a total of four *extinction positions* are shown by calcite during a 360° rotation of the stage. In fact, all anisotropic minerals, of which calcite is an example, have four extinction positions. The four extinction positions are seen in all grains of the mineral except those that are circular sections of the indicatrix (i.e., oriented exactly perpendicular to the optic axis). As stated before, such a grain will behave like an isotropic mineral and will always remain extinct when the polars are crossed.

The O-ray travels with a constant velocity in all directions; the E-ray does not. The O-ray follows Snell's law, but the E-ray does not. The refractive index with respect to the O-ray vibration direction is therefore constant; however, the refractive index with respect to the E-ray varies with direction. One other difference is that the vibration direction of the extraordinary ray is always contained within the plane, called an *optic axial plane*, which contains the optic axis.

Birefringence, Retardation, and Interference

The difference between the two extreme refractive indices of an anisotropic mineral is called *birefringence*. Calcite has a very high birefringence, which allows us to easily observe the double refraction of this mineral. On the other hand, quartz has a small birefringence, which makes it unsuitable for carrying out the double refraction experiment as described above.

When a thin section of an anisotropic mineral is placed between crossed polars (and it is not an isotropic section), it exhibits colors that change in intensity between the four extinction positions. These colors are called *interference colors* and result from interference between the two waves that pass through the anisotropic mineral grain and resolve into the vibration directions of the upper polar. When the incoming plane-polarized wave from the polarizer arrives at the mineral in a 45°off-extinction position, it splits into two waves because of the double refraction phenomenon (Figure 4.15). As the two daughter waves travel through the mineral with different velocities and wavelengths, one lags behind the other by a number of wavelengths or a fraction of wavelengths. This difference in distance between the fast and slow waves is known as *retardation* (Δ). Consider a slow ray (velocity = V_s) and a fast ray (velocity = V_f) traveling through a mineral grain of thickness d. Once the two rays exit the mineral, both must travel at the same velocity, say v, and will interfere with each other. However, the fast ray would have emerged first and covered an extra distance, Δ, by the time the slow ray emerged. If the slow ray takes time t to go through the mineral, then, applying the definition of velocity we obtain the following equation:

$$V_s = d/t$$

or

$$t = d/V_s \qquad \text{[Eq. 4.4]}$$

Over the same time t, the fast ray would have traveled a distance of d plus Δ. Note that the fast ray traveled with a velocity of V_f in covering the distance d at a velocity of v over the distance Δ. Therefore, we can write the following:

$$t = d/V_f + \Delta/v \qquad \text{[Eq. 4.5]}$$

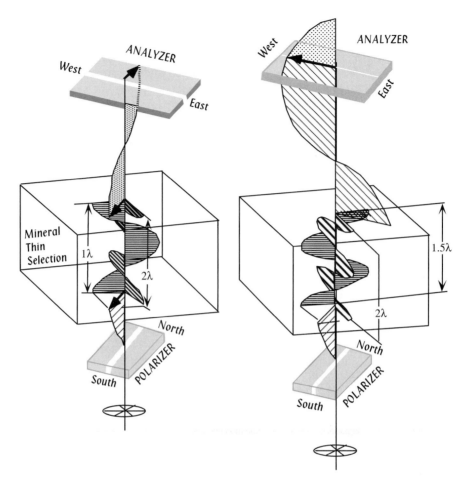

FIGURE 4.15   *The interference phenomenon is shown for an anisotropic mineral placed between the polarizer (with a N–S vibration) and an analyzer (with an E–W vibration). The incident plane-polarized light wave from below splits into two waves through the mineral, which recombine ("interference") at the top of the mineral section. On the left the path difference of the two waves traveling through the mineral is 1λ, and for the example on the right it is 0.5λ. For the case on the left, the light is blocked off at the analyzer and extinction occurs. The light is transmitted through the analyzer in the example to the right.*

From Equations 4.4 and 4.5, we obtain

$$d/V_s = d/V_f + \Delta/v$$

Rearranging, we get

$$\Delta/v = d/V_s - d/V_f$$
$$\Delta = d \cdot (v/V_s) - d \cdot (v/V_f)$$
$$\Delta = d \cdot [(v/V_s) - (v/V_f)] \qquad \text{[Eq. 4.6]}$$

By definition of refractive index, $n_s = v/V_s$ and $n_f = v/V_f$.

Therefore, from Equation 4.6 we get

$$\Delta = d \cdot [n_s - n_f]$$

That is,

$$Retardation = Thickness \times Birefringence.$$

When $\Delta = i\lambda$, where i is an integer $(1, 2, 3 \ldots n)$, the resultant wave formed by vector addition of the two daughter waves emerges from the upper surface of the mineral vibrating perpendicular to the preferred vibration direction of the analyzer, and is therefore extinguished (Figure 4.15a). On the other hand, when $i = 3/2, 5/2, 7/2 \ldots (n + 1)/2$, the resultant vector-added wave vibrates parallel to the analyzer's vibration direction and is allowed to pass through (Figure 4.15b). This wave will have an *interference color*, determined by its wavelength. The relationship between retardation and interference color is graphically portrayed in the Michel–Levy color chart (illustrated in the next section).

MICHEL–LEVY COLOR CHART

The *Michel–Levy chart* shows the relationship between birefringence, thickness, interference color, and retardation (Figure 4.16). Note that colors repeat in sequence at retardations of about 550, 1100, and 1650 nm. The colors with retardation less than 550 nm are called first order colors; those with retardation between 550 and 1100 nm are called second order colors, and so on. With increasing order, the color fades to grayish white.

Thin sections generally have a thickness of about 0.03 mm (30 microns). Minerals with low birefringence, such as quartz or feldspar, can be used to determine the

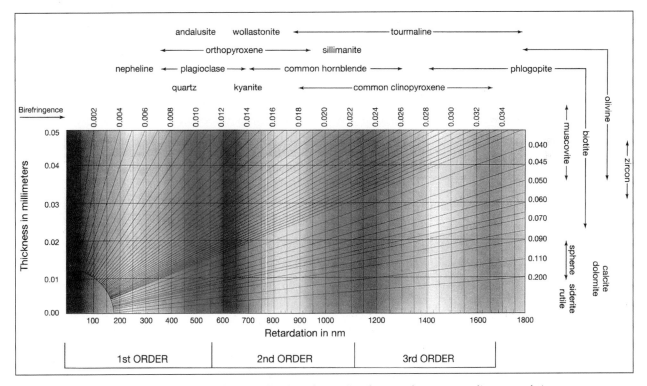

FIGURE 4.16 *The Michel–Levy color chart showing interference colors corresponding to retardation (bottom axis), birefringence (top and right axes), and thickness (left axis)* (from Perkins 1998). Some common minerals are shown around the top and right parts of the chart. Note that the standard thickness of a thin section is 0.03 mm. Therefore, if we want to determine which mineral thin section could give a 2° green color, we need to follow the 0.03 mm line to the box corresponding to the 2° green color, then follow the appropriate radial line to the top axis, where it shows the birefringence to be 0.025. The two likely minerals with such birefringence are tourmaline and hornblende. (From Perkins, 1998, *Mineralogy*, 1st ed., Prentice Hall)

thickness of the thin section while preparing it from a rock chip. As the chart shows, these two minerals should give a first order gray color for a thickness of 0.03 mm. However, if the thin section is too thick, the quartz or feldspar will be characterized by first order yellow or red interference color. For minerals with high birefringence, such as calcite, the interference color is of such high order that thickness of the section is not of much relevance. A mineral in a standard thin section (30 microns thick) with second order red/blue interference color should have a birefringence of 0.2 and retardation of 600 nm.

ACCESSORY PLATES AND RETARDATION

As noted earlier, the three types of accessory plates are the *gypsum plate*, the *mica plate*, and the *quartz wedge*. The gypsum plate produces a characteristic retardation of 550 nm. When a white light source is used, the gypsum plate produces a characteristic violet-to-red color that falls close to the upper limit of first-order colors in the Michel–Levy chart. Therefore, this plate has also been variously called the first-order red plate, sensitive-tint plate, or sensitive-violet plate.

The mica plate produces a retardation of 147 nm, which corresponds to the first-order white color. The quartz wedge is totally different from the other two accessory plates in terms of its wedge shape. Its thickness increases from 0 to 0.25 mm. Quartz itself has a birefringence of 0.0091, but the variable thickness produces a range of retardation that results in first to second to higher-order interference colors as the wedge is progressively inserted.

The following example shows how a gypsum plate may be used to determine the slow versus fast vibration directions in the thin section of a mineral grain. With the polars crossed, the mineral grain is first brought to an extinction position. Then, the stage is rotated clockwise through 45°. In this position, the grain exhibits maximum interference color. Although we know that the vibration directions of these fast and slow waves must be oriented at NW–SE and NE–SW directions, we do not know which is slow and which is fast. We note the interference color. We now insert the gypsum plate and note whether the color has risen or fallen in order in terms of the Michel–Levy chart. If the color increases, it means that the gypsum plate has added an extra amount of retardation (550 nm, to be exact) between the fast and slow waves. (This situation is called *addition*). Addition can happen only if the fast and slow directions of the gypsum plate match exactly with those of the mineral. On the other hand, if the slow direction of the mineral matches with the fast direction of the gypsum plate, the gypsum plate will actually reduce the amount of retardation (*subtraction*) and lower the order of interference color.

SECTION 3

UNIAXIAL AND BIAXIAL MINERALS

UNIAXIAL INDICATRIX

The Iceland spar experiment tells us that there must be two extreme refractive indices for calcite: one with respect to the vibration direction of the ordinary ray and the other for the extraordinary ray. Calcite is an example of a uniaxial mineral. All uniaxial minerals have two extreme refractive indices, conventionally represented as ω and ε corresponding to ordinary and extraordinary rays, respectively (Figure 4.17).

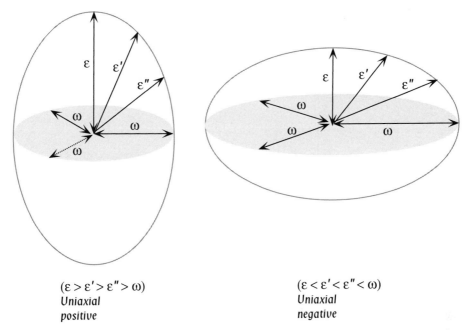

$(\varepsilon > \varepsilon' > \varepsilon'' > \omega)$
Uniaxial
positive

$(\varepsilon < \varepsilon' < \varepsilon'' < \omega)$
Uniaxial
negative

FIGURE 4.17   *Uniaxial indicatrices of positive and negative minerals.*

ω is constant for a given mineral regardless of the direction of propagation of light, but ε varies with the direction of propagation.

ε, ω, and the variation of ε in three dimensions are best illustrated in terms of a geometric figure, called the *uniaxial indicatrix* (Figure 4.17). The uniaxial indicatrix is an ellipsoid of rotation with semimajor and semiminor axes represented by the two extreme refractive index values, and the variable values of ε are represented as $\varepsilon', \varepsilon''$, and so on. In a positive uniaxial mineral, $\varepsilon > \omega$; in this case the semimajor and semiminor axes are equal to ε and ω, respectively. The opposite is true for a uniaxial negative mineral, which is defined by $\varepsilon < \omega$. The term *optic sign* is used to indicate whether a mineral is positive or negative.

The uniaxial indicatrix can be better understood in terms of three sections (or slices) through it: the circular section, the principal section, and the random section (Figure 4.18). The *circular section* (i.e., the "isotropic section") has a radius equal to ω and is perpendicular to the optic axis. Thus, any light propagating parallel to the optic axis in a mineral will vibrate in the circular section and give the refractive index ω. ε cannot be found from this section. As shown before, a plane-polarized light emerging from the polarizer will continue to propagate with the same vibration direction as it travels through the circular section, and it will be extinguished at the analyzer. Thus, the circular section of a mineral will always remain extinct between crossed polars. (That is, it will behave like an isotropic section.) Although the vibration direction of the light passing through the circular section remains the same, its wavelength must shorten to the extent required by the ω of the mineral.

A *principal section* will contain both ε and ω vibration directions; therefore, it will give the maximum birefringence (ε-ω) (Figure 4.18a). In a grain mount, the grains that show maximum interference colors are likely to be principal sections. A random section is one that is "cut" at an oblique angle to the optic axis and the circular section. As Figure 4.18c shows, such a section will give ω and ε'. $\varepsilon' > \varepsilon$ in the case of a negative mineral, and the opposite is true for a positive mineral (Figure 4.17). Random sections will always give some intermediate birefringence.

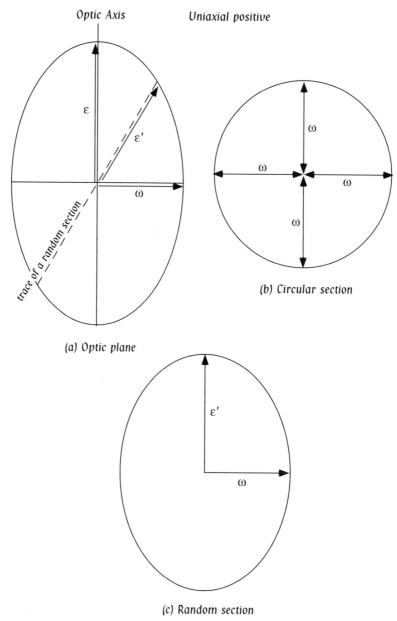

(a) Optic plane

(b) Circular section

(c) Random section

FIGURE 4.18   *Various planes through a positive uniaxial indicatrix.*

INTERFERENCE FIGURES

The interference figure of an anisotropic mineral is an image that can be seen through the microscope ocular using conoscopic illumination, crossed polars, a high-power objective (40×), and the Bertrand lens (explained later). It owes its origin to a complex set of interactions between interfering fast and slow waves as they pass through the mineral and undergo further changes before appearing at the ocular. Most importantly, the interference figure allows the determination of the mineral's optic sign. First, we examine some interference figures produced by uniaxial minerals. Biaxial interference figures will be discussed later. Figure 4.19 shows examples of interference

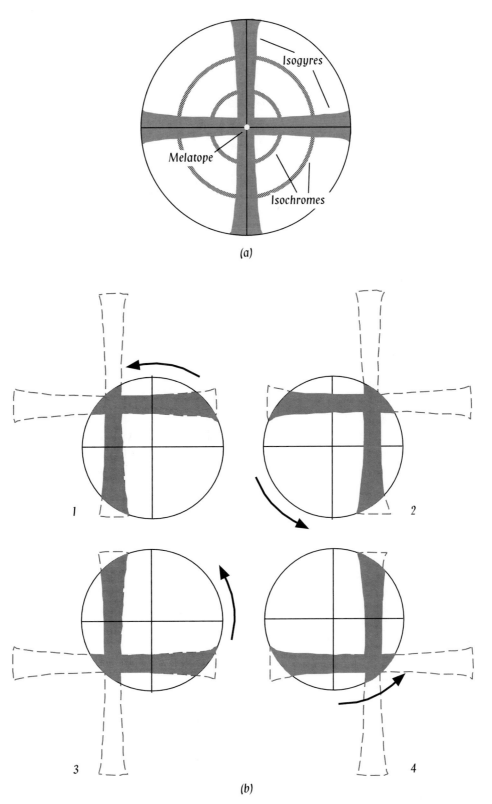

FIGURE 4.19   (a) Centered uniaxial interference figure. (b) A somewhat off-centered uniaxial interference figure. The four figures illustrate the progressive movement of the isogyres with rotation of the stage.

figures of a uniaxial mineral: 4.19a shows a centered optic axis figure and 4.19b shows somewhat off-centered figures as they change position continuously during rotation of the stage. The dark bands forming the cross are called *isogyres*. The "point" where the isogyres cross is known as the *melatope*. The melatope is the point where the optic axis emerges through the mineral section. The concentric color rings that appear are called *isochromes*. In the next two sections, we shall see how interference figures form and how to obtain them.

Origin of Interference Figure

The isogyres simply represent extinguished areas where the vibration directions of the mineral match with those of the polarizer and analyzer (Figure 4.20a). On the other hand, isochromes represent the *cones of equal retardation* emerging from the mineral on a horizontal plane (Figure 4.20b). These cones form due to the interference among the E- and O-rays as they travel out of the mineral. Figure 4.20c illustrates how each cone forms with the help of two parallel sets of convergent rays emerging off the condenser lens in the substage assembly. (Of course, there would be many more rays, the two sets chosen are for illustration purposes only.) Because these incident rays are not parallel to the optic axis of the mineral, each of them will split into O- and E-rays that will travel at different velocities and vibrate perpendicular to each other. As these rays emerge from the mineral, rays traveling along the same path must interfere; thus, the combined ray will be characterized by a specific retardation. For example, at point X the O1 and E2 rays coming off the pair of convergent rays shown in Fig. 4.20c will combine to form a single ray XY characterized by a path difference. If the path difference is an integer multiple of wavelength, darkness will result. Other resultant rays with path differences of fractional multiples of wavelength will produce color. (See the earlier discussion on retardation and interference color.) In three dimensions, the resultant rays with characteristic path differences will appear as cones with distinctive retardation (cones of equal retardation). The rings (i.e., cones) will be more closely spaced in a mineral with very high birefringence or with a thick grain.

Obtaining an Interference Figure

Grain selection is of paramount importance in obtaining an interference figure. An isotropic section (circular section) of an anisotropic mineral should be the best choice because the goal is to have the optic axis oriented perpendicular to the microscope's stage. However, such grains are generally impossible to obtain for most minerals; and the next best choice is to select a grain with minimum interference color (e.g., 1° gray). Such a grain would be close to being a circular section. In obtaining an interference figure, the following additional steps are necessary:

1. Polars must be crossed.
2. A higher power objective lens (40X) must be used.
3. Conoscopic illumination must be used by inserting the condenser lens in the substage.
4. The Bertrand lens must be inserted.

These steps should result in an interference figure visible through the ocular.

When a truly isotropic section is selected, the interference figure will appear as a dark "cross" (formed by the isogyres) overlying a set of concentric color and dark rings (isochromes). The center of the interference figure will be in the center of the field of view (Figure 4.19a). The figure will not change even when the stage is rotated

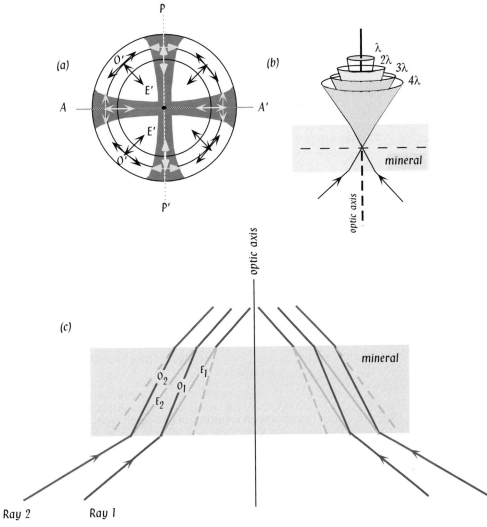

FIGURE 4.20 *These schematic diagrams show how isogyres (dark crosses) and isochromes (color rings) form in an interference figure of a uniaxial mineral.* (a) Vibration directions of ordinary and extraordinary rays, isogyres, and isochromes are shown. (b) Cones of equal retardation. (c) This diagram shows how individual incident rays split up into O- and E-rays as they travel through the mineral, and how these rays recombine to produce cones of equal retardation (bottom left).

(assuming that the objective lens and stage are perfectly centered). Melatope is the location where the optic axis perpendicularly "pierces" the circular section. For this reason, this type of interference figure is called an *optic axis interference figure*. The ε and ω vibration directions in this case are radially and tangentially, respectively, oriented (Figure 4.20a). These vibration directions are imposed by the requirement that the E-ray must vibrate in the plane that contains the optic axis, and the O-ray vibrates perpendicularly to it.

In the more likely event of selecting a grain with very low birefringence, an *off-centered optic axis figure* forms, in which the melatope does not occur at the center of the field of view (Figure 4.19b). This is because the optic axis is oriented at an oblique angle to the horizontal, and the melatope (the optic axis piercing point) obviously cannot be at the center. The degree off center depends upon the angle at which the optic axis is oriented to the plane of the stage. If the stage were rotated, the isogyres would move N–S or E–W but **would not rotate**. If the melatope lies outside the field of view, the isogyres may also go in and out of the field of view.

Finally, if a mineral grain showing maximum birefringence is chosen, one may actually obtain a *flash figure* (not shown). A flash figure is produced when the optic axis lies in the plane of the thin section, and it looks like a big dark cross that nearly fills the entire field of view.

OPTIC SIGN DETERMINATION

The optic sign of a mineral may be determined by evaluating whether the O-ray or E-ray is the fast ray. If the O-ray is the fast ray, it means that $\omega < \varepsilon$; therefore, the mineral is positive. In an interference figure the vibration directions of the O-ray and E-ray are known (as explained earlier). Which one of the two is fast may be determined by altering the retardation using an accessory plate. The following recipe may be followed in obtaining the interference figure first and then determining its optic sign:

Step 1. Use a 10× objective and select an appropriate grain, such as one that remains extinct or at least shows very low order interference color (1° gray) between crossed polars.

Step 2. Switch to a 40× objective and **adjust the focus very carefully**. Insert the convergent lens in the substage.

Step 3. Insert the Bertrand lens. An optic axis figure or an off-centered figure should be visible through the ocular at this stage.

Step 4. Identify the *quadrants* of the interference figure, which are the four illuminated areas, NE, NW, SW, and SE, containing the isochromes.

Step 5. Insert an accessory plate into the accessory slot (generally oriented NW–SE) while noting the vibration direction of the accessory plate (usually the slow direction, or γ, of the plate is oriented NE–SW).

Step 6. Notice the change in optical behavior of the isochromes in selected quadrants with reference to the vibration direction of the accessory plate. If the NE and SW quadrants experience a rise in the interference colors, the NW and SE quadrants will see a fall in interference colors. These interference color changes tell us whether the mineral is positive or negative (Figure 4.21).

For a gypsum plate, if the NW–SE quadrants become 1° yellow (which automatically means that the NE–SW quadrants become 2° blue), *subtraction* has occurred in NW–SE quadrants. Since the radial directions correspond to the E-ray vibration direction (discussed earlier), the subtraction means that the E-ray is slow and the mineral is positive (Figure 4.21a). And, if the NW–SE quadrants become 2° blue, the mineral is negative (Figure 4.21b). When a quartz wedge is used (when isochromes are distinct and numerous), the isochromes will move away from the center in the NW–SE quadrants for a positive mineral and in the NE–SW quadrants for negative mineral (Figures 4.21a,b).

DETERMINATION OF THE REFRACTIVE INDICES OF UNIAXIAL MINERALS

The ω and ε indices of a uniaxial mineral are determined using the liquid-immersion technique on grain mounts. The preparation of such grain mounts and the selection of oils of known refractive indices were discussed earlier. It is important to recall that

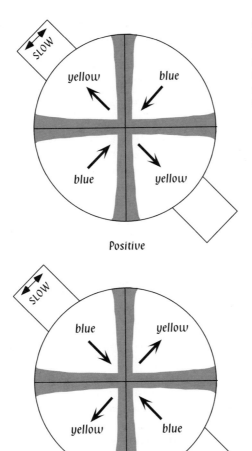

FIGURE 4.21 *Optic sign determination for uniaxial minerals.* The colors correspond to a gypsum accessory plate, whereas the arrows represent the directions of movement of the isochromes when a quartz wedge is inserted.

a uniaxial mineral has one constant refractive index (ω) (i.e., it does not vary with grain orientation) and the other (ε) varies with the optic orientation of individual grains. Therefore, not all grains are going to give ε although ω may be determined from any grain. In order to obtain tight constraints on ε, the goal should be to select those grains that show maximum birefringence, and to make many observations from several grain mounts with different oils to select the maximum ε' value for positive minerals and minimum ε' value for negative minerals. One must remember that the refractive index can only be determined with plane-polarized light emerging from the polarizer. Therefore, the desired vibration direction of the mineral (whether ω or ε) must be made parallel to the polarizer (N–S in the microscopes at most universities). Also, it is desirable to determine the optic sign of the mineral in consideration because knowing whether $\omega > \varepsilon$ or $\omega < \varepsilon$ at the outset makes the refractive indices determination much easier.

As we know already, parallelism of vibration directions with those of the polars results in four extinction positions during a complete 360° rotation of the stage. One may, therefore, perform the Becke line test by bringing the selected grain to an extinction position and pulling out the analyzer, then moving the stage down while noting the Becke line movement. The stage must then be turned until the next extinction position is attained; the logic of this is that if in the first extinction position, ω were parallel to the polarizer vibration direction, ε would be parallel to the polarizer vibration direction in the second extinction position. Once again the Becke line test must be performed by moving the analyzer out and moving the stage down (or up).

Let us suppose that the optic sign has been determined for the chosen mineral (from a grain showing isotropic behavior or 1° gray color), and it is negative ($\omega > \varepsilon$). To start, one may prepare a grain mount with an oil of $n = 1.50$. It is a good idea to determine ω first, because it does not vary with grain orientation. Let us say that a white light source is being used. We select a grain showing maximum birefringence and perform the Becke line test on two successive extinction positions (with the analyzer pulled out of the light path). With respect to both positions, we note that the BL moves into the mineral grain. That is both ε and ω are greater than n_{oil}. As in the case of isotropic refractive index determination, we need to make newer mounts with oil of successively greater refractive index until we see a perfect match (bluish and orangish BL's with equal intensity moving in opposite directions) with respect to one of the two successive extinction positions. Since we know (from the optic sign) that $\omega > \varepsilon, \varepsilon$ or ε' (which is close to ε) will be the first to be determined this way. A grain showing exactly ε will show a flash figure.

Once ε or minimum ε' has been satisfactorily determined, new grain mounts with oils of greater refractive indices will have to be prepared and similar BL tests performed until the greater refractive index (in this case ω) of the mineral is determined. Also, when choosing successive oils, one should first obtain a broad bracket on the ω or ε. As the mineral's ω or ε is narrowed down, one needs to obtain a perfect match by selecting oils with smaller differences in refractive index.

As a final reminder, note that a number of different grains showing maximum apparent birefringence must be used for ε measurement, then the maximum (if positive) or minimum (if negative) value should be taken as close as possible to the real ε.

BIAXIAL INDICATRIX

As the name suggests, biaxial minerals are characterized by two optic axes. By definition of optic axis, light traveling along each of these axes does not split into two separate rays; therefore, each of the two axes has a corresponding circular section perpendicular to it. Thus, a biaxial indicatrix must have 2 circular sections (Figure 4.22a,b). The presence of two circular sections requires the shape of the biaxial indicatrix to be a *triaxial ellipsoid* (Figure 4.22a,b). As opposed to the uniaxial indicatrix, which is an ellipsoid of rotation, the triaxial ellipsoid has three axes, and none of these axes coincide with the optic axes. The three axes coincide with three refractive indices, commonly named α, β and γ, where $\alpha < \beta < \gamma$. α, β and γ correspond to vibration directions X (fast), Y (intermediate), and Z (slow), respectively. Hence, there are three elliptical principal sections: the XY, YZ, and XZ sections (Figure 4.23). The XZ section contains the two optic axes and therefore is referred to as the *optic plane*. This is also the section that gives the maximum birefringence. Random sections are elliptical planes or sections that are not any of the three principal sections or circular sections. In the example shown in Figure 4.23e, the two axes are β and γ'''. (See Figure 4.22a as well). As Figure 22a shows, $\gamma > \gamma' > \gamma'' > \gamma''' > \beta > \alpha' > \alpha$.

The angle between the two optic axes is called the optic angle or 2V (Figure 4.22a,b). The axis (either X or Z) which bisects the acute 2V angle is called the *acute bisectrix* (Bx_a), and the one that bisects the obtuse 2V angle is called the *obtuse bisectrix* (Bx_o). The radius of both circular sections is equal to the intermediate refractive

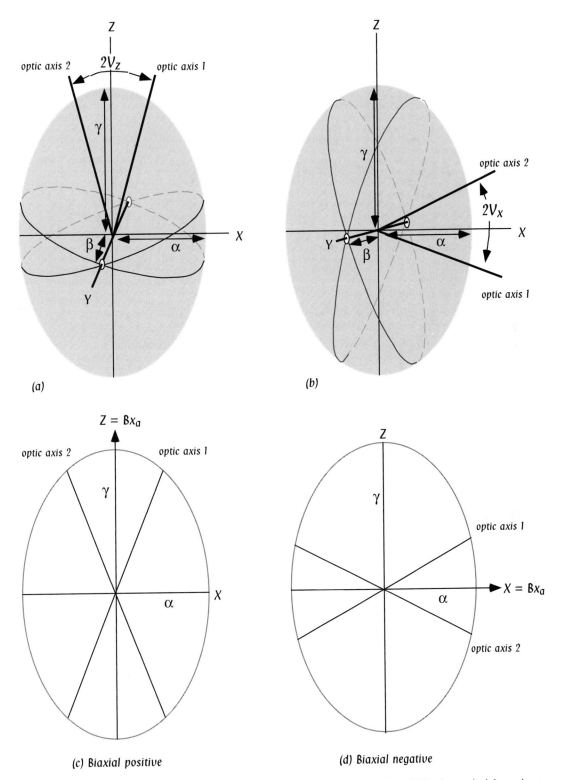

FIGURE 4.22 *Biaxial indicatrices for (a) positive and (b) negative minerals.* The three principle semiaxes are: γ—slow (Z-direction), β—intermediate (Y-direction), and α—fast (X-direction). (c) and (d) are the optic planes of (a) and (b) showing the orientation of the acute bisectrix and the two optic axes.

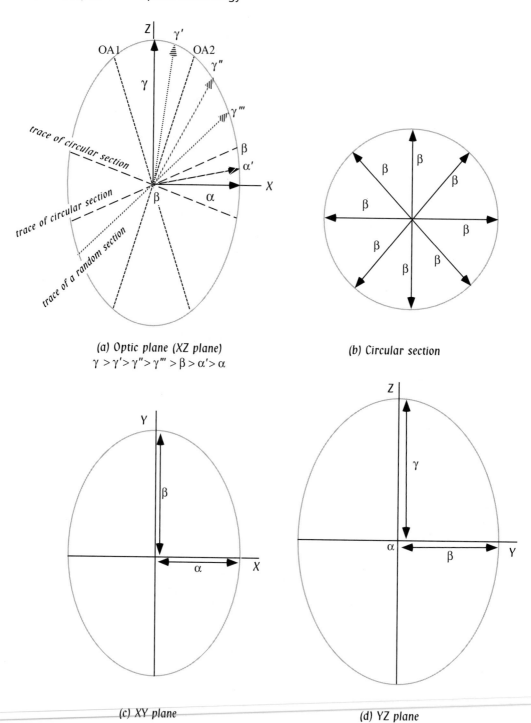

(a) Optic plane (XZ plane)
$$\gamma > \gamma' > \gamma'' > \gamma''' > \beta > \alpha' > \alpha$$

(b) Circular section

(c) XY plane

(d) YZ plane

FIGURE 4.23 *Different sections through a biaxial indicatrix.* (a) The XZ plane and how γ and α vary with orientation. Traces of the two circular sections (dashed lines) are also shown. OA1 and OA2 are the two optic axes. (b) A circular section showing only β vibration directions. (c) and (d) Two other principal planes showing the orientation of the three principal axes of the biaxial indicatrix. (e) A random section. *(continues)*

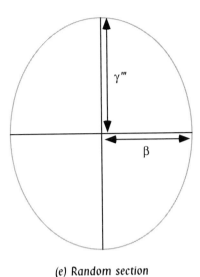

(e) Random section

FIGURE 4.23   *(continued)*

index, β (Figure 4.23b). Thus, β is constant for a biaxial mineral. γ and α vary according to grain orientation (Figure 4.23a). In terms of determining refractive indices, grains with maximum birefringence may give α and γ values or values close to them. Chances are that one may obtain several estimates for each from several grains. Therefore, similar to the uniaxial case discussed before, one must select the minimum α' value as α and maximum γ' value as γ. β can be determined from isotropic (circular) or near isotropic sections.

The optic sign of a biaxial mineral is defined on the basis of whether the Z or X direction coincides with the Bx_a: in a positive biaxial mineral, $Z = Bx_a$; and in a negative mineral, $X = Bx_a$ (Figure 4.22c,d).

BIAXIAL INTERFERENCE FIGURES

The same procedure used to obtain the uniaxial interference figure is followed for finding the biaxial interference figure (i.e., conoscopic illumination, 40× objective, crossed polars, and Bertrand lens). Biaxial interference figures are very different from uniaxial minerals in appearance and in the movement of isogyres, as will be seen later. Because biaxial minerals are characterized by two optic axes and three vibration directions, considerable variation can occur in their interference figures. (The interested reader should consult a proper optical mineralogy text for that purpose: see the Bibliography.) From a utilitarian point of view, however, only two types of biaxial interference figures are useful—the *acute bisectrix* (Bx_a) *figure* and *optic axis figure*. A biaxial interference figure is easily distinguished from a uniaxial figure on the basis of the nature of the isogyres and the way they move. Uniaxial isogyres always remain straight and move laterally with rotation of the stage, whereas biaxial isogyres curve.

Acute Bisectrix and Optic Axis Figures

Figure 4.24 shows examples of centered acute bisectrix and optic axis figures. In the centered acute bisectrix figure, the Bx_a is located at the center of the field of view, whereas in the centered optic axis figure, one of the two optic axes occurs at the center. Figure 4.24a shows how the acute bisectrix figure (centered) looks while the

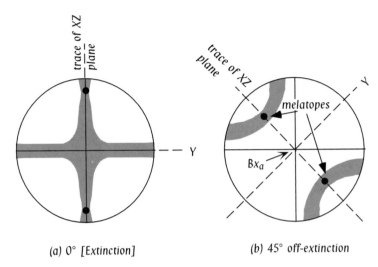

(a) 0° [Extinction] (b) 45° off-extinction

Centered acute bisectrix figures

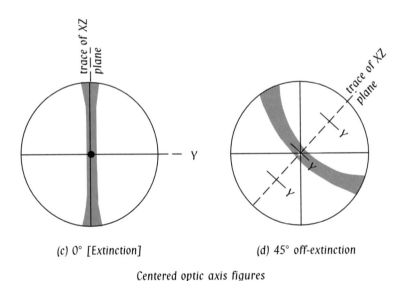

(c) 0° [Extinction] (d) 45° off-extinction

Centered optic axis figures

FIGURE 4.24 *Centered biaxial interference figures.*

mineral grain is at extinction. (That is, if the Bertrand lens were removed from the light path, the mineral grain should appear dark.) Figure 4.24b shows how the isogyres split and move away to their maximum separated positions when the stage is rotated by 45° off extinction. The melatopes are located at the most attenuated portions of the isogyres. The trace of the XZ plane (also called the optic plane or optic axial plane) is simply the line of intersection of the XZ plane with the horizontal plane of the field of view. Comparing this figure with the biaxial indicatrix will indicate that the XZ plane must contain the two optic axes (*melatopes* in the interference figure) and also that the Y direction must be perpendicular to this plane. With further rotation, the isogyres would approach each other and form a "cross" again at the 90° position. The difference between the crosses at the 0° (extinction) and 90° (second extinction) positions is simply that the XZ plane (and hence the Y direction) is rotated by 90° as well. Figures 4.24c and d show a centered optic axis figure during extinction and 45° off extinction, respectively.

The off-centered acute bisectrix and optic axis figures have their Bx_a and melatopes off the center of the field of view. Therefore, the isogyres may leave the field of view upon rotation of the stage, depending on the degree off the center and the 2V of the mineral. Choice of grains is once again a critical issue. As with uniaxial minerals, one may select an isotropic or near-isotropic grain if a centered or slightly off-centered optic axis interference figure is desired for optic sign and 2V determination (discussed later). For acute bisectrix figures, grain selection may not be easy and depends on the $2V_a$ (acute optic axial angle) because when the $2V_a$ is too large, the isogyres will leave the field of view even if the interference figure is a centered Bx_a figure. In such a case an optic axis figure should be more useful.

Figure 4.25a provides a three-dimensional perspective on how the Bx_a figure is related to optic axes, X, Y, and Z directions. Figure 4.25b shows isogyres and

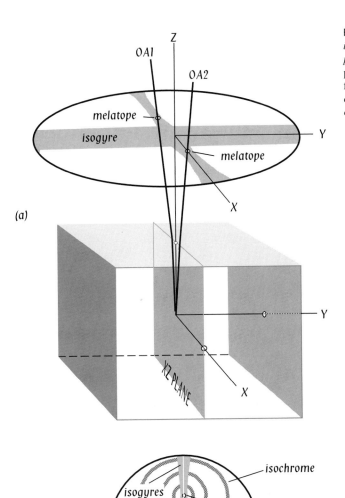

FIGURE 4.25   *An acute bisectrix interference figure in an extinction position.* (a) A three-dimensional perspective of the interference figure. OA1 and OA2 are the two optic axes. (b) Various components of the interference figure.

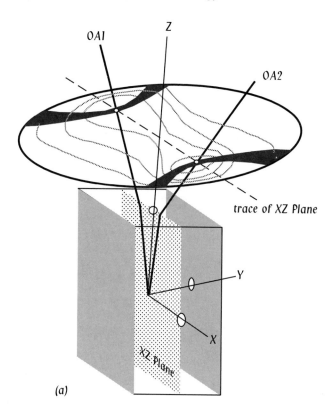

FIGURE 4.26 *Centered acute bisectrix figure in a 45° off-extinction position.* (a) A three-dimensional perspective of the isogyres, isochromes, optic axes, and vibration directions. (b) Determination of vibration directions at points Q and R using the Biot–Fresnel law.

(a)

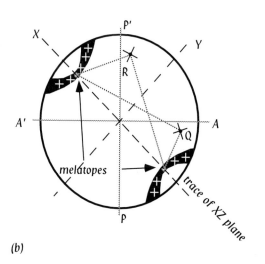

(b)

isochromes (isochromes are colored "rings") at an extinction position. The isochromes form broad pseudoelliptical color bands around the melatopes. Figure 4.26a provides a similar perspective at a 45° off-extinction position. In contrast to the uniaxial interference figure, locating vibration directions at various points within the biaxial interference figure is complicated. A useful method to determine the vibration

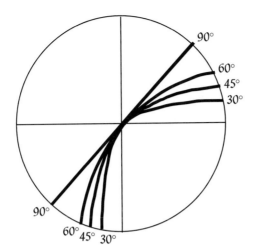

FIGURE 4.27   *The estimation of 2V (angles given in the diagram) from the curvature of isogyres by Kamb's method.*

directions at any location on a Bx_a figure is one that uses the *Biot–Fresnel law*, which is illustrated as follows. At any location, say Q, one may be able to locate the vibration directions by drawing a line from each of the two melatopes to Q; then, the vibration directions would be the angular bisectors between the two lines (Figure 4.26b). Figure 4.26b shows another example of vibration directions at a point R. If the Biot–Fresnel law is applied to various points on the darkest (middle) portions of each of the isogyres, one would observe that the two vibration directions are always N–S and E–W at those points.

As Figures 4.24c and d show, in a centered optic axis figure, a single isogyre is visible with its melatope at the center of the field of view. The isogyre is straight in an extinction position, but it curves and rotates around the center as the stage is rotated. Maximum curvature is obtained when the grain is in a 45° off-extinction position. The extent of curvature depends on 2V, and it is therefore possible to estimate 2V from the curvature of the isogyres by using Kamb's method (Figure 4.27). Although there are other methods that determine 2V more accurately, such methods are unnecessary for routine mineral identification purposes. Note that in a centered optic axis figure with 2V = 90°, the isogyre remains straight even when the stage is rotated; but its melatope stays pinned at the center.

It is impossible to confuse the straight isogyre (2V = 90°) in a centered biaxial optic axis figure with a seriously off-centered uniaxial figure. In the former, the isogyre remains straight and does not leave the field of view but only revolves around the center of the field of view. As we know from prior discussion, the single isogyre in an off-centered uniaxial figure moves laterally (while remaining straight) across the field of view.

Flash Figure

A flash figure appears as a huge dark cross-type feature that covers almost the entire field of view (not shown). It is obtained from the XZ section (i.e., when the optic axial plane is horizontal). The only utility of having a flash figure from such a mineral grain is that exact α and γ refractive indices may be determined.

OPTIC SIGN DETERMINATION

The easiest approach to optic sign determination is to use a mineral grain that shows the optic axis figure (centered, if possible). An ideal grain should stay extinct between crossed polars (i.e., an isotropic section). However, the next best thing is a grain that shows very low birefringence. Bring that grain to extinction, then rotate the stage through 45°. Inserting the condenser lens to obtain conoscopic illumination, then inserting the Bertrand lens into the light path should give an optic axis figure that would be expected at a 45° off-extinction position (Fig. 4.24c,d). If the 2V is less than 60°, a second isogyre may also be visible.

With the isogyre(s) occurring in NW–SE quadrants, insert an accessory plate of known orientation, for example a gypsum plate with slow direction parallel to NE–SW, through the accessory slot. The interference color on the convex side of the isogyre (or between isogyres in the case of two visible isogyres or a Bx_a figure) will decrease for a positive mineral (to first order yellow for gypsum plate because in this case Bx_a is slow) and increase for a negative mineral (to 2° blue for gypsum plate; Figure 4.28).

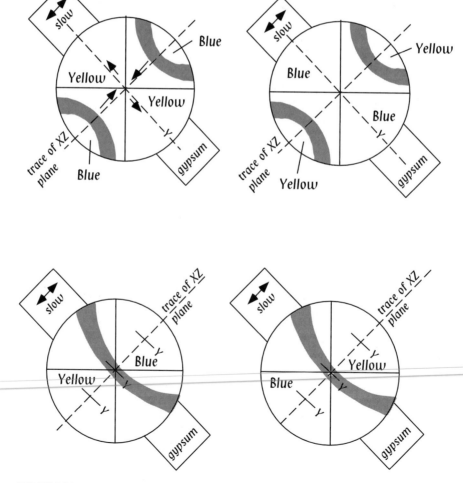

FIGURE 4.28 *Optic sign determination with the use of a gypsum plate.* In the top left diagram, the arrows show the movement of isochromes when a quartz wedge is inserted.

DETERMINATION OF REFRACTIVE INDICES OF BIAXIAL MINERALS

The most important thing to remember in determining the three refractive indices of a biaxial mineral is that β does not vary, whereas α and γ vary (i.e., $\gamma > \gamma' > \gamma'' > \gamma''' > \beta > \alpha'' > \alpha' > \alpha$; Figure 4.23a) depending upon the optic orientation of a grain (i.e., how a mineral grain is oriented with reference to the optic axes; see Figure 4.23a and the accompanying discussion in an earlier section).

Determination of β

If a truly isotropic grain (i.e., circular section) of the biaxial mineral is found, it is a simple matter to perform the Becke line test with different oils of known refractive indices in order to determine β. However, such grains are rarely found in actual practice. A more realistic alternative is to use a low birefringent grain for β determination. Because β is constant for a given mineral, this choice is just as good. Using conoscopic illumination, a high power objective lens, and a Bertrand lens, an acute bisectrix interference figure may be obtained on such a grain. As indicated in a previous section, this grain will give β and α or γ, depending on the optic sign.

Since the polarizer's vibration direction in most microscopes is N–S, β must be placed parallel to N–S when this oil immersion test is to be performed. The stage may be rotated such that the isogyres form a cross and the two melatopes, which mark the trace of the XZ plane, are oriented E–W. In such a case, β must be oriented N–S (i.e., parallel to the polarizer vibration direction). (However, if the microscope's polarizer is oriented E–W, the trace of the XZ plane must be made parallel to N–S.) The exact matching of the Y vibration direction with the polarizer may be checked by moving the Bertrand lens out and crossing the analyzer because in such a case, the grain should be extinct. The analyzer must be uncrossed and the Becke line test performed on the grain oriented in this way. Once again, an exact match will produce orange and bluish Becke lines of equal intensity.

Determination of α and γ

After determining β, fresh grain mounts with liquids of lower refractive index (n) value may be prepared to determine α. Maximum birefringent grains may be chosen, for when such grains are extinct, they will have α and γ parallel to the two polars at an extinction position. The reader is reminded that because a grain goes extinct when its two vibration directions are exactly parallel to those of the two polars, any refractive index determination must first necessitate bringing the grain to extinction and then uncrossing the analyzer, so that Becke line test may be performed with simple plane-polarized light emerging through the polarizer.

Once β has been determined, similar Becke line tests with oils of progressively lower refractive index should be performed on highly birefringent grains to determine α. These tests will eventually lead to finding a refractive index match (bluish and orangish Becke lines) between an oil with respect to the mineral. (Remember that this should be an extinction position, but the analyzer will have to be uncrossed for the Becke line test). Because α may vary from grain to grain, several grains must be examined, and the lowest refractive index value should be taken as α. Such a grain should yield a flash figure. Similarly, γ can be determined from highly birefringent grains by selecting oils of higher n than β and finding a match between the oil and mineral corresponding to an extinction position. In this case, the highest γ' value may be accepted as the true γ.

PREPARATION AND MICROSCOPIC DESCRIPTION AND IDENTIFICATION OF MINERALS

Thin Section Preparation

A mineral or a rock (which is an aggregate of minerals) needs to be cut into a very thin slice before it may be viewed under the microscope. Most geology departments have the facilities with which such "thin sections" of minerals can be made. Alternatively, the mineral or rock may be sent out to a commercial laboratory. The following paragraphs are a very general description of the steps involved and assume that the reader has access to such a mineral/rock thin-section-making facility and some departmental guidance on how to make a thin section.

In the first step, the rock or mineral must be cut with a diamond saw into a 0.5-cm-thick small slab measuring 2×1.5 cm^2. If the sample is very porous or fragile, it should be "impregnated" with epoxy-type resin under a vacuum and hardened before being cut into a small slab. The slab is then glued to a slide with an epoxy-type resin, making sure that air bubbles do not get in between the slide and the slab because such bubbles can later lead to peeling of the thin section. Preheating the slide often helps in this endeavor. Once glued and hardened, a significant portion of the extra thickness of the slab may be sliced off, again with a diamond saw. At this stage, the specimen should appear somewhere between transparent and translucent.

After cutting, the specimen should be washed with soap in an ultrasonic cleaner. In fact, at every step of the way, the specimen needs to be cleaned as well as possible. After cleaning and rinsing, the thin section may then be ground first with a 600-grit paper or powder (water saturated) on a glass (or 400 grit if the specimen is too thick) to a thickness where the individual grains become visible under a microscope. While grinding and polishing at this and all other succeeding stages, the thin section should be moved in a figure-eight pattern because such a movement prevents formation of any unidirectional fractures or scratches. At this stage and after ultrasonic cleaning, the specimen should be ground much thinner with a 1200-grit paper followed by a flattening with a 1200-grit diamond paste on a diamond lap. At this stage, the thin section should already have reached its destined thickness (approximately 30 microns), which may be verified by checking a quartz or plagioclase grain (or some other appropriate mineral, using the Michel–Levy chart), which should give a first order yellow or gray interference color. At this stage the thin section may be given a final polish (for about 5 minutes) with a 1-micron alumina powder (and water) on a polishing lap. Some technicians like to attach a cover slip (a thinly sliced glass wafer) on top of the thin section; however such sections cannot be used for electron microprobe or scanning electron microscope work.

Microscopic Description and Identification

It is important for the student to remember that a sample description must be so clear that a reader may have a mental picture of the rock or mineral without even looking at it under the microscope. A well-written description is best accompanied by sketches or photomicrographs. Under the petrographic microscope, a mineral or rock is first described without the analyzer (i.e., only with the polarizer in place), then with polars crossed along the following guidelines:

IN PLANE-POLARIZED LIGHT

Grain Size, Form and Shape

Exact sizes of mineral grains are generally not given in a routine description of a mineral, and terms such as *coarse* (>0.5 mm), *medium* (0.2–0.5 mm), and *fine* (<0.2 mm) are used to describe the relative sizes of mineral grains. Some additional terms are used to indicate how well developed the crystal faces are. The term *euhedral* (syn. idiomorphic and idioblastic—used only in metamorphic rocks) is used to describe grains with well-developed crystal faces. If none of the crystal faces are developed in a grain, it is said to be *anhedral* (syn. xenomorphic or xenoblastic—metamorphic terms). A *subhedral* (syn. subidiomorphic or subidioblastic—metamorphic terms) grain is one that is intermediate between anhedral and euhedral.

The shape or habit of a crystal is described in specific geometric terms such as *acicular* (needle-like), *equant* (when all dimensions are roughly equal), *bladed*, *lath-like*, *tabular*, *prismatic*, and *subprismatic* (Figure 4.29). Common examples of these shapes are as follows:

lath-like or tabular (as a coarse grain)—plagioclase

bladed—kyanite

acicular—sillimanite

equant—quartz, garnet

prismatic—amphibole

subprismatic—pyroxenes

Relief

The relief of a mineral is its relative "height" amidst the surrounding materials (minerals and epoxy), and is a reflection of the contrast between the refractive indices of the mineral and the surrounding materials. Relief can be a very good criterion in mineral identification, and a few examples are as follows:

very high relief—garnet, spinel, zircon, corundum, kyanite, sphene

high relief—pyroxene, olivine

medium relief—amphibole

low relief—plagioclase, alkali feldspars, quartz, nepheline

relief changing from low to high during stage rotation—calcite

Color and Pleochroism

Under this heading the color of a mineral in plane-polarized light is described. If, however, the mineral appears colorless, it should be noted as such. In the case of a colored mineral, one must note whether it is pleochroic (i.e., its color or intensity varies as the stage is turned) or nonpleochroic (i.e., its color does not vary). Some people prefer an exact description of which vibration direction in the mineral corresponds to what color (or, the pleochroic scheme) of a mineral. The following are some useful examples:

Colorless (quartz, feldspars, olivine, muscovite)

Colored and *nonpleochroic* (some garnet, some pyroxenes)

Colored and *pleochroic* (amphiboles, biotite, phlogopite, staurolite)

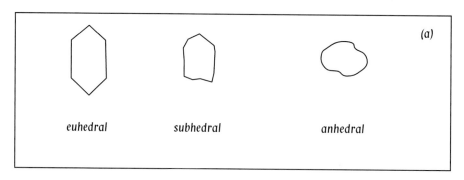

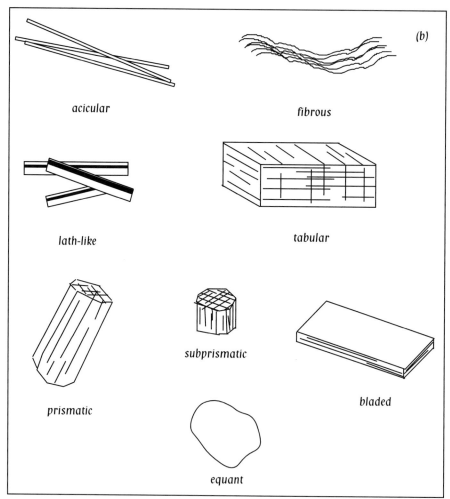

FIGURE 4.29   (a) Terms used to describe the relative development of crystal faces. (b) Some habits of crystals.

Cleavage

Cleavages in minerals generally show up as dark, fairly evenly spaced, straight cracks under the microscope. It is important to note the number of cleavage directions and their angular relationships. Two notable examples are pyroxenes and amphiboles; the two sets of cleavage in both minerals are best seen in their basal (001) sections. The

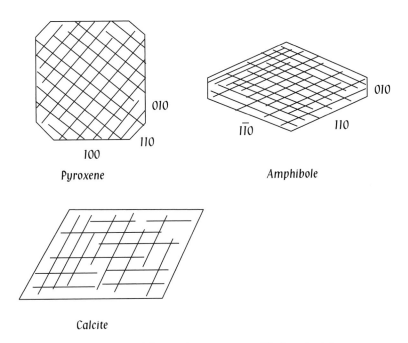

FIGURE 4.30   *Examples of cleavage in pyroxene, amphibole, and calcite (rhombohedral cleavage).*

angle between the two sets of cleavage is 87° in pyroxenes and about 56° in amphiboles (Figure 4.30).

Inclusions, Associations, and Alterations

It is sometimes important to describe inclusions within a mineral grain. What they are and how they occur (for example, whether they are oriented) should be described. In describing rocks, it is also important to note what the associated minerals or materials are, and their intergrain relationships (whether enclosed in one another or juxtaposed). Alterations are important to note by indicating whether the grains are entirely or partially altered. Sometimes the alteration phase keeps the original crystal form of the mineral it replaced, in which case it is called a *pseudomorph*.

BETWEEN CROSSED POLARS

Isotropic/Anisotropic

The first thing to note after crossing the polars is whether the mineral is isotropic or anisotropic. As stated before, if the mineral is isotropic, all of its grains will remain extinct. If the mineral is anisotropic, the interference colors, extinction, and optic sign need to be described.

Interference Colors

Using the Michel–Levy chart, the mineral's interference color and order must be written. If the color varies significantly within a grain, the term "variegated color" may be used. Zircon shows a variegated mix of high-order colors. The following are some typical interference colors, assuming a thin section having a normal thickness of 0.03 mm:

1° gray color—quartz, alkali feldspars, plagioclase, nepheline, apatite

2°–3° colors—pyroxene, olivine, amphibole, micas

High-order colors—calcite, dolomite, sphene

Extinction

As pointed out before, a mineral goes extinct when its vibration directions are parallel to those of the two polars of the microscope. Many different types of extinction may be observed. The nature of extinction depends on the crystallographic orientation of the mineral grain.

One should first note whether the mineral's grains exhibit *uniform extinction* when the whole grain goes extinct upon reaching an extinction position or *nonuniform extinction* when different parts of the grain go extinct at different angles as the stage is rotated. Nonuniform extinction may be due to twinning or zoning. Plagioclase typically shows lamellar twinned extinction in which alternating parallel lamellae within the grain go extinct at a given time (Figure 4.31). Plagioclase in igneous rocks commonly exhibits *compositional zoning* in which concentric zones of different chemical compositions develop within individual plagioclase grains (Figure 4.32). These zones go extinct at different times as the stage is rotated through 90°. Some pyroxene crystals in igneous rocks show *sector zoning*, in which different sectors within the grain go extinct at different times. A special variety of sector zoning is *hour-glass extinction* (hour-glass–shaped sectors go extinct) also seen rarely in pyroxene crystals.

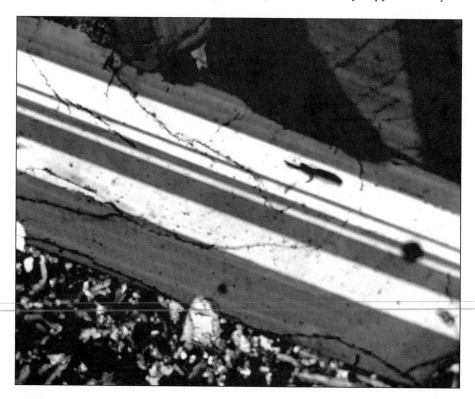

FIGURE 4.31 *Lamellar twinning in plagioclase. (Crossed polars, 80× magnification).* (Courtesy of Gautam Sen)

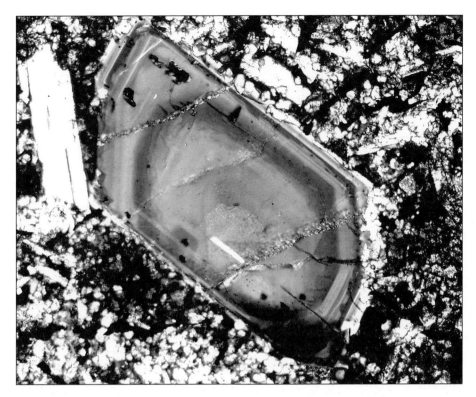

FIGURE 4.32   *Photomicrograph of a zoned plagioclase phenocryst in a basalt lava (36 ×
magnification, crossed polars).* (Courtesy of Gautam Sen)

The carbonate minerals (calcite and dolomite) and micas (muscovite, biotite) characteristically exhibit *mottled extinction*; that is, when the grain is extinct, light still comes through a large number of spots.

Uniform extinction may be of several different types: straight or parallel, inclined, no extinction angle, and symmetrical (Figure 4.33, Nesse 1991). A mineral is said to show *straight* or *parallel extinction* if it goes extinct when a cleavage set or its elongate (when cleavage is not distinct) grain boundary is parallel to one of the crosshairs. In this case the mineral's vibration directions are parallel to its cleavage (or elongation) and to those of the polarizer and analyzer.

If the mineral goes extinct when its cleavage or grain boundary is at an angle to the crosshairs, the extinction is *inclined extinction*. In reporting inclined extinction, it is customary to measure and report the extinction angle, which is the angle between one of the crosshairs and the position in which the grain goes extinct. It is measured using the graduations marked on the revolving stage. Using accessory plates, one may determine the X or Z direction of the grain and thus list the extinction angle as X $\wedge$ *c* or Z $\wedge$ *c* (Figure 4.33b). These angles are particularly helpful in the identification of the variety or "species" of amphiboles and pyroxenes.

In the case of anhedral mineral grains that do not lack any cleavage or other directional features, the extinction can simply be described as uniform extinction with no extinction angle (Figure 4.33c).

Symmetrical extinction refers to mineral grains that have two distinct directions, which may either be two sets of cleavage or two well-developed crystal faces. If the extinction angles measured with respect to two such directional features are the same, the mineral is said to exhibit symmetrical extinction (Figure 4.33d).

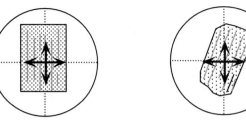

(a) Straight extinction

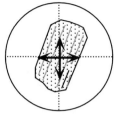

(b) Inclined extinction

FIGURE 4.33 *Some extinction types.* The arrows indicate vibration directions of the component waves. The parallel and subparallel lines in a, b, and d represent cleavage. In Figure 4.33d, 'x' and 'y' represent the angle between extinction position and two sets of cleavage.

(c) No extinction angle

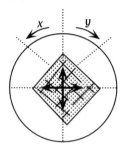

(d) Symmetrical extinction

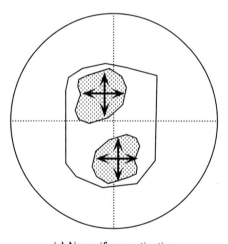

(e) Nonuniform extinction

Optic Sign

Finally, using conoscopic illumination, a high power objective, and the Bertrand lens, the *optic sign* (whether the mineral is uniaxial or biaxial and negative or positive) must be determined. In the case of a biaxial mineral, 2V should be estimated, if possible, based on Kamb's method. Precise measurements were a necessity some decades ago when fast quantitative methods were not available. However, such precise determinations of refractive indices and 2V are unnecessary today because fast, quantitative analyses of minerals have become routine using the electron microprobe or scanning electron microscope.

SUMMARY

[1] A light wave is an electromagnetic wave that can be made to vibrate in a single plane through plane polarization.

[2] A petrographic microscope makes use of two sets of polarizing plates that can be made to cross at 90°.

[3] The refractive index is an important property of a mineral that is dependent upon the wavelength of the light that passes through it. Isotropic minerals have a single refractive index, whereas anisotropic minerals have two (uniaxial) or three (biaxial) refractive indices. Refractive indices are generally determined using an oil immersion technique.

[4] Double refraction is the phenomenon by which a single light ray traveling through a mineral splits up into an ordinary and extraordinary rays. The optic axis is a unique direction within an anisotropic mineral along which a light ray does not split up into O- and E-rays. The difference between refractive indices is known as birefringence. Birefringence and thickness variations are related to interference colors that an anisotropic mineral may exhibit between crossed polars.

[5] Uniaxial minerals possess one optic axis, whereas biaxial minerals possess two optic axes. Uniaxial minerals exhibit two extreme refractive indices, namely ε and ω. Biaxial minerals possess three extreme refractive indices: α, β, and γ. ω and β do not vary with crystallographic direction.

[6] The indicatrix is an imaginary geometric figure that represents the variation in refractive index values of a mineral in three dimensions. The isotropic indicatrix is a sphere. The uniaxial indicatrix is an ellipsoid of rotation, and the biaxial indicatrix is a triaxial ellipsoid. A circular section is an imaginary circular plane (slice) through the indicatrix that is perpendicular to the optic axis. The uniaxial indicatrix contains one circular section. The biaxial indicatrix contains two circular sections corresponding to the two optic axes. 2V is the angle between the two optic axes of a biaxial mineral. The acute bisectrix bisects the acute 2V angle, whereas the obtuse bisectrix bisects the obtuse 2V angle.

[7] The principal sections are elliptical slices through the indicatrix whose semi-major and semi-minor axes represent the two extreme refractive indices of a mineral. The two refractive indices in the principal section of a uniaxial indicatrix are ε and ω. In a biaxial indicatrix three principal sections are possible—$\gamma\alpha$, $\beta\gamma$, and $\alpha\beta$.

[8] The optic sign of a mineral is an important and distinctive property. A uniaxial mineral is said to be positive if its ε refractive index is greater than its ω and negative if $\omega > \varepsilon$. A biaxial mineral is said to be positive if its slow direction (Z) coincides with its acute bisectrix (Bx_a). In a negative biaxial mineral X coincides with Bx_a.

[9] The refractive indices of a mineral are determined by the oil immersion method. Grain selection is of utmost importance in the determination of the refractive index (or indices) of a mineral. The interference figures are best obtained with isotropic or near isotropic grains of an anisotropic mineral. Conoscopic illumination and the Bertrand lens are required to obtain an interference figure.

[10] An interference figure is composed of isogyres, melatope(s), and isochromes. The two isogyres in the interference figure of a uniaxial mineral form a

"cross." The cross remains at the center of the field of view if the optic axis remains at the center during rotation of the stage (centered interference figure). However, in an off-centered figure, the isogyres remain straight and move laterally across the stage upon rotation.

[11] The optic axis figure and the acute bisectrix figure are most useful in the determination of optic sign of a biaxial mineral. 2V may be estimated from the curvature of the isogyre or from the maximum separation of the two isogyres in an acute bisectrix figure.

BIBLIOGRAPHY

Battey, M.H. (1981) *Mineralogy for Students* (2nd ed.). Longman (New York).

Blackburn, W.H. and Dennen, W.H. (1994) *Principles of Mineralogy* (2nd ed.). W.C. Brown (Dubuque, IA.).

Bloss, F.D. (1961) *An Introduction to the Methods of Optical Crystallography*. Holt, Rinehart, and Winston (New York).

Nesse, W. (1991) *Introduction to Optical Mineralogy* (2nd ed.). Oxford (New York).

Perkins, D. (1998) *Mineralogy*. Prentice Hall (Upper Saddle River, N.J.).

Shelley, D. (1985) *Optical Mineralogy* (2nd ed.). Elsevier (New York).

Stoiber, R.E. and Morse, S.A. (1994) *Crystal Identification With the Polarizing Microscope*. Chapman and Hall (New York).

Wahlstrom, E.E. (1979) *Optical Crystallography* (5th ed.). Wiley (New York).

C H A P T E R   5

Systematic Mineralogy

This chapter covers the following mineral descriptions:

SILICATES

Nesosilicates
    Olivines
    Aluminosilicates
    Garnet
    Zircon
    Staurolite
Sorosilicates
    Lawsonite
    Epidote
Cyclosilicates
    Tourmaline
Inosilicates
    Single Chain: pyroxenes
    Double Chain: amphiboles
Phyllosilicates
    Clay minerals
    Chlorite
    Micas: muscovite, phlogopite, biotite
    Serpentines
Tectosilicates
    Quartz
    Feldspars
    Feldspathoids

OXIDE MINERALS
    Spinels
    Ilmenite
    Hematite
    Chromite

CARBONATES
    $CaCO_3$ polymorphs
    Dolomite

SULFIDES
    Galena
    Chalcopyrite
    Sphalerite
    Pyrite

SULFATES
    Gypsum
    Halite

ABOUT THIS CHAPTER

A large number of minerals occur in nature, however, only a few have been found to be of interest to *most* geologists. In this chapter a short description of some of these minerals is presented.

SILICATES

Olivines

Chemical composition: $[Mg,Fe]_2SiO_4$

Olivines form a complete solid solution between their two end members: forsterite (Mg_2SiO_4) and fayalite (Fe_2SiO_4) (Figure 5.1; An explanation of the phase relations on binary solid solution is given in Chapter 7). Thus, a natural olivine may have 75% forsterite and 25% fayalite components and its composition may be written as $Fo_{75}Fa_{25}$ or $[Mg_{.75}Fe_{.25}]_2SiO_4$. Olivines generally contain less than 0.5% CaO and less than 0.5% Ni.

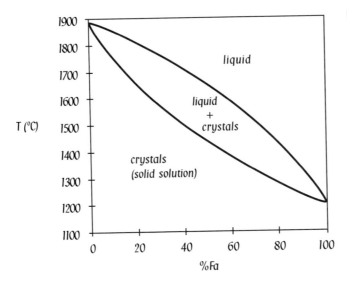

FIGURE 5.1 *Olivine solid solution.*

Crystallography: Orthorhombic.

Olivine is a nesosilicate with independent SiO_4 tetrahedra linked by M (Mg and Fe) cations (Figure 5.2a,b). The divalent cations form octahedra with surrounding oxygens belonging to different silicate tetrahedra. There are two types of *M sites*: M1 and M2. *M1 sites* are located at centers of symmetry, and *M2 sites* are located at mirrors within the structure. The *M1* and *M2* sites form parallel and alternate chains along the *c*-axis (Figure 5.2b).

Physical properties

Habit. Short prismatic.

Opacity and color. Generally transparent. Forsterite-rich olivines are colorless to olive-green colored. Fayalite-rich olivines are yellowish.

Luster. Vitreous.

Streak. Colorless.

Hardness (H). 7 (forsterite) to 6 (fayalite).

Density. 3.222 (forsterite) to 4.392 (fayalite).

Fracture. Conchoidal.

Optical properties

PLANE-POLARIZED LIGHT

Grain. Commonly equant; anhedral in most igneous rocks but can be euhedral in alkali olivine basalts.

Relief. High

RI	forsterite	fayalite
α	1.635	1.827
β	1.651	1.877
γ	1.670	1.880

[Note: The RI and $2V_z$ of all intermediate olivines can be calculated from these end members by assuming a linear relationship. For example, α of $Fo_{60}Fa_{40}$ should be $(0.6 \times 1.635 + 0.4 \times 1.827) = 1.7118$.]

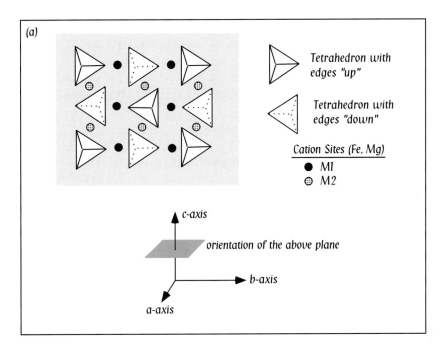

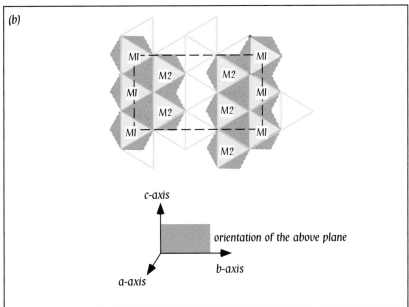

FIGURE 5.2 *Atomic structure of olivine as viewed (a) on a plane perpendicular to the c-crystallographic axis, and (b) on a plane containing the c-axis.*

Color and pleochroism. Generally colorless, however, fayalitic olivines are yellowish. Nonpleochroic.

Cleavage. None.

Alterations. Commonly alters to serpentine (colorless to green) or iddingsite (brownish).

BETWEEN CROSSED POLARS

> *Anisotropism.* Anisotropic with 2° interference colors.
>
> *Extinction.* Generally straight with respect to prismatic faces. Twin lamellae and sectors are common in olivines, in peridotites, and sometimes in basaltic rocks.
>
> *2V$_z$ and optic sign.* 82° (+) for forsterite, and 134° (−) for fayalite.

Distinguishing features: Their green color in hand specimens, lack of cleavage, 2° interference colors and equant shape under the microscope, and occurrence in peridotites and basaltic/gabbroic rocks are all quite distinctive. However, very often the olivine in basalts and peridotites is so severely altered to serpentine or iddingsite that it goes beyond recognition. In such cases, the original grain shape is sometimes retained.

Use: Olivine has a very high melting temperature and is therefore used in refractory bricks. It is also used as a gem (peridot) and as an abrasive.

Garnet group

Chemical composition:

Pyrope [$Mg_3Al_2Si_3O_{12}$]	Grossular [$Ca_3Al_2Si_3O_{12}$]
Spessartine [$Mn_3Al_2Si_3O_{12}$]	Andradite [$Ca_3Fe_2Si_3O_{12}$]
Almandine [$Fe_3Al_2Si_3O_{12}$]	Uvarovite [$Ca_3Cr_2Si_3O_{12}$]

Complete solid solution exists between pyrope, spessartine, and almandine in the dodecahedral cation site. Complete solid solution also exists between grossular, andradite, and uvarovite in the octahedral cation site in the garnet structure. Aside from the types of garnets mentioned previously, a special type of high-pressure garnet, called *majorite*, has been experimentally synthesized in the laboratory. (This is discussed extensively in later chapters.)

Natural garnets vary extensively in terms of pyrope, grossular, and almandine contents, and have been the focus of numerous petrologic studies. Figure 5.3a shows a triangular plot of such variations in garnets from a variety of rock types. An early attempt by Coleman (1965) to classify eclogites based on their garnet composition led to this famous diagram, although garnets from other rock types are also shown here. Recent studies of mantle rocks (principally eclogites and peridotites) have led to the recognition of garnets that have formed in ultrahigh-pressure conditions (discussed in a later chapter). As shown in Figure 5.3a,b, the garnets in such ultrahigh-pressure rocks are different from garnet peridotites in kimberlites; although there is considerable variation even among ultrahigh-pressure rocks from different places. (For example, note the difference between the western Alps and China, Figure 5.3a). The reader may refer back to these diagrams while going through the chapter in which ultrahigh-pressure rocks are discussed.

Crystallography: Isometric.

It is a nesosilicate with two types of cation sites that bond the tetrahedral sites; and these are distorted dodecahedral (called A-site; Coordination Number = 8) and octahedral (called B-site; CN = 6) sites (Figure 5.4). The A-sites are larger and are occupied by Ca, Mg, Fe^{2+}, Mn^{2+}, and the B-sites are occupied by Al, Cr^{3+}, and Fe^{3+}. Some garnets, called hydrogrossular [$Ca_3Al_2Si_2O_8(SiO_4)_{1-n}(OH)_{4n}$, where n ranges from 0 to 1], may contain small amounts of hydroxyl [(OH_4)] in solid solution.

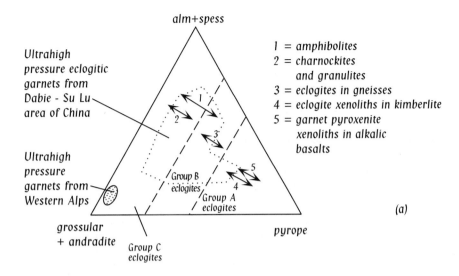

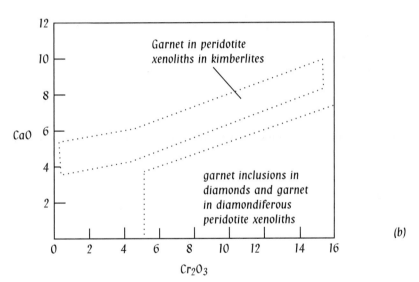

FIGURE 5.3 *Chemical composition of garnets.* (a) Composition of garnet in three type of eclogites as defined by Coleman et al. (1965). (b) Composition of mantle garnets.

Physical properties

Habit. Commonly dodecahedron and trapezohedron.

Opacity and color. Translucent to opaque. Common garnets (almandine–pyrope solid solutions) are deep brown to brown. Grossular-rich garnets can be colorless, pink, or yellowish; and uvarovite (extremely rare) is deep green.

Luster. Resinous.

Streak. Colorless.

Hardness (H). 6.5–7.5.

Density. 3.54 (pyrope), 4.33 (almandine), 4.19 (spessartine), 3.56 (grossular), 3.86 (andradite), 3.80 (uvarovite).

Fracture. Subconchoidal.

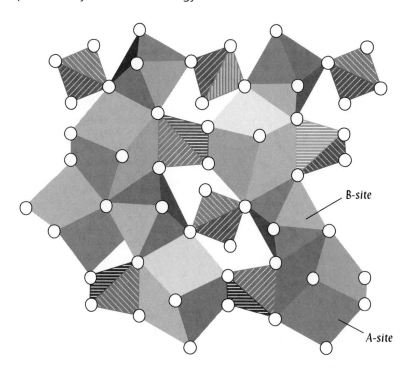

FIGURE 5.4   *Atomic structure of garnet.*

Optical properties

PLANE-POLARIZED LIGHT

Grain. Commonly equant, euhedral to subhedral, large grains.

Relief. Very High.

Species	RI
pyrope	1.71
almandine	1.83
spessartine	1.80
grossular	1.75
andradite	1.87
uvarovite	1.85

Color and pleochroism. Generally light brown to colorless. Nonpleochroic.

Cleavage. None.

Alterations. Commonly alters to clay along cracks.

BETWEEN CROSSED POLARS

Anisotropism. Isotropic.

Distinguishing features: Common garnets are pyrope almandine in composition and are brown in color. They are commonly found in a wide variety of metamorphic rocks. Additionally, lack of cleavage, dodecahedral (12-sided) or trapezohedral (trapezoid faces) crystal form in the hand specimen, and very high relief and isotropic character in the thin section all help to identify this mineral in a rock.

Use: Garnet is commonly used as a gem and as an abrasive.

Aluminosilicates

Chemical composition: [Al$_2$SiO$_5$]

	Kyanite	**Andalusite**	**Sillimanite**
Crystallography: (Figure 5.5)	Triclinic	Orthorhombic	Orthorhombic

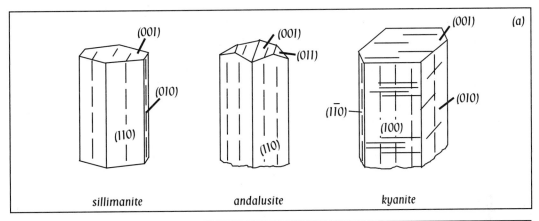

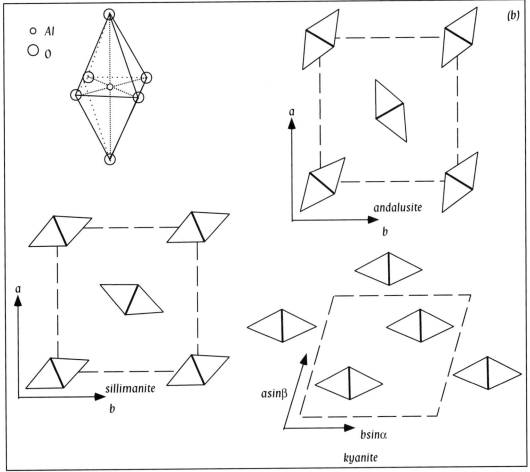

FIGURE 5.5   *Schematic drawings of aluminosilicate crystals.*

Physical properties

Habit.	Bladed	Prismatic, "tablets"	Fibrous or long prisms
Color.	Blue	Brown or white	White (brown)
Luster.	Vitreous, pearly on cleaved faces	Vitreous	Vitreous
Streak.	Colorless to white	Colorless	Colorless
Hardness.	Differential hardness: 4–5 along c-axis, 6–7 across.	7.5	6–7
Density.	3.6	3.18	3.23

Optical properties

PLANE-POLARIZED LIGHT

Grain.	Bladed, tabular prismatic		Fibrous or long prismatic
Relief.	Very high	High	High
RI			
α	1.712	1.632	1.658
β	1.720	1.640	1.662
γ	1.728	1.642	1.680
Color and pleochroism.	Colorless Nonpleochroic	Pinkish Feebly pleochroic	Colorless Nonpleochroic
Cleavage.	2 sets-(100) perfect, (010) good	2 sets {110} at right angles to each other	1 set (010) perfect
Alterations and inclusions.	All three alter to sericite.	*Chiastolite* variety has a cross-shaped pattern of carbonaceous inclusions	

BETWEEN CROSSED POLARS

Anisotropism.	1° colors	1°gray	1°–2° colors
Extinction.	Inclined	Straight with respect to the cleavage	Straight with respect to the cleavage
	$(Z \wedge c = 27\text{–}32°)$		
Optic sign.	−Biaxial	−Biaxial	+Biaxial
2V.	$2V_x = 78\text{–}84°$	$2V_x = 71\text{–}88°$	$2V_z = 20\text{–}30°$

Distinguishing features: Kyanite's bladed character, differential hardness along two different directions, very high relief, and inclined extinction are quite distinctive. Kyanite is often characterized by prismatic simple twinning. Sillimanite is distinguished by its long narrow prisms or fibrous character, moderately high relief, straight extinction, and optic sign. Andalusite is distinctive in the "cruciform" inclusions in its chiastolite variety. Its large 2V is also characteristic. (See the plate containing photomicrographs at the end of this chapter.) All three minerals occur in metamorphic rocks. Kyanite may sometimes occur in eclogites. The stabilities of these minerals are of great significance to the metamorphic petrologist and are dicussed in a later chapter.

Zircon

Chemical composition: $ZrSiO_4$

Crystallography: Tetragonal.
The principal structural unit is a chain of alternating edge-sharing SiO_4 tetrahedra and ZrO_8 triangular dodecahedra oriented parallel to the c-axis.

Physical properties

Habit. Small square prisms with a pyramidal termination.
Opacity and color. Generally transparent to translucent; brownish, but can be colorless, green or other colors.
Luster. Vitreous to nearly adamantine.
Streak. Colorless.
Hardness. 7.5.
Density. 4.6–4.7.
Cleavage. Rare {110}.
Fracture. Conchoidal or uneven.

Optical properties

PLANE-POLARIZED LIGHT

Grain. Distinctive small square prisms with pyramidal terminations; generally euhedral.
Relief. Very high. ($\omega = 1.922 - 1.96$; $\varepsilon = 1.961 - 2.015$)
Color and pleochroism. Colorless but with noticable dispersion; weak pleochroism (if present).

BETWEEN CROSSED POLARS

Anisotropism. 2–3° interference colors.
Extinction. Straight.
Optic sign. +Uniaxial.

Distinguishing features: It is hard to confuse zircon with anything else under a microscope. Its characteristic shape, very high relief, small grain size, very high birefringence, straight extinction, and uniaxial positive property are all good clues to its identification.

Special notes on zircon: Zircon is an extremely common accessory mineral in plutonic felsic and intermediate rocks and pegmatites. Zircon's real notoreity comes from its radioactivity due to U,Th radioactive isotopes and to its physical ability to withstand tectonic forces. Ion-probe techniques developed at Australian National University have led to the recognition of some of the oldest zircons in the world. It turns out that individual zircon crystals can be strongly zoned, with each zone recording a particular geological event. This allowed geologists to reconstruct the earliest history of the earth with the help of zircon crystals. The oldest zircon to be found is a 4.2 billion year old crystal embedded in Proterozoic sedimentary rock in Australia.

Staurolite

Chemical composition: $Fe_2Al_9O_6[SiO_4]_4(OH)_2$

Crystallography: Monoclinic. Penetration twins are common at 60° or 90° (Figure 5.6).

Physical properties

Habit. Prismatic crystals.
Opacity and color. Dark reddish brown.
Luster. Vitreous to resinous.
Streak. Colorless.
Hardness. 7–7.5.
Density. 3.7–3.8.
Cleavage. Poor (if any).
Fracture. Uneven.

Optical properties

PLANE-POLARIZED LIGHT

Grain. Commonly prismatic, elongate along *c*-axis; basal sections are generally six-sided.
Relief. Very High.

RI $\alpha = 1.736–1.747$; $\beta = 1.74–1.754$; $\gamma = 1.745–1.762$

Color and pleochroism. Pale yellow to deeper yellow; distinctly pleochroic.
Alterations and inclusions. May alter to sericite. Inclusions are generally abundant.

BETWEEN CROSSED POLARS

Anisotropism. Anisotropic. 1° yellow interference color.
Extinction. Straight.
$2V_z$ and optic sign. $2V_z = 80–90°$; Biaxial+.

Distinguishing features: Color, pleochroism, high relief, and 2V are all distinctive.

Additional notes: Staurolite commonly occurs as a porphyroblast in metamorphic rocks. Daniels et al. (1996) made an important discovery of staurolite inclusions in dia-

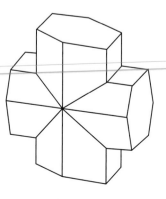

FIGURE 5.6   *Penetration twins in staurolite.*

monds. This staurolite is metastable because laboratory experiments tell us that the upper pressure limit of stability of staurolite is about 1.7 GPa (50 km; discussed in a later chapter), whereas diamond's stability field begins at pressures greater than 3.5 GPa (110 km) in a downgoing subducted slab. The logical inference is that prior to its incorporation in diamond, the staurolite formed as part of a metamorphic rock in the continental crust. This piece of continental crust was apparently subducted into fairly deep levels in the mantle where it was quickly enveloped by diamond. The temperature of the subducted slab never rose too high, otherwise the staurolite would have broken down. What this means is that the subduction was extremely rapid. Therefore, the occurrence of staurolite as a diamond inclusion is significant evidence that the recycling of continental crust into at least the deep upper mantle (or even lower mantle) does occur.

Lawsonite

Chemical composition: $CaAl_2Si_2O_7(OH)_2 \cdot 2H_2O$

Crystallography: Orthorhombic.

Physical properties

Habit. Tabular crystals; basal sections are rectangular or rhomb shaped.

Opacity and color. Colorless, bluish gray, bluish green, or white. Transparent to translucent.

Luster. Vitreous to greasy.

Streak. Colorless

Hardness. 8.

Density. 3.1.

Cleavage. Two sets (100) and (010).

Fracture. Uneven.

Optical properties

PLANE-POLARIZED LIGHT

Grain. Commonly prismatic elongate along *c*-axis; basal sections are generally squares or rhombs.

Relief. Moderately High;

RI $\alpha = 1.665$; $\beta = 1.674$; $\gamma = 1.685$

Color and pleochroism. Generally colorless; sometimes weakly pleochroic—colorless to blue-yellow-green.

Alterations and inclusions. May alter to sericite. Inclusions are generally abundant.

BETWEEN CROSSED POLARS

Anisotropism. Anisotropic. 1° yellow-to-red interference color.

Extinction. Straight or symmetrical when lamellar twinning is present.

$2V_z$ and optic sign. $2V_z = 76$–$86°$; Biaxial+.

Distinguishing features: Color, pleochroism, high relief, and 2V are all distinctive. Lawsonite is a common mineral in low-grade blueschist-facies metamorphic rocks.

Epidote

Chemical composition: $Ca_2(Al,Fe)Al_2Si_3O_{12}(OH)$

Crystallography: Monoclinic.

Physical properties

Habit. Small sand-sized grains or fibrous aggregates.
Opacity and color. Translucent with distinct pistachio-green color.
Luster. Subvitreous.
Streak. Colorless.
Hardness. 6–6.5.
Density. 3.4–3.5.
Cleavage. One set well-developed basal cleavage.
Fracture. Not a diagnostic feature.

Optical properties

PLANE-POLARIZED LIGHT

Grain. Commonly small granular aggregates; sometimes elongate prismatic or fibrous.
Relief. High;

RI $\alpha = 1.715$–1.751; $\beta = 1.725$–1.784; $\gamma = 1.734$–1.797.

Color and pleochroism. Some colorless, but most are distinctly pleochroic: light to deeper yellowish green.

BETWEEN CROSSED POLARS

Anisotropism. Anisotropic. 2° and 3° anomalous colors. Zoning is common.
Extinction. Straight with respect to cleavage.
$2V_z$ and optic sign. $2V_x = 0$–$25°$; Biaxial−.

Distinguishing features.   Color, pleochroism, high relief, crystal form, and optic sign are all distinctive. Epidote can occur in a wide variety of metamorphic rocks, particularly low- to medium-grade metamorphosed calcareous and mafic rocks.

Tourmaline

Chemical composition: $(Na,Ca)(Mg,Fe,Li,Al)_3Al_6(BO_3)_3Si_6O_{18}(OH)_4$

Crystallography: Hexagonal (rhombohedral). Six-fold rings of silicate tetrahedra interclated with BO_3 along the *c*-axis.

Physical properties

Habit. Typically trianguloid cross sections and columnar to acicular crystals forming radiating patterns.
Opacity and color. Transparent. Color can vary greatly from black to pink, green, and red.
Luster. Vitreous.
Streak. Colorless.

Hardness. 7.

Density. 3.03–3.25.

Cleavage. Absent. Longitudinal striations on crystal faces formed due to growth may be confused with cleavage or twinning.

Fracture. Conchoidal.

Optical properties

PLANE-POLARIZED LIGHT

Relief. Moderately high;

RI $\omega = 1.631–1.698$; $\varepsilon = 1.610–1.675$.

Color and pleochroism. Strongly pleochroic—light to deeper shades of black, green, brown, etc.

BETWEEN CROSSED POLARS

Anisotropism. Anisotropic. 2° interference colors; the absorption color generally masks the interference color.

Extinction. Straight.

Optic sign. Uniaxial−.

Distinguishing features: Color, pleochroism, crystal form, absence of cleavage, and optic sign are all distinctive. Tourmaline can be found in felsic-plutonic igneous rocks and pegmatites.

Pyroxenes

Chemical composition and crystallography: Pyroxenes can be crystallographically and compositionally diverse due to the nature of the sites where divalent and trivalent cations occur, linking the single chains composed of tetrahedra (Figure 5.7). Crystallographically, pyroxenes may be broadly classified into orthopyroxenes (orthorhombic: enstatite-ferrosilite series) and clinopyroxenes (monoclinic: everything else). Their general composition is written as $[(M2)(M1)(T)_2O_6]$, where M2 and M1 are cation sites that link the individual chains. M2 occurs between the bases of two adjacent chains, whereas M1 occurs between their apices (Figure 5.7). The following cations occupy the three sites:

Pyroxene end members	Smaller $\longrightarrow$				
	M2	M1	T,	T	
Diopside	Ca	Mg	Si,	Si	
Hedenbergite	Ca	Fe^{2+}	Si,	Si	Quadrilateral
Enstatite (clinoenstatite)	Mg	Mg	Si,	Si	pyroxenes
Ferrosilite (clinoferrosilite)*	Fe^{2+}	Fe^{2+}	Si,	Si	
Aegirine	Na	Fe^{3+}	Si,	Si	
Jadeite	Na	Al	Si,	Si	
Johannsenite	Ca	Mn	Si,	Si	
Mg-Tschermaks*	Mg	Al	Al,	Si	
Ca-Tschermaks*	Ca	Al	Al,	Si	
Spodumene	Li	Al	Si,	Si	

*not stable in nature

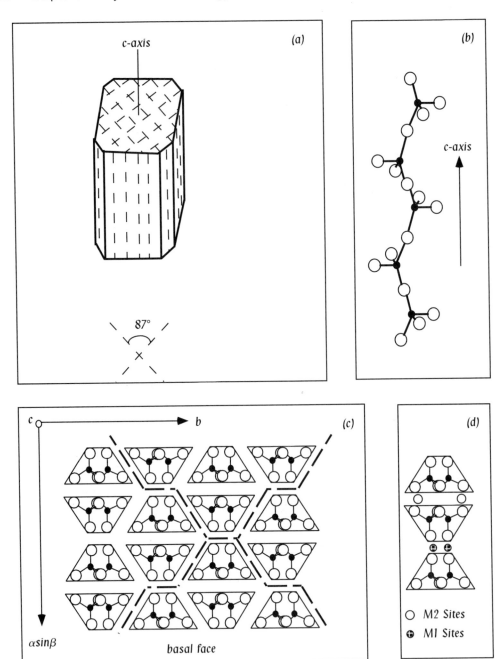

FIGURE 5.7 *Atomic structure of pyroxenes.*

The M1 site is an octahedral site with a coordination number of 6. The coordination number of the M2 site may be 6 (orthopyroxene, occupants—Mg, Fe^{2+}), 7 (pigeonite, Mg, Fe^{2+}, Ca), or 8 (Ca-rich pyroxenes, Ca, Na). Aside from the above listed cations, Fe^{3+}, Cr^{3+}, and Ti occur in M1 sites of natural pyroxenes. Na and Na–Ca pyroxenes are classified as omphacite (the pyroxene in eclogite), jadeite, aegirine, and aegirine augite depending on their content of jadeite, acmite, and (Ca + Mg + Fe) components (Figure 5.8a). Although not mentioned in the table above, it has been a

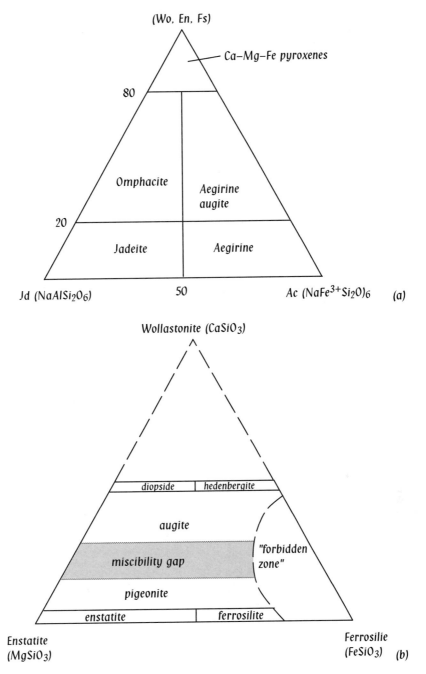

FIGURE 5.8   (a) Chemical composition of alkalic pyroxenes. (b) Composition of quadrilateral pyroxenes.

common practice to use the name hypersthene to refer to enstatite–ferrosilite solution minerals with about 10% to 50% Fs hypersthene.

Studies of clinopyroxenes that occur as inclusions in diamond (i.e., under ultrahigh pressure) have revealed that as much as 1.7 wt% K_2O can occur in them (Chopin and Sobolev 1995). This K presumably occupies M2 sites.

The common pyroxenes of mafic and ultramafic rocks can be described in terms of the so-called "pyroxene quadrilateral," which is a portion of the compositional triangle formed by $MgSiO_3$, $FeSiO_3$, and $CaSiO_3$. (Figure 5.8b; the diagram is slightly modified from the classification proposed by Morimoto 1988.) Orthopyroxenes, a solid solution of $MgSiO_3$ and $FeSiO_3$ components, can have up to 5% $CaSiO_3$ in solid solution, whereas pigeonite can have up to 15% $CaSiO_3$ component. Note that natural pyroxenes on earth are unstable in the area marked "forbidden zone." However, some Fe-rich lunar pyroxenes can be stable in that area. Also, a miscibility gap occurs between augite and pigeonite at low pressure. Metastable, groundmass pyroxenes in lavas can trespass this miscibility gap.

Figure 5.9 shows some general compositional ranges of natural pyroxenes that occur in basalts, diabases, lherzolites, eclogites, and in metamorphosed calc-silicate rocks. Diopsidic pyroxenes occurring in metamorphosed carbonate rocks are readily distinguishable from those in ultramafic rocks by the relatively high Cr-content (0.2–2 wt%) of the latter. Pyroxenes in tholeiitic sills and dikes can cover a broad range of compositions from augite to ferroaugite and from Mg to Fe-rich pigeonites. Ca-rich and Ca-poor pyroxenes simultaneously crystallizing from tholeiitic basalt magma in plutonic, layered intrusions (discussed in a later chapter) exhibit the maximum range of compositions (shown as thick lines in Fig. 5.9). The tie-lines show the composition of coexisting Ca-rich and Ca-poor pyroxenes. Figure 5.9 shows that in a magnesian, tholeiitic magma, a Ca-rich pyroxene first cocrystallizes with a magnesian orthopyroxene as the evolving magma becomes Fe-rich. Then, orthopyroxene stops crystallizing from the magma and pigeonite appears. Pigeonite continues to crystallize with augite up to the forbidden zone, at which point the pigeonite is no longer stable (and usually a fayalitic olivine and a SiO_2 phase appear). During subsolidus cooling, pigeonite ultimately breaks down into a orthopyroxene, and the excess Ca is exsolved as an augite.

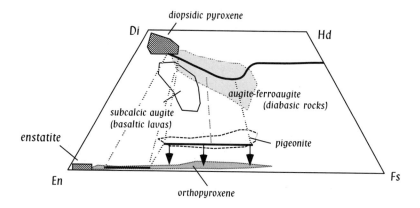

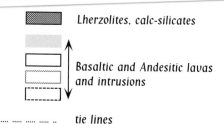

FIGURE 5.9   *Composition of pyroxenes in common tholeiitic and ultramafic rocks.*

Pyroxenes in mantle rocks are nearly pure diopsides and enstatites. Clinopyroxenes in mantle xenoliths have often been called Al-augite (occurs in pyroxenites) and Cr-diopside (occurs in peridotites). These high-pressure pyroxenes can contain as much as 9% Al_2O_3 and up to about 2% Cr_2O_3 in solid solution. The Na_2O content of such pyroxenes can be as high as 3%. McDonough and Rudnick (1998) compiled a large dataset on Cr-diopsides from upper-mantle peridotite xenoliths, and showed that Na_2O–Al_2O_3 trends are very different for Cr-diopsides in garnet peridotites (deeper, and continental in origin) versus spinel peridotites (shallower, continental as well as oceanic: Figure 5.10a). A large database also exists on the trace element composition of pyroxenes, particularly those from mantle peridotites. They vary widely, sometimes even within single grains (Figure 5.10b). Such variability provides clues to important petrologic processes such as the magma extraction process (fractional versus batch melting, Johnson et al. 1990), magma-wall rock interaction, and melt extraction rate as a function of depth from the lithosphere (e.g., Sen et al. 1993; Yang et al. 1998).

Properties of individual pyroxene solid solutions

Physical properties

Habit. Augite crystals are generally prismatic to subprismatic. Some enstatite crystals may be fibrous. Jadeite crystals occur in granular clusters.

Color, Luster, and Streak. In terms of the physical properties of hand specimens, color of pyroxenes can vary significantly, from colorless or white (Mg-rich end members) to brown, yellow, green, and black (natural augites and pigeonites are commonly black). Fe-rich varieties are generally brownish to yellowish green (in hedenbergite). Jadeite, omphacite, aegirine, aegirine augite, and Cr-rich diopside are all characteristically green, although their shades vary.

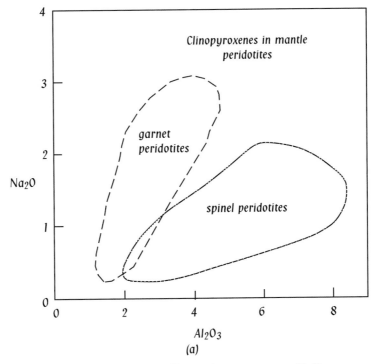

FIGURE 5.10   (a) Compositional difference between garnet peridotites versus spinel peridotites. (b) Rare earth element zoning in a single clinopyroxene crystal in a Hawaiian xenolith. (Yang et al., 1998)

(continues on next page)

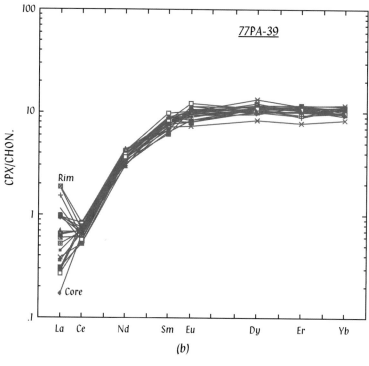

FIGURE 5.10   *(continued)*

Spodumene is "ash" colored. The luster of pyroxenes is generally vitreous to subvitreous, and the streak is colorless.

Hardness and cleavage. Orthopyroxenes: H = 5.12–6. Clinopyroxenes: H = 6–7. All pyroxenes are characterized by two sets of prismatic cleavage. (See Chapter 3.)

Density. 3.2–3.5

Optical properties

PLANE-POLARIZED LIGHT

Color and pleochroism. Most of the pyroxenes are colorless or have a pale brown color (those with greater Ti or Fe). They are usually nonpleochroic to feebly pleochroic.

Relief. High.

BETWEEN CROSSED POLARS

Extinction. Straight extinction with respect to prismatic cleavage sets for orthopyroxenes; inclined extinction with respect to the cleavage for monoclinic pyroxenes. The extinction angle, $Z \wedge c$, varies as follows (Gribble and Hall 1985):

Mineral	$Z \wedge c$
spodumene	24–28°
jadeite	30–40°
omphacite	36–48°
pigeonite	36–44°
augite	35–48°
aegirine and aegirine augite	70–100°

2V, optic sign, and refractive indices. The 2V and optic sign of orthopyroxenes can vary a great deal (Figure 5.11), with more Mg-rich and Mg-poor compositions being

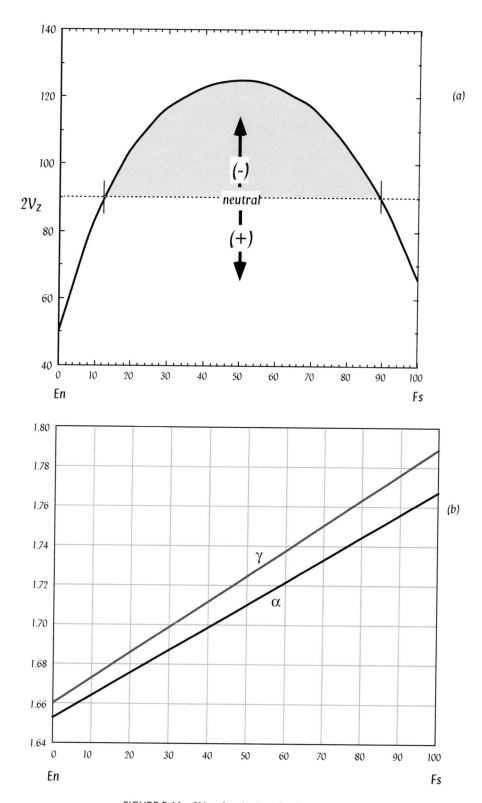

FIGURE 5.11   *2V and optic sign of orthopyroxenes.*

+biaxial, and others being −biaxial. Pigeonite has a small $2V_z$ (0–25°, commonly 0–10°) and is thus easily distinguishable from augite. 2V and β variations in quadrilateral pyroxenes are directly related to composition, as was originally shown by Hess (1949; see Figure 73 in Deer, Howie, and Zussman 1992). Prior to the popularity of modern analytical instruments, particularly the electron probe, virtually all authors (including this author) used this 2V–R.I.–composition relationship to estimate where pyroxenes would plot in the quadrilateral. However, this method is now obsolete.

Additional notes: No discussion of pyroxenes would be complete without mentioning two very important aspects: their exsolution characteristics and their compositional sensitivity to temperature. The mineralogy of the exsolved phases provides important clues to their depth of formation; and their growth and concentration profiles (a subject well beyond the scope of this book) allow the mineralogist to calculate the time required for the exsolved phases to grow. As for the second aspect, using exsolution growth characteristics, many scientists have calculated the cooling ages of meteorites. (See, for example, Grove et al.). The temperature sensitivity of compositions of coexisting pyroxenes (primary orthopyroxene–clinopyroxene or pigeonite–clinopyroxene pairs, or exsolved two pyroxene pairs) has made pyroxenes a very popular geothermometer for rocks of mafic and ultramafic compositions (discussed in a later chapter).

Experimental studies (e.g., Carlson 1988) of pyroxene phase relationships at atmospheric pressure show that pigeonite can only be stable in the presence of a melt over a narrow temperature range (Figure 5.12). Pigeonite has been found to occur as a phenocryst and groundmass in tholeiitic lavas from Japan (Kuno and Nagashima

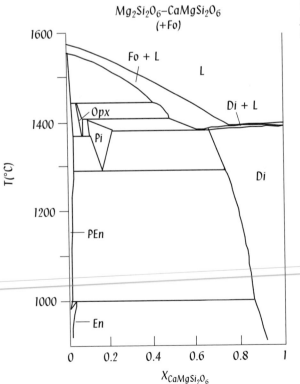

FIGURE 5.12   *Phase relations of pyroxenes at atmospheric pressure.* (Carlson, 1983)

1952) and India (Deccan Traps; Sen 1986). In the $MgSiO_3$–$CaMgSi_2O_6$ system, the maximum pressure of pigeonite stability is about 2 GPa (Lindsley 1983).

Studies of pyroxene exsolution in layered intrusions and thick tholeiitic intrusions have shown that orthopyroxene can exsolve augite (and vice versa), and pigeonite can exsolve augite (and vice versa) (e.g., Poldervaart and Hess 1951). Augite exsolution from pigeonite (or vice versa) characteristically occurs parallel to the (001) face of the host. On the other hand, augite–orthopyroxene exsolution occurs parallel to the (100) face of the host phase. As mentioned earlier, during slow cooling in plutonic layered intrusions, pigeonite eventually breaks down ("inverts") to orthopyroxene and augite. This inversion is generally accompanied by the development of twin planes parallel to the (100) face of the host pyroxene. Therefore, in an inverted pigeonite (now an orthopyroxene) exsolved augite occurs along both (001) and (100) of the host orthopyroxene. The twinning associated with it results in the development of the so-called "herringbone" exsolution (Figure 5.13).

At upper mantle pressures, clinopyroxene and orthopyroxene in mafic and ultramafic rocks may dissolve significant amounts of Al_2O_3 in the form of Ca-Tschermaks and Mg-Tschermaks components. Slow cooling at pressures less than about 1 GPa forces the pyroxene crystals in such rocks to expel their dissolved Al_2O_3 in the form of plagioclase. At pressures of about 1–3 GPa, the exsolved aluminous phase is a green spinel; and at P > 1.6 GPa, garnet may be exsolved from the host pyroxene (Figure 5.14). Whether garnet or spinel would exsolve depends on the bulk composition of the rock. Thus, in a garnet pyroxenite, a garnet may exsolve at, say, 2 GPa, but a Cr-spinel would exsolve from a Cr-diopside in a lherzolite at same pressure.

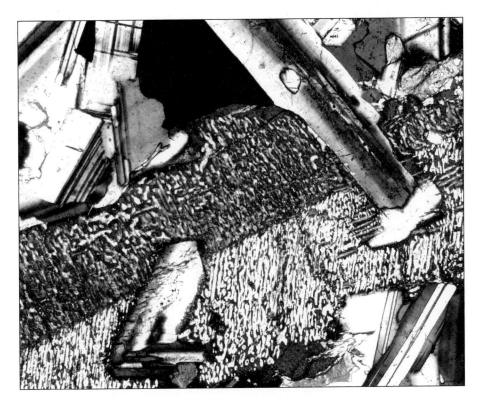

FIGURE 5.13   *Photomicrograph of herringbone exsolution in an inverted pigeonite.* (Courtesy of Gautam Sen)

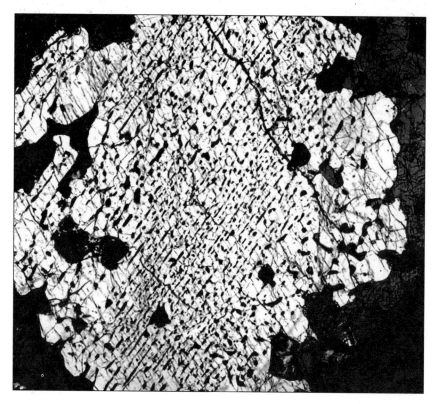

FIGURE 5.14   *Photomicrograph of exsolved garnet in clinopyroxene.* (Courtesy of Gautam Sen)

Amphiboles

Chemical composition and crystallography: The amphibole group can be thought of as a "mineralogical wastebasket" (a phrase attributed to Professor W. Gary Ernst of Stanford University and formerly of UCLA). This is because of their complex crystallography and composition. As we learned in an earlier chapter, all amphiboles are characterized by double chains formed of silicate tetrahedra. As a result, the chain width for amphiboles is greater than that for pyroxenes. The double chains also result in a different angle, 56° instead of 87°, between the two sets of cleavages (Figure 5.15). Like pyroxenes, amphiboles crystallize in orthorhombic and monoclinic systems and are characterized by six-fold to eight-fold coordinated cation sites. However, portions of these sites are very different for the two mineral groups. Some other notable differences, resulting from the double versus single chain structures, are the greater number of different cation sites (M1, M2, M3, M4) that laterally link the double chains, and the presence of the hexagonal "holes" within the double chains, which generate two types of additional sites in amphiboles—an anion site where $(OH)^{-1}$ is usually located, and a "vacant" 10- or 12-coordinated cation site, called an A-site (size = 1–1.4 Å), where the cation Na^+ (and sometimes, K^+) may or may not occur (Figure 5.15).

The general amphibole formula may be written as follows:

$$W_{0-1}X_2Y_5Z_8O_{22}(OH)_2$$

where W represents the A-site generally occupied by Na and K. This site is coordinated with O and (OH) in 10- or 12-fold coordination. X is the M4 site (CN = 6, 8) occupied by Ca, Mg, Fe^{2+}, and Na. Y represents the M1, M2, and M3 sites where Mg, Fe, and Al

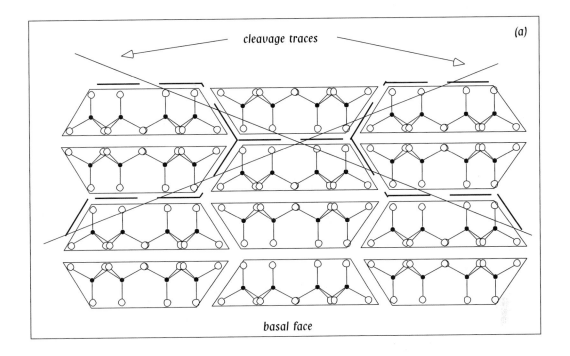

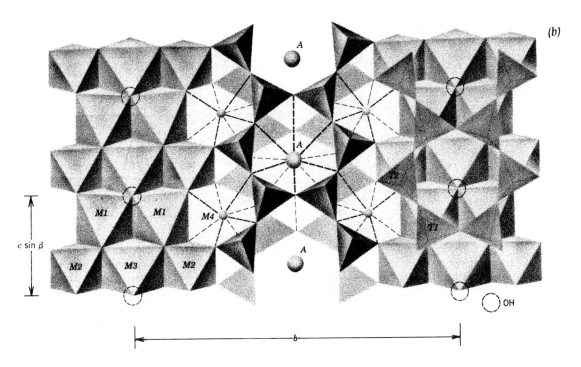

FIGURE 5.15 *Atomic structure of amphibole as viewed (a) perpendicular to the c-axis, and (b) parallel to the c-axis.* (From Hurlbut and Klein, 1999, 21st ed., *Manual of Mineralogy*, John Wiley and Sons Inc.)

cations are located. All of the three sites are in six-fold coordination. Z is the tetrahedral site T where Si and some Al are located (Leake 1978). Amphiboles are classified into four types based on their M4 site occupancy and chemical composition as follows (Figure 5.15):

Amphibole Type		
(Important types)	M4	Composition
Fe, Mg amphiboles	Mg, Fe	
anthophyllite (orthorhombic)		$Mg_7Si_8O_{22}(OH)_2$
cummingtonite–grunerite series (monoclinic)		$(Fe,Mg)_7Si_8O_{22}(OH)_2$
Calcic amphiboles	Ca	
tremolite–actinolite series (monoclinic)		$Ca_2(Mg,Fe)_5Si_8O_{22}(OH)_2$
hornblende (monoclinic)		$(Na,K)_{0-1}Ca_2(Mg,Fe^{2+},Fe^{3+},Al)_5(Si,Al)_8O_{22}(OH)_2$
kaersutite (monoclinic)		$NaCa_2(Mg,Fe^{2+})_4TiSi_6Al_2O_{22}(OH)_2$
Sodic–calcic amphibole	Na, Ca	
richterite (monoclinic)		$Na(Na,Ca)(Mg,Fe)_5Si_8O_{22}(OH)_2$
Sodic amphibole	Na	
glaucophane (monoclinic)		$Na_2(Mg,Fe)_3Al_2Si_8O_{22}(OH)_2$
riebeckite		$Na_2Fe_2^{2+}Fe_2^{3+}Si_8O_{22}(OH)_2$

Similar to pyroxenes, the Ca, Mg, and Fe amphiboles have a limited solid solution (miscibility gap) between the Ca-rich and Ca-poor types (Figure 5.16). Thus, two amphiboles may occur in the same rock, and they may have exsolved each other.

In general, it is sufficient for the undergraduate student to be able to distinguish between a clinoamphibole and an orthoamphibole; and between an amphibole and a pyroxene. Because of immense variability in amphibole compositions, it is often impossible to name them accurately without proper chemical (electron probe or wet chemical analysis) or structural (X-ray) characterization. Therefore, making the distinction between orthoamphibole and clinoamphibole under a petrographic microscope is generally sufficient. However, some amphiboles, such as glaucophane, have special tectonic significance, and therefore it is important to be able to identify them in rocks. With these thoughts in mind, a rather general set of criteria is presented below for the identification of amphiboles. Their potential petrogenetic significance will become apparent in later chapters.

Physical properties

Habit. Prismatic, columnar (mostly), and some are fibrous or needlelike (actinolite).

Color.   anthophyllite   "clove" brown
            hornblende   black

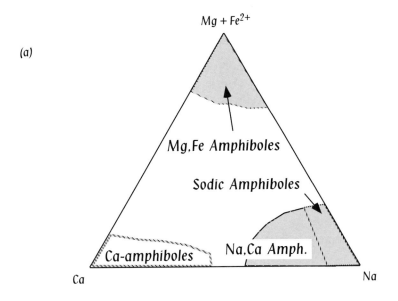

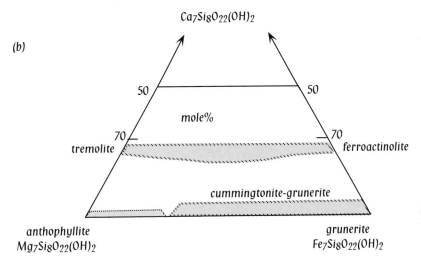

FIGURE 5.16   (a) A generalized classification of common amphiboles. (b) Amphibole quadrilateral.

tremolite	very light green
actinolite	deep green
glaucophane	blue

Luster. Vitreous.

Streak. Colorless.

Hardness. 5–6.

Density. 2.9–3.5.

Cleavage. Two well-developed prismatic sets (discussed earlier).

Optical properties   (These vary tremendously from one amphibole to the other. Only some common characteristics are outlined below.)

PLANE-POLARIZED LIGHT

Grain. Prismatic, needlelike, or fibrous (as stated before). Anthophyllite forms slender prisms or may be fibrous. Cummingtonite–grunerite amphiboles are generally fibrous, often forming radiating clusters. Tremolite–actinolite amphiboles are generally acicular or fibrous. Hornblende, glaucophane, and riebeckite form well defined prismatic crystals.

Relief. Moderately High

RI	hornblende	tremolite–actinolite	glaucophane
α	1.60–1.70	1.599–1.688	1.606–1.701
β	1.61–1.71	1.612–1.697	1.622–1.717
γ	1.62–1.73	1.622–1.705	1.627–1.717

Color and pleochroism. The color of amphiboles can vary from colorless to deep green, lavender, blue, or brown colors. Colored amphiboles are always strongly pleochroic. Hornblende's pleochroism is medium to deep green. A special orange-red type of hornblende, known as oxyhornblende, occurs as phenocrysts in some intermediate composition lavas. Glaucophane's pleochroism goes from lavender to sky blue. Riebeckite's pleochroism goes from light to deep blue. Magnesian tremolite is colorless, whereas the more Fe-rich actinolitic amphiboles are deep green.

BETWEEN CROSSED POLARS

Anisotropism. Anisotropic, 2° colors

Extinction, optic sign and 2V. Straight extinction for orthoamphiboles; inclined for all clinoamphiboles. Most amphiboles are biaxial negative with a large 2V.

	$Z \wedge c$	Optic sign	$2V_x$
anthophyllite	0°	biaxial + or −	65–148°
hornblende	12–34°	biaxial + or −	35–130°
tremolite–actinolite	10–21°	biaxial −	75– 88°
cumm.–grunerite	10–21°	biaxial + or −	82–145°
kaersutite	0–19°	biaxial −	66– 84°
glaucophane	10–21°	biaxial + or −	10– 45°

[Note: Crossite, a member of the glaucophane–riebeckite series, shows a much greater variation of $2V_x$ (0–180°)].

Additional notes: Amphiboles are quite common in metamorphic rocks and in intermediate to felsic volcanic rocks. Their composition can be used to decipher tectonic setting. For example, glaucophane can only be found in plate-convergent-boundary metamorphic rocks. The chemistry of amphiboles found in granitoid batholiths of specific characteristics can be used to infer their depths of emplacement. Although not as abundant in ultramafic rocks, amphibole does occur in mantle-derived ultramafic xenoliths. Their study has given significant insight into the role of water in mantle processes and volcanism.

Phyllosilicates (sheet silicates)

All phyllosilicates have one thing in common about their atomic structure: they have laterally extending sheets made of SiO_4 tetrahedra that share three of their corner oxygens with neighboring tetrahedra (Figure 5.17). When viewed down the *c*-axis,

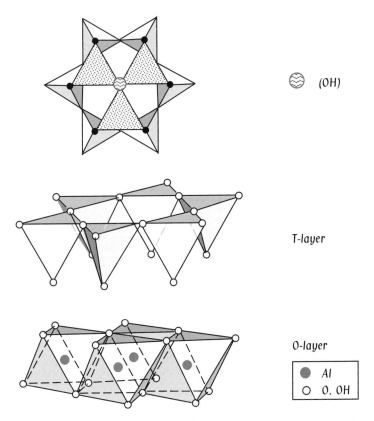

FIGURE 5.17 *Atomic structure of phyllosilicates is composed of tetrahedral (T-layer) and octahedral (O-layer) layers. T- and O-layers and the location of (OH)⁻(as viewed perpendicular to the T-layer) are shown.* (From Hurlbut and Klein, 1999, 21st ed., *Manual of Mineralogy*, John Wiley and Sons Inc.)

these tetrahedral sheets appear to form hexagonal rings (Figure 5.17). (OH) occurs at the center of each ring on the same plane as the unshared oxygens of the tetrahedral layer. The tetrahedral sheets are bonded by octahedral layers, each octahedra having apical O or (OH) and Al or other cations at its center (Figure 5.17). Micas and clay minerals are two important members of the phyllosilicates. In a generalized way, their structures are all a combination of octahedral layers (O-layers) and tetrahedral layers (T-layers).

Micas

Chemical composition and crystallography: Micas are an important group belonging to the phyllosilicate class. Their crystal structure is composed of parallel sheets of tetrahedra whose apices are linked by octahedral layers (Figure 5.18). Al, Fe, and Mg cations occur in this octahedral site. Some of these may remain vacant. The bases of adjacent sheets are separated by large sites where K^+ (Na^+ rare) occurs in 12-fold coordination.

A general formula for micas is $K(Al,Mg,Fe)_{2-3}(AlSi_3O_{10})(OH)_2$. Mica compositions and names can be described in terms of three end members: *muscovite* [or "white mica," $KAl_2(AlSi_3O_{10})(OH)_2$], *annite* [it is only an "ideal" end—member composition, $KFe_3(AlSi_3O_{10})(OH)_2$], and *phlogopite* [$KMg_3(AlSi_3O_{10})(OH)_2$]. The name *biotite* is given to the mica formed by solid solution between annite and phlogopite. There is very little solubility between muscovite and biotite.

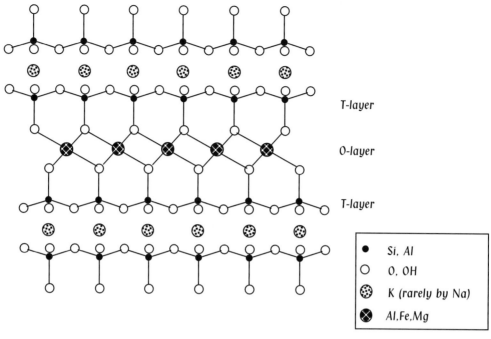

●	Si, Al
○	O, OH
⊛	K (rarely by Na)
⊗	Al,Fe,Mg

FIGURE 5.18   *Atomic structure of muscovite.*

Physical properties

Habit. Flaky, peels off easily.

Color. Muscovite—colorless; biotite—deep brown; phlogopite—light to reddish brown.

Luster. Vitreous, pearly on the flakes.

Streak. White.

Hardness. 2.5–3.

Density. 2.7–3.3 (high for Fe-rich biotites).

Fracture. Smooth.

Optical properties

PLANE-POLARIZED LIGHT

Grain. Usually as flakes.

Relief. Moderately low.

RI	muscovite	biotite	phlogopite
α	1.552–1.58	1.522–1.625	1.52
β	1.582–1.62	1.548–1.672	1.55
γ	1.587–1.623	1.549–1.696	1.57

Color and pleochroism.

	muscovite	biotite	phlogopite
	colorless nonpleochroic	deep to light brown, strongly pleochroic	light brown to clear, strongly pleochroic

Cleavage. In all cases, 1 basal set.

Anisotropism. Anisotropic, 2° colors

Extinction, optic sign and 2V. Straight extinction.

	Optic Sign	$2V_x$
muscovite	Biaxial−	30–47°
biotite	Biaxial−	0–25°
phlogopite	Biaxial−	0–15°

Additional notes: Micas occur in metamorphic as well as felsic and intermediate igneous rocks. In metamorphic rocks, their presence and composition have important implications for the source-rock composition, metamorphic grade, and thermobarometry. Phlogopite-rich micas are found in marbles and ultramafic xenoliths from the upper mantle. In the latter case, they provide evidence for the presence of structurally bound water at such depths. Muscovite is used in electrical insulators.

Chlorites

Crystallography and composition: The general composition of the chlorite group minerals may be written as

$$Y_6(Z_4O_{10})(OH)_8$$

where Z is the tetrahedral site upon which Si and Al are located. Y is an octahedral site where Fe, Mg, Al, and Mn are located. They are a typical sheet silicate with octahedral layers sandwiched between the apices of two tetrahedral sheets. Hydroxyl layers occur between the bases of the tetrahedral sheets. Chlorites are generally monoclinic, but some may be triclinic, and their chemistry can vary a great deal.

Physical properties

Habit. Typically sheets.
Color. Translucent. Color can vary, but generally a dull green color.
Luster. Subvitreous to greasy.
Streak. White.
Hardness. 2–3.
Density. 2.6–3.3.
Cleavage. One distinct basal set.
Fracture. Smooth.

Optical properties

Relief. Low to moderate;

RI α = 1.55–1.67; β = 1.55–1.69; γ = 1.55–1.69.

Color and pleochroism. Light to slightly deeper green. Pleochroic.

BETWEEN CROSSED POLARS

Anisotropism. Anisotropic. 1° gray; some have anomalous 1° blue-brown.

Extinction. Straight or inclined ($Z \wedge c = 0-30°$).

Optic sign. Biaxial − or +.

Distinguishing features, occurrence: Color, pleochroism, crystal form, and cleavage are all distinctive. Chlorite is generally a prominent mineral in low-grade metamorphosed mafic to intermediate and pelitic rocks. This mineral also occurs as an alteration product in many other rock types.

Clay Minerals

Clay minerals include a group of very fine-grained sheet silicates that are extremely diverse in crystal structure and chemical composition. In a general sense, clays are made of various configurations of tetrahedrally and octahedrally coordinated layers. They vary in size from amorphous materials to barely recognizable particles under a petrographic microscope. In general, proper identification of a clay mineral requires X-ray diffraction work, and optical microscopy does not really do much for their individual species identification. Clay mineralogy is a subdiscipline in itself and is of importance in environmental geology. It is well beyond the scope of this book to cover the clays in any detail, and the interested reader is advised to consult proper clay mineralogy reference and text books.

Structurally, two types of clays are distinguished: *single-sheet clays*, characterized by T–O layers (Figure 5.17), and *double sheet clays* with T–O–T layers. *Kaolinite* [$Al_2Si_2O_5(OH)_4$], popularly known as chalk, has a T–O type structure. Kaolinite is generally recognizable by its white color and dull luster. They are commonly an alteration product of feldspar generated under nonalkaline conditions (cf. Gribble and Hall 1985).

The *smectite* (of which *montmorillonite* is an example) and *illite* group of clays have a T–O–T structure. Smectites are simple aluminous clays [montmorillonite: $(Na,Ca)_{0.3}(Al,Mg)_2Si_4O_{10}(OH)_2 \cdot nH_2O$, where the n represents variable amounts of water]. Smectite is the dominant component formed by the alteration of bentonite clay, which is altered volcanic ash.

Illites are potassic clays, which turn into muscovite when recrystallized at higher temperatures. Illite is different from smectite because it is a *nonexpanding* clay, which is defined as a "potassium micalike mineral that occurs in clay-size ($<4\mu m$) fraction" (Srodon and Eberl 1988). Illite is a common mineral of mudrocks. It is produced as an alteration product of feldspars under alkaline conditions and high K and Al availability. Glauconite and sericite are two minerals related to illite. Glauconite is of importance in sedimentary petrology.

Talc [$Mg_3(Si_4O_{10})(OH)_2$] is a monoclinic, soft, commonly tubular clay that forms from the alteration of ultramafic rocks and thermal metamorphism of siliceous dolostones. Its color is usually some form of silky white to grayish white and may have a light greenish hue. Talc has a greasy feel and a pearly to greasy luster. The compact, massive variety is known as *soapstone*. Talc is often associated with serpentine. Under the microscope, it is colorless. Talc's optic sign is biaxial negative ($2V_x = 0-30°$). It shows strong birefringence, with $\alpha = 1.592, \beta = 1.622$, and $\gamma = 1.623$ (Nesse 1991).

The *serpentine* [$Mg_3Si_2O_5(OH)_4$] group of minerals are clay minerals that form through the alteration of olivine and orthopyroxene. The three common polymorphs of serpentine are chrysotile, antigorite, and lizardite: chrysotile is fibrous, while antigorite and lizardite (which is the matrix material) form very fine-grained massive bodies. They range from light green to deeper green and have a characteristic greasy luster. In thin section, antigorite appears to be nearly isotropic. The other two show 1° gray or white interference colors.

Quartz

Crystallography and composition: SiO_2. It is a tectosilicate with the tetrahedra sharing all their apical oxygens with the other tetrahedra. Hexagonal (rhombohedral).

Physical properties

Habit.   Prismatic, well-formed crystals, or massive.

Opacity and color.   Generally transparent. Quartz crystals may vary extensively in color, though the colorless variety is most common. Other coarse-grained varieties include rose quartz, "smoky" quartz, amethyst (purple), and so on. Microcrystalline quartz varieties can be many: jasper (Fe-rich, red color), agate (banded, colloidally precipitated), and onyx (gray and white bands).

Luster.   Vitreous to subvitreous.

Streak.   White.

Hardness.   7.

Density.   2.65.

Cleavage.   Absent. Horizontal striations occur.

Fracture.   Conchoidal.

Optical properties

PLANE-POLARIZED LIGHT

Relief.   Low.

RI $\omega = 1.544$; $\varepsilon = 1.553$

Color and pleochroism.   Colorless. Nonpleochroic.

BETWEEN CROSSED POLARS

Anisotropism.   Anisotropic. 1° interference colors—yellow, gray. Thick sections give 2° colors.

Extinction.   Straight with respect to prismatic faces in euhedral or subhedral grains. In metamorphic and sedimentary rocks, deformed and polycrystalline quartz grains exhibit wavy and sectoral extinction.

Optic sign.   Uniaxial +.

Distinguishing features, occurrence: Quartz can occur in a wide range of igneous, metamorphic, and sedimentary rocks. It can be confused with alkali feldspar and nepheline under the microscope. However, alkali feldspar commonly undergoes partial to complete alteration to *sericite*, which is clay. Also, alkali feldspar is biaxial. Nepheline is uniaxial negative and has a clouded appearance with clay; its relief is higher than that of quartz.

Feldspars

Composition and crystallography: The chemical composition of feldspars is relatively straightforward (Figure 5.19a): *plagioclase feldspars* form an extensive solid solution between anorthite and albite end members, and *alkali feldspars* form a solid solution between albite and a potash feldspar end member. Although the plagioclase feldspars form a continuous solid solution series, individual names are given to individual members of this solid solution series based on their composition ranges: albite,

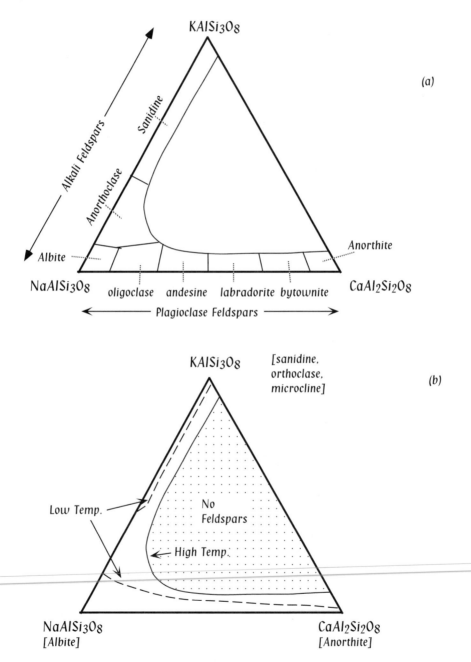

FIGURE 5.19 (a) Chemical composition of feldspars. (b) Limit of solid solubility in feldspars at high and low temperatures.

Ab_{90-100}; oligoclase, Ab_{90-70}; andesine, Ab_{70-50}; labradorite, Ab_{50-30}; bytownite, Ab_{30-10}; anorthite, Ab_{10-0} (Figure 5.19a). Alkali feldspar solid solutions are also classified into two types: sanidine and anorthoclase. As Figure 5.19b shows, both plagioclase and alkali feldspar solid solutions contain a limited amount of the third ternary component in solid solution. The extent of solubility is directly correlated with the mode of occurrence: in plutonic rocks the solubility is limited (as shown by the dashed fields), and in volcanic rocks the solubility is more extensive. Aside from these common rock-forming feldspar solid solutions, there is a rare feldspar, called Celsian ($BaAl_2Si_2O_8$), which will not be discussed here.

The $KAlSi_3O_8$ end member occurs as three distinct polymorphs: *sanidine* (stable at highest temperature), *orthoclase* (intermediate temperature polymorph), and *microcline* (lowest temperature variety). Microcline is triclinic; whereas sanidine and orthoclase are monoclinic. Plagioclase is triclinic, except that a high temperature form of albite, called monalbite, is *monoclinic*. Feldspars that form at high temperatures are *disordered* in the sense that tetrahedral Al is randomly distributed through the structure; whereas at a low temperature, *ordered* structure Al is located only in one of four tetrahedral sites. The disordered versus ordered varieties are generally given the prefixes "high" and "low," respectively, for example, high sanidine and low sanidine.

Physical properties

Habit. Well-formed crystals, or massive.

Opacity and color. Opaque. Plagioclase feldspar is generally white; orthoclase is generally "flesh" colored; and microcline is light green.

Luster. Subvitreous.

Streak. Colorless.

Hardness. 6–6.5.

Density. K-feldspars, 2.65; albite, 2.65; anorthite, 2.75.

Cleavage. Two very good sets roughly perpendicular to each other.

Fracture. Uneven.

Optical properties

PLANE-POLARIZED LIGHT

Relief. Low.

Color and pleochroism. Colorless, nonpleochroic.

BETWEEN CROSSED POLARS

Anisotropism. Anisotropic. 1° gray interference colors. Thick sections give 2° colors.

Twinning. Polysynthetic twinning is characteristic in plagioclase. Orthoclase generally shows simple twinning, whereas microcline is well known for its "cross-hatchered" twinning. (See plates at the end of this chapter.)

Extinction. Inclined extinction with respect to cleavages in alkali feldspars ($X \wedge a = 18°$ for microcline, 5–13° for orthoclase.)

Plagioclase shows symmetrical extinction with respect to twin lamellae (Figure 5.20). In such an extinction one side of a twin plane goes extinct during a clockwise rotation of the stage; and the other side of the twin plane goes extinct when the stage is rotated counterclockwise. The extinction angle can be used to obtain an estimate of chemical composition (Figure 5.21). In the Michel–Levy method, a symmetrical extinction angle is determined on the (010) twin plane. An appropriate grain for such measurement will have the following characteristics: it will show vertical albite twin lamellae (i.e., the twin plane should appear as thin sharply defined parallel lines, Figure 5.20); the entire grain should show the same interference color when the twin planes are oriented N–S, the difference in value of the measured extinction angles obtained from

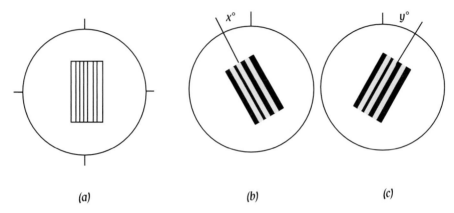

(a) (b) (c)

FIGURE 5.20 *Symmetrical extinction phenomena in plagioclase.* (a) Twin lamellae are N–S. (b) The grain is rotated so that one set of lamellae go extinct, and the extinction angle (x^0) is noted. Then the grain is turned the other way until the other lamellae go extinct, and the extinction angle is noted (y^0). The difference between (x^0) and (y^0) should be $\leq 4°$.

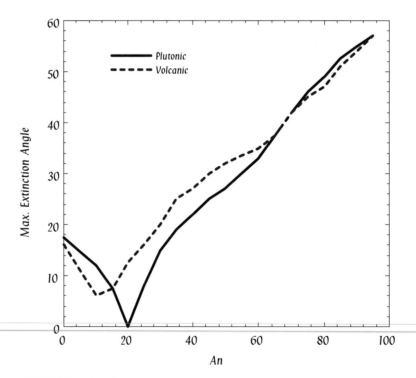

FIGURE 5.21 *Correlation between extinction angle and composition of plagioclases in plutonic and volcanic rocks.*

clockwise versus counterclockwise rotation of the stage will not exceed 4°. Many such measurements need to be made, then the maximum extinction angle is selected and compared with Figure 5.21 to estimate the composition. If the extinction angle is less than 16–18°, two compositional possibilities exist for the volcanic or plutonic plagioclase: for example, if the angle is 10°, the plagioclase could be An_{12} or An_{26} if it is in a plutonic rock. However, a final decision may be made by examining the refractive indices or optic sign of the plagioclase (e.g., + for < An_{20} and − for > An_{20}).

Optic sign. Biaxial + or − for plagioclase, but − for alkali feldspars.

Distinguishing features, occurrence: Feldspars are an abundant mineral in the crust. Polysynthetic (or lamellar) twinning is an important distinguishing property of plagioclase.

Feldspathoids

The common feature of all feldspathoids is that they are "like" feldspars in crystal structure and composition. There are four minerals that fall under this group; however, below the discussion is entirely on nepheline, the most common feldspathoid in alkaline igneous rocks.

Nepheline

Crystallography and composition: Hexagonal. $NaAlSiO_4$.

Physical properties

Habit. Generally short tabular or prismatic crystals. Sometimes massive.
Opacity and color. Generally translucent. Colorless or grayish or yellowish.
Luster. Greasy.
Streak. White.
Hardness. 5–6.
Density. 2.6.
Cleavage. One set common.
Fracture. Subconchoidal to uneven.

Optical properties

PLANE POLARIZED LIGHT
Relief. Low.

RI ω = 1.530–1.545; ε = 1.527–1.541

Color and pleochroism. Colorless, nonpleochroic.

BETWEEN CROSSED POLARS
Anisotropism. Anisotropic. 1° bluish gray.
Extinction. Straight.
Optic sign. Uniaxial −.

Distinguishing features, occurrence: Nepheline may be confused with feldspars and quartz under the microscope; however, optic sign is a key criterion in distinguishing them. Nepheline occurs in alkaline igneous rocks.

OXIDES

Oxide and sulfide minerals are often called "opaques" or "ore minerals," and their study under the microscope requires a different type of light source—a reflected light source rather than a transmitted light source. Their optical properties in reflected light are beyond the scope of this book, and only a list of physical properties of these minerals is given in Table 5.1.

TABLE 5.1 *Principal physical properties of some oxide minerals.*

Mineral	System	Habit	Color	Luster	Streak	H	Density	Special Properties
Magnetite ($FeO.Fe_2O_3$)	Isometric	Octahedra	Black	Metallic	Black	5.5–6.5	5.2	Magnetic
Spinel ($MgAl_2O_4$)	Isometric	Octahedra	Variable: (red, green, colorless, black)	Vitreous	Colorless	7.5–8	3.6	
Chromite ($FeO.Cr_2O_3$)	Isometric	Sugary	Black	Metallic	Brown	5.5	4.6	May be weakly magnetic
Hematite (Fe_2O_3)	Hexagonal	Many diff. habits	Reddish brown to black	Metallic	Brick red	5–6.5	5.3	
Ilmenite ($FeO.TiO_2$)	Hexagonal	Tabular	Black	Metallic	Black	5–6	5.0	

Among the oxides, the spinel group minerals are most common among rocks of igneous and metamorphic origin. The compositions of spinel group minerals are best visualized with the "spinel prism" (Figure 5.22). This group exhibits several different solid solutions and miscibility gaps (Haggerty 1976). The two solid solutions that are of considerable interest to the petrologist are the magnetite–ulvöspinel series and the chrome–aluminum spinels. Most geologists are interested only in the spinels (including spinel, magnetite, and chromite). Aside from ulvöspinel–magnetite solid solution, hematite–ilmenite solid solution series, and rutile are also of interest.

In mafic rocks (and metamorphosed rocks) two oxide phases generally coexist: one of these is a ulvöspinel–magnetite solid solution (often referred to as titanomagnetite because of its high Ti content); and the other is an ilmenite–hematite solid solution (Figure 5.23a). Figure 5.23a shows some schematic tie lines indicating the nature of the coexisting ilmenite–hematite and magnetite–ulvöspinel solid solutions. The significance of these coexisting oxide phases in natural rocks is that their compositions record temperature and redox state of formation of the rocks.

Chrome–alumina spinels, whose compositions fall on or very close to base of the spinel prism [that is, chromites ($FeCr_2O_4$), picrochromite ($MgCr_2O_4$), spinel ($MgAl_2O_4$), and hercynite ($FeAl_2O_4$)], occur in a variety of rocks—from mantle

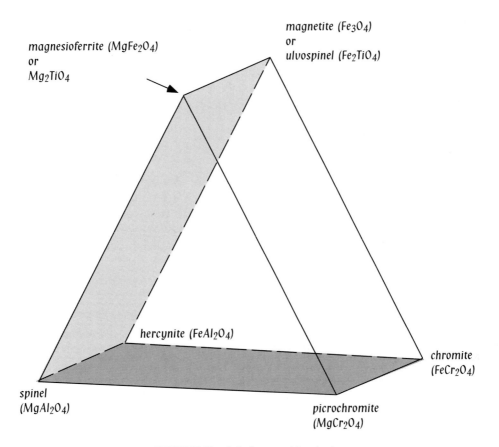

FIGURE 5.22   *Spinel compositional prism.*

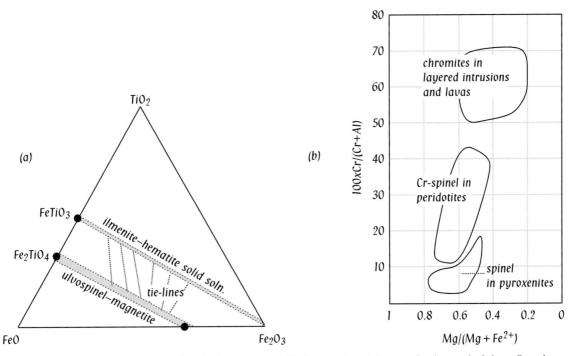

FIGURE 5.23   (a) Composition of Fe–Ti oxides in mafic rocks. (b) Composition of chrome–aluminum spinels in mafic and ultramafic rocks.

peridotites to mantle-crystallized pyroxenites, gabbros and basaltic lavas. The composition of the Cr–Al spinel is also very distinctive for each of these three major rock types (Figure 5.23b). As we will see in later chapters, mantle peridotite is a source or residue of partial melting that produces basalt magma. The $Cr/(Cr + Al)$ ratio of spinels that occur in mantle peridotites is a sensitive indicator of the degree of melting in that it increases (as does $Mg/(Mg + Fe^{2+})$).

CARBONATES

Carbonate minerals react easily with HCl and are therefore easily identified in the field. Although many different carbonate minerals are possible, from a geologist's point of view, only a few are of great interest. These have compositions within the $CaCO_3$–$MgCO_3$–$FeCO_3$ triangle (Figure 5.24): calcite, aragonite, dolomite, ankerite, magnesite, and siderite. Figure 5.24 shows the rough extent of solid solubility that occurs between the Ca–Mg–Fe carbonate minerals.

Calcite and Aragonite

Crystallography and composition: As mentioned earlier, calcite and aragonite are two polymorphs with the same chemical composition of $CaCO_3$. In their atomic structures (as with all carbonates), the strongest, partially covalent bonds occur between the carbon atom and three, coplanar, closest-neighbor oxygens (Figure 5.25). Each $(CO_3)^{2-}$ anion complex is bonded ionically to two Ca^{2+} cations in the case of calcite

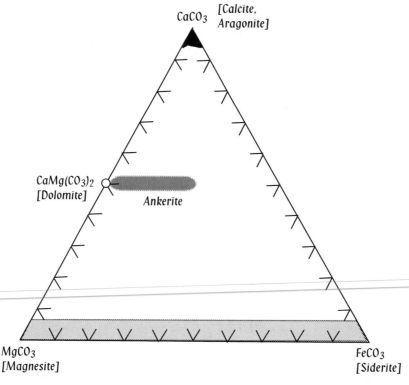

FIGURE 5.24   *Ca–Mg–Fe carbonate minerals and their mutual solid solubilities.*

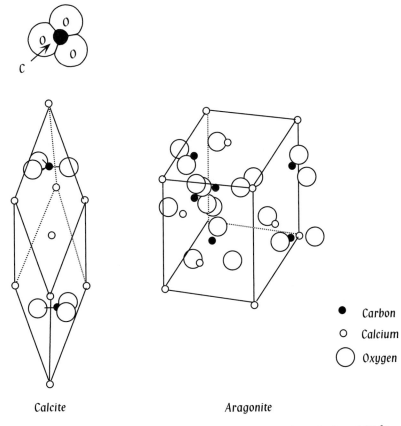

CO₃ anion group

Calcite

Aragonite

● Carbon

○ Calcium

◯ Oxygen

FIGURE 5.25   *Atomic structures of calcite and dolomite and a schematic view of* CO_3^{2-} *unit.*

and three Ca^{2+} in the case of aragonite. The resulting unit cell for calcite is rhombohedral and aragonite is orthorhombic.

Physical properties

Habit. Calcite can occur in many different habits but most commonly as rhombohedra (each face having the shape of a rhomb) or scalenohedra (each face having the shape of scalene triangle). Aragonite is acicular, tabular, or fibrous.

	calcite	*aragonite*
Opacity and color.	Transparent. White, colorless or many different colors	Transparent. Colorless or white, brownish
Luster.	Vitreous	Vitreous
Streak.	White	White
Hardness.	3	3.5–4
Density.	2.71	2.94
Cleavage.	three sets, rhombohedral	two prismatic sets

Optical properties

PLANE-POLARIZED LIGHT

Relief. Shows extreme variation with orientation of the grain because of its extreme birefringence. This is called the "Twinkling effect." Similar to calcite

Color and pleochroism. Both are colorless. Nonpleochroic.

BETWEEN CROSSED POLARS

Anisotropism. Anisotropic. Both show very high order creamy white interference colors.

Extinction. Mottled extinction is shown by both

Twinning. Three sets of twin lamellae common in calcite

Optic sign. Uniaxial− Biaxial−

Additional notes: Experimental studies show that aragonite is a higher pressure polymorph of calcite and is therefore not a stable phase at near-surface conditions (Hacker et al. 1992; Figure 5.26). However, aragonite occurs as a primary precipitate or as a biogenic component in marine environments: for example, in shallow marine reefal environments, aragonite is secreted by algae that occur sumbiotically with corals.

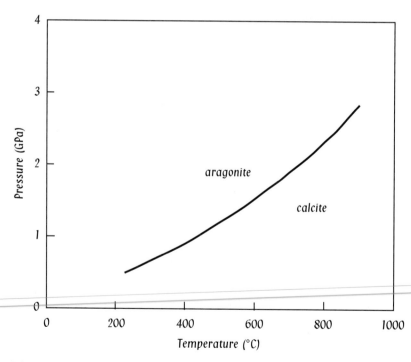

FIGURE 5.26 *Pressure-temperature stability of calcite and aragonite.* (Hacker et al., 1992)

(The skeletal material of coral is calcite.) Once deposited, aragonite needles eventually turn into calcite through diagnesis (a process that takes a few million years). Aragonite is also found in convergent-plate-margin metamorphic rocks, such as blueschist-facies metamorphic rocks. Figure 5.26 suggests that if chemical equilibrim is maintained, all aragonite made in the host rock during high-P metamorphism should turn back into calcite during exhumation of the rock. However, the common persistence of aragonite in these rocks suggests that reaction kinetics must be very important in the aragonite↔calcite transformation process. (See the review by Carlson 1983.) The exhumation (uplift and erosion) rate of aragonite-bearing metamorphic rocks must be sufficiently rapid in order to prevent the aragonite↔calcite transformation from going to completion (Hacker et al. 1992).

Dolomite–Ankerite

Crystallography: Hexagonal (rhombohedral).

Physical properties

Habit. Generally, rhombohedra often with curved surfaces and "saddle"-like twinning.

Opacity and color. Generally translucent; colorless or white or pink.

Luster. Greasy.

Streak. White.

Hardness. 3.5–4.

Density. 2.86–3.1

Cleavage. Rhombohedral cleavage.

Fracture. Subconchoidal.

Optical properties

PLANE-POLARIZED LIGHT

Relief. Low or high, depending on grain orientation.

Color and pleochroism. Colorless, nonpleochroic.

BETWEEN CROSSED POLARS

Anisotropism. Anisotropic. High-order creamy white.

Extinction. Straight and mottled.

Optic sign. Uniaxial−.

dolomite	$\omega = 1.679–1.690$; $\varepsilon = 1.500–1.510$
ankerite	$\omega = 1.690–1.750$; $\varepsilon = 1.510–1.548$

Additional notes: Dolomite is a common mineral in the sedimentary rock record; however, it is extremely rare in modern sedimentary environments. Dolomite and calcite form a complete solid solution at high temperature but decompose into two separate solid solutions due to the presence of a miscibility gap (Figure 5.27). The coexistence of calcite$_{ss}$ and dolomite$_{ss}$ is sensitive to temperature, and therefore may be used as a geothermometer for rocks containing both phases. (Geothermometry is discussed in a later chapter.)

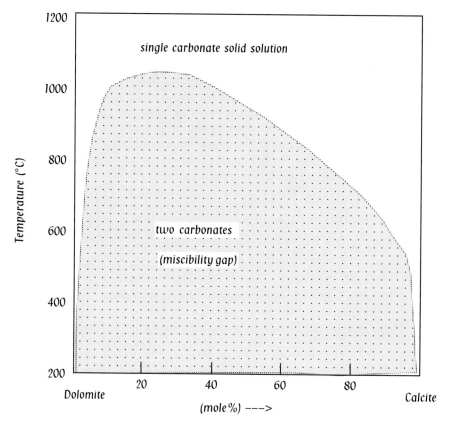

FIGURE 5.27   *Solid solution relationship between calcite and dolomite.* (Goldsmith, 1983)

SULFIDES AND SULFATES

Sulfides are metallic minerals characterized by metallic luster. Beyond hand specimens, they are normally described with a reflected light microscope. Table 5.2 gives a quick "run down" of the optical properties of some common sulfides.

Gypsum

Crystallography and composition: Monoclinic. $CaSO_4.2H_2O$

Physical properties

Habit.   Generally tabular or diamond shaped. Can be fibrous.
Opacity and color.   Transparent, colorless.
Luster.   Vitreous, also silky or pearly on cleaved surfaces.
Streak.   White.
Hardness.   2.0.
Density.   2.32.
Cleavage.   Good {010} cleavage.

TABLE 5.2 *Principal physical properties of some sulfide minerals.*

Mineral	System	Habit	Color	Luster	Streak	Cleavage	Fracture	H	Density	Special Properties
Sphalerite (ZnS)	Isometric	Combination of tetrahedra, cube, and dodecahedra	Brown to black	Resinous	Brown	Perfect {110}	Conchoidal	3.5–4	4.0	Polysynthetic twinning
Chalcopyrite ($CuFeS_2$)	Tetragonal	Generally	Brass yellow	Metallic	Grayish black	Poor	Uneven	3.5–4	4.2	Polysynthetic and penetration twins
Galena (PbS)	Isometric	Cubes	"Lead" gray	Metallic	Black	Perfect {100}	Subconchoidal	2.5	7.6	Penetration twins
Pyrite[*] (FeS_2)	Isometric	Pyritohedra, cubes, octahedra	"Gold" yellow	Metallic	Black	Poor	Subconchoidal	6–6.5	5.1	Striated faces, penetration twins

Note: Pyrite is often called "fool's gold" because it may appear to be gold.

[*]

Optical properties

PLANE-POLARIZED LIGHT

Relief. Low.

Color and pleochroism. Colorless.

BETWEEN CROSSED POLARS

Anisotropism. Anisotropic. $1°$ gray interference color.

Extinction. Inclined ($Z \wedge c = 52°$).

Optic sign. Biaxial +; $\alpha = 1.519–1.521$; $\beta = 1.522–1.526$, $\gamma = 1.529–1.531$

$2V_z = 58°$

Additional notes: Gypsum is a common evaporite mineral used to make plaster of Paris, which is further utilized in the manufacturing of wallboard and casts.

Anhydrite

Crystallography and composition: Orthorhombic. $CaSO_4$

Physical properties

Habit. Crystals are rare. Commonly massive. Radiated fibrous varieties.

Opacity and color. Transparent to translucent. Colorless to bluish, gray to dark gray. Sometimes discolored due to foreign inclusions.

Luster. Vitreous, also pearly on cleaved surfaces.

Streak. White to grayish white.

Hardness. 3.5.

Density. 2.98.

Cleavage. Three sets: {010} perfect, {100} very good, {001} good.

Optical properties

PLANE-POLARIZED LIGHT

Relief. Low.

Color and pleochroism. Colorless.

BETWEEN CROSSED POLARS

Anisotropism. Anisotropic. $1°$ gray interference color.

Extinction. Parallel with respect to cleavage.

Optic sign. Biaxial +; $\alpha = 1.570$; $\beta = 1.575$, $\gamma = 1.614$

$2V_z = 44°$

Additional notes: Anhydrite occurs in lower parts of evaporite sequences and often acts as a caprock for petroleum reservoirs. It is used to make fertilizer (ammonium sulfate) and sulfuric acid.

5.1 MEDICAL MINERALOGY

Minerals form the basis of technology and society. Minerals, in one form or another, are used by everyone in every facet of life—silicon ships, insulators, conductors, ceramics, glass, and many other things. There is also a negative side to minerals, which may be thought of as the "nuisance factor" as far as their effect on human health is concerned. Study of minerals in a health context has been emerging as a very critical area of interdisciplinary investigation, a subject that is called "Medical Mineralogy." Most people are more aware of the diseases caused by exposure to arsenic, lead, and asbestos minerals. Arsenic and lead were common ingredients of household paints until legislative efforts stopped their use. Asbestos was used as an insulation material in old buildings (and many old buildings still contain these materials); however, that has changed because of thorough studies of the effects of asbestos.

Many interdisciplinary studies of the health effects of minerals have been initiated by geologists, biologists, and medical professionals, as attested by several recent publications (e.g., Guthrie and Mossman 1993; Pasteris et al. 1999).

As an example, let us consider the effects of asbestos minerals. The term "asbestos" is generally applied to certain extremely fibrous mineral varieities belonging to the amphibole and serpentine groups, particularly *chrysotile* (in the medical literature, it is called crocidolite). The fibrous varities of the amphiboles tremolite–actinolite and anthophyllite have also been called asbestos. The effect of asbestos inhalation is *asbestosis*, which is "a fibrotic lung disease that is … characterized by the inability of the lung to oxygenate blood or to eliminate carbon dioxide and a decrease in the ability to expand or to respond to the action of the diaphram" (Guthrie and Mossman 1993, p. 557).

BIBLIOGRAPHY

Battey, M.H. and Pring, A. (1997) *Mineralogy for Students* (3rd ed.). Longman (London, U.K.).

Blackburn, W.H. and Dennen, W.H. (1994) *Principles of Mineralogy* (2nd ed.). W.C. Brown (Dubuque, IA.).

Bloss, F.D. (1961) *An Introduction to the Methods of Optical Crystallography*. Holt, Rinehart and Winston (New York).

Carlson, W.D. (1988) "Subsolidus phase equilibria on the forsterite saturated join $Mg_2Si_2O_6$–$CaMgSi_2O_6$ at atmospheric pressure." *Am. Mineral.* **73**, 232–241.

Chopin, C. and Sobolev, N.V. (1995) Principal mineralogic indicators of UHP in crustal rocks. In R.G. Coleman and X. Wang (eds.), *Ultrahigh Pressure Metamorphism*, pp. 96–131. Cambridge University Press (Cambridge, U.K.).

Coleman, R.G., Lee, D.E., Beatty, L.B., and Brannock, W.W. (1965) "Eclogites and eclogites; their differences and similarities." *Geol. Soc. Am. Bull.* **76**, 483–508.

Daniels, L.R.M., Gurney, J.J., and Harte, B. (1996) "A crustal mineral in a mantle diamond." *Nature* **379**, 153–156.

Deer, W.A., Howie, R.A., Zussman, J. (1992) *The Rock-forming Minerals*. Longman (London, U.K.).

Ernst, W.G. (1969) *Earth Materials*. Prentice-Hall (Englewood Cliffs, New Jersey).

Goldsmith, J.R. (1983) Phase relations of rhombohedral carbonates. In R.J. Reeder (ed.), Carbonates: mineralogy and chemistry. *Reviews in Mineralogy* **11**, 49–76, Mineralogical Society of America (Washington, D.C.).

Gribble, C.D. and Hall, A.J. (1985) *A Practical Introduction to Optical Mineralogy*. G. Allen and Unwin (London, U.K.).

Guthrie, G.D. and Mossman, B.T. (1993) Health effects of mineral dusts. Reviews in *Mineralogy* **28**, Mineralogical Society of America (Washington, D.C.).

Hacker, B.R., Kirby, S.H., and Bohlen, S.R. (1992) "Metamorphic petrology. Calcite→ Aragonite experiments." *Science* **258**, 110–112.

Johnson, K.T.M., Dick, H.J.B., and Shimizu, N. (1990) "Melting in the oceanic mantle: an ion microprobe study of diopsides in abyssal peridotites." *J. Geophys. Res.* **95**, 2661–2678.

Klein, C. and Hurlbut, C.S. (1992) *Manual of Mineralogy* (21st ed.), John Wiley (New York).

Kuno, H. and Nagashima, K. (1952) "Chemical compositions of hypersthene and pigeonite in equilibrium in magma." *Am. Mineral.* **37**, 1000–1006.

Leake, B.E. (1978) "Nomenclature of amphiboles." *Mineral. Mag.* **63**, 1023–1052.

McDonough, W.F. and Rudnick, R.L. (1998) "Mineralogy and composition of the upper mantle." In R.J. Hemsley (ed.), *Ultrahigh-pressure Mineralogy*, pp. 139–164, Mineralogical Society of America (Washington, D.C.).

Morimoto, M. (1988) "Nomenclature of pyroxenes." *Mineral. Mag.* **52**, 535–550.

Nesse, W.D. (1991) *Introduction to Optical Mineralogy* (2nd ed.). Oxford University Press (Oxford, U.K.).

Pasteris, J. D. and four others (1999) "Medical mineralogy as a new challenge to the geologist: silicates in human mammary tissue?" *Am. Mineral.* **84**, 997–1008.

Perkins, D. (1998) *Mineralogy.* Prentice Hall (Upper Saddle River, N.J.).

Phillips, W.R. and Griffen, D.T. (1981) *Optical Mineralogy, the Nonopaque Minerals.* W.H. Freeman (San Francisco).

Poldervaart, A. and Hess, H.H. (1951) "Pyroxenes in the crystallization of basaltic magmas." *J. Geol.* **59**, 472–489.

Sen, G., Frey, F.A., Shimizu, N., and Leeman, W.P. (1993) "Evolution of the lithosphere beneath Oahu, Hawaii: rare earth element abundances in mantle xenoliths." *Earth Planet. Sci. Lett.* **119**, 53–69.

Sen, G. (1986) "Mineralogy and petrogenesis of Deccan Traps lava flows around Mahabaleshwar, India." *J. Petrol.* **27**, 627–663.

Shelley, D. (1985) *Optical Mineralogy* (2nd ed.). Elsevier Science Publishers (New York).

Yang, H.–J., Sen, G., and Shimizu, N. (1998) "Mid-Ocean Ridge Melting: Constraints from Lithospheric xenoliths at Oahu, Hawaii." *J. Petrol.* **39**, 277–295.

C H A P T E R   6

Quantitative Analysis of Minerals

The following topics are covered in this chapter:

Introduction to the quantitative analysis of minerals
Electron microprobe and analytical scanning electron microscope
    Guiding principles
    Generation of X-rays
    Critical excitation energy
    Beam-sample interactions
    Quantitative analysis
    Matrix correction
    EDS and WDS analysis
Characterization of minerals: X-ray powder diffractometer

ABOUT THIS CHAPTER

The optical identification of minerals is a routine and important procedure in the study of minerals and rocks. However, a critical evaluation of the origins of minerals and rocks requires precise chemical analysis and high-magnification imaging of them. In most modern laboratories, X-ray techniques are routinely used in order to simultaneously obtain high resolution images of individual mineral grains and their chemical compositions. Three most commonly used instruments are the electron microprobe (EMP), the scanning electron microscope (SEM), and the X-ray powder diffractometer (XRD). SEM is primarily suited for the imaging of samples at micron scales and EMP is mainly used for chemical analysis and some imaging. XRD is routinely used to characterize minerals, especially when they are very fine grained.

All modern instruments come equipped with excellent, user-friendly, computer-driven automation systems, which render their operation so simple that only a functional knowledge of these instruments is necessary. In the case of EMP and analytical SEMs, however, the quality of the mineral or material analysis may vary considerably, even with such high level of sophistication. Therefore, the user must have some ways to distinguish between "good" and "bad" analysis. Excellent textbooks and journals on the topic of imaging and analysis are available. This chapter provides only a broad overview of these instruments. Finally, a short section is added on how to evaluate the analyses of common rock-forming minerals.

QUANTITATIVE ANALYSIS OF MINERALS: EMP AND ANALYTICAL SEM

The techniques involved in the chemical analysis of minerals may be grouped into two types: *destructive* and *nondestructive*. In destructive analysis the mineral sample is dissolved in solution or powdered and analyzed for its elemental abundances. Because the sample must be "destroyed" in such a process, such a technique does not allow detailed analytical study of such important features as zoning in individual mineral grain or reaction zones between grains. Zoning and reaction boundaries between adjacent mineral (or glass and mineral) grains can be examined with a very small electron beam employed in the nondestructive technique in an EMP and a SEM. In most laboratories, the SEM is used for imaging biological and fossil samples. However, in some laboratories, the SEM is fitted with analytical equipment so that chemical analysis of minerals/materials may be obtained. This type of SEM is called an *analytical* SEM.

An EMP or analytical SEM are only capable of analyzing for major (abundance greater than 3 weight%) and minor elements (abundance 0.1–3 wt%). Although some authors have used these instruments to obtain trace element abundances (at parts per million levels), that is not a routine practice. Trace element compositions on single crystals are more commonly obtained with an ion microprobe; and a discussion of this instrument is beyond the scope of this book. An additional handicap of most EMPs and analytical SEMs is that they cannot distinguish between polymorphs with identical chemical compositions. However, this is commonly not a problem because optical mineralogical techniques can usually tell us which polymorph it is. But when the mineral is extremely fine grained, as is the case with clay minerals, X-ray diffraction methods are used to identify a crystalline phase. Most laboratories house an X-ray powder diffractometer (XRD).

The advantages of EMP or SEM analysis are as follows: they are non-destructive; they are extremely rapid; sample preparation is relatively simple, and any well-polished thin section of a rock or a mineral may be analyzed; micronsized domains within individual mineral grains may be analyzed for major and minor elements with excellent accuracy; with the EMP, a microscope allows viewing of the sample as it is being analyzed; and changing samples is rapid and simple. There are certain things that these instruments are unable to do. They cannot distinguish between valence states of a particular element: for example, the amount of Fe in a mineral analyzed by an EMP is assumed to be in a ferrous state. This often results in a low total in a mineral analysis. This is particularly true of Fe-Ti oxide minerals, which often contain a significant amount of Fe_2O_3 component. The abundance of H_2O, an important constituent of hydrous minerals and glasses, cannot be determined with an EMP.

Since the basic premise of all X-ray techniques is X-ray generation, we first present a brief outline of how X-rays are generated.

X-RAYS AND THEIR GENERATION

X-rays are a type of electromagnetic radiation whose wavelengths fall in the range of 10^{-8} to 10^{-12} meters, which is shorter than the wavelengths of a different type of electromagnetic radiation—visible light (10^{-6} to 10^{-7} m). In the *ground state* of an atom, electrons occupy the lowest energy configuration possible. In SEM or EMP analysis, such an atom within the target mineral is "excited" by knocking off an electron from an inner shell, such as the K shell, with an energetic electron from the beam of electrons generated from the filament of the instrument. That is, the atom becomes ionized. An electron from one of the outer shells drops in to take the place of the missing electron in the K shell in that excited atom. Because the K-shell electron has lower energy than the outer shell electron, this electron transition releases the excess energy in the form of *X-ray photons* (or quanta of energy). The X-rays so produced have wavelengths that are characteristic of the elements present in the mineral. These characteristic X-ray lines are called $K\alpha$, $K\beta$, $L\alpha$, $L\beta$ etc. depending upon the orbitals

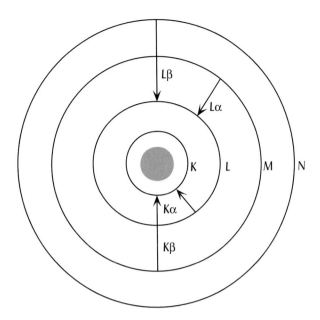

FIGURE 6.1   *Electron transitions (indicated by arrows) and the generation of Kα, Lα, and Mα X-rays.* (From Goldstein et al., 1992, 2nd ed., *Scanning Electron Microscopy and X-Ray Microanalysis*, Plenum Publishers)

involved in the electron transition: for example, when an electron jumps from L to K shell, the excess energy release results in a characteristic line called Kα (Figure 6.1).

Inner shell electrons and X-rays typically have energies of several kiloelectron volts (keV). This may be compared to the few electron volts (eV) of energy of light quanta. In generating a characteristic X-ray photon, the incident electron in an EMP or SEM must have sufficient kinetic energy $\left(E_o\right)$ that will exceed a certain amount of energy, called the *critical excitation energy* $\left(E_c\right)$, of, for example, the K-shell such that it can knock off a K-shell electron. The E_c's of the X-ray photons depend on the nature of the electron transition.

X-rays behave as both waves and particles (photons). The energy (E) and wavelength (λ) of an X-ray is related by the expression:

$$E\lambda = 12398 \qquad\qquad \text{[Eq. 6.1]}$$

where E is in electron volts and λ is in Angstrom units (Å; $1\text{Å} = 10^{-10}$ meters).

The basis for a qualitative analysis of elements is the relationship that the E_c of an X-ray "line" (e.g., $K\alpha_1$) varies approximately as the square of the atomic number (Z) of the emitting element (*Moseley's law*). Figure 6.2 shows energy and wavelengths of energetic K, L, and M lines as a function of Z. For most minerals we need only to be concerned with energies up to 10 keV. Also, silicate minerals and glasses are commonly analyzed for the following elements: Na, Mg, Al, Si, K, Ca, Ti, Cr, Mn, Fe, and Ni (in olivine). Na (Z = 11) and Ni (Z = 28) cover the atomic numbers of most elements that are commonly analyzed. Figure 6.2 shows that Kα lines with maximum energy of less than 10 keV will be exhibited by these elements.

Figure 6.3 shows the energy dispersive spectra of three silicate minerals (olivine, orthoclase, and clinopyroxene) and an oxide mineral. The energies of the lines representing various elements are shown at the bottom of each spectrum. Olivine spectrum shows two strong Kα peaks for Si and Mg and a small Kα peak for Fe. The relative intensities of these peaks are directly related to the abundance of these elements in the mineral. Actual analysis of this olivine is as follows: 41.93 wt% SiO_2,

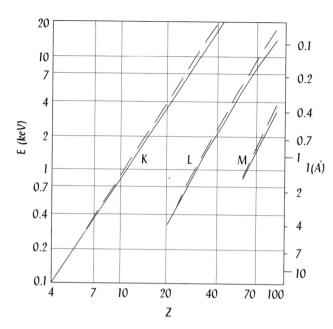

FIGURE 6.2 *Energy, wavelength, and atomic number (Z) relationships.* Energy (E) and wavelength (λ) of Kα, Lα, and Mα lines are shown as solid lines and their excitation energies are shown as dashed lines. (From Reed, 1996, *Electron Microprobe Analysis and Scanning Electron Microscopy in Geology*, Cambridge University Press)

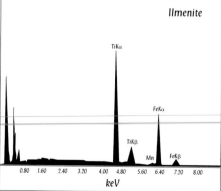

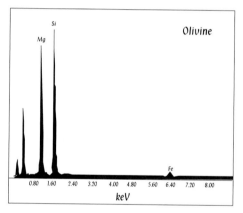

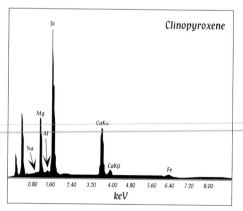

FIGURE 6.3 *Energy dispersive spectra of olivine, clinopyroxene, orthoclase, and ilmenite.* The energies are marked at the bottom of each spectrum.

49.91% MgO, and 8.16% FeO. In contrast, the clinopyroxene spectrum has additional Na, Al, and Ca peaks. Both Kα and Kβ peaks of Ca are visible. Relative to the olivine spectrum, both Fe and Mg peaks are smaller for the clinopyroxene. These peak intensities translate to the following composition of the clinopyroxene: 56.88 wt% SiO$_2$, 1.37% Al$_2$O$_3$, 2.26% FeO, 18.96% MgO, 20.09% CaO, and 0.44% Na$_2$O. Similarly, note the sharper Al Kα peak in the orthoclase and the prominent Ti Kα and Kβ peaks in the ilmenite sample.

In order to obtain characteristic K lines of various elements, it is clear that E$_o$ must exceed E$_c$. It turns out that in order to have the highest probability of K-shell ionization, the E$_o$/E$_c$ ratio must be about 2.5. It is common practice to use an accelerating voltage of 15 keV in EMP or SEM quantitative analysis of silicate minerals, which produces enough ionizations for useful quantitative analysis of materials.

THE ELECTRON MICROPROBE AND ANALYTICAL SCANNING ELECTRON MICROSCOPE

Basic Instrumentation

The theory and design of the EMP instrument was developed in 1949 by a French graduate student named R. Castaing. Based on this design, the first commercial instrument was built in 1956 by the French instrument maker, CAMECA. CAMECA and many other companies continue to manufacture the SEM and EMP. In both EMP and SEM, high velocity electrons are generated under high vacuum condition from a filament (usually made of tungsten). These electrons are focused through a series of electromagnetic lenses into a very narrow beam. As this beam impacts a target sample, characteristic X-rays and various types of electrons are generated from the sample. These are used to obtain images of the mineral (or any other) sample and to obtain quantitative chemical analysis of it by comparing the X-ray signals against a standard of known composition. Sample preparation is a major endeavor in EMP analysis. The sample must be well polished and be coated with carbon of a certain thickness.

The basic designs of EMP and SEM are similar: they are both high-vacuum instruments and have an electron optical column, a sample chamber, various detectors, and associated instrumentation (e.g., amplifiers) and a computer. The electron optical column has an electron gun, two or more electromagnetic lenses, mechanical apertures, spectrometers and detectors (for wavelength and energy dispersive spectrometry, secondary and back-scattered electron detectors), and a sample chamber (Figure 6.4a,b,c). Figure 6.4a shows a cross section of an electron optical column with the electron gun at the top and the sample at the bottom. Two electromagnetic lenses, a condenser lens above and an objective lens below, are also shown. Although one aperature and only two lenses are shown, most modern instruments have more lenses and more complicated designs. Also, EMPs commonly have an optical microscope for sample viewing. The number of electromagnetic lenses and spectrometers varies from instrument to instrument. This figure also shows a schematic wavelength dispersive spectrometer (WDS) in which the characteristic X-rays from a sample are diffracted by a crystal on to the detector. Figure 6.4b shows a schematic representation of an energy dispersive system in which X-rays from the sample are collected by an energy dispersive spectrometer (EDS). An amplifier and a computer process these X-rays and then generates an energy dispersive spectrum of the sample. Figure 6.4c shows a similarly detailed WDS system. An "average" EMP has four wavelength dispersive spectrometers (WDS) and a single energy dispersive spectrometer (EDS).

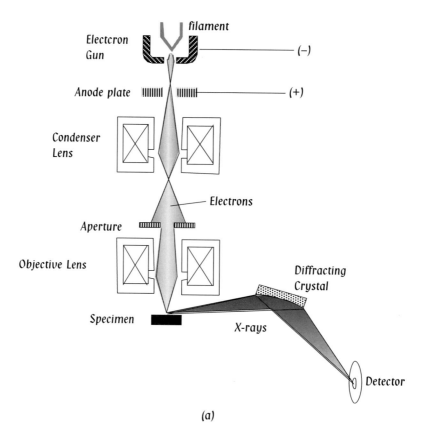

(a)

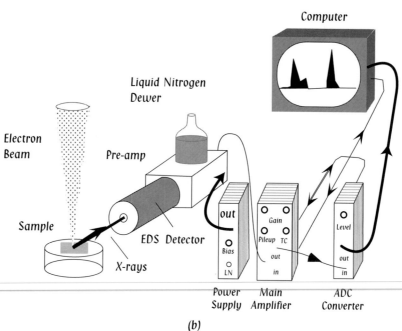

(b)

FIGURE 6.4   *Schematic diagram showing*: (a) the basic elements of an EMP or analytical SEM (EDS detector and scanning coils are not shown), (b) energy-dispersive spectrometer and associated electronics, and (c) wavelength-dispersive spectrometer and associated electronics. (Modified from Goldstein et al., 1992, 2nd ed., *Scanning Electron Microscopy and X-Ray Microanalysis*, Plenum Publishers)

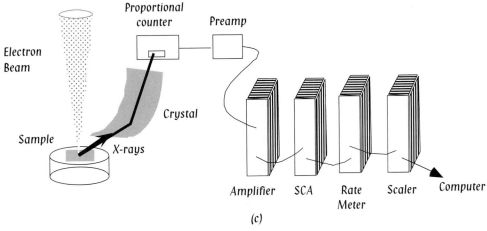

(c)

FIGURE 6.4   *(continued)*

The *electron gun* (Figure 6.5) houses a *filament*, which is heated to very high temperatures such that electrons are accelerated out of it. The current passed through the gun is referred to as *accelerating potential*. The filament is housed inside a negatively charged *Wehnelt cylinder*; and a positively charged *anode plate* is located at the base of the electron gun. The basic function of the Wehnelt cylinder and the anode plate is to create a field of equipotentials that force the electrons to form a beam. The *emission current* (the current used to control the emission of electrons from the filament) is the current sent through the anode plate and is generally kept at 100 microamperes for mineral analysis. Several *apertures* are used to control the beam diameter, and the magnetic electron *lenses* are used to focus the beam to a fine spot (1 μm or less in diameter) on the sample. Usually three or more such lenses are used: the lens closest to

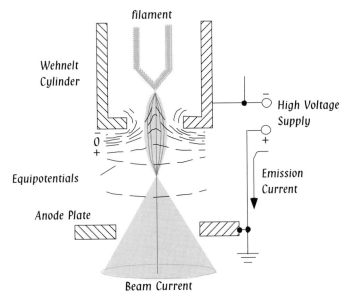

FIGURE 6.5   *Electron gun and beam formation.* (After Goldstein et al., 1992, 2nd ed., *Scanning Electron Microscopy and X-Ray Microanalysis*, Plenum Publishers with permission)

the sample is called an *objective lens*, and the upper lenses are called *condenser lenses*. Electric current flowing through the objective lens may be controlled to obtain a finely focused electron beam on the sample; on the other hand, the condenser lens current can be controlled to obtain a higher or lower beam current. As a matter of first order importance, the finer the focus of the beam the greater the resolution of the image. On the other hand, higher beam current often translates to higher photon counts from a sample and therefore better chemical data in a statistical sense.

BEAM–SAMPLE INTERACTIONS

When a focused electron beam hits the mineral sample (target) with sufficient energy, secondary electrons, backscattered electrons, Auger electrons, cathodoluminescence, fluorescence, and "X-ray continuum" are produced in addition to characteristic X-rays (Figure 6.6a,b,c). In general, mineralogists make use of characteristic X-rays, secondary electrons, and backscattered electrons. *Secondary electrons* are simply target electrons that are let loose by bombarding electrons from the electron beam generated by heating the filament in the electron gun (Figure 6.6b,c). Secondary electrons are generated near the surface of the mineral. Conventionally, secondary electrons are defined as those having <50 eV energy. Because of their low energies and origin near the surface of the target, secondary electrons are useful in imaging the topography of the mineral surface. *Backscattered electrons (BSE)* are incident beam electrons that are deflected back from inside the sample (Figure 6.6b,c). These come out of the sample with much higher energies than secondary electrons, sometimes reaching the energies of the incident electrons. The energies of BSEs increase with increasing atomic number of the element, and therefore, BSEs can be used to obtain images of zoning and reaction features in crystals. Beam electrons can decelerate as they travel through the coulombic field of adjacent atoms in the sample. Such deceleration results in the emission of X-rays with a wide range of energies that are called *X-ray continuum* or *Bremsstrahlung*. Bremsstrahlung creates a background noise and is therefore more of a nuisance to the analyst who is interested in characteristic X-rays. It affects the minimum concentration level of the element to be determined (i.e., its *detection limit*).

Depending upon E_c and Z of the element, characteristic X-rays may be produced from a substantial, commonly bulbous, volume (called *interaction volume*; Figure 6.7). Figure 6.7 shows two examples of beam-sample interactions in a copper-bearing alloy. The target specimen in Figure 6.7a has a much lower density (and hence lower atomic number or Z) than the one in Figure 6.7b. In both cases the electron beam is of the same diameter, and the sample current and E_o are also the same. We note that the shapes and sizes of the interaction volume in the two examples are very different. The depth of X-ray generation range, R, is much greater in the first case (lower Z). Although not shown here, greater E_o also results in a greater R, all other conditions being equal. Notice that the value of R for copper Lα X-rays is greater than that for the Kα X-rays in both examples. We conclude that R depends on Z, accelerating voltage (E_o), and the E_c of the X-ray line concerned (Figure 6.7). X-rays are not generated uniformly across the interaction volume. Also, even when all conditions are identical, the ranges for two elements with different Z's are different; the one with a higher Z will have a smaller range because its E_c is greater. The ranges for Kα *versus* Lα X-rays for the same element are different because Lα X-rays with lower E_c require less E_o and, therefore, can be generated from greater depth where the E_o is lesser than at a shallower depth.

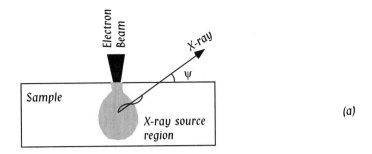

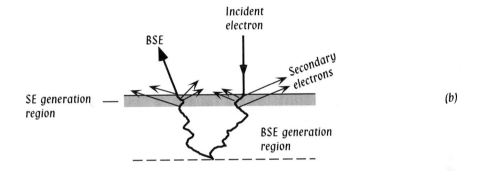

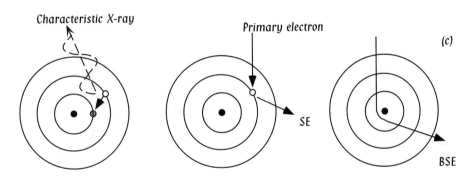

FIGURE 6.6 (a) Interaction volume showing the X-ray generation region and the X-ray take-off angle (ψ). (b) Generation of backscattered electrons (BSE) and secondary electrons (SE). BSEs are simply the electrons of the incident beam itself that go through the sample and are reflected back with minimal loss of energy. SEs are electrons with weak energies that are knocked off target atoms. (c) The generation of characteristic X-ray by inner shell ionization (left), the generation of secondary electrons (SE) by knocking off a shell electron by an incident (primary) electron (center), and (c) the generation of back-scattered electron (BSE) by scattering back of a beam electron (right).

The range is usually expressed in terms of a parameter called mass-depth (ρz), in which ρ is the density and z (as opposed to Z, which is the atomic number) is the actual depth. The function $\phi(\rho z)$ is used to describe X-ray generation as a function of mass-depth. Note that $\phi(\rho z)$ is not an absolute X-ray intensity (which is very difficult to measure or even calculate) but a normalized intensity (Goldstein et al. 1992). Figure 6.8 shows $\phi(\rho z)$ curves for the Kα X-rays for pure copper, titanium, and aluminum. All three curves have the same shape: note that the maximum amount of X-rays are generated somewhat below the surface which is also the region where most of the BSE's come from. The ranges are, however, different (as would be expected: discussed earlier) and the production of X-rays also decreases in a nonlinear

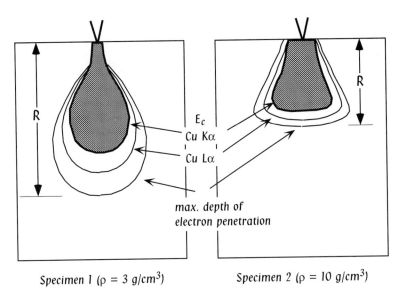

FIGURE 6.7 *Range (R) for Cu Kα and Cu Lα in two different specimens with different densities (left, 3 g/cm³ and right, 10 g/cm³).* Analytical conditions are the same for both cases. (Redrawn with permission from Goldstein et al., 1992, 2nd ed., *Scanning Electron Microscopy and X-Ray Microanalysis*, Plenum Publishers)

fashion in all three cases. Note that Kα intensity of Cu (Z = 29) is overall less than that for Ti (Z = 22) and Al (Z = 13). The lesson to be taken from Figure 6.8 is that one should be careful while analyzing fine grained minerals, zoned mineral grains, and mineral grains with fine foreign or exsolved inclusions. Because of the analytical volumes from which different X-rays may be produced during the analysis of a sample, the obtained analysis may include unwanted contributions from such material different overlapping mineral grains at depth within the target sample. Also, not all elements may be affected equally owing to the different ranges of their X-ray lines. Note that in practice, stray signals of "foreign" minerals may be filtered out by evaluating the structural formula of the mineral that is visible on the surface of the sample.

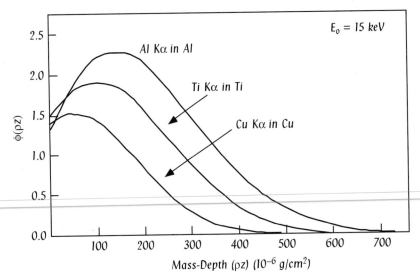

FIGURE 6.8 *Calculated ϕ(ρz) curves for Kα X-rays for pure Al, T and Cu at 15 keV* (Redrawn with permission from Goldstein et al., 1992, 2nd ed., *Scanning Electron Microscopy and X-Ray Microanalysis*, Plenum Publishers)

ENERGY DISPERSIVE SPECTROMETRY (EDS) AND WAVELENGTH DISPERSIVE SPECTROMETRY (WDS)

X-ray signals are processed by an energy-dispersive and wavelength-dispersive spectrometer. EDS uses a solid-state detector, commonly a lithium-drifted silicon or "Si(Li)" detector, which produces electronic pulses whose heights are proportional to the intensities of the X-ray photon energies. It is a fast method of simultaneously collecting an entire spectrum of energies of photons of different elements. On the other hand, WDS uses Bragg's principle (discussed later) and measures individual characteristic wavelengths of each element, one at a time. While EDS does "parallel" processing of energies of the photons, WDS does a "sequential analysis" since a spectrometer may be 'tuned' to only one element's characteristic wavelength at a time. The advantages of EDS are that it allows a fast determination of the proportions of all elements that are present in a mineral and thus is very helpful in a quick identification of the mineral. Quantitative analysis is done better with WDS, especially when it comes to the analysis of elements with Z < 15, because the background due to X-ray continuum is higher on the energy spectrum. Also, WDS is sometimes the only way to identify elements whose energy peaks overlap those of another element that is present in greater abundance.

WDS is based on *Bragg's law*, which states that when X-rays of wavelength λ and multiples of λ are 'reflected' from identical layers of atoms in a crystal that are separated by a constant distance d (known as d-spacing), λ and d are related by the angle of reflection 'θ' in the following way:

$$n\lambda = 2d \sin \theta$$

where n is $1, 2, 3, \ldots$ etc. and reflects the order of reflection (Figure 6.9). In a so-called "fully focusing" WD spectrometer, a crystal of known d-spacing and a detector are placed on the perimeter of an imaginary circle on which the sample surface is also located (Figure 6.10). This imaginary circle is known as the *Rowland circle*. The crystal's inner face is curved to match the curvature of the circle. For different wavelengths of different elements, the Rowland circle must be moved along a linear path such that the take-off angle (ψ, the angle at which X-rays take off the surface of the sample and reach the detector) is held constant. In Figure 6.10 ψ is known for an instrument, L is the distance between the sample and the crystal at any given point, R is the

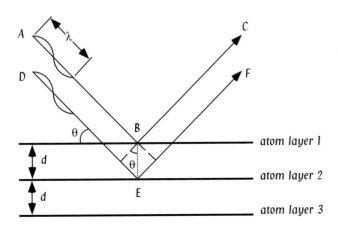

FIGURE 6.9 *Bragg's law: Two X-rays are reflected back from two parallel layers of atoms and are said to be in phase because their paths differ by only an integral number of wavelengths.*

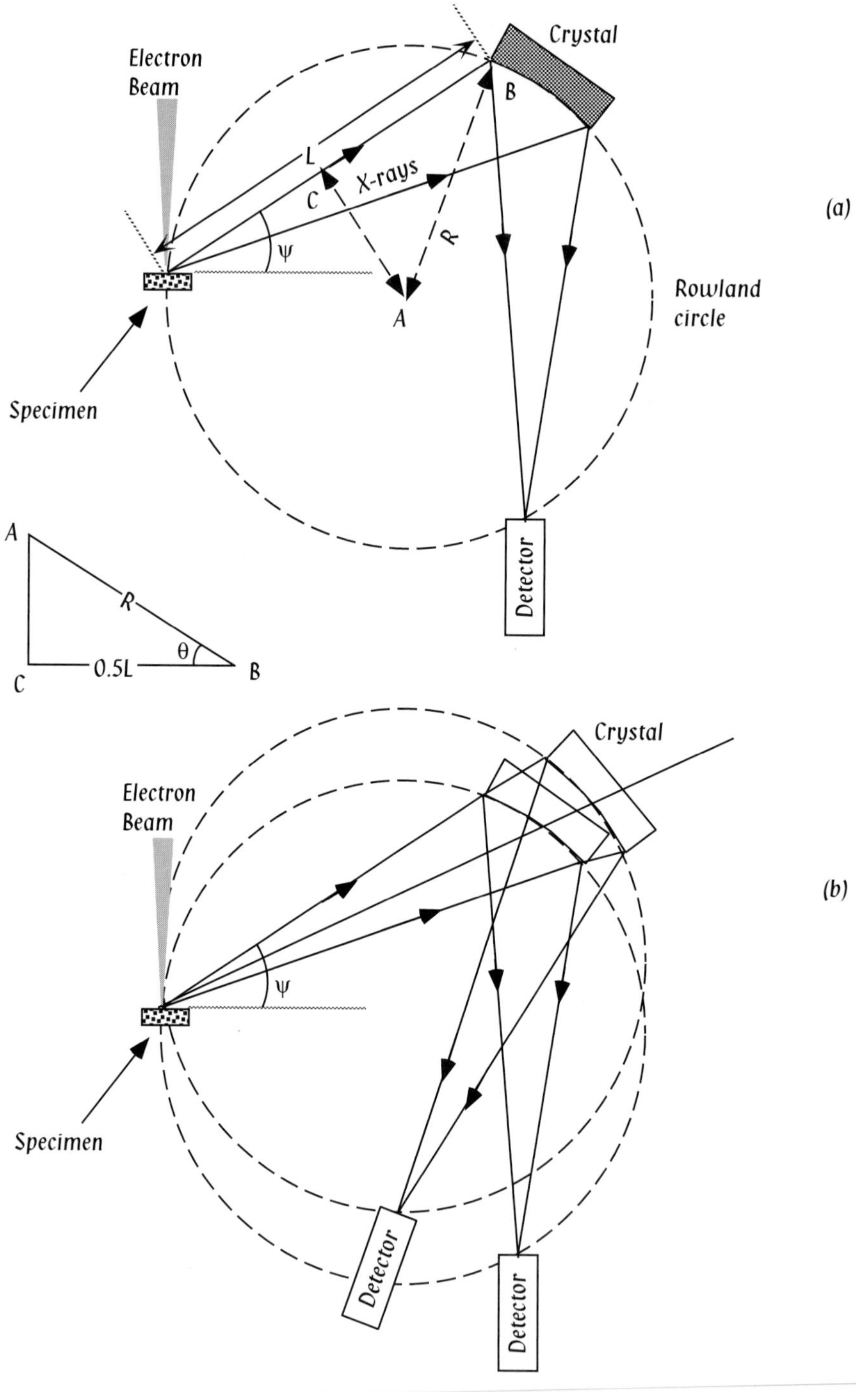

FIGURE 6.10   *Schematic drawings showing how wavelength-dispersive spectrometry works:* (a) The electron beam hits the sample and generates characteristic X-rays, which are reflected by a curved crystal toward the detector, where each photon is received and passed through an amplifier and other processing units. The radius of curvature of the crystal, the centroid of the sample, and the detector all lie on the perimeter of an imaginary circle, known as the *Rowland circle*. θ is the diffraction angle. ψ is the take-off angle. The triangle ABC (inset) shows the relationship between R, L, and θ which is utilized in the WDS method (see text). (b) In a fully focusing spectrometer, the positions of the detector and the crystal are moved for different X-ray lines, although the take-off angle remains constant. (Modified from Goldstein et al., 1992, 2nd ed., *Scanning Electron Microscopy and X-Ray Microanalysis*, Plenum Publishers)

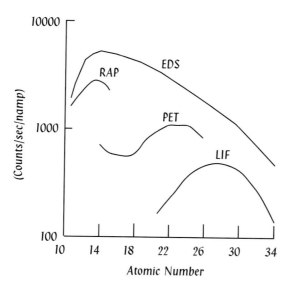

FIGURE 6.11   *Count rate (counts/sec/namp) and atomic numbers of elements whose wavelengths are covered by RAP, PET, and LIF crystals.* (Redrawn with permission from Reed, 1996, *Electron Microprobe Analysis and Scanning Electron Microscopy in Geology*, Cambridge University Press)

radius of the Rowland circle, and θ is the Bragg angle. From the triangle ABC in Figure 6.10, $\sin \theta = L/2R$. Transferring this value of $\sin \theta$ to Bragg's law ($n\lambda = 2d \sin \theta$) we obtain the following:

$$n\lambda = dL/R$$

The above relationship forms the basis of the functioning of WD spectrometers. Because d-spacing of the crystal is an important factor in resolving wavelengths of X-rays of different elements, several different crystals with different d-spacing and different wavelength coverage (Figure 6.11) are needed to analyze most elements in a mineral. This is done with the use of four or more spectrometers, each fitted with one crystal. More commonly, each spectrometer contains two crystals that can be flipped one at a time into the path of X-rays. Geologists commonly use ADP (or PET), LIF, and RAP (or TAP) crystals. Each of these can detect different wavelengths of different elements (Table 6.1). Table 6.1 lists the most commonly analyzed elements by their atomic numbers, $K\alpha$ or $L\alpha$ lines, their intensities, and peak positions on four of the more commonly used WDS crystals—LIF (Lithium Fluoride), ADP (Ammonium dihydrogen phosphate), RAP (Rubidium acid phthlate; try pronouncing this name), and PET (Pentaerythritol). The modern WDS spectrometers are commonly equipped with a greater variety of analyzing crystals that increase peak intensity of some of the more difficult elements. The analyst may a combination of crystals in different spectrometers to analyze a whole range of elements in a mineral: for example, let us consider the analysis of a clinopyroxene. The routinely analyzed oxides in a clinopyroxene are SiO_2, TiO_2, Al_2O_3, Cr_2O_3, FeO* (* represents all Fe assumed to be ferrous and analyzed as such), MnO, MgO, CaO, Na_2O. Based on Table 6.1 one may choose the following set up for analysis.

Spectrometer#	Crystal	Elements for Analysis
1	RAP	Na, Mg
2	ADP (or PET)	Al, Si, Ca
3	LIF	Cr, Ti, Mn, Fe

TABLE 6.1   *Atomic number and peak positions of elements of geological interest.*

El	Line	Z	I	LIF	RAP	ADP	PET
Na	Kα	11	100		1.8360		
Mg	Kα	12	100		1.5246		
Al	Kα	13	150		1.2856	3.1562	3.8415
Si	Kα	14	150			2.6969	3.2824
P	Kα	15	150			2.3304	2.8364
S	Kα	16	150			2.0334	2.4749
K	Kα	19	150			1.4163	1.7238
Ca	Kα	20	150	3.35948		1.2714	1.5474
Ti	Kα	22	150	2.74973			1.2665
Cr	Kα	24	150	2.2910			
Cr	Lα	24	100		3.3358		
Mn	Kα	25	150	2.10314			
Mn	Lα	25	100		2.9983		
Fe	Kα	26	150	1.93735			
Fe	Lα	26	100		2.7115		
Ni	Kα	28	150	1.65919			
Ni	Lα	28	100		2.2446		
Cu	Kα	29	150	1.54184			
Cu	Lα	29	100		2.0558		

The actual choice of crystals and spectrometers is dependent on many factors, such as the peak position relative to the limits of the spectrometer motors, peak overlap problems, counting statistics, etc. It is beyond the scope of this book to discuss these factors in any detail.

QUANTITATIVE ANALYSIS

The basic concept of quantitative analysis of minerals (or any solid material) is based on the fact that the intensities of X-rays for different elements are proportional to their concentrations in the sample. The concentration of an element in an unknown sample may be determined from its X-ray intensity if the X-ray intensity versus composition relationship is obtainable from a known standard whose composition had been determined by other methods (wet chemistry, for example). If $[I_i]^u$ and $[C_i]^u$ are X-ray intensity and composition, respectively, for an element 'i' in an unknown sample 'u', and $[I_i]^s$ and $[C_i]^s$ are those of the same element in the standard 's,' then:

$$[C_i]^u/[C_i]^s = [I_i]^u/[I_i]^s$$

The term on the right hand side is conventionally known as the *k-ratio*. The concentration of an unknown element with a $Z > 16$ determined from the above equation is generally in error by about 10%; and for light elements it can be much worse. Figure 6.12 demonstrates the effect of matrix on chemical analysis in a series of Fe-Ni alloys between pure Fe and pure Ni compositions. The lines marked "ideal" show a one-to-one correspondence of the k-ratio and concentration. The "measured" curves reflect the actual concentrations as determined from intensity ratios measured on the EMP. Note that the intensities of Ni X-rays are always too high whereas those of

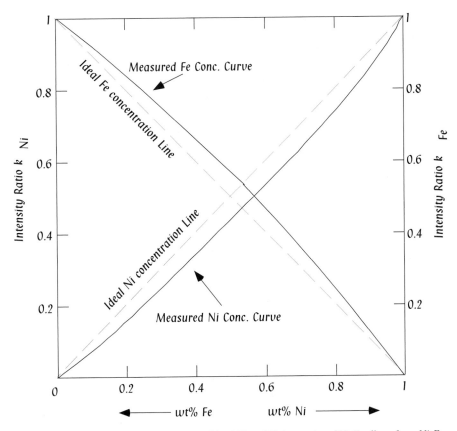

FIGURE 6.12   *Actual concentrations ("Ideal") of Ni and Fe in a series of Ni-Fe alloys, from Ni_0Fe_{100} to $Ni_{100}Fe_0$, are compared with measured concentrations as determined from intensity ratios of Ni and Fe.* (Redrawn with permission from Goldstein et al., 1992, 2nd ed., *Scanning Electron Microscopy and X-Ray Microanalysis*, Plenum Publishers)

Fe are always too low relative to their ideal values. This is because the Fe Kα line has a lower E_c than the Ni Kα line and therefore contributes extra photons from a region that is below the range for Ni. The opposite is true for the Fe Kα line. This difference is due to errors contributed by the presence of other elements in the same sample, and some nominal correction must be made to the raw intensity data. These corrections are known as *matrix correction factors*, and they stem mainly from three things: atomic number (Z), X-ray absorption (A) and fluorescence (F). Detailed discussion of these factors is beyond the scope of this book and may be found in the reference listed. A simple way to express the concentration and intensity ratios and matrix correction factors is as follows:

$[C_i]^u/[C_i]^s = [I_i]^u/[I_i]^s \times [ZAF]_i$, where ZAF is the correction factor for the same element. These correction factors depend to a significant degree on the beam current, nature of the sample, and the take-off angle (ψ) of the instrument. ψ is the angle contended between the mineral surface and the detector. For example, the smaller the take-off angle, the more sample X-rays will have to travel through and therefore, more absorption corrections will need to be made. The take-off angle of modern EMP's is fixed at 40°.

All modern instruments come with matrix correction software that allows three different options for matrix corrections—ZAF, $\phi(\rho z)$, and Bence–Albee corrections. The first two are based on metal alloys made in the laboratory, and the third is based on pure oxides synthesized in the laboratory. Most geologists have found the Bence–Albee corrections to be most useful for mineral analysis because in geological samples, errors resulting from the fluorescence factor are small. Note that matrix corrections would be minimal if the standard were chosen to be chemically and structurally close to the unknown; that is, if a diopsidic clinopyroxene is used to analyze another similar pyroxene in the unknown sample, the corrections would be few, if any. Mineral standards are often hard to come by, however, when they are available, one should use them for most of the major elements present. According to Goldstein et al. (1992), $\phi(\rho z)$ correction method is superior to the other two methods and does not require the use of a whole range of standards, as is the case for the Bence–Albee method.

EMP AND SEM ANALYTICAL CONDITIONS

Geological samples are mainly silicates and are poor conductors. Therefore, they need to be coated with a conducting material; and carbon is usually the conducting material of choice. The quality of the analysis of minerals and glasses depends upon many factors, perhaps the most important of which are accelerating potential, counting statistics, standards, nature of the material being analyzed, and instrument drift (see Goldstein et al. 1992 for a detailed discussion of these factors). For geological samples, the accelerating potential is commonly set at 15 or 20 kV. However, for high resolution imaging purposes, one may use a voltage of 25–30 kV. Such higher accelerating potential is also needed when the goal is to analyze elements of high atomic number (e.g., Uranium). The number of photon counts per second that may be obtained for a particular element depends on the abundance of the element in the mineral, the beam current and sample current, and the detector (and associated electronics). An optimum set of numbers for routine silicate mineral/glass analysis are as follows: accelerating potential, 15 kV; emission current, 100 microamps; sample current, 30 nannoamps. A minimum of 10,000 total counts is necessary for most major elements.

The nature of the target sample is an important factor in the analysis: the greater the concentration of "volatile" elements (i.e., elements with atomic number of <12, such as Na), the greater the risk of a problem due to volatile loss during analysis. This problem is generally manifested in low oxide totals, and can be minimized by using low sample current, a defocused beam (diameter ~ 10 μm), or a shorter counting time.

The electron beam may "drift" considerably (i.e., the beam current may fluctuate) during the course of an analysis. Such drift is common in old instruments and may result in problematic analysis. In order to check for such problems, it is important to analyze a known standard every 3–4 hours during the course of analysis.

The quality of the crystal used in a detector and associated electronics can pose a problem for old microprobes but is usually not a factor for new SEMs or EMPs. Over the years these crystals become bad, and when this happens the count rate goes down sharply, at which point they must be replaced.

The "goodness" of a mineral analysis is checked by analyzing a standard of similar composition, the oxide totals, and the structural formula (see Box 6.1). When dealing with minerals or glasses containing structural $(OH)^-$ or $(CO_3)^{2-}$, it is not possible to use the oxide totals criterion above. For hydrous minerals, often with structural vacancies, it may be hard to use the structural formula as well. At any rate, for most geological purposes, the analysis of hydrous or carbonate minerals is generally acceptable.

6.1 CHEMICAL ANALYSIS OF MINERALS

The quality of chemical analysis of materials is generally excellent with the electron microprobe or the analytical scanning electron microscope. However, the analyses are not always good, and the analyst must know how to separate the good from the poor analyses. One way to do this is to check the *structural formula*, which is simply the proportion of cations relative to a fixed number of oxygens in the mineral's formula.

Let us consider the structural formula calculation of the mineral clinopyroxene, which has six oxygens and four cations in its formula. For example, diopside ($CaMgSi_2O_6$) has six oxygens and four cations ($Ca^{2+} + Mg^{2+} + Si^{4+}$). Furthermore, clinopyroxene has primarily two types of cation sites—tetrahedral (or T site) and octahedral (M site: Chapter 5). Ideally, in a *real* pyroxene composed of 9 or 10 major and minor elements, the total number of cations of Si^{4+} and Al^{3+} filling up its tetrahedral site (or T-site) should be two, and the rest of the cations should then add up to two—filling up the M1 and M2 cation sites in pyroxene's atomic struc-

ture. (See Chapter 5.) In general, a pyroxene analysis is acceptable if the *total* cations add up to any value between 3.98 and 4.02. However, some authors may impose more stringent acceptability "filters."

The sample calculation shown in Table Bx 6.1 is an actual EMP analysis on a clinopyroxene in a Hawaiian xenolith, which shows that the analysis is good, based on the above mentioned criterion, because the total number of cations per 6 oxygens is 3.997. Similarly, one may use "total cation filters" for olivine (cation sum per 4 oxygens = 2.98–3.01) and feldspars (cation sum per 8 oxygens = 4.98–5.02). However, because EMP analysis cannot distinguish between the valence states of an element, the calculation of structural formula may become a tricky issue for minerals like spinel, garnet, and amphibole, which can contain a significant amount of Fe^{3+}, in addition to Fe^{2+}. Hydrous or other volatile-bearing mineral formula calculation is further complicated by the presence of elements that are not normally analyzed for.

TABLE BX 6.1   *Mineral structural formula calculation.*

A	B	C	D	E	F	G	H	I	J
Oxide	Cpx	Mol. Wt	Moles	Cation	Cations in Oxide	Real Cations	Oxygens per Cation	Real Oxygens	Cations Normalized to 6 Oxygens
SiO_2	52.92	60.084	0.881	Si^{4+}	1	0.881	2	1.762	*1.915*
TiO_2	0.69	79.9	0.009	Si^{4+}	1	0.009	2	0.017	*0.019*
Al_2O_3	6.96	102	0.068	Al^{3+}	2	0.136	1.5	0.205	*0.297*
Cr_2O_3	0.02	152	1E-04	Cr^{3+}	2	0.000	1.5	0.000	*0.001*
FeO*	5.1	71.85	0.071	Fe^{2+}	1	0.071	1	0.071	*0.154*
MgO	14.19	40.3	0.352	Mg^{2+}	1	0.352	1	0.352	*0.766*
CaO	17.68	56.08	0.315	Ca^{2+}	1	0.315	1	0.315	*0.686*
Na_2O	2.28	62	0.037	Na^+	2	0.074	0.5	0.037	*0.16*
Total	99.84		1.733			1.838		**2.759**	*3.997*

Column **B** gives the **actual Wt% analysis** of the clinopyroxene
Column **C** lists the molecular weights of the oxides
Column **D** gives the number of **moles in the analysis** = Column B/Column C
Column **F** gives the number of cations present in the oxide formula
Column **G** gives the number of **cations** in the analysis = Column D × Column F
Column **H** lists the number of oxygens per cation in each oxide formula
Column **I** gives the number of **real oxygens** in the analysis = Column G × Column H
Column **J** gives the calculated number of cations normalized to 6 oxygens = (6/2.759) × Column G, 2.759 being the total number of "real" oxygens calculated

X-RAY POWDER DIFFRACTOMETER (XRD)

The utility of X-rays in determining crystal structure is well known. This is one area of study where 14 Nobel prizes in Physics have been awarded (Battey and Pring 1997). Sophisticated single crystal structure determination requires different types of X-ray diffractometers, which cannot be discussed here. Instead, we will briefly discuss the most commonly used X-ray machine, an X-ray powder diffractometer, for routine mineral identification work.

The guiding principle on which a XRD operates is Bragg's law. In X-ray powder diffractometers, the source of X-rays is an evacuated X-ray tube (Figure 6.13), in which a beam of electrons is generated from a tungsten filament by applying an accelerating voltage of 20–100 kV. These electrons hit a metal target (usually copper for geological samples) and produce X-rays. These X-rays then collide with a target sample and get diffracted and collected by a detector. The signals received by the detector are then passed on to a computer, which identifies the 'reflections' and their peak intensities with reference to the angle 2θ and identifies the sample.

The procedure is simple: a mineral (or rock) is powdered to fine size and thinly and evenly spread, along with a bonding agent (acetone is commonly used), on a glass slide. This glass slide is then clamped onto a stage. The X-ray beam and the sample are lined up along a straight path. The detector sits on the other side of the sample from the X-ray source and moves along a circular arc (Figure 6.14) in order to catch the X-ray photons diffracted over a 2θ value of 2–60° (for most routine work).

The basic idea is that although the powdered sample will contain mineral grains of all kinds of orientation, reflections from a particular set of (hkl) planes of many different grains will obey Bragg's law. These may be matched with diffraction patterns of known minerals, and the unknown mineral may thus be identified. Most instruments can routinely identify minerals with automated systems that are attached to the actual instrument.

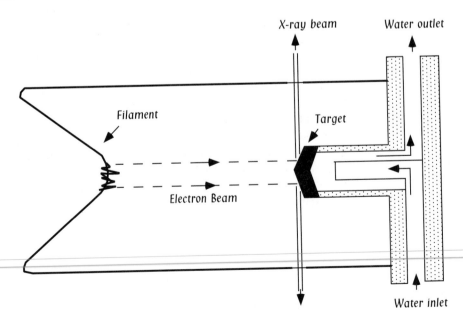

FIGURE 6.13 *Schematic drawing of an X-ray tube.* (Redrawn from Battey and Pring, 1997, 3rd ed., *Mineralogy for Students*, Addison Wesley Longman)

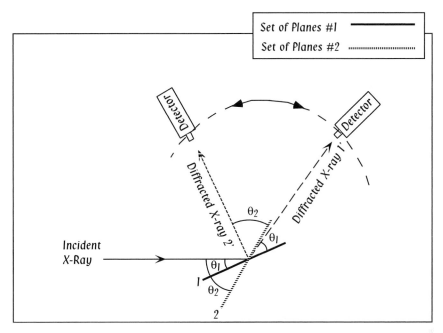

FIGURE 6.14   *The principle of X-ray powder diffraction.* The X-ray beam diffracted by a particular plane of atoms in a crystal is detected by a detector, which revolves around the sample in such a way that it can detect X-rays coming off the sample at many different angles.

SUMMARY

[1]   Three instruments that are commonly used in mineralogical laboratories are the scanning electron microscope (SEM), the electron microprobe (EMP), and the X-ray powder diffractometer (XRD). SEM and EMP are used for imaging and chemical analysis of minerals, whereas XRD is used to identify crystalline phases, particularly fine-grained ones.

[2]   In all three instruments, a high-energy electron beam is used to generate X-rays from a target. In the case of SEM and EMP, the target is the sample to be analyzed; but in XRD, the target is simply the source of the X-rays used to hit a sample and investigate its crystalline structure.

[3]   Two types of analytical procedures are used in SEM and EMP: wavelength and energy dispersive spectrometry, or WDS and EDS. WDS and XRD methods are based on Bragg's law. WDS analysis is highly quantitative, whereas EDS analysis can be semiquantitative to quantitative.

BIBLIOGRAPHY

Battey, M.H. and Pring, A. (1997) *Mineralogy for Students* (3rd ed.). Longman (Hong Kong).

Goldstein, J.I. and others (1992) *Scanning Electron Microscopy and X-Ray Microanalysis* (2nd ed.). Plenum Press (New York).

Lyman, C.E. and others (1990) *Scanning Electron Microscopy, X-Ray Microanalysis, and Analytical Electron Microscopy: A Laboratory Handbook.* Plenum Press (New York).

Reed, S.J.B. (1995) *Electron Microprobe Analysis and Scanning Electron Microscopy in Geology.* Cambridge University Press (Cambridge, U.K.).

C H A P T E R   7

Chemical Mineralogy: Mineral Phase Relations

The following topics are covered in this chapter:

Elementary thermodynamic principles
    Definitions
    Chemical equilibrium, chemical potential and free energy
    Clausius–Clapeyron equation and calculation of phase diagrams
Phase rule and phase equilibria of important mineral systems
    Melting, crystallization and melt mixing processes in binary and ternary systems

ABOUT THIS CHAPTER

Minerals and rocks form in a variety of temperature and pressure conditions within the earth. An understanding of their origin requires some elementary knowledge of thermodynamic principles. The objective of this chapter is to introduce the student to such concepts; however, a rigorous thermodynamic treatment of mineral phase equilibrium is beyond the scope of this book. The student will also be exposed to geometrical techniques that can be easily used to understand the conditions of mineral crystallization in simple mineral-melt systems that serve as analogues of crystallization and magma formation. Appropriate examples that reveal some important aspects of our earth "system" are used to demonstrate the power of phase diagrams in relating mineralogy to earth processes. The phase diagrams discussed in this chapter refer almost exclusively to igneous systems, that is, the crystallization of minerals from melts. Chemical equilibria involved in the formation of minerals in sedimentary and metamorphic rocks will be discussed in other chapters.

ELEMENTARY THERMODYNAMICS

The subject of thermodynamics deals with the stability of minerals and other materials in a variety of pressure and temperature conditions. Phase diagrams are a graphic means of portraying the stability of phases in pressure-temperature-composition space. Where thermodynamic stability of minerals and materials are concerned, an adequate description of them requires an understanding of such terms as system, phase, degrees of freedom, and components. These terms are first explained in the following paragraphs.

SYSTEM.   The definition of a *system* depends on how one defines its boundaries! While this statement may seem abstract, it is nonetheless the most appropriate definition. We may define the system to include an entire rock or a single mineral in the rock. If our problem is to examine the behavior of atoms within the mineral, the mineral can be "isolated" (in an abstract sense) from the rock and be considered as a system that is contained within itself. Consideration of all other minerals in that rock may be totally irrelevant to the mineral's internal atomic behavior; in such a situation the mineral, rather than the rock, is the system. On the other hand, if our interest is in the crystallization or melting behavior of the mineral, the entire rock must be considered as the system because in such a case, the mineral's crystallization and melting behavior are tied to other minerals in the rock.

In a *closed system* the system does not exchange materials with its surroundings, whereas in an *open system* such exchange occurs. In both cases, however, the system can exchange energy with the surroundings. Note that open systems are not considered at all in our discussion in this chapter.

PHASE.   A phase is a physically distinct and mechanically separable part of a system. Consider, for example, the rock *gabbro*. Gabbro is principally composed of only two minerals—pyroxene and plagioclase. Each of these two minerals (but not each grain of these minerals) is distinct by virtue of its compositional and physical characteristics and therefore is a *phase*. A phase may exist in solid, liquid, or gaseous states. If we are dealing with an immiscible mixture of oil and water, the two are distinct phases although both are in a liquid state because they are "physically distinct and mechanically separable".

COMPONENTS.   Each phase in a system is composed of chemical components. The number of chemical components in a system is the minimum number of chemical constituents required to define the compositions of all phases in the system. For example, consider the system forsterite-quartz. Forsterite's composition may be written as Mg_2SiO_4 or $2MgO.SiO_2$; and quartz is SiO_2. Although three elements occur in this system—Mg, Si, and O—compositions of phases in this system can be entirely described with just two components (i.e., Mg_2SiO_4 and SiO_2 or MgO and SiO_2) rather than three (i.e., Mg–Si–O) components. Therefore, the system forsterite-quartz is considered to be a two-component and not a three-component system.

One-, two-, three- and four-component systems are called *unary, binary, ternary*, and *quaternary systems*. It turns out that although rocks contain a large number of components, much of their phase relationships can often be adequately described with systems containing as few as three components. Therefore, in a later section of this book, a rigorous geometrical analysis of phase relationships in some important but simple one-, two-, and three-component systems is carried out.

CHEMICAL POTENTIAL AND GIBBS FREE ENERGY

The *chemical potential* (μ) may be viewed as an inherent tendency of a material to react. A chemical reaction within a system may occur when reactants have higher chemical potential than the products. For example, consider the polymorphic transformation reaction between diamond (chemical potential—μ_d) and graphite (chemical potential—μ_g). Figure 7.1 shows that at a temperature of 1000°C and pressure of 6 GPa (= 60 kbar), diamond is the stable mineral, and that is because $\mu_d < \mu_g$ at such a pressure(P)–temperature(T) condition. By similar reasoning, at approximately 1000°C and 2 GPa, graphite is stable (i.e., $\mu_g < \mu_d$). At equilibrium $\mu_g = \mu_d$, or, the chemical potential difference, $\Delta\mu = \mu_d - \mu_g = 0$.

When a phase is composed of more than one component, it is necessary to consider the *Gibbs free energy*, or G, which is the sum of the chemical potential of each component times the number of moles of each of the component in the phase. Thus, Gibbs free energy, or simply the *free energy*, may be expressed as:

$$G = \Sigma_i \mu_i n_i$$

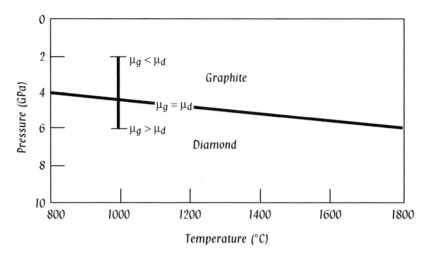

FIGURE 7.1   *Stability fields of diamond and graphite in pressure-temperature space.*
(From W.G. Ernst, 1969, *Earth Materials,* Mcmillan and Co.)

where μ_i and n_i are the chemical potential and the number of moles of component 'i'.
For a pure phase in a one-component system, $G = \mu$. When individual phases are not
pure but are solid solutions, one has to calculate G by considering free energy contributions made by each end member in the solid solution series. This may be illustrated
using the olivine$_{\text{solid solution}}$. Olivines form a complete solid solution between forsterite
(Mg_2SiO_4) and fayalite (Fe_2SiO_4) end members. The free energy of any intermediate
olivine (G_{ol}) with x moles of forsterite and $(1 - x)$ moles of fayalite may be given as:

$$G_{ol} = x\left(\mu_{fo}\right)^{ol} + (1 - x)\left(\mu_{fa}\right)^{ol} \qquad \text{[Eq. 7.1]}$$

It will be shown later that a solid solution will have lower free energy than the corresponding simple (mechanical) mixture of two end members.

The total free energies of the reactants and of the products are equal when they
are in chemical equilibrium. (That is, $\Delta G = G_{products} - G_{reactants} = 0$ at equilibrium.)
A reaction may proceed only when $G_{reactants} > G_{products}$ (i.e., $\Delta G < 0$).

FREE ENERGY, ENTHALPY, AND ENTROPY

The Gibbs free energy of a substance or system is also given as:

$$G = H - TS \qquad \text{[Eq. 7.2]}$$

where H is the *heat content* or *enthalpy*, and S is its *entropy*, which is a measure of the
degree of disorder (further discussed later). H is mathematically defined as

$$H = E + PV \qquad \text{[Eq. 7.3]}$$

where E is the *internal energy* of a system, and P and V are pressure and volume,
respectively.

Heat capacity $\left(C_p\right)$ is defined as the heat required to change the temperature of
1 mole of a substance by 1°C at constant pressure. The temperature of a substance increases roughly proportionally as it is heated at a constant pressure. C_p is generally
expressed as a proportionality constant as follows:

$$d\text{H} = C_p d\text{T},$$

or

$$C_p = dH/dT \qquad \text{[Eq. 7.4]}$$

Tables of data on enthalpy, heat capacity, volume, and other thermodynamic parameters of minerals can be found in many publications (listed in the references). Generally, these values are experimentally measured or calculated at a *reference temperature* of 298.15 K and 1 bar pressure. This reference state is generally expressed as a superscript "*o*" on the property in consideration (e.g., G^o, H^o etc., unless stated otherwise elsewhere in the text).

Determination of the values of these parameters at temperatures and pressures other than the reference state requires additional calculations. For example, H at a particular temperature T can be integrated from H^o (i.e., H at 1 bar pressure and 298 K temperature) as follows:

$$\int_{298}^{T} dH = \int_{298}^{T} C_p dT$$

or

$$H_T - H^o = \int_{298}^{T} C_p dT \qquad \text{[Eq. 7.5]}$$

The C_p of a mineral or glass is generally a polynomial function of temperature and may be given as

$$C_p = a + bT - c/T^2 \qquad \text{[Eq. 7.6]}$$

where a, b, and c are experimentally determined constants for the mineral or material of interest. Thus, H_T can now be obtained by combining Equations 7.5 and 7.6:

$$H_T = H^o + \int_{298}^{T} [a + bT - c/T^2] dT$$

or

$$H_T = H^o + [(a + bT^2/2 + c/T)dT]_{298}^{T} \qquad \text{[Eq. 7.7]}$$

Entropy may be thought of as a degree of disorder. In a gaseous state the molecules of a substance are much more dispersed (i.e., in greater disorder) than when the substance occurs in liquid state because in a liquid, the molecules are closer to each other and it is easier (in terms of probability) to "find" the molecules. Thus,

$$S_{gas} > S_{liquid} > S_{solid}$$

The entropy of a mineral can be estimated from relations between heat capacity and entropy, as laid down by the laws of thermodynamics (discussed in the next section):

$$dS = dH/T$$

or

$$dS = [C_p dT]/T$$

The entropy of any phase at a temperature T may be calculated from S^o and C_p as follows:

$$\int_{298}^{T} dS = \int_{298}^{T} (a/T + b - c/T^3) dT,$$

or

$$S_T = S^o + \left[a \ln T + bT + \left(c/2T^2\right)\right]_{298}^{T}$$

$$S_T = S^o + a \ln(T/298) + b(T - 298) + (c/2)\left(1/T^2 - 1/298^2\right) \quad \text{[Eq. 7.8]}$$

Similarly, the free energy of any substance can be calculated at a high P and T from the above equations. In a reversible reaction,

$$\Delta G^{P,T} = \Delta G^o + \int_1^P V^o dP,$$

which is the same as

$$\Delta G^{P,T} = \Delta H^o - T\Delta S^o + P\Delta V^o$$

Therefore, from Equations 7.7 and 7.8 above,

$$\Delta G^{P,T} = \left[H^o + \int_{298}^T C_p dT\right] - T\left[S^o + \int_{298}^T (C_p/T)\, dT\right] + \int_1^P V^o dP \quad \text{[Eq. 7.9]}$$

LAWS OF THERMODYNAMICS AND THE CLAUSIUS–CLAPEYRON EQUATION

The *first law of thermodynamics* states that the *internal energy* (E) of an isolated system is constant. The first law is best understood by the mathematical representation:

$$dE = dQ - dW \quad \text{[Eq. 7.10]}$$

where dQ represents the amount of heat added to the system. The system expends this heat by performing a certain amount of work (dW). The usual example used in thermodynamics textbooks to explain the first law is that of heating a gas in a cylinder fitted with a frictionless piston. When this gas is heated, its volume expands and displaces the piston, (i.e., the system performs work). Normally, work is defined as the force multiplied by the distance. In the example given, the work done is pressure–volume work, or

$$dW = PdV$$

where P is the pressure that causes the volume to change by dV. Thus, the first law can be rewritten as

$$dE = dQ - PdV \quad \text{[Eq. 7.11]}$$

In a reversible reaction, the *second law of thermodynamics* is expressed as

$$dS = dQ/T \quad \text{[Eq. 7.12]}$$

where T is temperature.

Combining the first and second laws (Equations 7.11 and 7.12) yields,

$$dE = TdS - PdV$$

or

$$dE - TdS + PdV = 0 \quad \text{[Eq. 7.13]}$$

Recall that

$$G = H - TS,$$

or,

$$G = E + PV - TS$$

Differentiating and rearranging,

$$dG = (dE + PdV - TdS) + VdP - SdT$$

Equation 7.13 shows that $dE + PdV - TdS = 0$, therefore we have the following:

$$dG = VdP - SdT \qquad \text{[Eq. 7.14]}$$

At equilibrium, $dG = 0$, and therefore

$$VdP - SdT = 0 \qquad \text{[Eq. 7.15]}$$

In order to understand the utility of the above relationship, consider the equilibrium transition between diamond and graphite as an example, where diamond is a higher pressure polymorph of carbon,

$$\text{Graphite} = \text{Diamond}$$

At equilibrium, the free energies (chemical potentials) of diamond and graphite must be equal. Based on Equation 7.14 therefore,

$$V_g dP - S_g dT = V_d dP - S_d dT$$

or

$$V_d dP - V_g dP = S_d dT - S_g dT$$

where the subscripts "d" and "g" stand for diamond and graphite, respectively.

The above equation can be rewritten below as

$$\Delta V dP = \Delta S dT$$

or

$$dP/dT = \Delta S/\Delta V \qquad \text{[Eq. 7.16]}$$

where ΔV and ΔS are volume and entropy differences, respectively, for the graphite = diamond reaction.

Equation 7.16 is known as the *Clausius–Clapeyron equation*. This equation is quite useful in that the slope of any equilibrium reaction boundary between any number of minerals can easily be calculated if entropy and volume data are known for them. Such data on minerals are easily available (see references). Below is an example in which the famous alumino-silicate phase diagram has been calculated from entropy-volume data.

Table 7.1 gives a set of thermodynamic data at 500°C and 1 bar for the alumino-silicate phases. The Clapeyron slope (dP/dT) for each of the reactions may be calculated as follows:

Reaction: kyanite $\leftrightarrow$ andalusite
$dP/dT = \Delta S/\Delta V = (245.1 - 236)/(52.29 - 44.69) = 1.197 \text{ MPa/K} = 11.97 \text{ bar/K}$

TABLE 7.1   *Thermodynamic data for the alumino-silicate phase diagram.*

Mineral	$V(cm^3/mol)$	$S°(J/mol/K)$	$H°(kJ/mol)$
Kyanite	44.69	236	−2519.31
Sillimanite	50.23	246.9	−2512.78
Andalusite	52.29	245.1	−2515.15

Similarly, we get the following dP/dT's for the following reactions: kyanite $\leftrightarrow$ sillimanite is 19.8 bar/K and andalusite $\leftrightarrow$ sillimanite is −8.7 bar/K.

We now need to calculate at least one point in P, T space for each of the previous three reactions in order to completely draw the phase diagram. We know from Equation 7.2 that at equilibrium, $\Delta G = \Delta H - T\Delta S = 0$ for each one of the three reaction boundaries. Therefore, equilibrium temperature $(T_e) = \Delta H/\Delta S$. As an example, we calculate the T_e at 1 bar for the reaction:

kyanite $\leftrightarrow$ andalusite:

$$T_e = [\Delta H/\Delta S]_{ky=and} = [(-2515150 + 2519310)/(245.1 - 236)] = 457 \text{ K}$$

Similarly, we obtain for kyanite $\leftrightarrow$ sillimanite, $T_e = 599$ K, and for andalusite $\leftrightarrow$ sillimanite, $T_e = 1317$ K.

We know from published reports that the kyanite $\leftrightarrow$ sillimanite boundary does not extend to 1 bar pressure. Therefore, the T_e for this reaction at 1 bar cannot be correct. Without going into much additional detail, note that this temperature at 1 bar pressure is the location of the metastable extension of the equilibrium boundary. Based on the calculated values of T_e and the slopes of the curves for the reactions kyanite $\leftrightarrow$ andalusite and andalusite $\leftrightarrow$ sillimanite the two curves are drawn first in P, T space. Then, the kyanite $\leftrightarrow$ sillimanite boundary is constructed with the calculated slope (19.8 bars/K) through the point (which is the invariant point on this diagram; discussed in a later section) where the other two curves meet (Figure 7.2).

In passing it is worth mentioning that even this apparently simple Al_2SiO_5 system is not so simple in the sense that large uncertainties exist in experimental and calculated data on each of the three reaction boundaries. Metastable persistence of these phases way beyond their stability field has also been demonstrated in many field studies of metamorphic rocks.

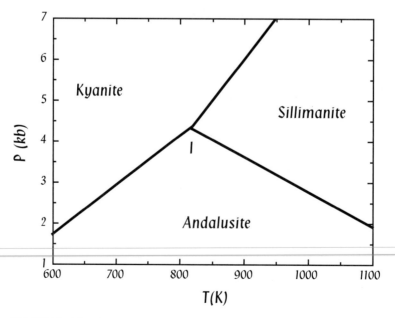

FIGURE 7.2 *The alumino-silicate phase diagram calculated here.* 'I' is the invariant point where all three phases coexist. (From M.J. Holdaway, Am. J. Sci.)

GIBBS PHASE RULE AND PHASE DIAGRAMS

The phase rule or the Gibbs phase rule was perhaps one of the most outstanding contributions to science made in 1876–1878 by J. Willard Gibbs, an American chemist. The phase rule states that in a system of c components and ϕ phases, the variance (f) or degrees of freedom,

$$f = c - \phi + 2.$$

"f," or degrees of freedom, represents the number of intensive variables (temperature, pressure, and composition) that can be changed independently without the loss of a phase. The numeral '2' represents pressure and temperature; when one of these is held constant, the numeral reduces to '1'. At constant pressure (or temperature) the phase rule thus reduces to $f = c - \phi + 1$. The implications of the phase rule may be understood with the following example of the Al_2SiO_5 system (Figure 7.2).

Unary Systems: Al_2SiO_5 Phase Diagram

In this one-component or *unary* system, three phases occur—kyanite, sillimanite, and andalusite. In the phase diagram there are three lines along which two phases are stable (i.e., kyanite = sillimanite, andalusite = sillimanite, and kyanite = andalusite). The three lines meet at a single point where all three phases are stable. Let us consider the kyanite = sillimanite boundary. At any point on this line, the degrees of freedom, $f = 1$ (# of components) $- 2$ (# of phases) $+ 2 = 1$. Because the kyanite = sillimanite boundary line has a variance of 1, it (and the other two lines as well) is called a *univariant line*. It is evident that along any univariant line, pressure and temperature can be varied but dependently of each other if both phases are to occur in equilibrium; and hence the degree of freedom (to vary pressure or temperature) is 1. At the triple point where all three phases are stable, $f = 0$; and the implication is that there is no freedom to vary either pressure or temperature otherwise, one or two phases may be lost. This point is called an *invariant point*. Within the stability fields of each of the three minerals, one is free to vary both temperature and pressure independently or dependently and yet will have a single mineral phase in stable existence. Because $f = 2$, each of the three stability fields is called a *divariant field*.

It is noteworthy that the phase rule does not say anything about the maximum number of phases that can occur in a system. For example, consider the system SiO_2. A large number of phases occur in this system, for example: liquid SiO_2, α-quartz, β-quartz, tridymite, cristobalite, coesite, and stishovite (Figure 7.3). Note that not all of these phases are stable around a single invariant point in this diagram, because the phase rule allows only a maximum of number of three phases at any single invariant point in this (or any other) unary system.

In addition to obeying the phase rule, the pressure-temperature phase diagram of SiO_2 shows some other features that are generally common to many other systems. First, note that the melting temperature of silica increases with pressure (i.e., the solidus in the pressure-temperature space has a positive slope); and the slope changes as a new polymorph becomes stable (i.e., the slope of the solidus becomes steeper at each invariant point). Also, the slopes of all polymorphic transformation reactions, except the cristobalite = tridymite reaction, are positive. When volatiles, such as H_2O and CO_2, are introduced in a system, the slope of the solidus becomes negative up to a certain pressure.

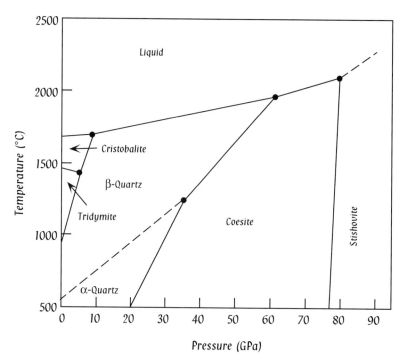

FIGURE 7.3 *Phase diagram for the unary system Silica.* (From Zoltai and Stout, 1984, *Mineralogy: Concepts and Principles*, Burgess).

BINARY SYSTEMS

Graphical Representation

Two component systems at a constant pressure (*isobaric* phase diagram) are best viewed in terms of a temperature-composition (or T-X) graph, in which temperature is plotted on the ordinate and composition is plotted on the abscissa. One has a choice of plotting molar versus weight percentages of the two components on the composition axis. The choice of molar versus weight percent makes a big difference in the nature of the phase diagram; therefore, it is important to understand how mineral compositions may be calculated on a molar versus weight basis from their chemical constituents.

Consider the examples of forsterite and enstatite, both of which are composed of MgO and SiO_2. Their formulas can be written as $2MgO \cdot SiO_2$ and $MgO \cdot SiO_2$, respectively; that is, the ratio of MgO to SiO_2 moles is $2:1$ in forsterite and $1:1$ in enstatite. In terms of mole percent, therefore, forsterite is composed of 67% MgO and 33% SiO_2 and enstatite is composed of 50% MgO and 50% SiO_2. If the molecular weights of MgO and SiO_2 were identical, then the weight% compositions of forsterite and enstatite, in terms of MgO and SiO_2, would be identical to their respective mole% compositions. However, the molecular weights of MgO and SiO_2 are different (MgO = 40.3 and SiO_2 = 60.08); therefore, weight% composition of both forsterite and enstatite are different from their mole% compositions, as calculated below.

100 wt% forsterite contains

$$100 \times [2 \times MgO \text{ mol. wt}/[(2 \times MgO \text{ mol. wt}) + SiO_2 \text{ mol. wt}]]$$

$$= [(2 \times 40.3)/[(2 \times 40.3) + 60.08]]\%$$

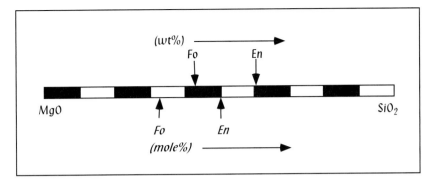

FIGURE 7.4   *The location of forsterite and enstatite in mole versus weight percent diagrams.*

or 55.3 wt% MgO. SiO_2 wt% is $(100 - 55.3)$ or 42.7%.

100 wt% enstatite similarly contains $100 \times [MgO$ mol. wt$/(MgO + SiO_2$ mol. wt$)]$

or, 40 wt% MgO and 60 wt% SiO_2.

The numbers calculated in the previous example may be plotted on the composition axis of a binary diagram in which one end represents 100% SiO_2 and the other end is 100% MgO (Figure 7.4). Each tick mark in Figure 7.4 represents a 10% increment. The upper and lower parts of this figure show where forsterite and enstatite plot in terms of weight versus mole percent, respectively. In this particular example, the differences between mole percent and weight percent plots are clearly large and demonstrate how important it is to note whether a particular phase diagram is a mole% or weight% diagram. All phase diagrams in this chapter are weight% diagrams unless otherwise specified.

Phase Relations

Binary systems exhibit a wide range of mineral-melt phase relationships—from eutectic crystallization (melting) to solid solution to incongruent melting (peritectic reaction). We will explore these different types of phase behavior with a number of binary systems that are of relevance to geology. In each system, we will consider two extreme conditions of crystallization and melting of liquids; in one of these (*equilibrium crystallization* and *equilibrium melting*), chemical equilibrium is always maintained between coexisting phases, and in the other (*fractional crystallization* and *fractional melting*), rapid physical separation of melt and crystals results in complete disequilibrium between phases. In nature, the crystallization of magmas and the melting of rocks likely follow some intermediate paths between the two extremes mentioned.

Geologic processes associated with the formation, transportation, and crystallization of magma ("molten rock") are discussed in the sections that follow with reference to specific phase diagrams. These processes include crystallization, melting, wall-rock assimilation, and mixing of different magmas. Where appropriate, textural aspects of such processes will also be considered. Application of phase diagrams in the study of igneous rocks will become clearer as we delve into the chapters that deal with igneous rocks.

Solid Solution: The System Forsterite–Fayalite at 1 Atmosphere Pressure

Plagioclase and olivine are important examples that exhibit extensive solid solution between the two end members. Olivine forms solid solution between forsterite (Mg_2SiO_4) and fayalite (Fe_2SiO_4). The system forsterite-fayalite plays an important role in our understanding of a number of phenomena—from the crystallization of basalts to the formation of the lithosphere.

At atmospheric pressure, relevant to the crystallization of shallow basaltic intrusions and lavas, the olivine phase diagram forms a "loop" in a temperature-composition (T-X) diagram, with the melting points of forsterite and fayalite defining the two ends of the loop (Figure 7.5). The higher temperature and the lower temperature curves of the loop are known as the *liquidus* and the *solidus*, respectively. Above the liquidus only a liquid phase is stable, and below the solidus a solid phase is stable. Between the liquidus and solidus, a solid phase coexists with a liquid phase. During crystallization (or melting) the path traversed by a liquid is called the *liquid path*: for example, (1) → (4) is the liquid path during equilibrium crystallization of the liquid "BC" (discussed in the next paragraph). The corresponding path on the solidus is the *solid path*.

It is important to note that Figure 7.5 shows phase relationships at atmospheric pressure. An increase of pressure causes the melting points of both forsterite and fayalite end members to rise, and therefore, the entire loop is shifted to higher temperatures.

Equilibrium Crystallization and Equilibrium Melting

Consider the case of a melt of bulk composition "BC" (i.e., 60% fayalite, 40% forsterite, or, $Fo_{40}Fa_{60}$). This melt will begin to crystallize when it is cooled to its liquidus temperature of 1590°C (stage 1). The composition of the first crystal that will form in equilibrium with this melt can be determined by drawing a line, called a *tie line*, from the liquidus intersection point to the solidus at 1590°C. In this example, the coexisting solid phase will have a composition $Fo_{75}Fa_{25}$. During equilibrium crystallization, the liquid and solid will continuously react with each other and undergo compositional change as the temperature drops. At any given temperature the composition of the coexisting solid and liquid phases are given by the solidus and liquidus: for example, at 1400°C the coexisting solid and liquid compositions are $Fo_{47}Fa_{53}$ and $Fo_{20}Fa_{80}$, respectively.

The *lever rule* is an important tool that allows calculation of proportions of all phases in a system at a particular temperature and for a given bulk composition. The "lever" is a tie line connecting the coexisting phases at a given temperature, and the bulk composition is the "fulcrum" of the lever. How the lever rule works may be best illustrated with the following example of equilibrium crystallization of liquid "BC". At 1525°C (stage 2 in Figure 7.5), a "lever" $x - z$ is drawn, where x and z are coexisting solid and liquid compositions, respectively. The "fulcrum" y of this "lever" is the bulk composition. Since y is the bulk composition, it follows that it can be defined by the sum of some proportions of the solid x and liquid z. That is, $y = mx + nz$, where m and n are proportions of x and z, respectively. How can one calculate m and n (or convert to percentages by normalizing to 100 percent)? The answer is easily obtained by comparing "levers" 1 through 4 in Figure 7.5: In stage (1), when the first olivine crystals have just formed, the right-hand side of the lever is almost nonexistent. In successive stages, the right-hand side of each lever becomes more prominent at the expense of the left-hand side of each lever; and at 1380°C the left side of the lever becomes virtually nonexistent. Since more solid must form with progressive cooling, it is apparent that the right-hand side of each lever must represent solid proportion.

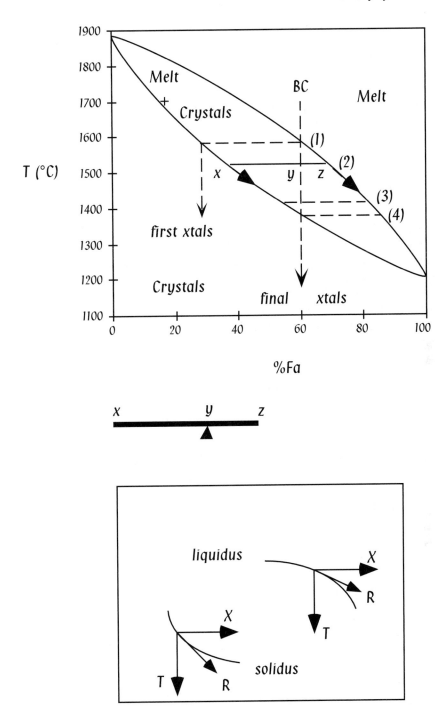

FIGURE 7.5   *Equilibrium crystallization of a liquid of composition "BC" in the system forsterite-fayalite at 1 atmosphere pressure.*

In our example, then, the percentages of the liquid and solid ($Fo_{64}Fa_{36}$) at 1525°C can be calculated from the following equations:

$$\%\text{liquid} = 100 \times (\text{distance } xy)/(\text{distance } xz)$$

$$\%\text{solid} = 100 \times (\text{distance } yz)/(\text{distance } xz)$$

When stage 4 (i.e., the bulk composition) is reached, all the remaining liquid is used up as this is the temperature where the last lever can be drawn. The final solid will have the same composition ($Fo_{40}Fa_{60}$) as the starting melt composition as long as equilibrium is maintained.

In *equilibrium melting*, the solid and liquid paths are exactly reversed: for example, an olivine of composition $Fo_{40}Fa_{60}$ will not melt until the temperature reaches approximately 1380°C (Figure 7.5). As the temperature is raised further, more melting will occur. At each temperature the proportion of liquid and solid phases can once again be calculated from the lever rule. Complete melting will take place at 1590°C, that is, the liquidus temperature of the $Fo_{40}Fa_{60}$ olivine.

Fractional Crystallization and Fractional Melting

In fractional crystallization crystals are removed from the melt as soon as they form; thus, a new bulk liquid composition is created, which is more fayalite-rich than the liquid that would form by equilibrium crystallization. Such perfect removal of crystals makes it theoretically possible for the liquid path to be extended to pure fayalite composition. Two types of solid paths are recognized in fractional crystallization: the *instantaneous solid path* (ISP)—one that tracks the compositions of solids that are in equilibrium with the liquid at each *instant* of crystallization, and the *total solid path* (TSP), which tracks the *integrated* compositions of all solid fractions. In fractional crystallization the ISP falls on the solidus and can reach pure fayalite composition, but the TSP follows a curved path below the solidus and reaches the bulk composition when the last drop of liquid runs out (Figure 7.6).

In fractional melting, the melting begins at the same temperature as in equilibrium melting; however, because each melt fraction is removed as soon as it forms, a new

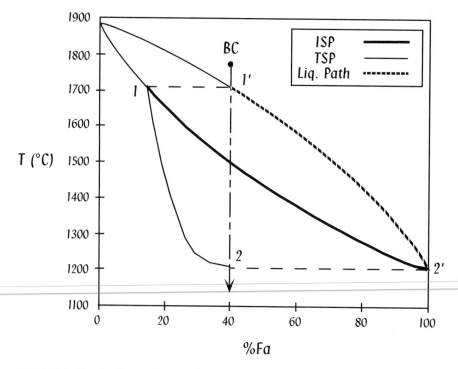

FIGURE 7.6 *Fractional crystallization of a liquid of composition* $Fo_{40}Fa_{60}$ *in the forsterite-fayalite system.*

bulk solid source composition is created. Such efficient melt extraction could lead to a final solid residue of pure forsterite. Further details on these types of phenomena are beyond the scope of this book, but note that in nature, perfect fractional crystallization or melting is highly unlikely. Imperfect fractional crystallization can produce zoned olivine crystals such as those found in many alkali-olivine basaltic lavas.

Melt Mixing

Consider a case where two melts, M_1 and M_2, each carrying olivine crystals (phenocrysts) of composition X_1 and X_2, respectively, are mixed in a 50:50 proportion such that the mixed melt falls on a line connecting M_1 and M_2 (Figure 7.7a). As Figure 7.7a shows, the new mixed melt is situated below the liquidus. Therefore, it will be forced to crystallize crystals of composition X_n and reach the liquidus at that temperature (approximately 1530°C). Creation of the new mixed melt makes X_1 and X_2 crystals unstable

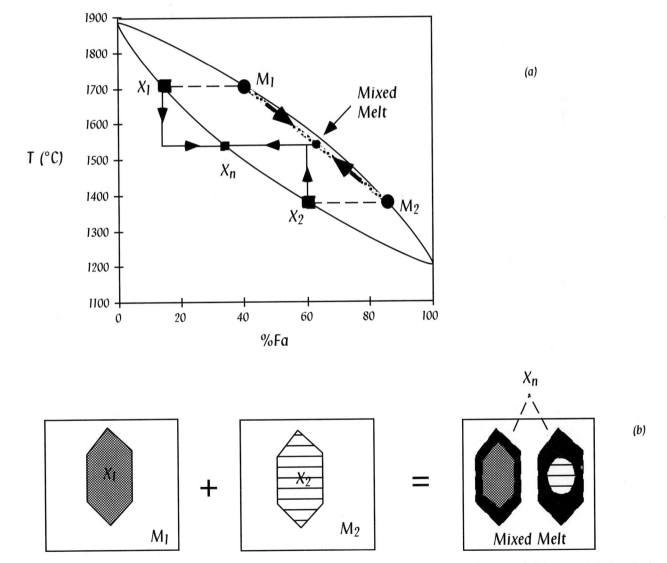

FIGURE 7.7   (a) Mixing of melts M_1 and M_2 in the system. (b) The lower panel shows textural development of olivine crystals in the mixed melt.

in the new melt, since the only solid phase that can coexist with this melt is of composition X_n. Crystals (X_1) in the hotter melt will be supercooled to 1530°C, and a euhedral rim of olivine composition X_n will grow on them (Figure 7.7b). On the other hand, olivine crystals (X_2) in the cooler melt M_2 will be superheated and mostly dissolved (resorbed) until 1530°C is reached, then a rim of X_n composition will grow on them (Figure 7.7b). *(Can you calculate what percentage of the crystals must dissolve?)* In igneous rocks, mixed and differently zoned (both normal and reversed zoning) olivines are sometimes seen and are a good indicator of magma mixing.

EUTECTIC SYSTEM: THE SYSTEM DIOPSIDE–ANORTHITE AT 1 ATMOSPHERE

Minerals in a binary system sometimes exhibit *eutectic* behavior, in which the compositions of the minerals remain fixed (i.e., do not vary like a solid solution) but any mixture of the two minerals begins to melt at a much lower temperature point, called the *eutectic point*. The temperature and composition of the eutectic point are referred to as the *eutectic temperature* and *eutectic composition*, respectively. One familiar (though incorrect in detail—we will ignore such details here) example of a binary eutectic system is the *diopside–anorthite* system at 1 atmosphere pressure (Figure 7.8). In this system pure diopside or anorthite crystals melt to a liquid of diopside or anorthite composition at their respective melting points. (This is not strictly true because diopside has a small melting range rather than a congruent melting point, Kushiro 1973). However, when the two are mixed, no matter what the proportion of diopside and anorthite crystals in the starting mixture, melting will always begin at the *eutectic point* "E". For example, the mixtures $An_{55}Di_{45}$, $An_{10}Di_{90}$, $An_{90}Di_{10}$ will all begin melting at E.

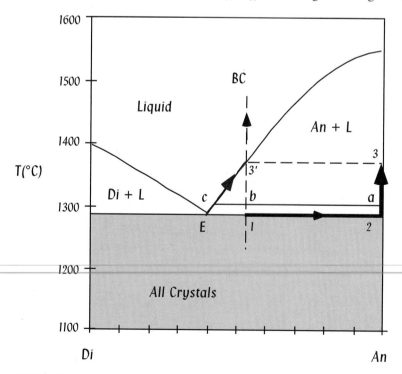

FIGURE 7.8   *The system diopside-anorthite at 1 atmosphere pressure with the equilibrium melting of a solid mix containing $Di_{48}An_{52}$.*

As pressure is increased under water-saturated conditions, the position of the eutectic point shifts to more anorthitic compositions (not shown here). That is, with increasing pressure more aluminous melts will be produced from any mixture of diopside and anorthite under hydrous conditions. Basaltic lavas erupting over subduction zones are often extremely aluminous. Based on the phase equilibrium just described, it can be concluded that water must play an important role in their genesis.

It is important to note that at any single fixed pressure, the phase rule becomes $f = c - \phi + 1$, because pressure is no longer a variable, and the "1" in right-hand side of the equation represents variations in temperature. At a eutectic point at a fixed pressure, f is zero (i.e., an invariant point) because $c = 2$ and $\phi = 3$ (two crystalline phases and a melt phase). When both pressure and temperature variation are considered, the phase rule becomes $f = c - \phi + 2$, and the eutectic point at a single pressure now becomes part of a univariant line along which all three phases can occur in P,T space. (This is not shown here, but it should look like the solidus in Figure 7.3).

ADVANCED READING: CALCULATION OF EUTECTIC SYSTEMS USING THERMODYNAMIC DATA

It is possible to calculate a binary eutectic diagram from thermodynamic data using what is called a *cryoscopic equation*. As an example, let us calculate the diopside-liquidus in the Di–An system. At any temperature T on the liquidus, the liquid and solid diopside are in equilibrium. In the following example, the chemical potential of diopside in the solid is equal to that of diopside in the liquid:

$$\mu_{Di}^{s} = \mu_{Di}^{L}$$
$$\mu_{Di}^{s} = \mu_{Di}^{o,s} + RT \ln a_{Di}^{S}$$

and

$$\mu_{Di}^{L} = \mu_{Di}^{o,L} + RT \ln a_{Di}^{L}$$

where μ_{Di}^{s} and μ_{Di}^{L} are the chemical potentials of diopside in solid and liquid phases, and a_{Di}^{S} and a_{Di}^{L} are the actvities of diopside in solid and liquid, respectively. $\mu_{Di}^{o,s}$ and $\mu_{Di}^{o,L}$ are the chemical potentials of pure solid and molten diopside, respectively. R is the universal gas constant.

At T then,

$$\mu_{Di}^{o,s} + RT \ln a_{Di}^{S} = \mu_{Di}^{o,L} + RT \ln a_{Di}^{L}$$

or

$$\mu_{Di}^{o,s} - \mu_{Di}^{o,L} = RT \ln a_{Di}^{L} - RT \ln a_{Di}^{S}$$

The left side is the ΔG of melting of pure diopside; therefore, the above equation can be rewritten as follows:

$$-\Delta G^{o} = RT \ln a_{Di}^{L} - RT \ln a_{Di}^{S}$$
$$-\Delta H^{o} + T \Delta S^{o} = RT \ln a_{Di}^{L} - RT \ln a_{Di}^{S}$$

ΔH^{o} and ΔS^{o} are the enthalpy and entropy of fusion of pure diopside.

$$-\Delta H^{o} + T \Delta S^{o} = RT \ln a_{Di}^{L}$$

or

$$-\Delta H^{o}/RT + \Delta S^{o}/R = \ln a_{Di}^{L}$$

Above, note that we ignored the second term (i.e., RT ln $a_{Di}{}^{S}$) on the right-hand side of the equation since the activity of pure diopside crystal is 1. In the melting of pure diopside, $\Delta S^o = \Delta H^o / T^m$, where T^m is the melting point of pure diopside. Therefore, the previous equation becomes

$$-\Delta H^o / RT + \Delta H^o / RT^m = \ln a_{Di}{}^{L}$$

or

$$\ln a_{Di}{}^{L} = \Delta H^o / R \left[1/T^m - 1/T \right]$$

$a_{Di}{}^{L}$ is generally written as $X_{Di}{}^{L} \cdot \gamma_{Di}{}^{L}$, where X is the mole fraction and γ is a parameter known as the *activity coefficient*, which is 1 when the solution behaves ideally. In the present example, the melt's activity coefficient is assumed to be 1, thus, activity = mole fraction. Therefore, the equation can be written as

$$\ln X_{Di}{}^{L} = \Delta H^o / R \left[1/T^m - 1/T \right].$$

This equation is known as the *cryoscopic equation*. Since ΔH^o, T^m, and R are known, T can be determined at any given $X_{Di}{}^{L}$ or vice versa. From several different T and $X_{Di}{}^{L}$, the liquidus curve may be drawn.

Equilibrium Crystallization and Melting

During *equilibrium melting*, melt and crystalline phases remain in contact at all times; and (depending upon the bulk composition of the starting mix) with greater extents of melting, the liquid path climbs up the diopside or anorthite liquidus. This can best be illustrated with the example of melting of a mixture of composition $An_{55}Di_{45}$. Note that this composition falls on the anorthite side of the eutectic point E. This starting mix will begin melting at E. The melt will stay at E until all crystals of diopside have been melted. When all diopside crystals are dissolved in the melt phase, the solid path would reach the anorthite composition axis (the solid line in Figure 7.8). Since no diopside crystals remain, the melt will be free to move up the liquidus of anorthite from E to 3′ and the solid path will move up from 1 to 3. At any given temperature one can determine the proportions of melt and crystals by applying the lever rule. For example, when the solid path reaches "*a*," the lever would be *a-b-c*, and

$$\%\text{anorthite crystals} = 100 \times \text{distance } ab / (\text{distance } ac)$$

Once again, successive levers can be constructed to see how the proportion of melt increases as a function of temperature or degree of melting. It can be seen that when the liquid reaches 3′ (i.e., bulk composition), all the crystals would have melted.

Equilibrium crystallization can be thought of as exactly the reverse of equilibrium melting. Liquid and solid paths follow the same paths as in equilibrium melting except that the directions (arrows in Figure 7.8) are reversed (Figure 7.9). Consider the liquid "BC" in Figure 7.9. This liquid will crystallize anorthite when it reaches its liquidus at *t*. Continuous cooling and crystallization will result in the movement of the residual liquid to E. At this point diopside crystals will form, and eventually the liquid will completely crystallize. At any point on the liquid path a tie line may be constructed to calculate proportions of liquid and solid phases. For example at *c*, the tie line is *a-b-c* such that

$$\%\text{liquid} = 100 \times \text{distance } ba / \text{distance } ca$$

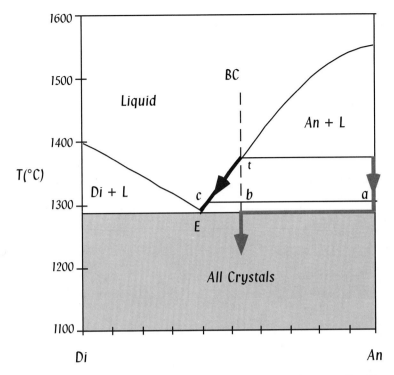

FIGURE 7.9   *The equilibrium crystallization of liquid "BC" in the system diopside-anorthite.*

Crystallization at the eutectic point often produces a blebby intergrowth texture (Figure 7.10), called *eutectoid* or *granophyric texture*. This texture is most common in a type of igneous rock called *granophyre*. In these rocks such texture originates from eutectic crystallization of quartz and alkali feldspar.

Fractional Crystallization

In binary eutectic systems, liquid and solid paths for any particular liquid composition are the same for fractional and equilibrium crystallization; however, a much greater amount of residual liquid crystallizes at the eutectic point in the case of fractional crystallization.

Fractional Melting

The only thing that fractional melting has in common with equilibrium melting is that the melting begins and continues at the eutectic point until one of the solid phases is exhausted from the residue. For example, a "rock" composed of $An_{55}Di_{45}$ will begin to melt at the eutectic point. Through progressive melting and extremely rapid extraction of the melt from the source (which is fractional melting), the solid residue will ultimately be composed only of anorthite. That is, the solid path will reach the anorthite axis. Because melts were quickly extracted as soon as they formed, at this point no melt coexists with the residue made of pure anorthite. Naturally, this pure anorthite residue will not melt until it is heated to the melting point of anorthite! Therefore, in fractional melting, the solid path is still a continuous path whereas the liquid composition 'hops' from the eutectic point to the congruent melting point of anorthite (Figure 7.11).

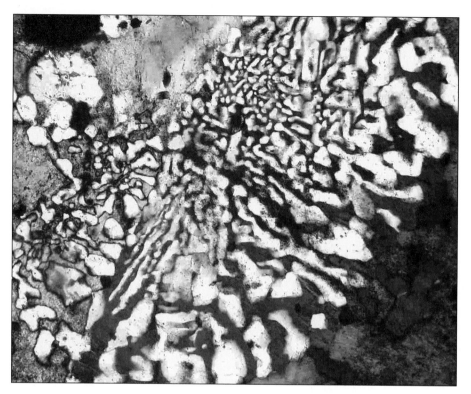

FIGURE 7.10 *A photomicrograph showing granophyric texture (also called eutectic or eutectoid texture) in a granophyre.* (Courtesy of Gautam Sen)

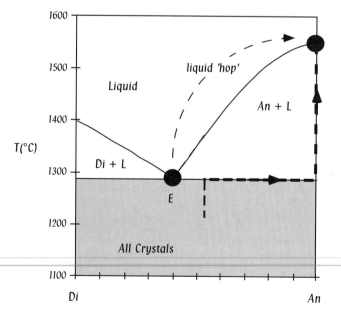

FIGURE 7.11 *The fractional melting of a solid mix of composition* $Di_{48}An_{52}$.

Some Additional Comments

Eutectic melting has always been a subject of great interest to igneous petrologists. In the 1950's and 1960's, before significant modern analytical tools were developed and before a great many details of compositional variations in basalts were discovered, voluminous basalt lavas on continents (e.g., the Deccan Trap flood basalts of India) and on the ocean floor (e.g., the Ontong–Java plateau in the Pacific) were thought to be compositionally uniform. Because eutectic melting provides a way to generate compositionally uniform magmas from a diverse variety of source rocks, many authors proposed that these magmas were generated in the upper mantle by an eutectic-type melting process.

In the late 1980's and early 1990's, some authors noted that the rock *lherzolite*, which is the most abundant rock in the earth's upper mantle, melts directly to its own composition at a high pressure, much like an eutectic mixture melts to its own composition (i.e., eutectic composition). Based on this eutectic or near-eutectic behavior of the mantle peridotite, they hypothesized that the earth's upper mantle may have once been molten like a giant *magma ocean* that formed by eutectic melting (Figure Bx 7.2). Examples like these show how simple observations on phase diagrams can lead to the interpretation of global scale phenomena!

DOUBLE EUTECTIC SYSTEMS: THE SYSTEM NEPHELINE–SILICA AT 1 ATMOSPHERE PRESSURE

In the system nepheline–silica, *albite* (which is the sodic end member of the plagioclase solid solution series) forms an intermediate compound with its independent melting point (Figure 7.12). Formation of albite creates two separate eutectic points (E1 and E2) as in the following equations and in the nepheline–silica diagram:

$$\text{albite (ab)} + \text{nepheline}_{ss}\ (\text{ne}) = \text{L (E1)},$$

and

$$\text{ab} + \text{tridymite (tr)} = \text{L (E2)}.$$

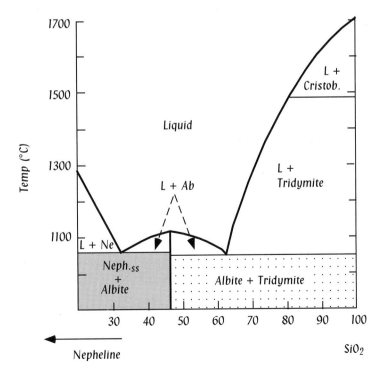

FIGURE 7.12   *Only a part of the system nepheline-silica at 1 atmosphere pressure is shown here.* Nepheline actually forms solid solution with carnegeite.

Crystallization of melts with lower silica than albite will drive the final liquid to E1 where nepheline and albite can coprecipitate; whereas those with higher silica than albite will end up at E2 where tridymite and albite can crystallize simultaneously from the liquid. Thus, like a watershed or a mountain divide that forces streams on one side to flow one way and streams on the other side to flow in an opposite direction, the albite "line" acts as a compositional barrier or a "thermal divide" between liquids that crystallize "free silica" (i.e., a silica polymorph) and those that crystallize nepheline.

This diagram explains why nepheline and a free silica phase are never found together in the same rock, an observation that was made by S. J. Shand from field studies of rocks long before this phase diagram was determined experimentally. In fact, based on this finding, Daly classified magmas (molten rocks) into two genetic classes: alkalic magmas (those that can potentially crystallize nepheline) and subalkalic magmas (those that can potentially crystallize free silica polymorph). Not surprisingly, such a genetic classification is still valid today!

INCONGRUENT MELTING BEHAVIOR AND PERITECTIC SYSTEMS: THE SYSTEM FORSTERITE—SILICA AT 1 ATMOSPHERE PRESSURE

In contrast to the systems we have considered so far, some binary systems contain an additional mineral that cannot melt to its own composition (i.e., congruent melting) but breaks down to another mineral and a liquid when heated to a certain temperature. Such melting behavior is called *incongruent melting*. In the system forsterite–silica at atmospheric pressure, the mineral enstatite exhibits such incongruent melting behavior as in the following equation:

$$enstatite = forsterite + liquid.$$

In Figure 7.13, when enstatite crystals are heated to 1557°C (T_p), they will melt to forsterite crystals and a liquid whose composition will be at "P", called the *peritectic point*. The temperature T_p is the *incongruent melting point* of enstatite.

Equilibrium Crystallization and Melting

In the system forsterite–silica, enstatite and a silica polymorph (cristobalite) exhibit eutectic behavior with a eutectic point located at E in Figure 7.14. Any crystalline starting mix with a bulk composition falling on the right-hand side of En will first melt at E. Also, liquid compositions falling on the right-hand side of the dotted line CB-En (e.g., the bulk composition "QTh") will, upon crystallization, always end up at the eutectic point E. We will not consider the eutectic part of the diagram any further and instead will focus on the peritectic point P. The three starting liquid compositions, OTh, Th, and QTh, are selected as analogues of the three types of common basalt magmas: olivine–tholeiite, tholeiite, and quartz–tholeiite, respectively in the following discussion.

Liquid OTh:

A liquid with a bulk composition of OTh will start crystallizing forsterite crystals at approximately 1710°C (i.e., its liquidus temperature), and progressive cooling will drive the liquid composition to migrate down the liquidus toward P (Figure 7.14). Once again, at any given temperature the lever rule may be used to calculate the proportions of crystals and liquid. For example, when the liquid composition reaches "n",

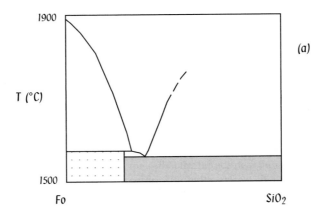

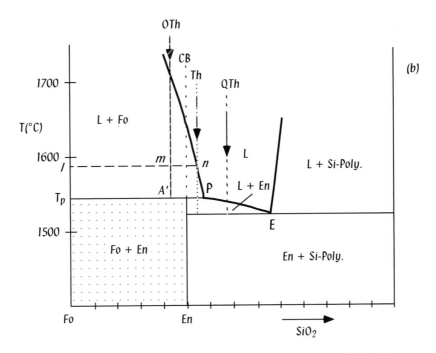

FIGURE 7.13   (a) The system forsterite-silica at 1 atmosphere pressure. (b) P and E are peritectic and eutectic points, respectively.

$$\text{liquid}\% = 100 \times \text{distance } lm/\text{distance } ln$$

$$\text{forsterite}\% = 100 \times \text{distance } mn/\text{distance } ln.$$

With further cooling the liquid will reach P, and at this point forsterite crystals will react with the liquid P and produce crystals of enstatite owing to the following reaction:

$$\text{forsterite} + \text{liquid} = \text{enstatite}$$

Texturally, the forsterite crystals will appear resorbed with rounded grain boundaries and sometimes enveloped by crystals of enstatite. This reaction will continue until all of the liquid is used up, and that will happen when the solid path, which is composed of mixtures of forsterite and enstatite crystals, reaches the bulk starting composition at A′.

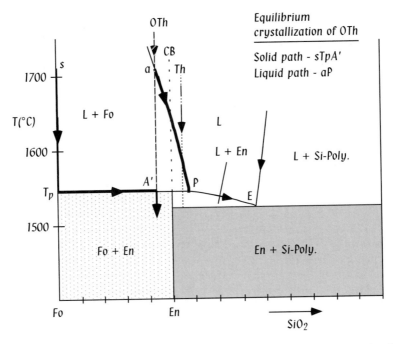

FIGURE 7.14   *The equilibrium crystallization of a liquid "OTh" in the system forsterite-silica.*

Note that as long as equilibrium is maintained throughout, the liquid *cannot* move down to E because that would mean that *all* forsterite crystals would be dissolved, which cannot happen since the bulk composition constraint forces the final crystalline assemblage to be composed of forsterite and enstatite. Table 7.2 shows the crystallization history of liquid OTh.

TABLE 7.2   *Equilibrium crystallization history of starting composition OTh (Figure 7.13(b)).*

Temperature (°C)	Phase(s)
>1710	Liquid
$1710 - 1557(T_p)$	Liquid + Forsterite
1557	Liq. + Fo = Enstatite
1557 − 1543	Liq. + Enstatite
1543	Liq. + En + SiO_2-polymorph
<1543	Enstatite + SiO_2-polymorph

Liquid Th:

Consider another bulk melt composition "Th" that falls between the P and CB-En lines (Figure 7.15). This melt will begin crystallizing at approximately 1580°C, its liquidus temperature, and will move down to P with a lowering of the temperature. At P, the reaction Fo + L = En will ensue, and more and more enstatite crystals will form at the expense of the reactants forsterite and the liquid. Note that this bulk composition falls in an area where the solidus assemblage must be composed of enstatite and a silica polymorph. Therefore, in this case all of the forsterite crystals will be used up while the temperature remains fixed at the pertitectic point, and, when all of the

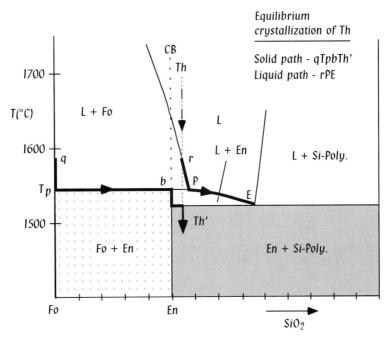

FIGURE 7.15   *The equilibrium crystallization of "Th" in the system forsterite-silica.*

forsterite is lost, the liquid will be free to move down to E while crystallizing enstatite only. At E, both enstatite and a silica polymorph will crystallize.

Equilibrium melting paths for different bulk compositions in this system are exactly the reverse of equilibrium crystallization paths. Any solid assemblage composed of enstatite and a silica polymorph (i.e., the bulk composition to the right side of CB-En line) will generate first melts at E; and those with forsterite and enstatite crystals will generate the first melt at P.

> [**Question:**   Can you tell where melting would begin and end for a pure "enstatite rock"?]

Fractional Crystallization and Fractional Melting

In perfect fractional crystallization, the bulk composition does not impose a control on the liquid path because a new instantaneous "bulk" composition is produced every time the crystals are separated from the liquid as soon as they form. Therefore, for all bulk melt compositions, including OTh, the final liquid will end up at E, and the final solid assemblage will have a eutectic composition.

As we noted earlier in the case of eutectic systems, fractional melting produces continuous solid residue paths but distinct liquid compositions or "liquid hops". In the present case, fractional melting behavior may be understood with reference to the bulk solid composition Th (Figure 7.16). Th will begin to melt at E (1543°C) and will continue until the residue becomes pure enstatite. Once the residue becomes pure enstatite, further melting will not occur until the temperature is raised to T_p (1557°C). At T_p, the forsterite crystals and a melt of peritectic composition P are produced as enstatite crystals melt. Melting will stop once all enstatite crystals are melted and the residue becomes pure forsterite. Melting will resume only when the temperature is raised to the melting point of forsterite (Figure 7.16).

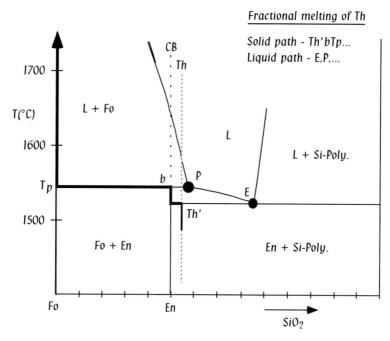

FIGURE 7.16   *The fractional melting of a solid of composition "Th" in the system forsterite-silica.*

Melt Mixing

Mixing of melts, which is of paramount importance in igneous processes, may produce some interesting results in this system. Vigorous mixing between olivine-saturated melts, such as P_i in Figure 7.17, and the ol + en-saturated melt P, will always result in a new bulk melt composition that will be on the line P_iP and be situated within fo + L

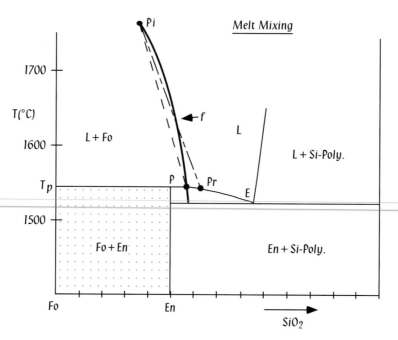

FIGURE 7.17   *Mixing of melts in the system forsterite-silica.*

liquidus area. Such a melt will be supersaturated with olivine since its composition will not be on the liquidus but below it; and enstatite crystals in melt P will either dissolve or be completely surrounded (i.e., isolated from melt) by forsterite crystals (a disequilibrium situation). The new melt will move from its location on the mixing line toward the fo + L liquidus curve at a constant temperature while crystallizing forsterite crystals. With further cooling the new mixed melt will then move down the liquidus.

A very different result may be obtained if mixing occurs between an olivine- and an enstatite-saturated melt. For example, an interesting phenomenon occurs when the new mixed melt's composition falls on the line between f and P_r. Such a mixed melt will be superliquidus, that is, it cannot contain any crystals. Small amounts of the crystals of forsterite and enstatite in the parent magmas P and P_i will dissolve in such a mixed melt as will any physical evidence of mixing of the two magmas because the resultant magma will be fully molten. In igneous petrology today, an assumption is commonly made that phenocryst-free melts may approximate primary magmas generated directly from the source rock. The point of the present exercise is to draw attention to the fact that even aphyric magmas themselves may be derived by the mixing of two very different magma types but may not be recognized as such.

[**Question:** Can you determine what will happen if Pi and Pr carried as much as 50% fo and en crystals, respectively, and they mixed in a 50:50 proportion?]

The system Fo–Silica is highly relevant to basalt and andesite magma generation and crystallization. Some preliminary aspects of melting and crystallization in this system at high pressure conditions (both "dry" and "wet") are discussed in Box 7.2.

BINARY SYSTEMS WITH LIMITED SOLID SOLUTION AND EXSOLUTION: THE JOIN ALBITE–ORTHOCLASE

Some minerals show only limited solid solution between their end members (e.g., alkali feldspars and pyroxenes). Figure 7.18 shows the phase relationships of the "system" albite–orthoclase at a number of pressures—from 1 atmosphere to 5 kbar

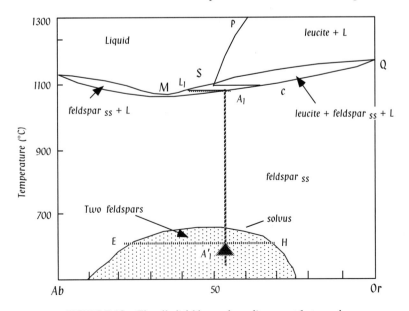

FIGURE 7.18   *The alkali feldspar phase diagram at 1 atmosphere.*

pressure under water-saturated conditions. (In reality, the albite–orthoclase is not a true system but a "join" in the ternary system albite–orthoclase–silica, because liquids on this compositional join leave the join upon crystallization or melting and enter the albite–orthoclase–silica triangle.) For our purpose here, however, we will assume that the albite–orthoclase join behaves like a binary system, and all phase compositions are defined by the two components at all times.

Let us first consider the albite–orthoclase "system" at 1 atmosphere pressure. For the time being, we will ignore the complications caused by the appearance of the mineral leucite and focus on the fields marked as Feldspar$_{ss}$ ('ss' is the abbreviation for solid solution) + L (liquid). The liquidus and solidus touch each other in the middle of the diagram producing what is called a "minimum" (M). Point M is not a binary eutectic because only a single feldspar$_{ss}$ can crystallize at this point from a corresponding liquid. We have noted earlier that exactly *two* solid phases crystallize at a binary eutectic point at constant pressure. Note that the two loops in Figure 7.18 is such that liquids more enriched in the "Or" component than M would crystallize a last feldspar$_{ss}$ that is more Or-rich than M. The same condition will apply to liquids that are more Ab-rich than M.

An interesting complication is the existence of a *miscibility gap* or a *solvus* below the solidus at 1 atm. Within this gap a single feldspar$_{ss}$ cannot be stable and must decompose into two feldspar$_{ss}$. For example, a feldspar$_{ss}$ crystal A$_1$ (Or$_{53}$Ab$_{47}$) crystallizing at approximately 1080°C from a liquid L$_1$ will begin decomposing to two feldspar$_{ss}$ as it cools extremely slowly (i.e., maintaining chemical equilibrium) through the miscibility gap (around 680°C). One of these will be "exsolved" as microscopic blebs of a separate feldspar phase within the other (host) feldspar phase (Figure 7.18). This phenomenon is called *exsolution*. In our example, when the feldspar$_{ss}$ A$_1$ is cooled to 600°C blebs of a feldspar$_{ss}$ of composition E (around Ab$_{78}$Or$_{22}$) will be exsolved from the host feldspar$_{ss}$ H (around Or$_{68}$Ab$_{32}$). The proportion of the exsolved blebs at this temperature may be easily calculated using the lever rule since the bulk feldspar composition is fixed at A$_1'$:

$$\%\text{Exsolved phase E} = 100 \times \text{A}_1'\text{H}/\text{EH, and}$$

$$\%\text{Host phase H} = 100 - \text{E (since it is a binary solution)}$$

An increase of pressure under water-saturated conditions depresses the solidus to lower temperature and elevates the solvus to a higher temperature. The diagram at 5 kbar water pressure (Figure 7.19) shows that the solvus "penetrates" the solidus and liquidus so that the minimum is now replaced by a eutectic point (E). A liquid of eutectic composition will crystallize two feldspar$_{ss}$ crystals—S$_1$(Ab$_{80}$Or$_{20}$) and S$_2$(Or$_{54}$Ab$_{46}$). This is in contrast to the single feldspar$_{ss}$ crystal that crystallizes from any liquid in the Ab-Or system at low pressure. The existence and importance of one versus two feldspars in igneous rocks of granitic composition have been recognized for a long time. Large, tens-of-kilometer–sized, plutons of granite (called batholiths) often occur in the core regions of folded mountain belts. They have been "traditionally" classified into two groups, *hypersolvus* and *subsolvus granites*, depending on whether they contain a single or two feldspars, respectively. In the context of the solvus versus pressure effects described above, it is clear that two-feldspar granites must solidify at a deeper level in the crust than one-feldspar granites.

At 5-kbar partial water pressure (i.e., pH$_2$O = 5 kbar) the two feldspars (Ab$_{80}$ and Ab$_{46}$) crystallizing from the eutectic liquid E will both exsolve as the temperature is lowered slowly enough so that equilibrium is maintained. By the time these

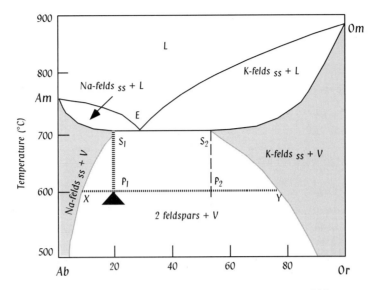

FIGURE 7.19   *The alkali feldspar phase diagram at $p_{H_2O} = 5$ kbar.*

crystals reach a temperature of 600°C, one of these feldspars would have reached X and the other Y. Both X and Y crystals would contain exsolved Y and X, respectively, in proportions dictated by the lever rule:

In crystals of X composition, %Y (exsolved) $= 100 \times XP_1/XY$, and

in crystals of Y composition, %X (exsolved) $= 100 \times YP_2/XY$

Based on the albite-orthoclase diagram one can obtain depth (pressure) and temperature of melt crystallization temperature and subsolidus temperature from coexisting exsolved and host phase compositions in alkali feldspars.

TERNARY SYSTEMS

The addition of a third component requires a different type of representation of information on variations in temperature and in the proportions of three components involved. Triangular graph papers (Figure 7.20), available in university bookstores, are used to portray all such relationships. A triangular graph paper contains grid lines parallel to each of its three sides. The three components are placed at the apices of such a graph. Each of the three sides of the triangle is binary, and proportions of the two components along each of these lines can simply be plotted as before. Any composition, rock or liquid, composed only of the three components must have certain percentages of each component, and the total of these must add up to 100%. For example, consider the hypothetical system A-B-C (Figure 7.20). A liquid containing only A and B will plot on the A-B boundary. Another liquid composed of A, B, and C must plot inside the triangle. Let us assume that our liquid has 30% A, 20% B, and 50% C. For simplification, we can write its composition with subscripts (as before)—$A_{30}B_{20}C_{50}$. It is clear that contents (percentages) of two of the components automatically define the content of the third component in the liquid. This liquid may be plotted by locating any two of the grid lines corresponding to the percentages of two of the components; and the intersection of the two lines

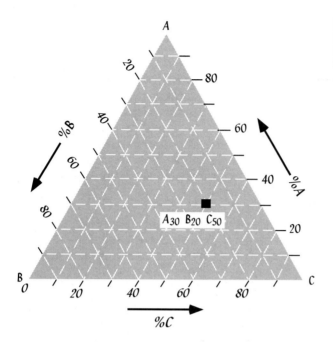

FIGURE 7.20 *An example of a ternary graph showing how the composition of a solid mix or liquid is plotted.*

gives the location of this liquid composition (Figure 7.20). Since the apex A of the triangle is 100% A, A must be 0% at the base BC. Thus, % A must gradually increase along lines parallel to BC, going from 0% at BC to 100% at A. Using the same logic, lines for the contents of B and C can be constructed on a triangular graph paper.

Consideration of temperature requires the addition of a third dimension, as shown in Figure 7.21. The use of such 3-D diagrams is clearly cumbersome in quan-

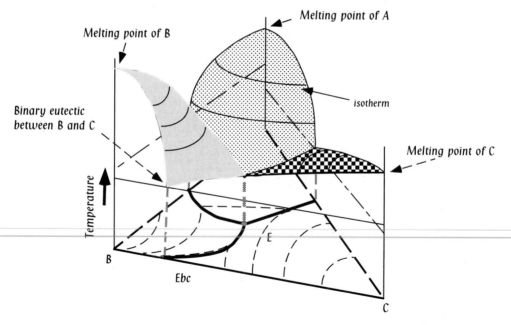

FIGURE 7.21 *The hypothetical system A-B-C at 1 atmosphere in 3-D perspective, in which the compositional base (A-B-C triangle) and the vertical temperature axis are shown.*

tifying phase relationships. Therefore, such liquidus surfaces are "projected" onto the triangular base ABC from the temperature axis, and temperatures on each surface are shown by *isotherms*. An isotherm is a constant-temperature curve on a diagram. It may be thought of as analogous to a topographic contour, which represents a constant altitude contour in a topographic map. Similar to topographic contours, the spacing between adjacent isotherms indicates the steepness of the slopes of the liquidus surfaces. For example, in Figure 7.21, B's liquidus surface slopes most steeply because the isotherms are most closely spaced in the liquidus field $B + L$.

One must also clearly state whether weight or molar compositions of phases are plotted in the triangular diagram for reasons stated earlier. Note that in all cases discussed later, the phase diagrams are weight% diagrams.

Ternary systems (i.e., systems with 3 components) can be of many different types—from eutectic to peritectic and solid solution. Here we consider a few important examples of ternary phase relationships of mineral-melt systems.

TERNARY EUTECTIC SYSTEMS

In an effort to illustrate ternary eutectic behavior, we consider a hypothetical example of the hypothetical system A-B-C (Figures 7.22–7.25). Note that each side of the triangle ABC is a binary eutectic: E_{bc} is the eutectic point between B and C, E_{ca} is the eutectic point between C and A, and E_{ab} is the binary eutectic point between A and B. Figure 7.21 shows that when all the binary sides of ABC are considered, in three dimensions the *liquidus surfaces* for each phase would appear like a mountainside, and a junction between two adjacent surfaces would be a like a valley. Such valleys are called *cotectic curves* and a melt on any such curve in a ternary system will precipitate two crystalline phases while moving "downslope" toward the lowest temperature point in the diagram, that is, the *ternary eutectic point* "E" (Figure 7.22).

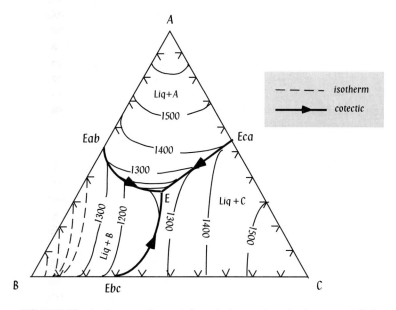

FIGURE 7.22 *Isotherms and cotectic boundaries are shown in the system A-B-C.*

Equilibrium Melting and Crystallization

Consider the equilibrium melting of rock X ($A_7B_{30}C_{63}$; Figure 7.23). Because the ternary eutectic point E is the lowest temperature point in this diagram, the first melt must be of eutectic composition that will form at the eutectic temperature (around 1000°C). The formation of such a melt will drive the remaining solid composition directly away from the eutectic point to Y. All crystals of A will disappear into the melt when the solid residue reaches Y. Because A crystals are no longer present, the liquid will be free to move up the cotectic E → E_{bc} while the solid residue moves toward C from Y. At any instant, a tie line, such as any of the dashed lines shown in Figure 7.23, may be constructed and the lever rule applied in order to calculate the proportions of liquid and crystals. For example, consider the tie line or lever 1-X-C, in which X is the bulk composition (the "fulcrum" of the "lever"). As before, the liquid side of the lever is on the other side of the fulcrum X, and therefore,

$$\%\text{liquid} = 100 \times \left[XC/(XC + 1X) \right]$$

Note that when the liquid reaches 1, all of the B crystals would be dissolved. With further melting, the liquid will move toward X; and once it reaches X the rock would be totally molten because that is the bulk composition of the rock!

The equilibrium crystallization path is exactly the opposite of the melting path (not shown here). If X were a liquid, it would begin to crystallize crystals of C at around 1400°C because it sits on the 1400°C isotherm on C's liquidus surface. It would continue to crystallize C until the cotectic E_{bc} − E is reached at point 1. With further crystallization the residual liquid will reach E while the solid path goes from C to Y to X.

Fractional Melting

Perusal of phase relations in binary systems showed that during fractional melting the residual solid composition changes continuously, whereas the liquid composition path is a discontinuous one—"hopping" from invariant point to invariant point. This behavior persists in ternary systems as well, as illustrated by the example below.

Consider the example of a "rock" X in Figure 7.24. It will begin to melt at the ternary eutectic point E (approximately 1000°C), and the solid residue will move toward Y. Because fractional melting would have extracted all the melts, the residual solid at Y will simply be composed of B and C. Melt being absent, the solid Y will no longer behave like a ternary but as a binary BC; and melting will resume only when the temperature is raised to the binary eutectic temperature of E_{bc} (around 1170°C). Prompt removal of the melt of composition E_{bc} will drive the composition of the solid residue toward C. When the solid residue reaches C, it will behave like a unary system being composed entirely of C crystals; thus, it will not melt again until the melting temperature of C (around 1670°C) is reached. Once again the TSP (BC → Y → C) is continuous, and the liquid path $\left(E \ldots E_{bc} \ldots C \right)$ is discontinuous in fractional melting.

Fractional Crystallization

Consider a liquid of bulk composition X (Figure 7.25). As with equilibrium crystallization, this liquid will begin to crystallize C at approximately 1400°C, which is the liquidus temperature of X. Prompt removal of C crystals from the melt, as required

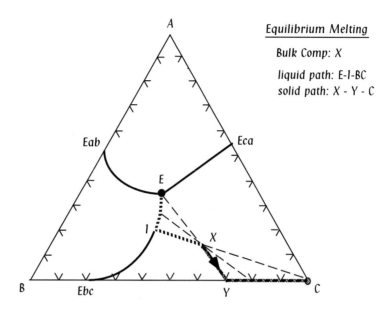

Equilibrium Melting

Bulk Comp: X

liquid path: E-I-BC
solid path: X - Y - C

FIGURE 7.23   *The equilibrium melting of a solid mix "X" in the system A-B-C.*

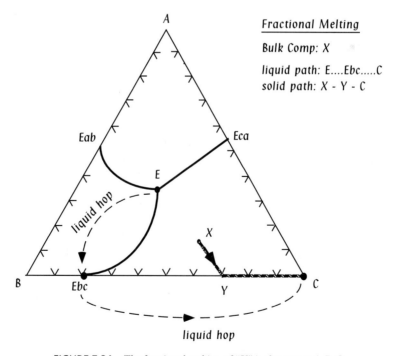

Fractional Melting

Bulk Comp: X

liquid path: E....Ebc.....C
solid path: X - Y - C

FIGURE 7.24   *The fractional melting of "X" in the system A-B-C.*

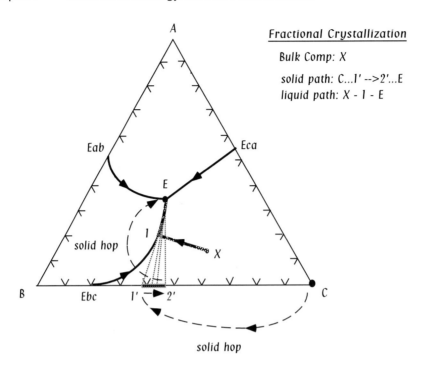

FIGURE 7.25 *The fractional crystallization of "X" in the system A-B-C.*

by fractional crystallization, will drive the residual liquid composition directly away from C toward point 1, and at any given instant the liquid composition is a new bulk composition. Once the liquid reaches 1, it will start crystallizing B and C. The instantaneous mixture of B and C crystals will have its composition at the point where, 1–1′, which is a tangent to the cotectic E-E_{bc} at 1, meets the binary side BC of the triangle. The proportions of B and C in the crystalline precipitate can be determined using the lever rule: the B proportion in solid is equal to C1′/BC, and the C proportion in solid is equal to B1′/BC.

The liquid at 1 will continue to move down the cotectic toward E; and a series of tangents from the liquid on the cotectic to the base BC may be constructed in order to track the TSP (1′ → 2′). When the liquid reaches E, the TSP would have reached 2′. At E, a crystalline assemblage containing eutectic proportions of A, B, and C will crystallize directly from the eutectic liquid. Therefore, in fractional crystallization, the liquid path (X → 1 → E) is once again continuous, and the solid path (C … 1′–2′ … E) is discontinuous.

The "System" Anorthite-Forsterite-Diopside as an Example

The system anorthite-forsterite-diopside at atmospheric pressure shows ternary eutectic behavior between the three crystalline phases—anorthite, forsterite, and diopside (Figure 7.26). However, the appearance of a spinel liquidus field introduces some complexities because the crystallization of spinel, whose composition does not plot within this triangle, drives the residual liquid composition out of the ternary system. Thus, any liquid plotting in the shaded region in the figure will eventually crystallize spinel, and the liquid and solid paths cannot be fully treated within the confines of this ternary "system."

Liquid and solid paths during equilibrium and fractional crystallization of any melt within the unshaded area can be easily dealt with, as shown with the examples

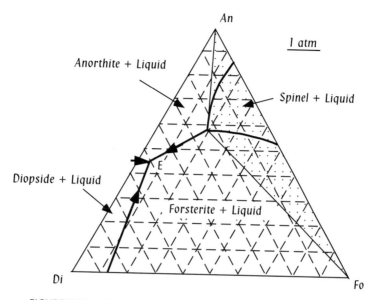

FIGURE 7.26   *The system anorthite-forsterite-diopside at 1 atmosphere.*

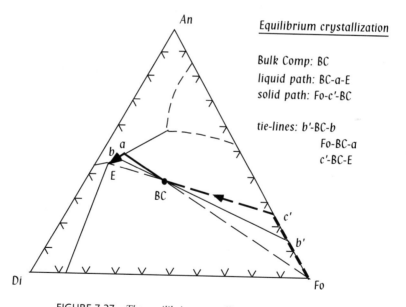

FIGURE 7.27   *The equilibrium crystallization of a liquid "BC".*

in Figures 7.27 and 7.28. For example, equilibrium crystallization of a liquid BC in Figure 7.27 will generate a liquid path BC → a → E. At any given point, a lever rule may be constructed to extract information on the percentages of the melt and solid phases in equilibrium. An example of this is shown by the tie line $b \ldots BC \ldots b'$:

$$\text{solid\%} = 100 \times b\text{BC}/bb'$$

$$\text{liquid\%} = 100 - \text{solid\%}.$$

The proportion of An and Fo in the solid at b' is of course given by its position on the An-Fo line.

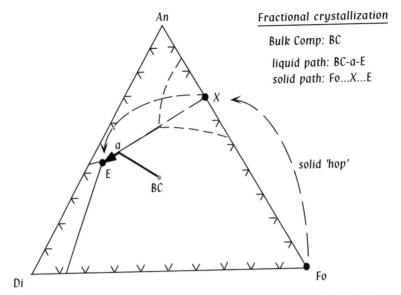

FIGURE 7.28 *The fractional crystallization of liquid X: liquid and solid paths.*

Equilibrium melting is also easy to deal with in the unshaded area of the diagram because the liquid and solid paths in such a case will be exactly opposite of those for equilibrium crystallization.

The fractional crystallization of BC is portrayed in Figure 7.28 and described as follows. Crystallization of the melt BC will drive the residual liquid directly away from the Fo corner (the crystallizing phase) toward *a*. At this point a tangent drawn to the cotectic curve and on to the base An-Fo will give the solid extract composition (X in Figure 7.28). Because the cotectic appears almost as a straight line, the solid extract composition is unlikely to change too much as the liquid composition evolves toward E. Once the melt reaches E, all three crystalline phases will cocrystallize in ternary eutectic proportions. The solid "hops" for BC composition are shown in Figure 7.28.

Fractional melting of a rock of composition BC gets more complicated because of the spinel field. Initial melting of BC will occur at E (Figure 7.29). Extraction of the eutectic melt will continue until the solid residue reaches the An-Fo boundary. Beyond this stage, how the melting path will proceed cannot be depicted from this diagram because of complications involving the spinel liquidus field and the absence of a binary eutectic between anorthite and forsterite.

TERNARY SYSTEM WITH SOLID SOLUTION: THE SYSTEM DIOPSIDE–ALBITE–ANORTHITE AT 1 ATM PRESSURE

Basalts and gabbros are a group of igneous rocks that forms the oceanic crust. They are also a major component of the continental crust. The two abundant minerals in these rocks are clinopyroxene and plagioclase. The system diopside–albite–anorthite at 1 atmosphere pressure (Figure 7.30) serves as an analog of natural basalt magmas because the crystallization behavior of clinopyroxene (diopside) and plagioclase$_{ss}$ from basaltic magmas may be illustrated in some detail with this system. The three binaries that comprise the boundaries of this system are a solid solution between albite and anorthite (i.e., plagioclase$_{ss}$) and two eutectic systems—diopside–anorthite and diopside–albite.

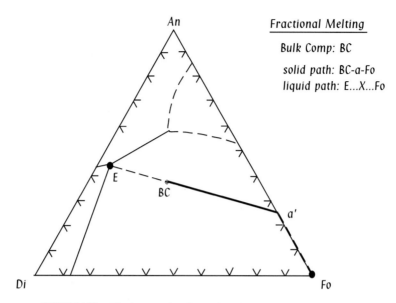

FIGURE 7.29   *The fractional melting of "rock" X: liquid and solid paths.*

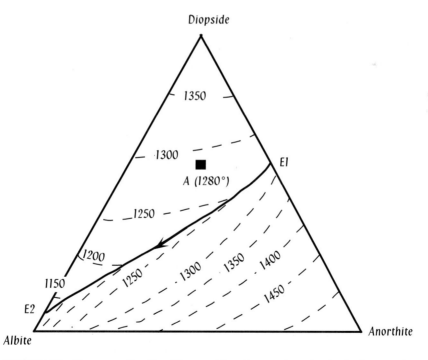

FIGURE 7.30   *The system diopside-albite-anorthite at 1 atmosphere showing isotherms and the location of a liquid of composition "A". The crystallization behavior of "A" is discussed further in the text.*

E1-E2 is the only cotectic boundary that occurs in this system, and it is like a thermal valley whose high temperature terminus is the eutectic point E1 between anorthite and diopside, and the lowest temperature point is the eutectic point E2 between albite and diopside. (There are some interesting complications at E2, which are avoided in our discussion). Liquids on this curve will evolve by moving downward E2 while crystallizing diopside and plagioclase$_{ss}$. An important handicap of this diagram is that the plagioclase composition coexisting with any liquid on the cotectic cannot be directly read off the diagram but requires additional knowledge of the Ab-An binary diagram. A detailed discussion of the paths taken by solid and liquid phases during fractional crystallization in this system is beyond the scope of this book, except to note that the liquid will reach E2 and the final solid will be a eutectic mixture of diopside and albite. In the next section, we will only discuss *equilibrium crystallization paths* of liquids and solids in the diopside and plagioclase$_{ss}$ liquidus fields in some detail. Fractional crystallization or melting paths are briefly discussed, mostly in the form of a passing reference.

Equilibrium Crystallization

Consider the liquid A located in the Di + L field on the 1280°C isotherm. This liquid will begin to crystallize diopside crystals at 1280°C, its liquidus temperature. Crystallization of diopside will drive the residual liquid composition directly away from diopside corner (Figure 7.31a); and the liquid will eventually reach point B on the cotectic. At this point a plagioclase$_{ss}$ (An$_{80}$, based on plagioclase diagram, which is not

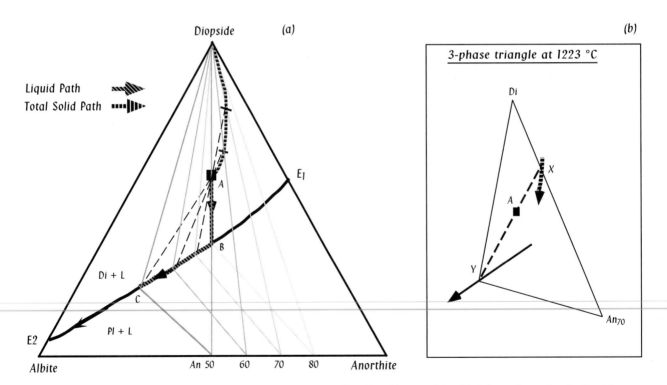

FIGURE 7.31   (a) The equilibrium crystallization of "A" in the system diopside-albite-anorthite. (b) A three-phase triangle diopside-Y-An$_{70}$ at 1223°C. The tie line X (solid)—Y (liquid) can be used as a "lever" to calculate proportions of liquid and solid at 1223°C. The bulk composition "A" is the fulcrum. The dashed arrow is the trace of the solid path, and the solid arrow is the trace of the liquid path. The proportions of diopside and An$_{70}$ crystals in the solid assemblage at 1223°C can also be calculated from the Di-An$_{70}$ line using "X" as the fulcrum. (See the text for further discussion.)

shown here) will crystallize together with diopside. A triangle, called a *three-phase triangle*, connecting the three coexisting phases (liquid B and crystals of diopside and An_{80} plagioclase) can now be constructed (Figure 7.31a). Several such triangles are constructed to illustrate solid and liquid paths as both liquid and plagioclase$_{ss}$ compositions change with cooling. Further cooling of liquid B will move it down the cotectic as the plagioclase$_{ss}$ composition continuously changes toward the albite corner. Because diopside composition does not change, *all* three-phase triangles will be anchored at the Di-corner while "swinging" from the right to the left of the diagram.

At any instant, the coexisting solid composition can be determined by drawing a line from the liquid composition through the bulk composition on to the base of the appropriate three-phase triangle. For example, note the three-phase triangle Di-An_{70}-Y at 1223°C (Figure 7.31b). When a line is drawn from the liquid composition Y through the bulk composition A to the base of this three-phase triangle, we obtain the solid composition X coexisting with liquid Y. Thus, XY is a tie line. When TSC is located this way for each successive three-phase triangle, a curved TSP emerges that starts at the diopside corner and terminates at A, that is, the bulk composition (Figure 7.31a). When TSP reaches A, the last liquid will be at C, and the coexisting plagioclase will be An_{50}. At this point the right arm of the last three-phase triangle will pass through the bulk composition. It is clear that one cannot construct any more down-temperature three-phase triangles beyond this point because such triangles will not contain the bulk composition (which is an impossibility). The proportions of crystals and coexisting liquid at any temperature can be determined using the appropriate tie line and lever. For example, at 1223°C, the tie line is X-A-Y; and

$$\%solid = 100 \times AY/XY \text{ and } \%liquid = 100 - solid\%$$

$$\%plagioclase (An_{70}) = 100 \times XDi/DiAn_{70} \text{ and } \%diopside = 100 - \%plagioclase$$

Liquids whose compositions fall within the plagioclase$_{ss}$ + L field take a curved path during crystallization because plagioclase$_{ss}$ continuously changes composition (Figure 7.32a). Consider the crystallization of a liquid P, for example (Figure 7.32,a,b). The first plagioclase$_{ss}$ to crystallize from it is S, which will drive the liquid directly away from S. As the liquid evolves slightly, the plagioclase$_{ss}$ will become more Ab-rich, which will force the liquid path to curve up. Several tie lines, S-P, Q-P-*q*, T-P-*t*, and R-P-M, are constructed to show how the liquid path curves up. Each of these tie lines connects coexisting solid and liquid phases and goes through the bulk composition P. Once the liquid reaches M, diopside will start crystallizing and the residual liquid will move down the cotectic E1-E2. Three three-phase triangles, Diopside-R-M, Diopside-N-*n*, and Diopside-O-*o*, describe the liquid path and coexisting solids. The liquid will be completely crystallized at O. Once again, we see that when equilibrium is maintained, the liquid solidifies long before it reaches the lowest temperature in the system.

The total solid path can be divided into two parts: from S to R when plagioclase$_{ss}$ is the only phase to crystallize; and R to P, which is constructed by drawing appropriate tie lines from the liquid through bulk composition to the three-phase triangle at each instant of liquid movement (Figure 7.32b). Several tie lines (dashed lines in Figure 7.32b) are constructed to demonstrate this: B-P-1 starts from liquid composition B, goes through P (bulk composition), and cuts the three-phase triangle B-*b*-Diopside at 1. The second tie line N-P-2 cuts its appropriate three-phase triangle N-*n*-Diopside at 2. The total solid compositions will be located at 1 and 2 when the liquids are at B and N, respectively. The final tie line connects the final liquid O with the bulk composition P, where the TSP must end. Connecting R, 1, 2, and P thus gives the part of the TSC path after diopside joins plagioclase in crystallization.

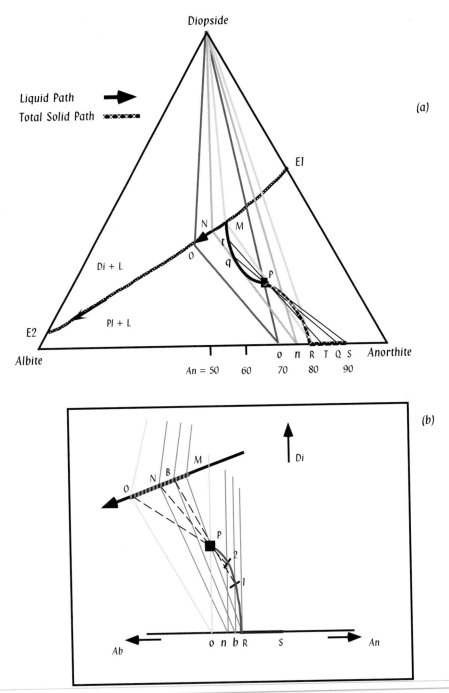

FIGURE 7.32 (a) The equilibrium crystallization of a liquid "P" in the liquidus field of plagioclase$_{ss}$ (Pl + L). The liquid path is marked by the thick solid curve with an arrow: P-M-O; and the liquid path is the dash-dot curve S-R-P. The tie lines are: P-S, q-P-Q, t-P-T; and the three-phase triangles are: M-R-diopside, N-n-diopside, and O-o-diopside. (b) Details of the construction of a portion (R → P) of the solid path S-R-P shown in this figure. For example, when the liquid is at "B," a dashed line B-P is drawn and extended to meet the appropriate three-phase triangle. This line intersects the diopside-b arm of the three-phase triangle at "1," which is where the coexisting solid composition must exist. Similarly, a second coexisting solid composition "2" corresponding to liquid "N" is located. The last tie line is the dashed line O-P, when the liquid is all used up as the solid reached the bulk composition. Joining R, 1, 2, P gives the curved portion of the solid path, which results from cotectic crystallization of diopside and plagioclase$_{ss}$.

Once again, at any instant an appropriate tie line can be used to calculate the percentages of liquid and solid. Consider the following example of the tie line B (liquid composition)–P (bulk composition)–1 (solid composition):

$$\text{Liquid\%} = 100 \times [(\text{distance P-1})/\text{distance B-1}]; \text{ and solid\%} = 100 - \text{liquid\%}$$

The proportions of diopside and plagioclase$_{ss}$ (composition at 'b') are given as follows:

$$\text{Di\%} = 100 \times \text{Di}/(b - \text{Di})$$

$$\text{Plagioclase}_{ss}\% = 100 - \text{Di\%}$$

During the *fractional crystallization* of liquid P, the liquid path will have less of a curvature than in equilibrium crystallization because of instantaneous separation of the crystals from the melt at each stage. The residual liquid will not end at O but will ultimately reach E2.

The first melt generated by either equilibrium or fractional melting of any solid assemblage lying within the Di-Ab-An phase diagram will be located somewhere on the cotectic curve E1-E2 as required by the appropriate three-phase triangle. In equilibrium melting, the solid and liquid paths will be exactly the reverse of the equilibrium crystallization paths. However, in fractional melting the final solid will either be pure diopside or anorthite depending upon the location of the original staring composition.

TERNARY SYSTEM WITH REACTION RELATIONSHIP: THE SYSTEM FORSTERITE–DIOPSIDE–SILICA AT 1 ATMOSPHERE PRESSURE

The system forsterite–diopside–silica at 1 atmosphere pressure serves as an important analogue for basalt magma crystallization at low pressure. Earlier in our discussion of the system forsterite–silica, we noted that enstatite, an orthopyroxene, forms as an intermediate mineral between forsterite and silica and exhibits the reaction relation Fo + L = En. This reaction relation persists even when diopside is added to the system (Figure 7.33). In addition to *protoenstatite*, a type of orthopyroxene, a Ca-poor clinopyroxene, called *pigeonite*, appears in this diagram and has a very small liquidus field. Both pigeonite and protoenstatite show a reaction relation with forsterite, and hence the univariant boundaries between liquidus fields of forsterite and of these two Ca-poor pyroxenes are *reaction curves* (p-$p1$ and $p1$-$p2$) along which the reaction Fo + L = Ca-poor pyroxene occurs. The boundary p1-e1 between liquidus fields of protoenstatite (Pen) and pigeonite (Pig) is also a reaction curve, where the reaction Pen = Pig + L occurs. All the other boundaries within the diagram are cotectic curves. It is standard practice to mark reaction curves with double arrows and cotectic curves with single arrows (Figure 7.33). The liquidus fields of silica polymorphs (tridymite and cristobalite) are not of particular interest here, and therefore, shall not be discussed except to note the existence of the two eutectic points e_1 and e_2.

It is important to note that the diopside that crystallizes from liquids (i.e., liquidus diopside, Di$_{ss}$) is not pure diopside but one that has some amount of an enstatite component in solid solution. Hence, it plots on the line between diopside and enstatite in Figure 7.33. Pigeonite and protoenstatite are solid solutions as well and plot on the Di-En line. Note that pigeonite actually shows a broader range of solid solution than shown in Figure 7.33; however, in order to keep complexities to a minimum, pigeonite's composition is shown as a point.

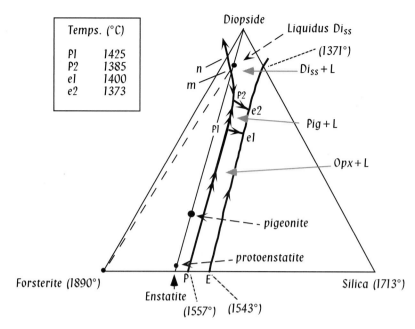

FIGURE 7.33 *The system diopside-forsterite-silica at 1 atmosphere.* There are three pyroxene phases—diopside$_{ss}$, pigeonite$_{ss}$, and protoenstatite$_{ss}$—that crystallize in this system. The extents of the solid solution shown by each of these three phases are not shown for brevity. P-P1 is a reaction curve along which the reaction Fo + L = Pen occurs. Along the curve P1-P2, the reaction Fo + L = Pig occurs. $e1$ and $e2$ are two ternary eutectic points where the respective assemblages are En + SiO_2-polymorph + Pig + L and Di$_{ss}$ + SiO_2-polymorph + Pig + L. The boundaries between the liquidus fields of the three pyroxene phases and that of the silica polymorphs are cotectic curves. The dashed line Fo-Di$_{ss}$ creates a thermal maximum on the cotectic curve between Di$_{ss}$ + L and Fo + L fields.

The dashed line connecting the liquidus diopside composition (Di$_{ss}$) and forsterite (Fo) produces a *thermal maximum* (T$_{max}$ in Figure 7.34) on the cotectic boundary between the diopside and forsterite liquidus fields. Figure 7.34 shows that three-phase triangles involving Fo, L, and Di$_{ss}$ (for example, Fo-1-Di$_{ss}$ and Fo-2-Di$_{ss}$) on the left-hand side of the dashed line Fo-Di$_{ss}$ point toward the Fo-Di boundary. This means that liquid compositions plotting on the cotectic Fo + Di$_{ss}$ + L to the left of the Fo-Di$_{ss}$ line will move toward the left, that is, toward the Fo-Di boundary (as shown by arrow). The three-phase triangles on the right side of the dashed line Fo-Di$_{ss}$ point toward P2 (i.e., L + Fo + Di$_{ss}$ + Pig invariant point); therefore, cotectic liquids plotting on the right side of T$_{max}$ will move toward P2.

Two fundamentally different basalt magma types occur in nature: alkali-olivine basalts (A) and olivine-tholeiites (OTh, Figure 7.34). Figure 7.34 illustrates some of the distinctive crystallization patterns of the two magma types: alkali-olivine basalts cannot crystallize a Ca-poor pyroxene, whereas olivine tholeiites can; olivine does not show a reaction relation with the magma in the former, whereas olivine tholeiites do (i.e., Fo + L = Ca-poor pyroxene); and residual liquids from alkali-olivine basalt crystallization become poorer in silica, whereas the opposite trend is shown by tholeiitic magmas.

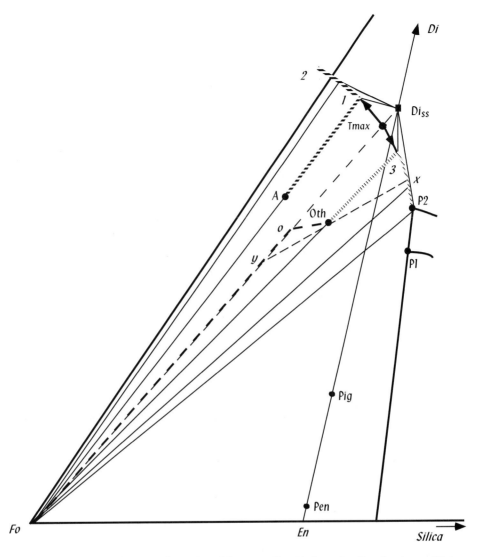

FIGURE 7.34 *An expanded view of a portion of the system diopside-forsterite-silica showing equilibrium crystallization liquid paths for two starting liquid compositions, "A" (A → 1 → 2) and "Oth" (Oth → 3 → P2).* In case of "A," the residual liquid moves out of the diagram, and hence only a portion of the solid path can be drawn here, and therefore, it is not shown. The solid path for "Oth" is Fo → o → Oth.

Equilibrium Crystallization of Liquids "A" and "OTh"

The crystallization of forsterite from liquid A will drive the residual liquid from A to 1 (Figure 7.34). At 1, diopside$_{ss}$ will join forsterite as a crystalline phase, and their cocrystallization will drive the residual liquid toward 2, out of this diagram! Thus, liquid A (an analogue for alkali-olivine basalts) ultimately becomes more silica depleted at advanced stages of crystallization and never crystallizes a Ca-poor pyroxene; and olivine does not show any reaction relation with pyroxenes.

In contrast, the crystallization of forsterite from a liquid OTh will force the residual liquid to move down to 3, at which point diopside$_{ss}$ crystals will join forsterite. Cotectic crystallization of forsterite and diopside$_{ss}$ will drive the residual liquid to P2 (i.e., Fo + Di$_{ss}$ + Pig + L invariant point). The final liquid will be used up at P2, and the end product will be a mixture of forsterite, diopside$_{ss}$, and pigeonite crystals.

> [**Question:**   What prevents the residual liquid from moving down from P2 to the invariant point Di$_{ss}$ + Pig + SiO$_2$ + L($e2$)?]

The TSC path for this bulk composition (OTh) is constructed as follows: when forsterite is the only phase crystallizing (i.e., the liquid is moving from OTh to 3 in Figure 7.34), the TSC is stuck at the Fo-corner. Once diopside$_{ss}$ starts cocrystallizing with forsterite (i.e., the liquid path: 3 → P2), the TSC will move up from the Fo-corner toward Di$_{ss}$; and the composition of the solid coexisting with any liquid can be obtained by drawing a tie line from the liquid through the bulk composition to the Fo-Di$_{ss}$ line. An example of one such tie line x-OTh-y is shown in Figure 7.34. The TSC path constructed this way moves up the Fo-Di$_{ss}$ boundary to o. When the TSC reaches o, the liquid must be at P2 (a tie line from o through OTh will reach P2). As pigeonite starts crystallizing together with diopside but at the expense of forsterite (remember: Fo + L = Pig), the TSC will move from o to OTh. Once the TSC reaches OTh, the liquid would be totally used up. The TSC path shows that although forsterite dissolves and pigeonite forms due to the reaction relation at P2, the liquid gets totally used up before all forsterite crystals are dissolved. Note that this is true to the extent that equilibrium is maintained within the system. By comparison, note that in fractional crystallization, the residual liquid would be free to leave P2 and move down the L + Di$_{ss}$ + Pig cotectic boundary because olivine crystals would be pulled out of the melt as soon as they formed and the bulk composition is not held constant.

Returning to equilibrium crystallization, once again note that the percentages of crystals and TSC at any instant may be calculated by constructing an appropriate tie line, as illustrated by the tie line x-OTh-y in Figure 7.34:

$$\%\text{Liquid } x = 100 \times y\text{-OTh}/xy \text{ and } \%y = 100 - x$$

The same concept can be used to calculate the proportion of Di$_{ss}$ and Fo crystals in the solid assemblage y.

> [**Question:**   Calculate the proportions of Di$_{ss}$ and Fo crystals in y.]

Equilibrium Crystallization of "Th"

We now consider the crystallization path of a tholeiite-analogue Th (Figure 7.35). Note that Th lies in the liquidus field of forsterite, and therefore it will crystallize forsterite and move toward 0. From 0 to P1, the reaction Fo + L = Pen will occur.

As an aside, note that from 0 to P1, one of the apices of the three-phase triangles falls on the reaction boundary, which can only happen when the curve is a reaction curve. In contrast, note that the three-phase triangles Fo-1-Di$_{ss}$ and Fo-3-Di$_{ss}$ in Figure 7.34 both "contained" the boundary curve between diopside$_{ss}$ + L and Fo + L, which is only possible when the boundary is a cotectic boundary. Thus, three-phase triangles allow a distinction between cotectic and reaction boundaries.

Returning to our discussion on the crystallization of Th, one needs to determine what happens to the residual liquid when it reaches P1—whether it will (1) continue to move up the boundary 1-P2 (where Fo + L = Pig reaction occurs); (2) completely freeze at P1; (3) move down the liquidus boundary Pen + L = Pig; or (4) move into the Pig + L field. Note that at P1, two sets of reaction will occur: Fo + L = Pig and Pen + L = Pig. Only one of these four paths is possible, and the only way to deduce this is to construct several three-phase triangles. Any three-phase

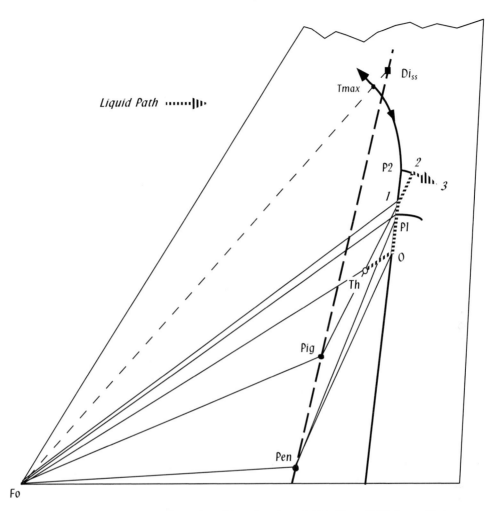

FIGURE 7.35   *The liquid path Th → O → P1 → 1 → 2 → 3 (e2 in Figure 7.33) for equilibrium crystallization of a liquid of composition "Th."* The three three-phase triangles, Fo-Pen-O, Fo-Pen-P1, and Fo-Pig-1, are shown to construct the liquid path. (See the text for further discussion).

triangle connecting Pig, Pen, and a liquid on the Pig/Pen liquidus boundary (not shown in Figure 7.35) will not include the bulk composition Th; and therefore, option (3) above can be ruled out. Similarly, a three-phase triangle connecting any liquid composition in the Pig + L field, Pig, and Fo will not include the bulk composition Th, which includes option (4). Three-phase triangles connecting certain liquids along P1-P2 (such as 1-Fo-Pig in Figure 7.37) will include the bulk composition Th; and therefore, the residual liquid will move from P1 to 1. Option (2) is automatically ruled out. Clearly, at P1, all protoenstatite crystals would be lost by reaction.

The tie line 1-Th-Pig shows that forsterite crystals would have all been dissolved when the liquid reaches 1, thereby allowing the liquid to move across the Pig + L field to 2, when diopside$_{ss}$ starts crystallizing with pigeonite. The liquid will then move along 2 to 3 and ultimately be finished up at the eutectic point e_2 in Figure 7.33.

The total solid path for Th can be constructed from appropriate tie lines. It is a simple matter to show that the TSP will be composed of several parts (Figure 7.36). When the liquid crystallizes only forsterite (the liquid path O → P1; Figure 7.35), the solid composition is pinned at the Fo-corner. Concomitant with the movement of

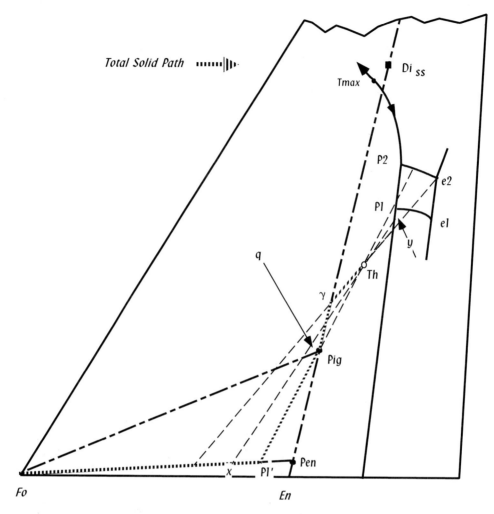

FIGURE 7.36 *The equilibrium crystallization of Th: The solid path, Fo → Pl′ → q → Pig → r → Th.* A tie line (lever) *x*-Th-*y* is shown to aid in the calculation of liquid and solid proportions when the liquid reaches "*y*." The solid assemblage at this instant can also be calculated: %Fo crystals = 100 × (Distance *x*-Pen/Distance Fo-Pen), and %Pen crystals = 100 − %Fo crystals. (See the text for further discussion.)

the residual liquid from O to P1, while the reaction Fo + L = Pen occurs, the TSC moves from Fo toward Pen. At any instant a tie line may be constructed to determine solid and liquid percentages. Consider the tie line *x*-Th-*y* in the following example:

$$\text{Solid}\% = 100 \times (\text{Distance Th-}y/\text{Distance } xy)$$

The proportion of protoenstatite and forsterite in the solid at this instant can also be calculated as follows:

$$\text{Forsterite}\% = 100 \times (\text{Distance } x\text{-Pen}/\text{Distance Fo-Pen})$$

$$\text{Protoenstatite}\% = 100 - \text{Forsterite}\%$$

The tie line P1-Th-P1′ shows that when the liquid is at P1, the coexisting solid is at P1′ (Figure 7.36). While all protoenstatite crystals are reacting out, the liquid stays at P1 and the solid composition moves to *q*. (Although not apparent in this figure, q is a point that is not exactly at pigeonite but on the Fo-Pig line.) When the solid composition reaches q, the liquid is depleted of all protoenstatite crystals; however, some forsterite crystals still remain along with pigeonite. The solid composition then moves from q to Pig, as the liquid composition moves from P1 to 1 in Figure 7.36. Once forsterite is entirely dissolved (liquid at 1, solid at Pig; Figure 7.35), the liquid moves across the Pig + L field and the solid composition stays pinned at Pig. As the liquid (now on the Di_{ss} + Pig + L curve) moves from 2 to 3 (Figure 7.35), the solid moves from q to r (Figure 7.36). When the liquid reaches the invariant point e_2 (Figure 7.35), the solid composition moves from r to Th; and when the total solid composition reaches Th, all the melt would be crystallized.

More complicated scenarios of crystallization paths can be constructed. Also, many different and useful magma-mixing and contamination models can be illustrated with this system, and the reader is encouraged to explore these. In the next section, we present only one example of how fractional crystallization may proceed from a liquid like X. The only reason we pick X is because it allows clear drawing of the solid and liquid paths during fractional crystallization and fractional melting.

Fractional Crystallization of X

Crystallization and immediate removal of Fo from the liquid X will move the residual liquid toward m (Figure 7.37). At m, the liquid will be free to leave the Fo + Pen + L boundary and move into the Pen + L field because the boundary is a reaction boundary Fo + L = Pen and Fo crystals would have been removed (a requirement in fractional crystallization). Inside the Pen + L field, the liquid will move

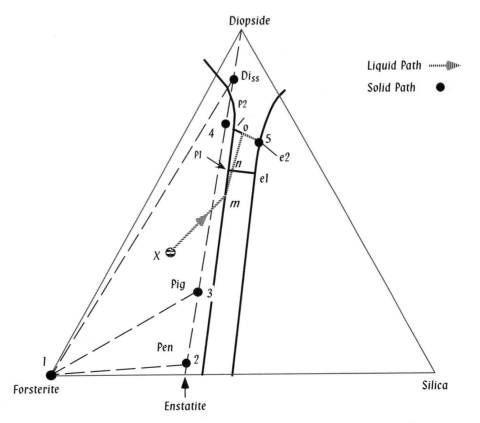

FIGURE 7.37   *The fractional crystallization of liquid X: Liquid and solid paths are shown.*

directly away from Pen composition because it is the only phase precipitating from the liquid. The residual liquid will eventually reach the Pen + L = Pig reaction curve at *n*. At *n* the liquid will be free to move across the Pig + L field, directly away from Pig composition, while precipitating pigeonite. The liquid will eventually reach *o* and move down the Di$_{ss}$ + Pig + L cotectic toward *e2*. The total solid path (TSP) and instantaneous solid compositions are relatively simple to construct in this case. First, the solid is pinned at the Fo-corner, then it hops to Pen and then to Pig. Dropping tangents from each point on the liquid path on the Di$_{ss}$ + Pig + L curve to the Di$_{ss}$-En line gives the next portion of the solid path. Once the liquid reaches *e2*, the solid hops to *e2* and is entirely crystallized there.

The example shown in Figure 7.37 is of course one of many possible paths, called *liquidus fractionation paths*. Consideration of such paths for a large number of liquids results in a diagram like Figure 7.38.

Equilibrium and Fractional Melting of X

As with other systems, equilibrium melting paths are the reverse of equilibrium crystallization paths (Figure 7.39). Because X is composed of diopside, pigeonite, and forsterite, its initial melting must occur at P2. The melt composition will stay at P2 until all diopside crystals melt and the total solid composition moves from X to 1. The total solid composition will then move from 1 to 2 as the melt composition moves from P2 to P1. Further melting will not move the liquid composition from P1 until the solid residue reaches 3 from 2. When the solid residue reaches 3, all pigeonite crystals would

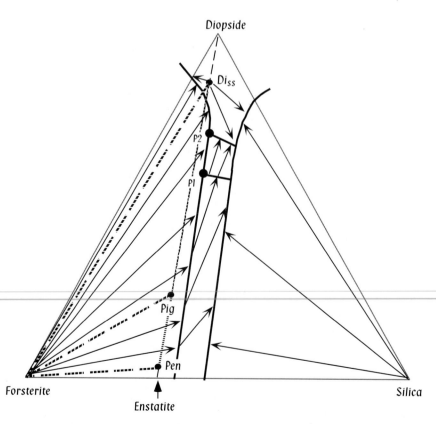

FIGURE 7.38   *The liquidus fractionation paths for each of the liquidus fields are shown.*

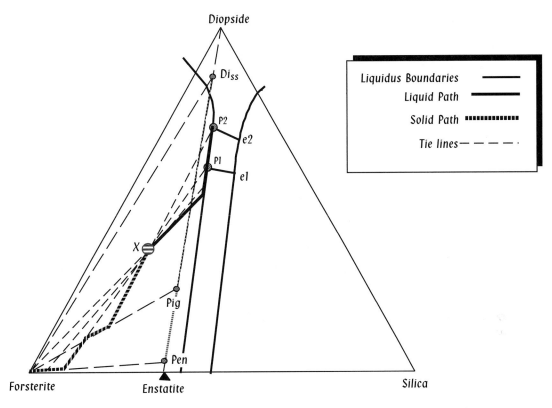

FIGURE 7.39 *The equilibrium melting of X: Liquid and solid paths and a few tie lines are shown.*

have been dissolved. The melt composition will then be free to leave P1 and move toward 2′, as the solid residue composition will move from 3 to the forsterite corner. Once all protoenstatite crystals melt and the residue is only composed of forsterite, the liquid composition will leave the reaction curve (Pen = Fo + L) and move toward X.

During fractional melting of any mixture of crystalline phases, the initial melt will be generated at the appropriate invariant point, and the solid residue will move directly away from that invariant point. Figure 7.40 considers two solidus surfaces—Fo-Di$_{ss}$-Pig and Fo-Pig-Pen. As a specific example, let us consider melting of X, a "rock" composed of Di$_{ss}$, Fo, and Pig crystals. The invariant point where these phases are involved with a liquid is P2. Therefore, melting will begin at P2, and the solid residue will move from X to 1 (i.e., directly away from P2). At this point all Di$_{ss}$ crystals would have disappeared from the residue, and melting cannot begin until the temperature is raised so that melting can resume at P1. Once this happens the solid residue would move from 1 to 2 (directly away from P1). When the solid residue reaches 2, there would be no pigeonite crystals left. Once again the temperature will have to be raised so that melting can resume again at the peritectic point En = Fo + L on the forsterite-silica binary. Melt production at this binary invariant point will drive the residue composition to the forsterite corner. Once the residue is composed entirely of forsterite crystals, melting must stop and cannot renew until forsterite's melting point is reached.

Many more solid residue paths like the one above can be constructed for a whole range of starting solid compositions. Within the solidus Di$_{ss}$-Fo-Pig, such paths must radiate away from P2; and within the solidus Pig-Fo-Pen, such lines would radiate away from P1 (Figure 7.41). These radiating lines are called *solidus fractionation paths*.

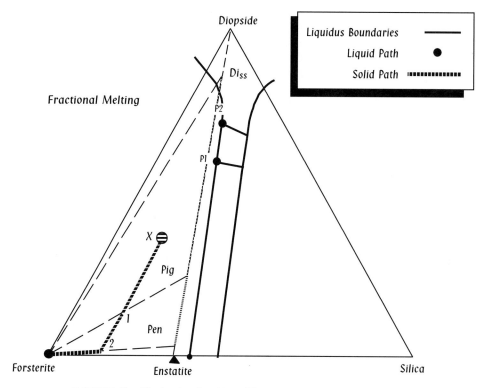

FIGURE 7.40 *The fractional melting of X: Liquid and solid paths are shown.*

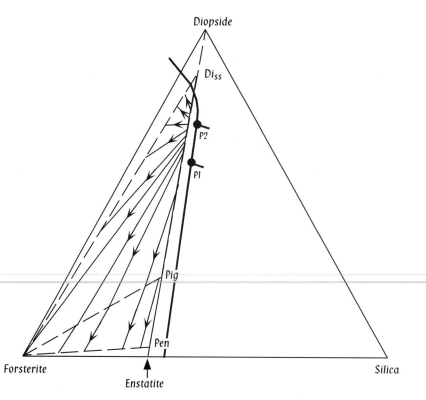

FIGURE 7.41 *The solidus fractionation paths for two of the solidus surfaces, Di_{ss}-Fo-Pig and Pig-Fo-Pen, are shown.*

7.1 THE OLIVINE PHASE DIAGRAM AND THE COMPOSITION OF THE EARTH'S UPPER MANTLE

Rocks and minerals from the earth's upper mantle are sometimes ripped off and brought to the surface by violent magmas. Such rocks and minerals are called xenoliths and xenocrysts, respectively, because they do not crystallize from the host magma itself but are accidentally picked up on the way. A familiar example of a xenocryst is the mineral diamond. We know that diamond can only be stable in the upper mantle (Figure 7.1); and therefore, it follows that all the world's diamond must have been transported to the earth's surface by magmas. Some of the deepest xenoliths (xenocrysts) come from as deep as the top of the transition zone!

Scientists have long been fascinated with xenoliths and xenocrysts because they provide the only direct window into the earth's mantle. Here, we consider an example of how a simple binary phase diagram can give us a fundamental piece of information about the composition of the earth's upper mantle based on a study of mantle xenoliths. In 1970, James L. Carter of the University of Texas at Dallas published a research paper on a very large number of xenoliths from an area in New Mexico, called Kilbourne Hole. Using textural and chemical criteria, he identified two classes of xenoliths—one of them principally lherzolites (dominantly olivine, ortho- and clinopyroxene, and minor spinel), and the other pyroxenites and wehrlites (composed largely of clinopyroxene; olivine and spinel are minor phases). Carter plotted the compositions of olivines in the two xenoliths suites (as shown in Figure Bx 7.1a) and noted a fundamental difference between the two: 90% of olivines in lherzolites are more forsteritic than Fo_{88}; whereas those in the other suite (Carter interpreted this group to have had an igneous origin, that is, they formed by magma crystallization) are $< Fo_{86}$. The olivine phase diagram (Figure Bx 7.1b) tells us that during the partial fusion of an olivine, the residues

of melting will be more forsteritic than the source, and the olivines that crystallize from the melts will generally be less forsteritic than the source. For example, when a source olivine "Y" (Fo_{87}) is melted, the residual olivine becomes progressively forsteritic (as shown by the arrows and the example "X"). Carter used this reasoning to conclude that the majority of lherzolites are residues of melt extraction and that the "undepleted source" lherzolite (undepleted means the magma has not been extracted at all from the source) of the earth's upper mantle must have an olivine composition of Fo_{87}, where there is a paucity of residual and magmatic olivine compositions. The latter conclusion allowed him to provide an overall estimate of the abundances of other elements in the upper mantle as well.

Similar studies from other areas by different authors (including those on Hawaiian xenoliths by the present author) have noted a very similar frequency distribution of olivine compositions. Therefore, Carter's conclusion regarding the composition of the earth's undepleted upper mantle remains valid today. Several other authors (particularly, the late Professor A.E. Ringwood of Australian National University) used different approaches to obtain constraints on the undepleted upper mantle composition. In a general sense, their calculations have not altered Carter's significant conclusion, which was based on a rather simple phase diagram!

Note that although the binary system Fo-Fa was used to reach such an important conclusion, it is highly unlikely that the temperature of melting in the shallow mantle ever reaches 1700°C! The presence of other components in a lherzolitic source rock drops the melting temperature by 100's of degrees. However, the compositional effects on olivine, as described previously, remain valid.

(*continues on next page*)

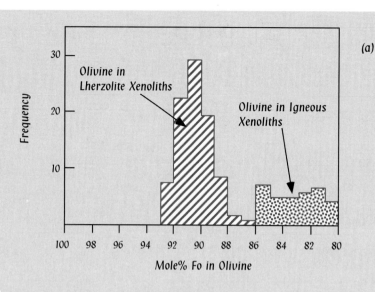

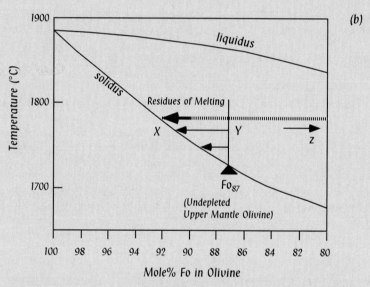

FIGURE BX 7.1 (a) Frequency distribution of olivines in two classes of mantle xenoliths from the Kilbourne Hole, New Mexico. (b) A part of the binary phase diagram Forsterite-Fayalite. In Carter's model, Fo_{87} is the undepleted (i.e., before any melt was extracted from it) upper mantle olivine. This olivine (represented by the bulk composition "Y"), upon melting and loss of melt, will become progressively richer in forsterite component (i.e., move toward "X", for example). The derived melts (not shown, but note the arrow toward z) will likely undergo crystallization-differentiation and ultimately crystallize olivines that are more Fe-rich than Fo_{87}, as seen in magmatic xenoliths and as phenocrysts in basalt lavas.

7.2 THE EFFECTS OF "DRY" VERSUS "WET" PRESSURE ON THE FO-SILICA PHASE DIAGRAM AND THEIR SIGNIFICANCE

Pressure of the surrounding rocks, also known as load pressure (referred to as "dry" pressure), can have a profound effect on the phase relationships in a mineral-melt system. Volatiles, such as water, may reside in pore spaces between grains in a rock and can generate as much pressure ("wet" pressure) on the rock as load pressure (i.e., load P = P_{H_2O}). The effect of P_{H_2O} on mineral phase relations is consid-

erably different from that of load P. This is explained with the help of Figure Bx 7.2. This diagram essentially combines three isobaric phase diagrams in one! The now-familiar 1 atmosphere phase relations in the system Fo–SiO_2 are shown in the middle; and the top and bottom diagrams are for "dry" versus "wet" phase relationships, respectively, at 2 GPa (after Chen and Presnall 1975).

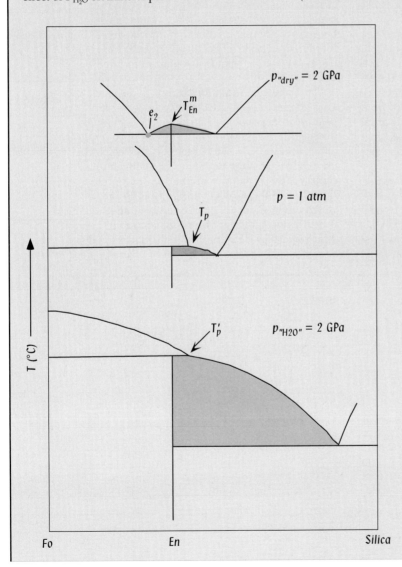

FIGURE BX 7.2 *Effect of "dry" vs. "wet" pressure in the system forsterite (Fo)—silica (SiO_2).* Enstatite (En) forms an intermediate compound with its incongruent melting point at T_p at 1 atmosphere pressure (the middle diagram). The enstatite + liquid field is shaded. Enstatite melts congruently at a pressure of 2GPa in a volatile-free environment (upper diagram). Again, enstatite + liquid field is shaded. In a water-saturated environment and at a high pressure of 2 GPa, enstatite melts incongruently at T'_p, and the enstatite + liquid field is much expanded (lower diagram).

(continues on next page)

Note the enstatite + liquid field is marked with gray shading. First, in the dry case, the 1 atmosphere peritectic point, Tp, moves toward "En" then toward "Fo" as "dry" pressure is increased. Due to this movement, at a certain pressure (probably around 1.1 GPa: Chen and Presnall 1975), this point (called "singular point") falls exactly on the En line; and with a further increase of pressure, it shifts further toward Fo, creating a second eutectic point (e_2: Fo + En = L at P_{dry} = 2 GPa) in this diagram, and thus the well-known low pressure peritectic reaction (Fo + L = En) disappears. At pressures above 2 GPa, this eutectic point moves further toward Fo, still showing the same eutectic behavior.

A direct application of this P-effect may be made to basalt magma, which is the material that forms the oceanic crust. As will be seen in a different chapter, mantle peridotite, composed largely of olivine and a substantially lesser amount of pyroxenes, is the predominant source rock from which basalt magmas are generated by partial melting. The system forsterite-silica is a good analogue for studying basalt magma generation and crystallization.

The P-shift of the invariant point Tp shows that basaltic magmas produced at increasingly greater pressure will become increasingly richer in the Fo (or olivine) component. Note that the melt extracted from a peridotite at 2 GPa will have the composition at "e_2". Because of the thermal barrier created by the En composition line at 2 GPa (high pressure in general), such a basaltic magma cannot fractionate to the second eutectic point at that pressure where both enstatite and a silica polymorph may crystallize from the eutectic melt. However, if this same magma rises to a shallow-level magma storage chamber (where the low pressure phase diagram is applicable), it may undergo fractional crystallization of large amounts of forsterite (olivine in a natural magma) and pyroxene and proceed toward the En + Silica = L eutectic point at low pressure. Unless natural magmas rise tremendously fast so that their pristine chemical composition is retained, the above relationship tells us that most basaltic lavas erupted on the surface are non-pristine and may be derived through shallow-level fractional crystallization of olivine from more primitive magmas generated at much higher pressures.

The three significant effects of P_{H_2O} are as follows: magma is generated at a much lower temperature from peridotite than under P_{dry} condition; the magma composition is silica-rich relative to that produced by "dry" melting; the field of En + L (pyroxene + liquid in natural magma) is significantly expanded at the expense of the Fo + L field. Hydrous melting of peridotite, in which olivine and orthopyroxene are important constituents, is relevant to the generation of andesitic and high-alumina basalt magma and will be discussed in a later chapter.

7.3 MAGMA OCEAN ON PRIMITIVE EARTH(?): CLUES FROM PHASE DIAGRAMS

Evolution of the earth since its birth 4.6 billion years ago is an exciting area of study. Unfortunately, the virtual absence of the record of the earth's earliest few million years in rocks has made it a rather difficult problem to unravel. Many fundamental questions continue to haunt scientists. For example, when and how did the earth's mantle differentiate into three layers—the lower mantle, the transition zone and the upper mantle! Was there ever a time when the earth's outer part was molten like a giant magma ocean?

In the mid-1980's, some scientists (most notably, Claude Herzberg of Rutgers University, and David Walker of the Lamont–Doherty Earth Observatory, Columbia University) noted that the solidus and liquidus of the earth's upper mantle rock seem to converge at a pressure of about 15 GPa (about 450 km) in pressure-temperature space. (Recall that the solidus of a particular rock is the curve at which the rock begins to melt; and liquidus is the curve above which it is entirely molten: Figure Bx 7.3). These scientists pointed out that the only way this convergence could happen is if the earth's upper mantle rock itself formed by eutectic-type crystallization from a giant magma ocean at that pressure range. Verification of this idea came when Eiji Ito

and Eichi Takahashi of Okayama University carried out melting experiments on an upper mantle xenolith up to a pressure of 25 GPa. They noted that indeed the solidus and liquidus come to a near-convergence at about 16 GPa. What is also interesting is that the liquids near the convergence zone are compositionally almost identical to the starting mantle rock. As discussed elsewhere in the chapter, this is exactly what would be expected when the starting rock composition coincides with the eutectic composition at a given pressure! Some authors have doubted the magma ocean hypothesis on the basis of other chemical arguments; however, Ito and Takahashi indicated that many or all of those chemical differences are explained by the fractionation of perovskite in the lower mantle. On the other hand, based on some simple oxide-oxide plots (discussed in a later chapter) Carl Agee and David Walker of Columbia University noted that simple fractionation of perovskite from a chondritic magma ocean (with iron taken out to form the earth's core) cannot form the peridotitic material that is supposed to be the dominant constituent of the earth's upper mantle. These authors noted that addition of olivine (by flotation in the magma ocean) would be necessary to form the upper mantle.

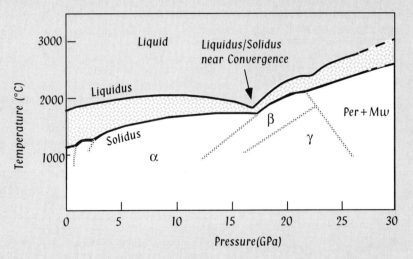

FIGURE BX 7.3 *Experimentally determined solidus and liquidus of an upper mantle rock. α-olivine, β-modified spinel, γ-spinel, Per-Mg-perovskite, and Mw-magnesiowustite.* (From Ito and Takahashi, 1987, *Nature* 328)

The debate about the existence of a primordial magma ocean on earth is perhaps not entirely over. For our purpose, however, the important

thing is to realize how far a simple observation on a phase diagram can go in deciphering the earth's history.

SUMMARY

[1]   Crystallization and melting behavior of rock-forming minerals are constrained by appropriate phase relationships. The concepts of phase rule, chemical potential, Gibbs' free energy, entropy, enthalpy, volume are all fundamentally important in understanding phase relationships.

[2]   Many different types of phase behavior are shown by mineral/melt systems, such as eutectic, reaction relationship, complete solid solution, limited solid solution, subsolidus unmixing, liquid immiscibility.

[3]   Both melting and crystallization process in any system can be of two extreme types—equilibrium and fractional. In most systems, liquid and solid paths during equilibrium versus fractional melting (crystallization) are very different. Liquid paths in fractional melting are different from liquid paths in fractional crystallization.

[4]   Lever rule is an important constraint on the proportion of crystals and melt at any given P, T condition.

[5]   Mixing of melts can result in extensive crystallization or partial or total resorption of crystals and crystallization.

[6]   The systems Fo-Fa, Di-Fo-Silica, and Di-An-Ab are all highly relevant to basalt magma crystallization and melting of peridotite.

[7]   Addition of volatiles lowers the melting temperature (solidus) relative to the "dry" solidus. The composition of the melt generated at the water (or carbon dioxide)-saturated solidus is also very different from that produced at the dry solidus: water tends to make the melt more silicic.

BIBLIOGRAPHY

Bowen, N.L. (1924) *The Evolution of the Igneous Rocks*. Princeton University Press (Princeton).

Carmichael, I.S.E., Turner, F.J. and Verhoogen, J. (1974) *Igneous Petrology*. McGraw-Hill (New York).

Carter, J.L. (1970) Mineralogy and chemistry of the earth's upper mantle based on the partial fusion-partial crystallization model. *Geol. Soc. Am. Bull.* **81**, 2021–2034.

Cox, K.G., Bell, J.D. and Pankhurst, R.J. (1979) *The Interpretation of Igneous Rocks*. George Allen and Unwin (London, U.K.).

Ehlers, E.G. (1972) *The Interpretation of Geological Phase Diagrams*. W.H. Freeman (San Francisco).

Herzberg, C.T. and O'Hara, M.J. (1985). Origin of mantle peridotite and komatiite by partial melting. *Geophys. Res. Lett.* **12**, 541–544.

Ito, E. and Takahashi, E. (1987) Melting of peridotite at uppermost lower mantle condition. *Nature* **328**, 514–517.

Kato, T., Ringwood, A.E., and Irifune, T. (1988) Experimental determination of element partitioning between silicate perovskites, garnets and liquids: constraints on early differentiation of the mantle. *Earth Planet. Sci. Lett.* **89**, 123–145.

Maaloe, S. (1985) *Igneous Petrology*. Springer-Verlag (New York).

McBirney, A.R. (1993) *Igneous Petrology*. Jones and Bartlett Publishers (Boston).

Morse, S.A. (1980) *Basalts and Phase Diagrams*. Springer-Verlag (New York).

Philpotts, A.R. (1990) *Principles of Igneous and Metamorphic Petrology*. Prentice-Hall (Englewood Cliffs).

Presnall, D. C. (1969) The geometrical analysis of partial fusion. *Am. J. Sci.* **267**, 1178–1194.

Takahashi, E. (1986) Melting of dry perdotite KLB 1 up to 14 GPa: implications on the origin of peridotite upper mantle. *J. Geophys. Res.* **91**, 9367–9382.

Walker, D. (1986) Melting equilibria in multicomponent systems and liquidus/solidus convergence in mantle peridotite. *Contrib. Mineral. Petrol.* **92**, 303–307.

Wood, B.J. and Fraser, D.G. (1978) *Elementary Thermodynamics for Geologists*. Oxford University Press (Oxford, U.K.).

Zoltai, T. and Stout, J.H. (1984) *Mineralogy: Concepts and Principles*. Burgess (Minneapolis).

PART II

Petrology: An Introduction

Petrology is the study of rocks, and the scientist who studies rocks is called a **petrologist**. A rock is defined as an aggregate of minerals. Most rocks are commonly composed of grains of different minerals; but some are *monomineralic*—composed of grains of a single mineral: e.g., the rock *dunite* contains 90% or more olivine. Rocks are of three types, namely, igneous, sedimentary, and metamorphic. Igneous rocks form by solidification of molten rock materials, called *magma*.

Sedimentary rocks form by conversion of sediments into rocks due to a process known as diagenesis. Diagenesis commonly takes place due to deep burial of the sediments where some type of cementing material bonds the loose grains. In rare cases, diagenesis may occur *at* the earth's surface—an example of which is *beach rock*, which is found on the beaches of central and north Florida. Metamorphic rocks form from preexisting igneous/sedimentary/metamorphic rocks (generally called *protoliths*) when they are subjected to prolonged heat, pressure, and highly reactive fluids that migrate along grain boundaries. Metamorphic reactions are generally considered to take place in solid state, although pore fluids participate in and/or enhance the formation of new metamorphic minerals by breaking down old ones.

In general, the distinction between the three rock types can be made on the basis of texture and mineralogical composition. *Petrography* is the study of descriptive aspects of rocks, including variations within individual grains and physical interrelationships between grains (i.e., texture). The term *petrogenesis* refers to interpretative aspects of rocks.

In the chapters that follow, we learn about what rocks are made of and how they form. Because it is not possible to cover every aspect of petrology in a book of this type, we focus on some of the more abundant types of rocks that compose our planet.

CHAPTER 8

Introduction to Igneous Rocks

The following topics are covered in this chapter:

ABOUT THIS CHAPTER

This chapter introduces the reader to igneous rocks. Igneous rocks form by the so-lidification of magma (or molten rock) and comprise the bulk of the earth, moon, and other terrestrial planets. The nature of magma and its formation is the topic of the chapters that follow. This chapter deals with some of the fundamental descriptive aspects of igneous rocks—primarily their mineralogy and textures. Finally, the reader is introduced to the concepts of igneous rock classification.

IGNEOUS ROCKS: MODE OF OCCURRENCE AND PETROGRAPHY

Volcanic versus Plutonic Rocks

The formation, segregation, and ascent of magmas are topics of great interest to the petrologist. Whether a magma erupts or completely solidifies beneath the surface depends on two competing factors: how fast it cools versus how fast it rises. Magma that actually erupts on the earth's surface is what we know as *lava*. Lavas that erupt from volcanoes do so because of their rapid rise to the surface. Once erupted, they quickly solidify/quench to form what are called *extrusive* or *volcanic rocks*. In general, how big the crystals of minerals will grow from a magma in forming an igneous rock depends on how slow the magma cools. Because extruded lavas cool very fast due to rapid loss of heat to the surroundings, they either quench to a **glass**, which is a solid without any ordered internal arrangement of atoms, or form very small crystals (see Box 8.1 for a more detailed discussion of crystal growth in magmas). Therefore, extrusive rocks are generally fine grained and/or glass rich. The terms *aphanitic* and *glassy* (or *holohyaline*) are used to describe the texture of such fine-grained and glass-rich rocks, respectively.

    At the other extreme, magmas that pond at great depths within the earth's crust cool very slowly because they lose heat much more slowly to the surrounding wall rocks, which are already quite hot. Crystals can grow slowly to sufficiently large sizes in such a deep-seated igneous intrusion (called *plutons*). Therefore, the igneous rocks that compose plutons are coarse grained and are referred to as *plutonic rocks*. Plu-

8.1 NUCLEATION AND CRYSTAL GROWTH FROM MAGMA

Phase relationships (Chapter 7) describe the equilibrium conditions under which a mineral may crystallize from a melt or magma. In reality, however, crystals do not form from a liquid until it has been supercooled—that is, the liquid has been cooled a few degrees below the equilibrium temperature at which they should have formed. This happens because an embryo (called a *nucleus*) of the crystal must first form and be stable enough to grow into a larger crystal. The process of nucleus formation is called *nucleation*. Nucleation can occur either randomly in the melt (*homogeneous nucleation*) or on other existing crystals or gas bubbles in the melt. Although the details of nucleation are not well understood, some basic facts about stability of nuclei and crystal growth are known.

In the case of nucleation from a liquid, the free energy change associated with its growth (ΔG_{gr}) is the sum of change in free energy of the volume (ΔG_{vol}) and the free energy change associated with the formation of a new interface between nucleus and liquid (ΔG_{interf}):

$$\Delta G_{gr} = \Delta G_{vol} + \Delta G_{interf}.$$

ΔG_{vol} has to be negative (i.e., G of crystalline phase must be less than that of the liquid). ΔG_{interf} is a positive number. For a nucleus to grow, then, ΔG_{interf} must be a smaller number than ΔG_{vol} so that ΔG_{gr} is negative.

As a matter of convenience, the nucleus may be assumed to be spherical, and the energy balance may be expressed as follows:

$$\Delta G_{gr} = 4/3\pi r^3 \Delta G_{vol} + 4\pi r^2 \Delta G_{interf}.$$

Figure Bx 8.1 graphically represents the form of this equation. For extremely small embryos with radii less than r_c in Figure Bx 8.1, the term $4\pi r^2 \Delta G_{interf}$ of the prior equation (i.e., interfacial energy contribution, which is positive) is larger than the other term (volume free energy contribution, which is negative), and therefore ΔG_{gr} is positive. Such nuclei are not stable and thus dissolve. The opposite is true when the nucleus has a radius greater than r_c.

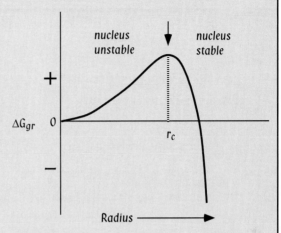

FIGURE BX 8.1 *Stability of nuclei during nucleation from a magma. Nuclei with a radius less than the critical radius (r_c) are unstable.*

Cooling rate exerts an important control over nucleation. Plagioclase is particularly known for its reluctance to nucleate. Experimental studies on basalts from the moon have shown that at high cooling rates plagioclase nucleation may be suppressed so much that a phase reversal may occur; that is, plagioclase may actually crystallize after a particular phase (such as ilmenite), although equilibrium phase diagram indicated that plagioclase should have crystallized before that phase.

During the growth of a crystal, certain faces may grow faster than the others because of lower interfacial energies associated with their growth. Extremely rapid cooling (quenching) leads to the formation of glass. Slow cooling leads to more equilibrium shapes of crystals, whereas fast cooling leads to the development of skeletal or acicular crystals. Also, two different phases may grow at very different rates, although both may start nucleating at the same time from a magma. The resulting texture in such a case may give a false impression that the phase forming the larger grains appeared first.

tonic rocks naturally lack glass. Rocks that compose shallow-level intrusions, called *hypabyssal intrusions*, have an intermediate grain size. Plutons are emplaced at depths >3 km, whereas hypabyssal intrusions are emplaced at <3 km. In the following section, the reader is introduced to the various forms of extrusive and intrusive bodies.

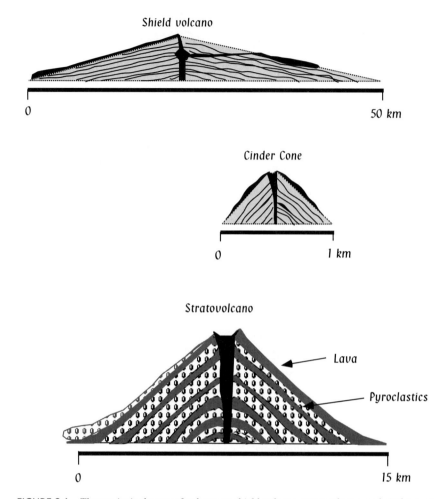

FIGURE 8.1   *Three principal types of volcanoes: shield volcano, stratovolcano, and cinder cone.*

Three kinds of material are extruded during a volcanic eruption: solid materials called *pyroclastic materials* (e.g., volcanic ash, bombs, blocks, etc.), lava, and gases. These can erupt through cracks or fissures in the crust or through well-defined volcanic conduits. Based on their geometric form and relative size, volcanoes can be fundamentally of three classes: shield volcanoes, stratovolcanoes, and cinder cones (Figure 8.1).

Shield volcanoes are truly giant volcanoes with gently dipping flanks. They are best exemplified by the earth's largest volcano—Mauna Loa (Hawaii). When its height is counted from the ocean floor to the peak, Mauna Loa is taller than Mt. Everest. Compared with stratovolcanoes, they are often called *nonexplosive* because they erupt very fluid basaltic lava without emitting gases in any significant amount.

Stratovolcanoes are typically associated with subduction zone volcanism. They are highly explosive and are built of alternate layers of pyroclastics and lava. Lavas are dominantly andesitic in composition, although basaltic and shoshonitic lavas may form a part of them as well. Cinder cones are really small and are found in all tectonic environments. They generally form by a single phase of volcanism.

Igneous intrusions can take many different forms as well (Figure 8.2). A first-order distinction is made between discordant and concordant intrusions. An intrusion is said to be *discordant* when it cuts across the regional structural planar/linear fea-

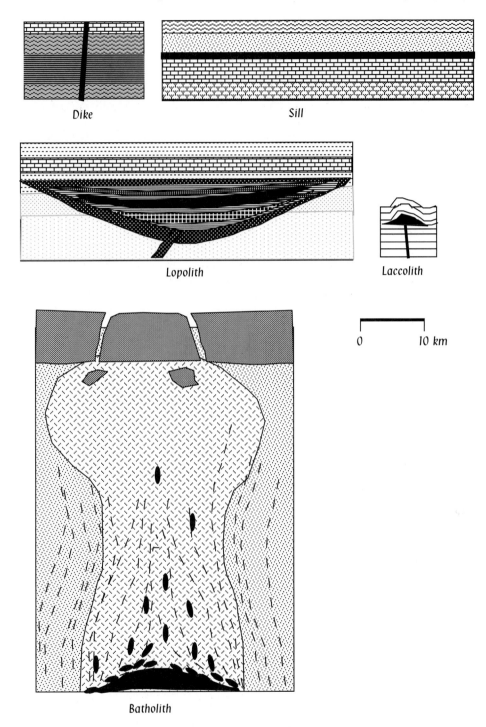

FIGURE 8.2   *Major types of intrusions.*

tures, such as sedimentary strata. In contrast, a *concordant* intrusion's contacts are parallel to the regional structural features. A *sill* is a type of tabular, concordant, hypabyssal intrusion. A tabular, discordant intrusion is referred to as a *dike*.

Really large mafic plutons (100s of km in diameter) can take hundreds of thousands of years to completely solidify. Such a long cooling history is conducive to crystal growth, and therefore plutonic rocks are generally coarse grained and free of glass. Hypabyssal rocks are generally medium grained. Although cooling plays a major role in crystal size, in some unusually coarse-grained intrusive rocks, called *pegmatites*, such crystal growth is attributed to crystallization in equilibrium with a vapor phase. Pegmatites are volumetrically almost insignificant when compared with basalts, andesites, and granites. However, they commonly host economically important ore minerals and gemstones such as emerald (scientific name—Beryl).

Geometric forms of plutons can vary a great deal (Figure 8.2). Plutons of mafic magmatic composition tend to have the form of a *lopolith*, which is a broad funnel- or keel-shaped intrusion whose exposed surface area could be as much as 64,000 km^2 (e.g., Bushveld intrusion of South Africa). Lopoliths are generally discordant. Felsic and intermediate magmas generally form *batholiths*, which are complex bodies of vast aerial extent. Smaller batholiths are called *stocks*. In some cases, forceful emplacement of a magma results in uplift of the overlying rocks, giving the igneous intrusion the geometric shape of a dome. Dome-shaped plutonic intrusions are called *laccoliths*. A system of parallel and cross-cutting dikes and sills, called a *dike swarm*, is often common in continental flood basalt provinces such as the Columbia River Basalt Group of northwestern United States. Dike systems forming ringlike structures (called *Ring dikes*) and cones (i.e., *Cone sheets*; Figure 8.2) are often associated with some minor alkalic magmas.

Elements of Petrography of Igneous Rocks

Physical description of the appearance of a rock is the first step of any petrologic study. Petrography deals with such descriptive or observational aspects of a rock. A good petrographic description should be such that it affords the reader a clear mental picture of what the rock looks like without even looking at the actual specimen. It is also helpful to draw sketches of the rock or take photomicrographs with a microscope.

There are two aspects of petrography: mineralogy and texture. A typical petrographic description must begin with the identification of the essential minerals (i.e., the dominant mineral components of the rock) and accessory minerals (i.e., the minor mineral constituents). The next step is to provide a visual estimate of the mode (i.e., proportions of minerals). If necessary, one may wish to obtain quantitative modes by point counting under a microscope or image processing methods from back-scattered electron images of the rock obtained via a scanning electron microscope. Normally a visual estimate is sufficient for routine descriptions.

In describing the texture of an igneous rock, the following aspects must be brought out:

1. Crystallinity.
   It refers to the extent to which a magma has crystallized. The terms *holocrystalline* (fully crystallized, generally medium to coarse crystals; somewhat analogous to the term *phaneritic*), *microcrystalline* (crystalline but crystals are very small [<1 mm] even under a microscope; also *aphanitic*), *merocrystalline* (mixture of crystals and glass), and *holohyaline* (glassy) are generally used (Figure 8.3).
2. Sizes, shapes, and habits of crystals of the essential minerals.
   In general, the terms *coarse* (>5 mm), *medium* (1–5 mm), and *fine* (<1 mm) are used to describe the overall grain size. In the case of very large crystals of pegmatites, the term *pegmatitic* is used to describe such crystals.
   The terms *euhedral*, *subhedral*, and *anhedral* are used to describe, in a relative sense, how well developed crystal faces are for each mineral phase. A min-

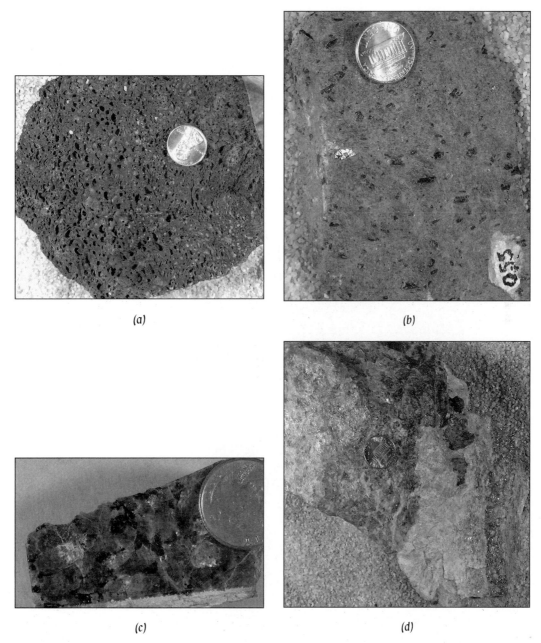

(a)

(b)

(c)

(d)

FIGURE 8.3   *Hand specimen photographs of textures of rocks various grain sizes.* (a) Fine-grained vesicular basalt. (b) Porphyritic andesite with amphibole phenocrysts (darker crystals). (c) Coarse-rained anorthosite cumulate (tabular crystals are plagioclase). (d) Pegmatite with extremely large crystals. (Courtesy of Gautam Sen)

eral grain is said to be *euhedral* when all/most of its faces are smooth and well developed, and anhedral is just the opposite—none of the faces is well developed. A subhedral grain has some of its faces well developed (see Chapter 4).

The shape of a grain is described by using such terms as *equant, subequant, prismatic, subprismatic, tabular, lath-shaped*, and so on (discussed in Chapter 4).

3. Specific Textures
Texture of a rock is essentially a description of the geometrical relationships between individual mineral grains (and/or glass or other materials) of a rock.

Several terms are used to describe textures of igneous rocks in the field: When a hand-sized rock sample is very fine grained so that individual grains are not discernible, the texture is called *aphanitic*. The term *phaneritic* is used to describe coarsely crystallized texture. In some volcanic and hypabyssal rocks, the texture may be *porphyritic*, defined by the presence of coarse phenocrysts set in a fine-grained groundmass (Figure 8.4). If the phenocrysts form clusters, then the texture is called *glommeroporphyritic*.

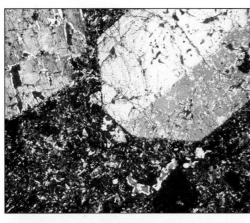

(a)

(b)

FIGURE 8.4 *Photomicrographs of textures.* [80× magnification, crossed polars.] (a) Porphyritic texture in an alkali basalt. (b) Aphanitic texture in a basalt. (c) Phaneritic texture in a gabbro (gabbroic texture). (Courtesy of Gautam Sen)

(c)

FIGURE 8.5   *Photomicrograph of flow texture in a rhyolite.* The arrow marks the flow direction. [80× magnification, plane-polarized light]. (Courtesy of Gautam Sen)

Textures, as viewed under the microscope, may be broadly divided into two groups: *inequigranular*, in which two or more distinct size groups of crystals occur, and *equigranular*, in which all grains are roughly equal in size. If distinct size groups do not occur and instead the grain size variation is a continuum between large and small sizes, then the term *seriate texture* is used. Special textures that do not fall into these groups are few; for example, rhyolites often exhibit flow texture (Figure 8.5), in which small elongate crystals are aligned like dead plants in a smoothly flowing river. Such a texture develops due to viscous flowing of the lavas while the crystals are growing in it. Another special texture is shown by graphic granites, in which quartz and alkali feldspar form hieroglyphicslike intergrowth (Figure 8.6).

Inequigranular texture may be of a number of different types. *Porphyritic texture*, in which coarse phenocrysts are embedded in a finer ground mass, is an example of inequigranular texture. In passing, it is worth mentioning that, although porphyritic texture is common among volcanic rocks, some volcanic rocks may be aphyric (i.e., devoid of phenocrysts). *Poikilitic texture* refers to a texture in which small crystals are entirely enclosed by larger crystals, and the term *subpoikilitic* is used when such enclosure

FIGURE 8.6   *Photomicrograph of graphic granite.* (Courtesy of Gautam Sen)

FIGURE 8.7 *Photomicrograph (80× magnified, crossed polars) of poikilitic and subpoikilitic inclusions of olivine in plagioclase.* (Courtesy of Gautam Sen)

is only partial (Figure 8.7). The terms *subophitic* and *ophitic* are used to describe the special textures of a type of igneous rocks called *gabbro*, in which coarse grains of augite partially to entirely enclose lath-shaped crystals of plagioclase (Figure 8.8). In

FIGURE 8.8 *Photomicrograph of subophitic texture.* Note the twinned plagioclase crystals partially enclosed by clinopyroxene. [30× magnified, crossed polars]. (Courtesy of Gautam Sen)

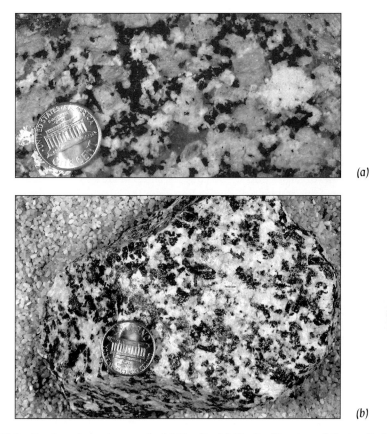

FIGURE 8.9   *Hand specimen photographs.* (a) Granite. (b) Diorite. (Courtesy of Gautam Sen)

gabbro and diabase, intergranular texture is also common, in which grains of pyroxene and oxide minerals occupy the intergranular space between coarse plagioclase laths. If glass (often hydrated or altered to clay, referred to as *devitrified glass*), instead of crystals, occupies the interstices, then the texture is called *intersertal*.

Equigranular textures are generally found in plutonic rocks. Such overall textural terms as *panidiomorphic granular, hypidiomorphic granular*, and *allotriomorphic granular* are used when the bulk of the constituent grains are euhedral, subhedral, or anhedral, respectively (Figure 8.9). In large mafic (defined later) plutonic intrusions, layered structure on field to microscopic scales is common. Such rocks are called *cumulates*, and a whole range of textural terms are used to describe such rocks (described in a later chapter).

CLASSIFICATIONS OF IGNEOUS ROCKS

A Simple Field Classification

Chemical composition of a magma dictates what minerals may crystallize from it. The rate at which the magma cools, which is related to its depth and mode of emplacement, controls the sizes and shapes of crystals and intergrain textural relationships. Therefore, it is only natural that naming an igneous rock must take into account its chemical composition, mineralogy, and texture (mode of emplacement). Table 8.1 shows such a general classification. A classification of this type is particularly useful in the field, where usually a pocket lens is the only tool one can use to identify a rock. There is a direct correlation between the hand specimen color of a rock and its content of mafic minerals (minerals with relatively high content of Mg, Fe, such as olivine and pyroxene): The greater the mafic mineral content, the darker the rock. Based on color of the hand specimen, igneous rocks are classified into *melanocratic* (mafic minerals >60%), *mesocratic* (mafic minerals 30%–60%), and *leucocratic* (mafic minerals <30%) types.

Let us consider an example to illustrate how to read Table 8.1 in naming a rock. The rock in question is coarse grained (i.e., plutonic) and is principally composed of plagioclase and clinopyroxene. Based on these facts, the third column in this table shows that this rock should be called a *gabbro*. However, if the pyroxene were not a clinopyroxene, but an orthopyroxene, then the rock would have been named a *norite* (further clarified in the IUGS classification that follows). In general, norite is rather uncommon, therefore in the field it is generally safe to call a plagioclase + pyroxene-bearing plutonic rock *gabbro*. If the rock is chemically and mineralogically identical to a gabbro but is volcanic (i.e., fine grained), then it should be called a *basalt* (third column, third from the bottom in Table 8.1), and so on.

TABLE 8.1   *A simple classification of common igneous rocks.**

Magma Type ⟶	Ultramafic (mafic mins. >90%)	Mafic (mafics 60-65%)	Intermediate (mafics 30-60%)	Felsic (mafics <30%)
Mineralogy ⟶	Olivine generally dominant, followed by pyroxenes	Pyroxene and plagioclase dominant	Plagioclase and pyroxene and/or amphibole dominant	Alkali feldspar and quartz dominant Biotite and amphibole in variable abundance
Volcanic (fine grained)	**Komatiite**	**Basalt**	**Andesite**	**Rhyolite**
Hypabyssal (medium grained)	**Komatiite**	**Diabase**	**Diorite-Granodiorite**	**Granite**
Plutonic (coarse-grained)	**Peridotite**	**Gabbro**	**Diorite-Granodiorite**	**Granite**

(Mode of Occurrence spans the Volcanic, Hypabyssal, and Plutonic rows)

*Rock names are in bold.

BRIEF PETROGRAPHY OF SOME COMMON IGNEOUS ROCKS

Peridotites and Pyroxenites

Peridotites are generally coarse-grained rocks, and their mode may vary a great deal. They are generally composed of ≥60% olivine (Fo_{88}–Fo_{92}), ≤25% aluminous (>5 wt% Al_2O_3) orthopyroxene, ≤15% diopsidic clinopyroxene (with 4–8 wt% Al_2O_3 and 0.5–1.2% Cr_2O_3), and an aluminous phase (plagioclase, spinel, or garnet). Textures of peridotites may vary tremendously from metamorphic types (foliated, cataclastic, etc.) to more hypidiomorphic granular (Figure 8.10). In a protoclastic texture, large deformed crystals (protoclasts) show granulation along grain boundary and are set in a ground mass composed of sheared, much smaller, grains. Exsolution in pyroxenes, triple-point junctions between grains, and subgrain boundaries (deformation lamellae) in olivine are generally common in these rocks. Most commonly, these rocks thus display metamorphic textures, which become apparent on comparison with metamorphic rock textures.

(a)   (b)

FIGURE 8.10   *Spinel lherzolite from the upper mantle.* (a) Photograph of a hand specimen showing a sheared textured spinel lherzolite on the left and a granular textured spinel lherzolite on the right. Both are embedded in a basanite (visible in the center). (b) Photomicrograph of a spinel lherzolite [8× magnification. Crossed polars]. (Courtesy of Gautam Sen)

Pyroxenites are very coarse-grained rocks that are composed of $\geq 90\%$ pyroxene (Figure 8.11). Clinopyroxenites are generally much more common than orthopyroxenites. Their grain size can vary from pegmatitic (extremely coarse) to coarse/medium. The very coarse-grained pyroxenites often have pyroxenes with spectacular exsolution structures. The exsolved phase may form lamellae to highly irregular, blebby shapes. The exsolved phases may be a pyroxene or an aluminous phase (garnet, spinel, or plagioclase). An aluminous phase, commonly a green-colored spinel, plagioclase, or garnet, is generally present in pyroxenites. Olivine is rare to absent. Phlogopite and a brown amphibole (kaersutite) are generally present in minor amounts. The textures of these rocks are commonly igneous types, mostly hypidiomorphic granular. However, strongly recrystallized metamorphic textures with triple-point junctions are more common among the smaller grained pyroxenites.

(a)   (b)

FIGURE 8.11   (a) Photograph of a garnet pyroxenite embedded in a basanite lava. (b) Photomicrograph of a garnet pyroxenite [36× magnification. Crossed polars.] (Courtesy of Gautam Sen)

Basalt, Gabbro, Diabase

These mafic igneous rocks are all dominantly composed of lath-shaped to tabular crystals of plagioclase (commonly An_{70}–An_{55}) and subprismatic to prismatic crystals of augitic clinopyroxene. These two minerals occur in roughly equal proportions in the rock. Pigeonite, and sometimes orthopyroxene, may also occur only in the tholei-itic/olivine-tholeiitic variety of these mafic rocks. Olivine's abundance usually varies between 0% and 15%. Olivine may often be altered (partially or wholly to an id-dingsite [an orange colored clay] or serpentine), but can be recognized by its relict shape. Magnetite and ilmenite are generally present, although their abundance is generally less than 5%. Besides these small globular forms of sulfides (usually chal-copyrite, pyrite-pyrrhotite, and pentlandite), formed due to sulfide-silicate liquid im-miscibility, are also ubiquitous (<1%). Tiny (generally <25 microns) euhedral crystals of apatite are also found in the interstitial spaces.

The difference between the basalt, diabase, and gabbro is principally one of tex-ture. Basalt is a fine-grained or glass-rich rock and is often porphyritic in nature (Fig-ure 8.4). It may contain vesicles (round quenched bubbles) or amygdules (bubbles filled with secondary materials). The most common phenocryst assemblages are olivine or olivine + plagioclase or olivine + plagioclase + augite. Intergranular and intersertal textures are common in the groundmass of porphyritic basalt. Diabase is medium grained and has a characteristic ophitic-subophitic texture. Gabbros are coarse grained, and the plagioclase and pyroxene crystals generally form coarse hypidiomorphic gran-ular type textures. Intergranular texture, defined by the occurrence of pyroxene gran-ules in the interstitial spaces between plagioclase crystals, is also a common feature.

Andesite and Diorite

Andesites vary greatly in their modal composition. They generally exhibit porphyritic texture dominated by strongly zoned plagioclase phenocrysts, euhedral-to-subhedral red-brown amphibole phenocrysts, and/or pyroxene phenocrysts (Figure 8.12). The plagioclase composition varies tremendously due to zoning, but the mean composi-

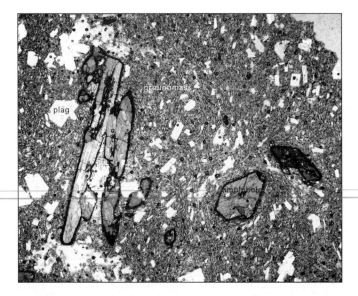

FIGURE 8.12 *Photomicrograph of an andesite (80× magnified) with amphibole and plagioclase phenocrysts (plane-polarized light).* (Courtesy of Gautam Sen)

tion is $\sim$ An$_{40}$. Cores of individual plagioclase crystals may be as a calcic as An$_{85}$. Amphibole phenocrysts are often surrounded by rims of pyroxene and opaque oxide mineral grains. One interpretation of these rims is that they form due to dehydration breakdown reaction of the amphiboles as the lavas rise toward the surface. Zoned pyroxene phenocrysts of subcalcic augite, pigeonite, or orthopyroxene are often found. Olivine, biotite, or glass are generally uncommon.

Diorites (broadly used here to include IUGS-defined diorite, monzodiorite, quartz diorite, and quartz monzodiorite) are composed of plagioclase, amphibole, pyroxenes $\pm$ quartz. They exhibit more equigranular, hypidiomorphic granular texture. Plagioclase generally tends to dominate the mode. Granodiorites and other somewhat quartz-rich rocks have similar textures.

Granite and Rhyolite

By definition, the two minerals that dominate the compositions of these rocks are quartz and an alkali feldspar. In hypabyssal intrusions and volcanic rhyolites, sanidine, a high-temperature polymorph of K-feldspar, is usually the dominant feldspar. The quartz phenocrysts in lavas are sometimes euhedral, but quartz grains in the ground mass are always anhedral and small. An extremely glassy, dark appearing (in hand specimen) equivalent of rhyolite is recognized as an *obsidian*. Rhyolites are generally vesicle free. *Flow texture* (Figure 8.5), defined by curvilinear arrangement of alkali feldspar and quartz crystals, may occur in hypabyssal intrusions or crystal-laden lavas.

Granite and granodiorite typically form batholiths. Their texture is generally coarse-grained hypidiomorphic granular (Figure 8.9). They are composed of an alkali feldspar and quartz. Subhedral to euhedral albitic plagioclase crystals may also occur. Whether one or two types of feldspars occur in a granite has significance in understanding their depths of origin (discussed further later). Alkali feldspar crystals often carry exsolutions of a more albitic plagioclase feldspars, known as *perthites* (Figure 8.13). Biotite and hornblende may occur, and their proportion may vary greatly between different granitic rocks. Blebby intergrowths of quartz and alkali feldspar, variously referred to as *granophyre*, *eutectic*, or *eutectoid*, are common in the interstices. Minor euhedral crystals of zircon, apatite, and sphene may also occur in the interstitial spaces.

FIGURE 8.13   *Photomicrograph of perthite.* (crossed polars; 45$\times$ magnified). (Courtesy of Gautam Sen)

Syenite and trachyte are similar to granite and rhyolite, respectively, except that they have lesser amounts of normative and modal quartz (note that nepheline syenite and phonolite are alkalic rocks and therefore have normative and modal nepheline). In general, syenites form small plutons and ring dike complexes.

MORE RIGOROUS CLASSIFICATIONS

Many different classification schemes have been proposed in the literature. Among them, the one proposed by the International Union of Geological Sciences (more popularly known as the *IUGS classification*, given in LeMaitre, 1976) for plutonic rocks is commonly used today. This classification is based on the observed proportion of mineral constituents (i.e., mode). The IUGS classification of volcanic rocks is not widely used because in such rocks minerals are often not discernible. In the case of volcanic rocks, classification schemes based on chemical or normative (discussed later) compositions are generally used. In this book, we adopt the IUGS classification of plutonic rocks and LeMaitre's chemical classification of volcanic rocks.

PLUTONIC ROCK CLASSIFICATION

The IUGS classification is based on the modal abundance of quartz [Q], alkali feldspar [A], plagioclase [P], feldspathoids [F], and mafic minerals [M]. If the proportion of M, also known as the color index, exceeds 90%, then it is an ultramafic rock. Rocks with a color index <90 are classified in terms of their proportions of Q, P, A, and F contents. This is done by recalculating the percentages of Q, P, A or F, P, A into 100% and plotting them in the appropriate triangle—either QPA or FPA (discussed later).

Ultramafic Rocks

Ultramafic rocks are dominantly composed of three minerals: olivine, orthopyroxene, and clinopyroxene. Figure 8.14a shows IUGS classification of ultramafic rocks in terms of the olivine–orthopyroxene–clinopyroxene triangle. A rock containing 60% olivine, 25% orthopyroxene, and 15% clinopyroxene would plot in the field labeled *lherzolite*, and therefore should be called a *lherzolite* (Figure 8.14a). Figure 8.14a also shows where >90% of the lherzolites worldwide plot (gray shaded area). Because an aluminous phase (plagioclase, spinel, or garnet) is usually present in a lherzolite, an additional qualifier is added as a prefix to indicate what type of aluminous phase is present: If the aluminous phase is a spinel, then the rock is called a *spinel lherzolite*. Similarly, the names *garnet lherzolite* and *plagioclase lherzolite* are used when garnet or plagioclase, respectively, occur. It is important to note that lherzolite is the predominant constituent of the earth's upper mantle. Over the past two decades, it has been recognized that the nature of the aluminous phase is an important indicator of the depth of origin of a lherzolite—plagioclase lherzolite can generally be stable at depths <30 km, garnet lherzolite at >75 km, and spinel lherzolite at intermediate depths (discussed in a later chapter). Therefore, there is important genetic significance to naming a lherzolite.

Harzburgites, which are dominantly olivine + orthopyroxene rocks (Figure 8.14a), are also quite abundant. Harzburgites can be found as xenoliths in basalts or in ophiolite sections, where they are interpreted to be residues of partial melting (discussed in a later chapter). They can also occur as layers in layered, mafic, plutonic in-

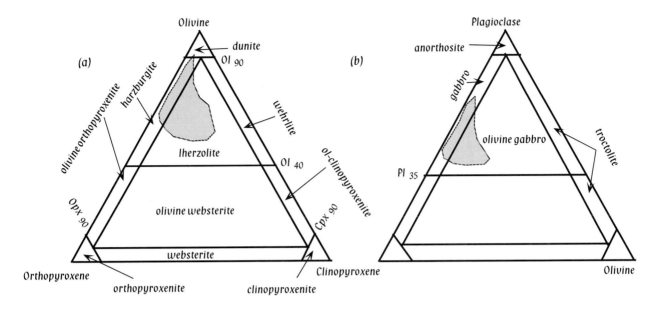

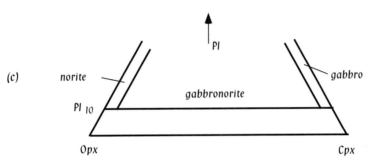

FIGURE 8.14   *IUGS classification of (a) plutonic, (b) ultramafic, and (c) mafic rocks.*

trusions, where they appear to form by crystallization processes (also discussed later). Wehrlites and clinopyroxenites occur as xenoliths in lavas and as veins in ophiolitic complexes, where they are inferred to have a magmatic crystallization origin.

The majority of upper-mantle lherzolites and harzburgites exhibit distinct metamorphic textures. Therefore, it is ironic to consider these important rock types as igneous rocks. Presumably this tradition has continued because lherzolite happens to be the source rock for basalt magmas, which form all of the oceanic crust and much of the continental crust. In such a case, eclogite, clearly a metamorphic rock composed of pyrope-rich garnet and an omphacitic clinopyroxene, must also be considered here because it too likely plays some role as a source rock for magmas.

In the literature, the terms *peridotite* and *pyroxenite* are used to refer to olivine-rich versus pyroxene-rich ultramafic rocks. Thus, peridotites broadly include lherzolite, dunite, harzburgite, olivine, websterite, and wehrlite, whereas pyroxenites include mainly clinopyroxenite, orthopyroxenite, and websterite.

Figure 8.14a does not include three other important ultramafic rock types—kimberlites, lamproites, and komatiites. *Kimberlites* are a class of rare potassic volcanic ultramafic rocks that contains phenocrysts of phlogopite, altered olivine, abundant xenoliths, and xenocrysts of the mantle, mixed with a very fine-grained material. Its popularity stems from two things: Many are diamond bearing, and they brought up the deepest mantle xenoliths ever recorded. Most kimberlites are highly altered. In reality, kimberlite magmas were far from being molten silicates, but are believed to have been gas-charged mixtures of xenoliths, xenocrysts, some phenocrysts, and small amounts of melt. Their eruptions were likely at supersonic velocities through typically pipe-shaped intrusions known as *diatremes*.

Lamproites are a rare group of volcanic to hypabyssal rocks that are unusually rich in K_2O and vary from ultramafic compositions through mafic to more evolved compositions. Leucite and sanidine are generally present in the highly evolved lamproites. Olivine lamproites are ultramafic and contain phenocrysts of olivine (Fo_{91-94}), diopside, phlogopite, potassic amphibole in a fine-grained ground mass dominantly composed of diopside and phlogopite. Kimberlites and ultramafic lamproites are thought to have been generated deep within the upper mantle in vapor-saturated conditions.

Komatiites are clearly quenched, dry ultramafic lavas that largely erupted during Archean. Phanerozoic komatiites are extremely rare. The single most important criterion used to identify these fossil ultramafic lavas is the presence of spinifex structure, which is due to the presence of long, feathery, skeletal crystals of olivine and pyroxene. Experiments in the laboratory have shown that such olivine morphology can only develop when olivine crystallizes rapidly from a rapidly quenching magma. The presence of komatiites in Precambrian attests to very high temperatures within the earth's shallow upper mantle.

Mafic Plutonic Rocks

In this book, a slightly modified version of the IUGS classification of mafic and felsic rocks is used. In this version, many of the nomenclature details (which are essentially descriptive jargon) have been skipped because the focus of this book is only on those most common rocks.

Mafic rocks are classified on the basis of modal contents of olivine, plagioclase, and pyroxenes (Figures 8.14b, 8.14c). Rocks containing >90% plagioclase are called *anorthosites*. Anorthosites occur as layers in plutonic-layered intrusions and as massive, km-scale bodies of Precambrian age. Note that the highlands on the moon are largely composed of anorthosites. Gabbros (plagioclase + clinopyroxene) and olivine gabbros (gabbros commonly with 5%–15% olivine) are the most abundant mafic plutonic rocks (Figure 8.14b). Troctolites (olivine + plagioclase) rocks occur as layers in layered intrusions. Figure 8.14c shows that if the dominant pyroxene is an orthopyroxene (instead of clinopyroxene), then the orthopyroxene-bearing equivalent of a gabbro is called *norite*. Norites and olivine gabbronorites generally occur in layered intrusions and may also be found associated with anorthosite massifs.

Felsic Plutonic Rocks

The modal proportions of quartz (Q), alkali feldspar (A), plagioclase (P), and feldspathoids (foid, F) are used to name various rock types (Figure 8.15). IUGS classification recognized a fundamental distinction between rocks that contain quartz and those containing feldspathoids. In Chapter 7, it was shown that feldspathoids and quartz are not compatible in the sense that crystallization processes do not allow them to occur in the same rock. Thus, IUGS classification has two triangles, QAP and FAP, which are shown to share the PA edge. In the field, more general terms are often used—for example, a rock containing more than about 20% quartz is referred to as a *granitoid*. An exact mode allows a more precise naming of the rock—for example, it is called a granite if it has 40% Q, 40% A, and 20% P because it plots within the

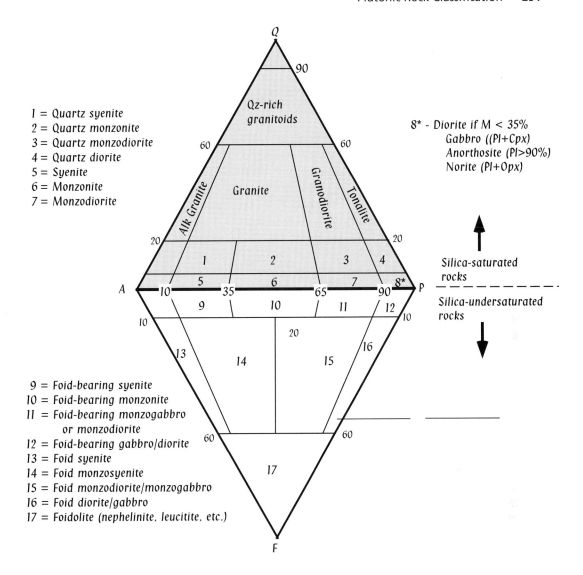

I = Quartz syenite
2 = Quartz monzonite
3 = Quartz monzodiorite
4 = Quartz diorite
5 = Syenite
6 = Monzonite
7 = Monzodiorite

8* - Diorite if M < 35%
    Gabbro ((Pl+Cpx)
    Anorthosite (Pl>90%)
    Norite (Pl+Opx)

 9 = Foid-bearing syenite
10 = Foid-bearing monzonite
11 = Foid-bearing monzogabbro
        or monzodiorite
12 = Foid-bearing gabbro/diorite
13 = Foid syenite
14 = Foid monzosyenite
15 = Foid monzodiorite/monzogabbro
16 = Foid diorite/gabbro
17 = Foidolite (nephelinite, leucitite, etc.)

Q = quartz, A = alkali feldspar and albitic plagioclase, P = plagioclase
F = feldspathoids (Foid), M = mafic minerals

This classification scheme is for plutonic rocks containing <90% M.
For rocks with >90%M, the classification for ultramafic rocks should be used.

FIGURE 8.15   *QAPF classification of felsic plutonic rocks.* The upper triangle (gray) is applicable to rocks with modal quartz (i.e., SiO_2-saturated). All silica-undersaturated rocks containing feldspathoids plot in the lower triangle (SiO_2-undersaturated).

granite field. The IUGS classification recommends the use of a complex system of pre-fixes based on detailed modes; however, such detailed naming is often unnecessary.

The P-corner of each of the triangles in the QAPF classification has more than one rock name in a given field, which obviously needs additional explanation. The distinction between gabbro and diorite can only be made based on the proportion and nature of mafic minerals (gabbro being generally pyroxene rich and diorite being amphibole rich) and plagioclase composition (gabbro >An_{55}; diorite <An_{50}).

CLASSIFICATION OF VOLCANIC ROCKS

There are a number of classifications of volcanic rocks available in the literature. Among these, LeMaitre's (1976; Figure 8.16) classification based on alkalis versus silica contents of volcanic rocks has been shown to be quite useful (Cox et al., 1979). However, even this classification scheme is not entirely free of criticism. For our purpose, we do not dwell on the subject, but accept this classification in naming intermediate and felsic volcanic rocks. However, as far as mafic volcanic rocks are concerned, the most useful classification is a genetic classification that was proposed by Yoder and Tilley (1962). This classification (popularly known as the *basalt tetrahedron*; discussed in a later chapter) is based on calculated mineral molecules (or norm) and has been known to be most useful in understanding the petrogenesis of basalt magmas.

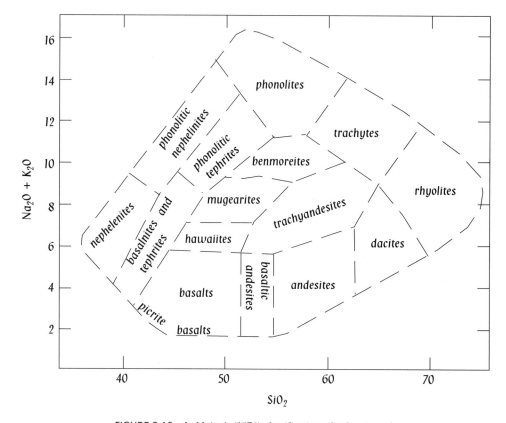

FIGURE 8.16   *LeMaitre's (1976) classification of volcanic rocks.*

CIPW NORM CALCULATION

In 1903, Cross, Iddings, Pirsson, and Washington realized that a way to treat volcanic rocks would be to calculate the proportions of a set of standard minerals (called *normative minerals*) from the chemical composition of a rock (detailed procedure given in appendix). This procedure, which may seem daunting, has almost become a standard practice over several decades, and today most students and researchers use a computer to do these calculations. The mineral percentages so calculated from a chemical analysis are referred to as *CIPW norm* as opposed to *mode*, which refers to actually counted mineral percentages from a thin section of the rock specimen using various microscopic techniques. It should also be noted that CIPW method calculates normative minerals in terms of weight percent. However, if molecular percent of such normative minerals is calculated, it is called a *Niggli norm*. The real advantage of calculating norm is that it allows a way to compare rock compositions to liquidus phase volumes and boundaries in appropriate phase diagrams (e.g., Presnall et al., 1979).

Commonly, a relatively smaller number of normative minerals are used, which generally are apatite (*ap*), ilmenite (*il*), magnetite (*mt*), feldspar (composition given in terms of orthoclase [*or*], albite [*ab*], and anorthite [*an*] components), diopside (*di*—composition is further broken down in terms of wollastonite [*wo*], enstatite [*en*], and ferrosilite [*fs*] components), hypersthene (*hy*—composition may be further broken down into *en* and *fs* components), olivine (*ol*—composition may be further broken down into forsterite [*fo*] and fayalite [*fa*] components), nepheline (*ne*), and quartz [*q*].

The following are important assumptions made in norm calculation:

1. Hydrous phases are ignored and the magmas are assumed to be dry.

2. Mafic minerals are assumed to be free of Al_2O_3. Alumina content is used to calculate the feldspars or feldspathoids.

3. Mg/Fe ratios of all Mg, Fe-containing minerals are assumed to be the same.

4. The procedure recognizes the incompatibility of quartz and feldspathoids (e.g., nepheline, leucite) based on the well-founded observation that quartz and feldspathoids can never occur in the same rock under equilibrium conditions. Thus, quartz and nepheline never appear together in the norm of any rock.

None of these assumptions is particularly off base. However, like any simplifying procedures, the norm calculation procedure also has its limitations—for example, ignoring hydrous phases renders this method to be of limited application to andesitic rocks. Perhaps the greatest value of norm calculation is in the classification of basaltic rocks, as is seen in a later chapter.

A Few Examples That Illustrate the Basic Principle of CIPW Norm

The underlying premise in developing the CIPW scheme of norm calculation is as follows: (a) a magma, if cooled slowly in a crustal environment, would form a rock containing proportions of crystals of certain chemical compositions; and (b) because minerals contribute to the chemistry of a rock, a rock's chemical composition can be redistributed to form mineral molecules of ideal formula compositions. This logic of norm calculation may be illustrated with a few examples as follows.

Consider first an example in which the chemical analysis of a natural titanomagnetite "rock" needs to be recalculated in terms of its normative magnetite (mt) and ilmenite (il) components. Column 1 in the following table gives the actual chemical analysis (weight%) of the "rock." Column 2 simply gives the molecular weights of each oxide. Column 3 lists the moles of each oxide, obtained by dividing Column 1 by

Column 2. At this stage, we are in a position to distribute the appropriate molecular amounts of each oxide to form the two mineral molecules—namely, the *mt* and *il* components. We have two equal choices: We can calculate either ilmenite or magnetite first.

We know that both magnetite ($FeO.Fe_2O_3$) and ilmenite ($FeO.TiO_2$) have FeO. Clearly FeO cannot be used to distinguish the contributions of magnetite versus ilmenite to the rock's composition. However, TiO_2 and Fe_2O_3 are contributed only by ilmenite and magnetite, respectively. Therefore, either of the two can be used to distinguish between the contributions of the two normative minerals. In calculating magnetite ($FeO.Fe_2O_3$), we have to allocate all of Fe_2O_3 and an equal amount of FeO to form magnetite. Similarly, calculation of ilmenite requires allocation of all TiO_2 moles

	(1)	(2)	(3)	(4)	(5)	(6)	(7)
				mol.no.		*CIPW norm*	
		mol. wt.	moles	mt	il	*mt*	*il*
FeO	39.78	71.846	.554	.215	.323	49.78	
Fe_2O_3	34.27	159.69	.215	.215			
TiO_2	25.81	79.9	.323		.323		49.01

and the same amount of FeO to form ilmenite. Columns (4) and (5) show such molecular distributions. In passing, note that at this stage we can recalculate the Niggli molecular norm by recalculating the molar numbers in terms of 100%:

$$\%mt = 100 \times \left[.215/(.215 + .323)\right] = 39.96\%$$

$$\%il = 100 \times \left[.323/(.323 + .215)\right] = 60.04\%$$

Columns (6) and (7) show calculation of CIPW wt% normative *mt* and *il*. These are calculated by multiplying the mole numbers of *mt* and *il* components by their respective molecular weights. For example, wt% *mt* = .215 × 231.536 = 49.78. As expected, molecular% and wt% norms give very different numbers.

As a short digression, note that the prior calculation does not use up all of the FeO molecules available. Also the calculated CIPW norm does not add up to 100%, but 98.79%. This discrepancy may reflect uncertainty in the chemical analysis or the fact that the actual mineral species that forms a solid solution with magnetite is ulvospinel ($2FeO.TiO_2$) and not ilmenite ($FeO.TiO_2$). However, ignoring these facts does not take away anything from achieving our basic goal, which is to understand why norm calculation is done the way it is done.

Consider, as a second example of slightly greater complexity, a rock composed only of normative olivine and hypersthene. The following table shows the wt% contents of FeO, MgO, and SiO_2 in the rock. As before, we first obtain molecular numbers of each oxide.

	wt%	÷	mol.wt.	=	mol.no.
SiO_2	43.62		60.3		0.723
FeO	11.40		71.85		0.159
MgO	44.82		40.3		1.112

After calculating the mol. no's, we are at a loss as to how to distribute the oxide molecules between normative olivine [2(MgO,FeO).SiO_2] and hypersthene [(MgO,FeO).SiO_2] since both are composed of MgO, FeO, and SiO_2. The easiest way would be a simple algebraic formula as follows. We assume that x = no. of hy molecules (which is also equal to the no. of SiO_2 molecules in hypersthene) and y = no. of ol molecules (which is the same as the no. of SiO_2 molecules in olivine). Let S = available SiO_2 and M = available MgO + FeO. We know from chemical formulae of the two minerals that hypersthene [(MgO,FeO).SiO_2] requires equal amounts of (MgO + FeO) and SiO_2, whereas MgO + FeO = 2 SiO_2 in olivine [2(MgO,FeO).SiO_2]. Therefore, we can write:

For hypersthene, $(SiO_2)^{hy} = x = (MgO + FeO)^{hy}$

For olivine, $(SiO_2)^{ol} = y$, and $(MgO + FeO)^{ol} = 2y$

Since S is the total SiO_2 available, it follows that $S = x + y$. Similarly, $M = x + 2y$. From these two equations, x and y may be calculated as follows:

$$x = 2S - M \quad \text{and} \quad y = S - x.$$

In our example, S = 0.723 and M = 1.271. Therefore,

number of hy molecules = x = 0.175, and

number of ol molecules = y = 0.548.

At this stage of calculations, we know that in hypersthene MgO + FeO = 0.175. However, we do not know MgO and FeO values individually, and we need to know those values if we are to obtain CIPW norms, which require multiplication by appropriate molecular weights. Here we make the assumption, as required by the CIPW calculation procedure, that the MgO/(MgO + FeO) ratios of both minerals are equal to that available:

$$[MgO/(MgO + FeO)]^{hy} = [MgO/(MgO + FeO)]^{ol} = [MgO/(MgO + FeO)]^{available}$$

We know from previous calculations that

$$[MgO/(MgO + FeO)]^{available} = 1.112/1.271 = 0.875.$$

Therefore,

$$[MgO/(MgO + FeO)]^{hy} = [MgO/(MgO + FeO)]^{ol} = 0.875$$

We already know $(MgO + FeO)^{hy} = 0.175$, and therefore easily calculate

$$(MgO)^{hy} = 0.875 \times 0.175 = 0.153 \quad \text{and} \quad (FeO)^{hy} = 0.175 - 0.153 = 0.022.$$

These numbers can now be used to obtain CIPW norms of enstatite and ferrosilite components in normative hypersthene in the same manner as in our first example:

	mol.no.	en	fs
SiO_2	0.175	0.153	0.022
FeO	0.022		0.022
MgO	0.153	0.153	

CIPW normative en = 0.153 × 100.38 (mol.wt. of enstatite) = 15.36

CIPW normative fs = 0.022 × 132.15 (mol.wt. of ferrosilite) = 2.91

Similarly, olivine has 0.548 SiO_2 and total $MgO + FeO = 1.096$. Using the prior ratio assumption, we calculate as follows:

	mol.no.	fo	fa
SiO_2	0.548	0.4795	0.0685
FeO	0.137		0.137
MgO	0.959	0.959	

$$\text{CIPW normative fo} = 0.4795 \times 140 \text{ (mol.wt. of fo)} = 67.13$$

$$\text{CIPW normative fa} = 0.0685 \times 204 \text{ (mol. wt. of fa)} = 13.97$$

Adding up all the normative components, en + fs + fo + fa, we get 99.37.

More detailed examples of norm calculation from chemical analysis of actual rocks are provided in the appendix. The objective of these examples is to give the student an idea of the science of norm calculation.

SUMMARY

[1]   Igneous rocks form by solidification of magma. They are broadly grouped into volcanic, hypabyssal, and plutonic rocks based on their mode of occurrence.

[2]   Volcanoes are fundamentally of three kinds—shield volcanoes, stratovolcanoes, and cinder cones. Hypabyssal intrusions are of two kinds—dikes and sills. Plutons can be of many types—lopolith, laccolith, batholith, ring dikes, etc.

[3]   A typical rock description must include a description of the minerals and their mode, and texture. Simple field classification of igneous rocks is based on mineralogy and texture.

[4]   IUGS classification presents a more rigorous approach to naming plutonic rocks based on their mode. Volcanic rocks are named on the basis of their chemical composition.

[5]   CIPW norm calculation method allows one to convert a chemical analysis of a rock to modal constitution in terms of some end member mineral molecules.

BIBLIOGRAPHY

Barker, D.S. (1983) *Igneous Rocks*. Prentice-Hall (Englewood Cliffs, New Jersey).

Carmichael, I.S.E., Turner, F.J. and Verhoogen, J. (1974) *Igneous Petrology*. McGraw-Hill (New York).

Cox, K.G., Bell, J.D. and Pankhurst, R.J. (1979) *The Interpretation of Igneous Rocks*. George Allen and Unwin (London, U.K.).

LeMaitre, R.W. (1976) *Numerical Petrology*. Elsevier (Amsterdam, The Netherlands).

McBirney, A.R. (1993) *Igneous Petrology* (2nd ed.). Jones and Bartlett (Boston, MA.).

Philpotts, A.R. (1990) *Principles of Igneous and Metamorphic Petrology*. Prentice-Hall (Englewood Cliffs, New Jersey).

Presnall, D.C. et al. (1979) *Generation of Mid-Ocean Ridge Tholeiites. J. Petrol.* **20**, 3–35.

Ragland, P.C. (1989) *Basic Analytical Petrology*. Oxford University Press (Boston, MA.).

Wilson, M. (1989) *Igneous Petrogenesis*. Unwin Hyman (Boston, MA.).

CHAPTER 9

Nature and Properties of Magma

The following topics are covered in this chapter:

Magma
  Chemical and physical nature
  Structure, viscosity and density of silicate liquids
  Magma formation, segregation, and ascent

ABOUT THIS CHAPTER

The starting point of all igneous rocks is magma (or molten rock). Magma is not often a simple liquid, but rather a complex mixture of liquid, solid materials, and dissolved vapor. Magma's chemical composition and physical properties, such as viscosity and density, are all important factors that control magma's overall behavior, including its mineralogy and texture. This chapter deals with an evaluation of magma's composition and behavioral aspects during its formation, transport, and emplacement.

MAGMA: FUNDAMENTALS

Chemical Composition of Magma

Most magmas that crystallize at depth or erupt on the earth's surface as lavas carry crystalline materials and dissolved vapor—mainly CO_2 and H_2O. In terms of chemical composition, all magmas (except rare carbonatites) are silicate magmas, in which the dominant component is silica (SiO_2), generally comprising 45% or more by weight. Alumina (Al_2O_3), with its abundance in common magmas somewhere between 13% and 18%, is a distant second (Table 9.1). SiO_2, TiO_2, Al_2O_3, Fe_2O_3, CaO, MgO, MnO,

TABLE 9.1 *Chemical composition of some igneous rocks.*

wt%	Komatiite	Basalt	Andesite	Rhyolite
SiO_2	46	48.9	60	75
TiO_2	0.35	2.45	0.6	0.3
Al_2O_3	7	14.1	16	13
Fe_2O_3	1.6	1.8	1.5	0.2
FeO	9.8	10.9	6	1.7
MgO	25	8	4	0.2
CaO	7.5	9.7	7	1.2
Na_2O	0.5	2.8	3.2	4.5
K_2O	0.05	0.4	2	4.2

FeO, Fe_2O_3, Na_2O, and K_2O comprise 99% of any igneous rock. There are many other elements (e.g., V, Sr, Ni, La, Ce, Nd, Sm, Eu, Yb, Hf, etc.) whose abundances are generally much lower (parts per thousand or parts per million levels) and are hence referred to as *trace elements*. These elements are nonetheless important because they provide signficant clues to the origin of igneous rocks.

As a starting point, it is useful to classify magmas into four broad groups—namely, *ultramafic, mafic, intermediate*, and *felsic* types (Table 9.2)—based on relative abundances of SiO_2, $MgO + FeO + Fe_2O_3$ (called the *mafic* component) and total alkalis ($Na_2O + K_2O$). It is apparent from Table 8.2 that the mafic component is minimal in felsic magmas and increases toward ultramafic magmas. Total alkalis and silica exhibit the opposite behavior. Naturally, what minerals may form from a magma directly depend on the magma's chemical composition (e.g., an ultramafic magma with its low SiO_2 and high $MgO + FeO$ should be expected to crystallize minerals like olivine and pyroxenes, which have high $MgO + FeO$). Quartz (SiO_2) would not be expected to form in such a magma because olivine and pyroxene crystals would use up all the available silica so that there would not remain any excess silica needed to form quartz. However, in a felsic magma, which is very high in silica and alkalis and poor in $MgO + FeO$, one may expect crystals of alkali feldspars ($[K,Na]AlSi_3O_8$) and quartz to form. Tables 9.2 and 8.1 (Chapter 8) show this direct correlation between magma's chemical composition and minerals that crystallize from it. This is easily anticipated based on the discussion of norm calculation in Chapter 8.

TABLE 9.2   *First-order division of magma types.*

wt%	Ultramafic	Mafic	Intermediate	Felsic
SiO_2	>43–49	46–53	60–65	>65
$MgO + Fe_2O_3 + FeO$	35–46	15–28	10–21	<3
$Na_2O + K_2O$	<1	2–3.5	3–6	5–10

Solid Component in Magma

A magma rises from its place of origin primarily due to its lower density (although other factors, such as volatile pressure, also contribute to magma's ascent) than that of the surrounding wall rocks. During its ascent, a magma cools such that crystals may form and separate from the rising magma. In erupted lavas, such crystals, often euhedral, may be recognized as phenocrysts by virtue of their large size relative to the surrounding glass or fine crystals that form due to rapid quenching of the lava.

A magma may also carry rock and mineral fragments broken off the conduit wall as it rises violently from the deep. Such wall rock and mineral inclusions are called *xenoliths* and *xenocrysts*, respectively (Figure 9.1). The study of xenoliths and xenocrysts has proved to be particularly valuable because many of them represent great depths in the earth's upper mantle, which cannot be reached by drilling. Xenoliths and xenocrysts are often identified on the basis of their chemical composition, mineral assemblages, and disequilibrium reaction textures (such as rounded or resorbed boundaries), which develop due to their being out of chemical equilibrium with the host magma.

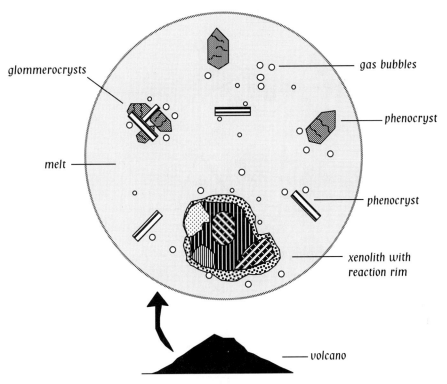

FIGURE 9.1 *A schematic drawing of a microscopic view (crossed polars) of a volcanic rock showing the presence of glass (isotropic), phenocrysts of two different minerals (olivine and plagioclase), cluster of phenocrysts (glomerocrysts), exsolved gas bubbles (preserved as vesicles), and a xenolith.* Because the xenolith was not in equilibrium with the host lava, a reaction rim composed of different minerals developed between the host lava and the xenolith.

Structure of Silicate Magma

Magma's viscosity and density are important factors that determine how fast it rises. These physical properties of a magma are fundamentally related to its internal arrangement of atoms and molecules. Magma's atomic structure may be understood by referring to the atomic structures of silicate minerals. We noted earlier that the fundamental structural unit of all silicate minerals is the $[SiO_4]^{4-}$ tetrahedron, in which a small Si^{4+} cation at the tetrahedron's center is covalently bonded to 4 O^{2-} anions occupying its apices. In most silicate groups (e.g., the chain silicates), $[SiO_4]$-tetrahedral units share some of the apical oxygens, whereas the nonshared oxygens are bonded to other cations (Mg, Fe, Na, Ca, etc.) present in the mineral's structure. The shared versus nonshared oxygens are called *bridging* and *nonbridging* oxygens, respectively. Si and other fourfold coordinated cations (principally Al) that form the backbone structure of a tetrahedral network are called *network formers*. The cations (e.g., Mg, Ca, etc.) that usually occur in higher coordination and link the networks are called *network modifiers*. Silicate magmas, like silicate minerals, are composed of extensive networks of $[SiO_4]$—tetrahedral units with neighboring tetrahedra containing bridging and nonbridging oxygens. During melting of a silicate mineral or rock, the covalent bonds between Si^{4+} and oxygens of each tetrahedron are generally too strong to break. However, the bonds between nonbridging oxygens and network modifiers do break, enabling these other cations to move around within the framework of polymerized $[SiO_4]$ chains in a molten silicate (Figure 9.2). As one would

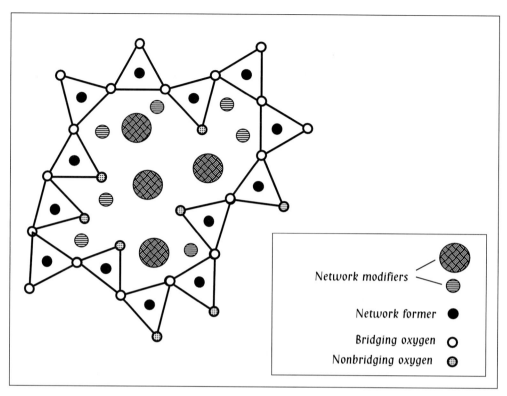

FIGURE 9.2 *Atomic structure of a silicate melt showing polymerized chains of silicate tetrahedra, network formers, and network modifiers.*

expect, SiO_2 content of a magma exerts strong control over the extent of polymerization of a melt—the greater the SiO_2, the more polymerized the melt.

Viscosity

Viscosity (η) of a magma is simply defined as its internal resistance to flow and is given as:

$$\eta = \sigma/\varepsilon,$$

where σ and ε are applied shear stress and rate of shear strain, respectively. A convenient way to understand what viscosity means is to compare the difference between how syrup and water flow: Syrup is said to be more viscous than water because it flows more slowly than the latter. Magmas or fluids in general can show a range of viscosity characteristics: Newtonian fluids exhibit a linear relationship between σ and ε and pass through the origin in a stress versus strain rate plot (Figure 9.3). Crystal-free basalt magmas show Newtonian behavior. In contrast, a basalt or andesitic magma containing abundant crystals may behave like a Bingham plastic—that is, they may possess some finite yield strength and thus flow only when a certain threshold value of stress has been exceeded. Rhyolitic magmas exhibit pseudoplastic behavior, in that they show a nonlinear relationship between stress and strain rate. As should be expected, the extent of polymerization (i.e., how many silicate chains occur in the magma, which is directly dependent on the abundance of SiO_2) exerts a strong control over viscosity. Thus, rhyolite magma, with its highest SiO_2 content, is more viscous than andesite and basalt magma (Figure 9.4). Temperature increase breaks down

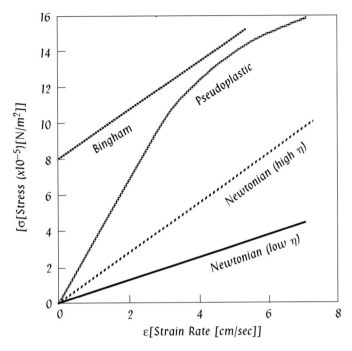

FIGURE 9.3   *Stress–strain relationships and types of magma viscosities.*

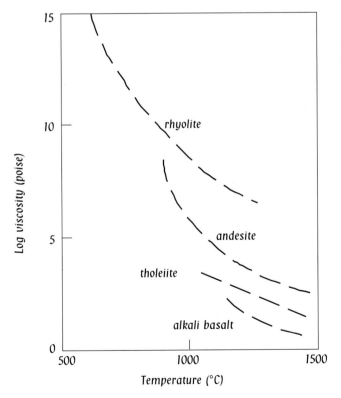

FIGURE 9.4   *Viscosities of four principal magma types at 1 atmosphere pressure and as a function of temperature.* (After McBirney, 1993, 2nd ed., *Igneous Petrology*, Jones and Bartlett, with permission)

more silicate networks in the magma and thus lowers its viscosity. However, increasing crystallization makes a magma progressively more viscous. Increase of dissolved H_2O, a network modifier greatly lowers the viscosity of a magma (Figure 9.5).

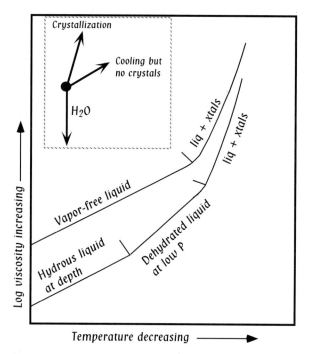

FIGURE 9.5 *Viscosities of hydrous versus anhydrous magmas as a function of devolatilization and partial crystallization (modified from Hess, 1989).* Hydrous magma has lower viscosity than anhydrous magma of broadly similar composition. As a hydrous magma rises and volatiles exsolve from it (due to lowering of pressure or temperature; discussed in a later chapter), its viscosity increases. Appearance of crystals sharply increases the viscosity of a magma as well.

Density

Densities of different magmas have been measured in the laboratory, mostly at atmospheric pressure, and they vary between 2.2 and 3.1 g/cm^3 (Figure 9.6). Density of a magma is directly related to the abundance of the mafic (i.e., Mg + Fe) component in the magma. Thus, rhyolite magmas are less dense than andesites, and andesites are less dense than basalts. A magma rises from its point of origin by buoyancy, much

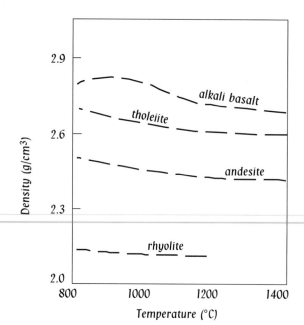

FIGURE 9.6 *1 atmosphere densities of magmas as a function of temperature.* (Redrawn with permission from McBirney, 1993, 2nd ed., *Igneous Petrology,* Jones and Bartlett)

9.1 VISCOSITY, DENSITY, AND VELOCITY OF A XENOLITH-BEARING MAGMA FROM HAWAII

The island of Oahu is composed of two large-shield volcanoes and numerous small-cinder cones. Lavas that erupted some 1 to 2 million years ago from the small cones contain mantle and crustal xenoliths. Using some simple viscosity–density values of the xenoliths and magma, it is possible to calculate a minimum ascent velocity of the magma. Note that these xenoliths are heavier (density = 3.4 g/cm^3) than the magma (density = 2.8 g/cm^3), and therefore should have precipitated (or settled) out of the magma, but they were brought up because the magma was rising faster than the xenoliths could settle. Using a simple law called Stoke's law, and assuming that the magma behaved as a Newtonian fluid, one can calculate the minimum ascent velocity of the magma that prevented the xenoliths from settling.

Stoke's law is given as:

$$V = \frac{2\, g r^2 (\rho_{cryst} - \rho_{liq})}{9\eta},$$

where

V = settling velocity of the xenolith in magma

g = acceleration due to gravity (assumed to be constant; 980 cm/sec^2)

r = radius of the xenolith (2.5 to 10 cm)

ρ_{cryst} = density of xenolith (assumed to be 3.4 g/cm^3)

ρ_{liq} = density of liquid (assumed to be 2.8 g/cm^3)

η = viscosity (varies between 200 and 500 poise for basalt magmas ± crystals)

Plugging these values into the prior equation, minimum ascent velocities of the host magma were calculated and plotted in Figure Bx 9.1 for three different sizes of xenoliths and three different viscosity conditions. It is clear that viscosity has no effect on ascent velocity for small (which are more common in Hawaii) xenoliths. However, for larger sizes, the effect is quite strong. Note that these estimates increase by twice as much if the magma behaved like a Bingham fluid. However, field observations (such as a thin glassy coat of the host lava around the xenolith) on some of these xenolith-bearing magmas of Hawaii suggest that they were extremely fluid and may have behaved like a Newtonian fluid (Spera 1980; Sen 1983).

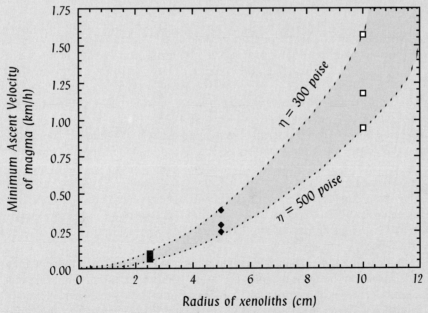

FIGURE BX 9.1   *Minimum velocities of xenolith-bearing magmas are calculated using Stoke's law for three different xenolith sizes (radius, r = 2.5, 5, and 10 cms) and a range of magma viscosities between 300 and 500 poise.*

like a helium-filled balloon rises through the atmosphere. This happens because the magma is lighter (i.e., less dense) than the surrounding wall rock.

Consider a simple example of buoyant rise of a basaltic magma from a depth of 60 km to the surface. We assume that the wall rock at ~60–40 km is peridotite with a density of 3.3 g/cm³. The magma's density is assumed to be constant at 2.9 g/cm³. The pressure on the magma and wall rock at any depth can be calculated from:

$$P = \rho g h$$

where P is pressure (in GPa), ρ is density, and g is acceleration due to gravity (assumed to be constant with a value of 980 cm/sec²).

$$P_{rock} \text{ at } 60 \text{ km} = (60,000,00*3.3*980)/10^{10} = 1.94 \text{ GPa}$$

$$P_{magma} \text{ at } 60 \text{ km} = (60,000,00*2.9*980)/10^{10} = 1.70 \text{ GPa}$$

Therefore, the pressure difference (ΔP) of 0.24 GPa between the magma and wall rock makes the magma sufficiently buoyant to rise to the surface. In reality, however, this pressure difference may disappear when the magma reaches the Moho because crustal rocks, especially in continents, often have a lower density (~2.7 g/cm³) than that of the magma. Therefore, the magma is not buoyant when it reaches the Moho. Instead, it may be trapped at the Moho, where it would start crystallizing. After some amount of crystallization, the remaining differentiated magma may rise again when its density becomes less than that of its surrounding crust. Stolper and Walker (1980) showed that the vast majority of midoceanic ridge basalts (MORB) undergo such density filtration and differentiation in magma chambers. Thus, they argued that erupted MORB are not the primary magmas formed directly by partial melting of the mantle, but are derived from them.

Density of a magma at atmospheric pressure can be calculated from its chemical composition and molar volume as follows:

$$\rho = \frac{\sum X_i M_i}{\sum X_i V_i}$$

where X_i, M_i, and V_i are mole fraction, gram molecular weight, and molar volume of an oxide. In a magma composed of many different oxide components (SiO_2, Al_2O_3, TiO_2, and so on), the mole fraction of any particular oxide—say SiO_2—is the number of moles of SiO_2 present divided by the total moles of all oxides (including SiO_2 moles) and may be represented as: X_{SiO_2} = moles of SiO_2/(moles of SiO_2 + moles of TiO_2 + moles of Al_2O_3 + ... etc.).

Expressed in more general terms, in a magma or crystal composed of chemical components a, b, c ... i, mole fraction of the component "i" may be given as: $X_i = (m_i)/(m_a + m_b + m_c + ... m_i)$, or $X_i = m_i/\sum m_{a...i}$ where $m_{a...i}$ are moles of $a, b, c, ... i$ components.

High-pressure measurements of density and viscosity have been very few (Figure 9.7). Kushiro's (1980) experiments show that the density of a basaltic magma rises from 2.95 g/cm³ at 1 atm pressure to about 3.5 at a pressure of 3 GPa, whereas viscosity decreases drastically.

Earlier we noted that densities of magmas are generally less than those of mantle minerals. However, because magmas are more compressible than minerals, density of magmas increases at a more rapid rate than minerals as pressure is increased. Some scientists have determined that ultramafic (komatiitic) magmas become denser than mantle minerals at 11–12 GPa (i.e., 330–360 km; Rigden et al., 1984; Agee and Walker, 1988; Ohtani et al., 1995). The occurrence of such a density cross-over implies

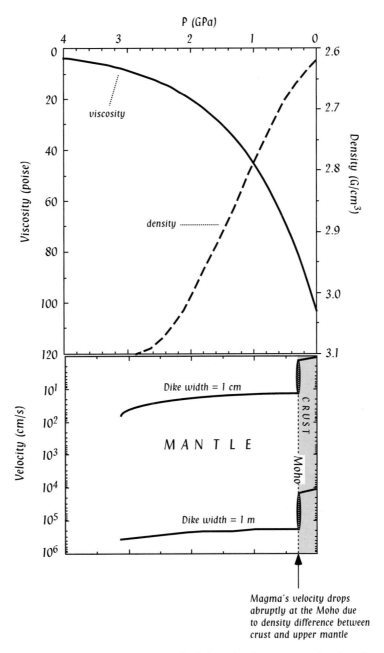

FIGURE 9.7 *Density and viscosity of a tholeiitic basalt magma as a function of pressure.* (Based on Kushiro, 1980)

that magmas generated in the earth's lower mantle will not leave its source area and instead will sink. Magmas produced from the mantle at such high pressures are likely to be highly magnesian in composition (ultramafic) (e.g., Takahashi, 1986; further discussed in a later chapter). However, because their density is likely to be greater than the surrounding mantle rocks (and therefore lack of buoyancy), it is highly unlikely that they will rise toward the surface. This probably explains why we do not have any modern eruptions of komatiite-like magmas.

MAGMA FORMATION, SEGREGATION, AND ASCENT

Under favorable conditions, tiny fractions of magma may form in the crust or mantle. Once formed, such magma must somehow segregate from the source rock and form larger pools that will ascend toward the surface. Many magma bodies may only rise to a certain level and freeze, whereas others may erupt through volcanoes. Next we consider the processes that are responsible for the generation and delivery of magma to the crust.

Magma Formation

Magma may form at any depth within the earth if the temperature at that depth is sufficent to melt the rock(s) present at such a depth. This may happen in one of two ways: by bringing a hot, deep rock to a shallow level where its temperature is higher than its solidus temperature (referred to as *decompression melting*) or by lowering the solidus temperature through the introduction of a volatile phase (called *volatile-aided melting*). In divergent plate margins (e.g., midoceanic ridges), as plates are pulled apart, hot asthenosphere passively rises and begins to melt at a shallow level (called *passive melting;* Figure 9.8). In hot spots, decompression melting happens due to the injection of a hot, deep plume or jet at a shallow depth and is referred to as *plume melting*.

ADVANCED READING

It is apparent from Figure 9.8 that decompression melting can happen because the solidus has a positive slope (i.e., the melting point of a rock increases with pressure) and the path followed by the rising mantle parcel has a shallower slope. The rise of deep, hot rock bodies is generally believed to follow an adiabatic gradient (i.e., they cool entirely due to expansion of volume as pressure decreases). The adiabatic gradient is expressed as:

$$\frac{d\mathrm{T}}{dz} = \frac{\alpha g \mathrm{T}}{\mathrm{C}_p},$$

where α is the volume coefficient of thermal expansion, T is temperature, z is depth, and C_p is heat capacity. The adiabatic gradient ($d\mathrm{T}/dz$), or change of temperature as depth decreases, for the upper mantle lherzolite is ~ 0.3 K/km, which is a much smaller number than the $d\mathrm{T}/dz$ slope of the mantle solidus (~ 8 K/km) at ~ 60–20 km. Therefore, the adiabatic gradient of a rising mantle parcel intersects the mantle solidus at a shallow angle, which results in partial melting of such a parcel (Figure 9.8). Once melting ensues, latent heat of fusion (~ 100 cals/g) is used up in the melting process and cools the mantle parcel: the resulting $d\mathrm{T}/dz$ gradient is then a steeper gradient ($d\mathrm{T}/dz \sim 3$ K/km). Note in the prior calculations the following numbers were used: $\alpha = 3 \times 10^{-5}$/K, $\mathrm{C}_p = 0.3$ cals/g.

A likely locus of volatile-aided melting is above a subduction zone, where hydrous minerals in the subducting crust break down and release H_2O into the overlying mantle wedge (Figure 9.9). This H_2O lowers the solidus of the rocks comprising

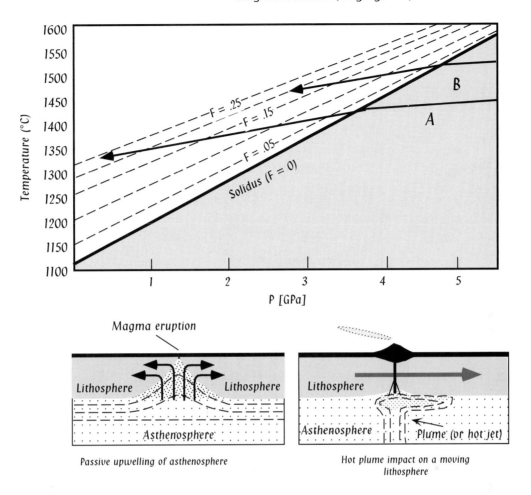

FIGURE 9.8 *Melting in a passive melting regime (bottom left panel) versus a plume with a lithosphere passing over it (bottom right panel).* The bottom panels: *black*—crust, *gray shading*—lithosphere, *light dots*—asthenosphere, *densely spaced dots*—partially molten asthenosphere, *lighter gray*—partially molten plume. Arrow in the bottom right panel indicates movement of the lithospheric plate. Arrows in the bottom left panel indicate directions of mantle flow. Dashed contours—isotherms.
Top panel shows the paths of mantle parcels in the plume versus passive melting regimes.
[1 GPa = 10 kb ≃ 30 km.] F represents degree of melting of the mantle rock: F = 0.1 means that 10% of the rock is molten. Note that the F lines are more closely spaced at higher pressures than at lower pressures [This diagram is principally based on data from Jaques and Green (1980), Hirose and Kushiro (1993) and Baker and Stolper (1994).]
Path A—Rise of a single asthenospheric parcel in a passive melting regime. It follows an adiabatic path until it crosses the solidus, where it begins to melt. These melts are extracted rapidly in a near-fractional melting process. Melting uses up latent heat and thus cools the parcel, and the slope of the adiabat changes slightly, but the parcel continues to melt as it continues to rise upward. Melting finally stops when a mineral phase (likely to be clinopyroxene) in the rock is used up or melting regime hits the cool lid (lithosphere) on top; as a consequence, the parcel cools rapidly and moves to the side becoming the lithosphere.
Path B—A much hotter parcel from the plume rises along an adiabat and begins to melt as it crosses the solidus at a deeper level than Path A. The Path B mantle parcel continues to melt as it rises until one or more mineral phases in it is/are exhausted or the plume impacts the cool lithosphere. Because the %F lines are more closely spaced at higher pressure, the melting rate (dF/dP, %melting as a function of pressure) is greater for Path B than Path A.

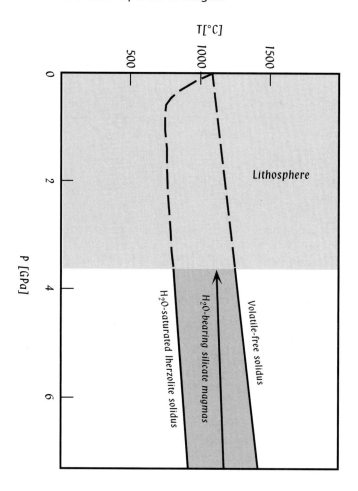

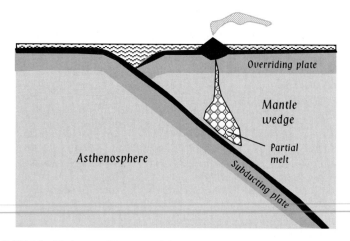

FIGURE 9.9 *Hydrous melting in a subduction zone.* Bottom panel shows a schematic cross-section of a subducting plate where H_2O-rich fluid is given off the subducting plate. This fluid lowers the melting temperature of the mantle wedge above the subducting plate, triggering melting. Top panel shows the extent to which H_2O lowers the mantle lherzolite solidus. It also shows the vertical temperature gradient in the wedge (arrow). Note that this path is lower than the volatile-free solidus; therefore, if H_2O were not added to the wedge, melting would not occur.

the mantle wedge, triggering melting if the ambient temperature of the wedge exceeds the new solidus temperature. A magma formed in the presence of volatiles is significantly different from the magma produced under volatile-absent condition from the same rock at the same pressure. For example, a lherzolite melting at ~1.5 GPa may yield an olivine tholeiitic partial melt, whereas the same rock under hydrous conditions at 1.5 GPa may form a high-alumina basalt or a magnesian andesite magma. Further discussion follows in a later chapter.

Magma Segregation

Experimental petrology and seismology indicate that most magmas erupted within the Cenozoic era have been generated within the top 200 km of the earth. Deep in the source region where melting begins, tiny pockets of magma are probably distributed along intergranular pore spaces of the source rock. Such partially molten rocks rise to shallower levels by diapirism (i.e., the entire mass of rock + melt may flow upward by deforming the very hot rocks that surround the partially molten zone; e.g., Philpotts, 1990). During this process, magmas leave behind their matrix by porous flow (i.e., they migrate along intergranular pore spaces). At some point, compaction must become a highly effective process. In compaction, the matrix deforms and flows downward (and sideways, in the case of oceanic lithosphere formation) and the melt is squeezed out to form a layer at the top. Eventually pipe flow takes over, in which magma pockets become sufficiently large and are transported in well-defined conduits. Magma may create its own conduit by stoping (i.e., by fracturing and incorporating wall rocks), or the conduit may be a preexisting fracture.

ADVANCED READING

To understand magma segregation via porous flow, one must know the geometry of pore spaces in the source rock. Significant progress has been made in recent years toward quantifying the segregation of basaltic magmas from their upper-mantle source rock, lherzolite. Melting appears to begin at the intersections (triple-point junctions) of grains of three different phases. Melting along corners occurs to minimize the interfacial energy differences between the crystalline surfaces and melt. In this process, the dihedral angle (θ) between two adjacent solid grains and melt plays an important role in controlling melt distribution in the pore spaces of the rock (Figure 9.10). If $\theta < 60°$, then melts in all corners are interconnected, which allows the melt to escape along grain boundaries even when the melt fraction is very small. Basaltic magmas and lherzolite matrix exhibit this behavior. When $\theta > 60°$, the melts at the pore spaces do not form an interconnected network, and thus melt escapes from the matrix only when a certain porosity threshold (melt fraction) is crossed. Granitic melts generally show this type of behavior and therefore do not escape from the source very easily.

Flow of magma through interconnected pores (i.e., porous flow) is governed by D'Arcy's law:

$$v = \frac{K}{\mu\phi} \cdot \frac{dP}{dz},$$

where v is magma velocity, K is permeability, ϕ is porosity, and μ is viscosity of the magma. dP/dz is simply the pressure gradient caused largely by the density difference between a magma and the solid residue. Permeability is affected

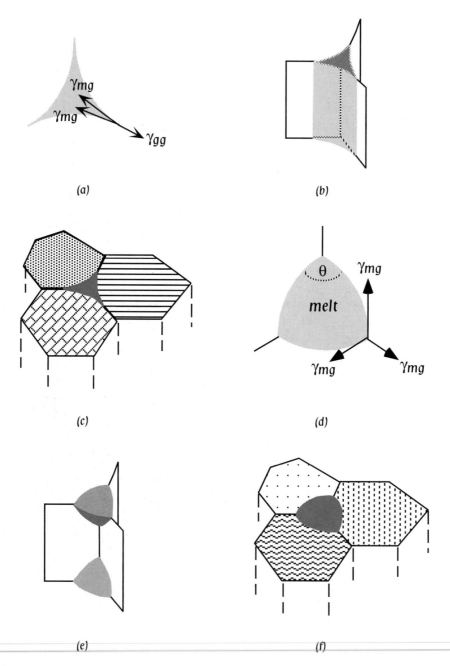

FIGURE 9.10 *Melt distribution in pore spaces between grains in a rock.* (a) Interfacial energy differences between two adjacent grains and between melt and grain are given as γ_{gg} and γ_{mg}, respectively. θ = Dihedral angle. (b) The first melt occupies triple-point junctions, where three grains of three different minerals meet. In cases of very low dihedral angles (applicable for basalt magma—lherzolite residue), the melt is expected to form melt channels through triple-point junctions as well as thin films along grain edges. (c) In case of greater θ applicable to highly silicic melts (rhyolitic), the melts do not interconect until a sufficiently high degree of melting has occurred (after Watson, 1982).

by grain size and porosity: Coarser grains have greater permeability. The greater the viscosity, the slower will be melt velocity. Therefore, a basalt magma will travel faster than a granitic magma because of its lower viscosity. D'Arcy's law indicates that melt velocity is inversely proportional to melt fraction (ϕ). Note that at higher ϕ, the dihedral angle may become greater than 60°, as a result of which the channels connecting the pores may get pinched off (Figure 9.10).

In compaction, a two-phase flow occurs in which the melt flows (segregates) upward from the solid matrix (residue) as the latter deforms and flows downward (Figure 9.11). This happens due to the sharp density contrast between melt and matrix. McKenzie (1984) provided a rigorous treatment of compaction and showed that, in the case of basalt magmas, a maximum of $\sim 3\%$ melt may occupy the pores to satisfy grain/melt interfacial energies. Once such a value is exceeded, the matrix deforms, expelling the melt.

Consideration of a slightly modified version of D'Arcy's law tells us whether compaction would occur or not:

$$v_{melt} - V_{matrix} = -[K/\phi\mu] \cdot [dP/dz + \rho_{melt}\, g],$$

where v_{melt} and V_{matrix} are velocities of melt and matrix, K = permeability, P = pressure, ρ_{melt} = density of melt, g = acceleration due to gravity, and z = height, which becomes positive upward. When the pressure is hydrostatic, melt movement will not occur:

$$dP/dz = -\rho_{melt}g \qquad\qquad \text{[Eq. 9.1]}$$

or

$$dP/dz + \rho_{melt}g = 0.$$

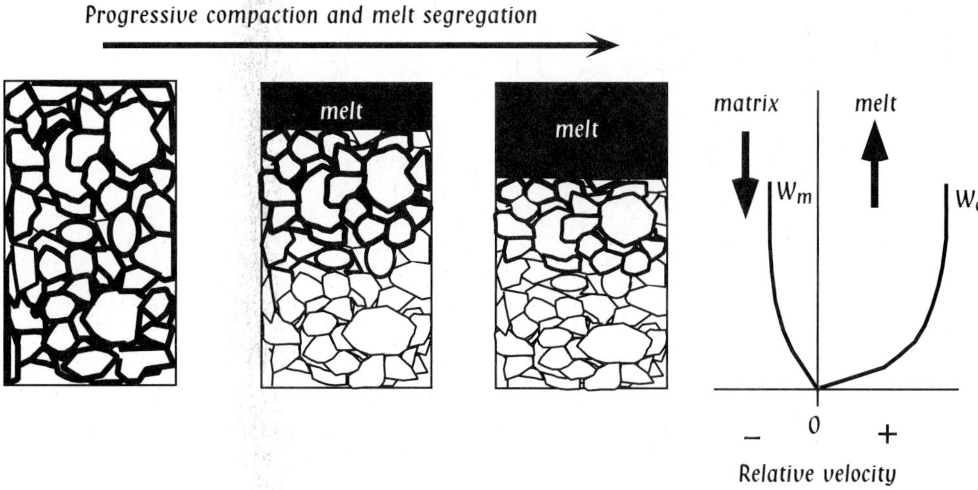

Progressive compaction and melt segregation

FIGURE 9.11 *The left three panels show schematic drawings of how compaction proceeds.* In the beginning (uncompacted), the melts are located along grain boundaries (left). With progressive compaction, the solid matrix settles to the bottom while the melt separates and collects at the top. The right panel shows relative velocities with which the melt $\left(\text{velocity} = W_m\right)$ and matrix (W_o) segregate from one another. The velocity signs represent the direction along which material is flowing: $(+)$ for upward flow and $(-)$ for downward flow. The melt velocity is high in the lower part of the compacting column and becomes constant above a certain height. The matrix velocity is lower than the melt velocity at all heights. (From Sparks, 1992, in *Understanding Earth* (Brown et al., eds.), Cambridge University Press)

ρ_{melt} can be related to the mean density of matrix $+$ magma assemblage as follows:

$$\rho_{mean} = (1 - \phi)\rho_{matrix} + \phi\rho_{melt}$$

or

$$\rho_{melt} = \left[\rho_{mean} - (1 - \phi)\rho_{matrix}\right]/\phi.$$

Matrix does not expand or compact when the pressure gradient

$$dP/dz = -\rho_{mean}g$$

or

$$dP/dz = \left[(1 - \phi)\rho_{matrix} + \phi\rho_{melt}\right]g \qquad \text{[Eq. 9.2]}$$

To prevent matrix deformation both Equations 9.1 and 9.2 need to be true, which can only happen if $\phi = 1$ (i.e., no matrix may be present). Since this is an impossibility, matrix must undergo compaction.

Compaction must be considered in terms of length and time scales (Figure 9.11). Since melt must form and segregate upward, the melt velocity should be zero at some depth (the lower boundary of the length scale). The upper boundary of this system may be defined to have zero stress on the matrix. Melt velocity (W_m) increases from the lower boundary until at some height it reaches a constant value ($w°$), at which it can balance the pressure gradient due to mean density. Therefore, compaction is greatest close to the lower boundary and diminishes exponentially upward, becoming zero at the upper boundary, which may be an impermeable layer or a permeable layer through which the melt may percolate (McKenzie, 1984). McKenzie (1984) used a dimensionless parameter, called *expansion rate* ($1/\phi \; \partial\phi/\partial t'$), to show this behavior: It is zero at the lower boundary, remains negative up to the upper boundary, but its rate decreases exponentially from bottom to top (Figure 9.11). The *compaction length* (δ_c) is defined as a dimensionless quantity:

$$\delta_c = \sqrt{\frac{\zeta + 1.33\eta}{\mu}} \cdot K, \qquad \text{[Eq. 9.3]}$$

where η and ζ are shear and bulk viscosities of the matrix, μ is the viscosity of the melt, and K is permeability.

Permeability may be expressed, as a first-order approximation, as:

$$K = a^2\phi^3/\left[k(1 - \phi)^2\right],$$

where a is the spherical radius of grains in the matrix, ϕ is porosity (or melt fraction), and k is a constant whose value may be about 1,000 (McKenzie, 1984).

As an example of a back-of-the-envelope calculation, consider the case of a basalt magma segregating from the upper mantle beneath a midoceanic ridge. We assume the values 10^{19} and 100 poise for matrix viscosity ($\zeta + 1.33\eta$) and melt viscosity μ, respectively. Table 9.3 shows permeability and compaction length values calculated for different melt fractions (ϕ) and two different sizes of grains ($a = 0.2, 0.3$ cm). Table 9.3 and Figure 9.11 show that the compaction length (δ_c) is likely to be $< \sim600$ m for melt fractions (F) of up to 3% in the upper mantle where the average grain radius (as estimated from xenoliths) is

~0.2–0.3 cm. Beyond this height, the magma must move by percolation or in dikes. Note that this compaction length is much smaller than the average length over which magma is believed to be generated, say, beneath a midoceanic ridge (10s of km). Thus, it seems likely that the magma segregates from the bottom to near the top of the melting zone through a series of compaction layers and not one single compacting column. In the lowermost column, expelled magma ponds for a short time near the top of the layer and percolates through pore spaces in the rock above, eventually saturating its pore spaces. This new matrix + magma layer undergoes another compaction expelling the melt upward, and so on. Thus, the magma may reach the crust through a series of magma waves resulting from compaction.

TABLE 9.3   *Calculation of compaction heights.*

a (radius) (cm)	porosity (cm^2)	k	K Permeability (cm^2)	Matrix Viscosity (poise)	Melt Viscosity (poise)	Compac. length (m)
0.2	0	1000	0	1.00E+20	100	0.00
0.2	0.01	1000	4.0812E−11	1.00E+20	100	63.88
0.2	0.02	1000	3.3319E−10	1.00E+20	100	182.54
0.2	0.03	1000	1.1478E−09	1.00E+20	100	338.00
0.2	0.04	1000	2.7778E−09	1.00E+20	100	527.05
0.2	0.05	1000	5.5402E−09	1.00E+20	100	744.32
0.2	0.06	1000	9.7782E−09	1.00E+20	100	988.85
0.3	0.01	1000	9.1827E−11	1.00E+20	100	95.83
0.3	0.02	1000	7.4969E−10	1.00E+20	100	273.80
0.3	0.03	1000	2.5826E−09	1.00E+20	100	508.20
0.3	0.04	1000	6.25E−09	1.00E+20	100	790.57
0.3	0.05	1000	1.2465E−08	1.00E+20	100	1116.48

Magma's flow through dikes (i.e., pipe flow; Figure 9.12) can be described with the following equation (Sparks, 1992):

$$v = \frac{d^2}{64\eta} (\Delta\rho)g, \qquad \text{[Eq. 9.4]}$$

where v = magma velocity, d is the width of the dike, η is the viscosity of the magma, $\Delta\rho$ is the density difference between the rock and magma, and g is acceleration due to gravity. In the case of a tholeiitic basalt magma ($\eta = 300$ poise) moving along a dike ($d = 10$ cm) within the lherzolitic lithosphere, the $\Delta\rho = 0.6$ g/cm^3; and the flow velocity may be calculated from the prior equation:

$$v = \left[10^2/(64*300)\right]*0.6*980$$
$$v = 3.06 \text{ cm/s}$$

Volatile-saturated magmas can ascend much faster and more explosively as the vapor exsolves from the magma during ascent. A classic example of this is kimberlite magma, which appears to have erupted at supersonic velocities (McGetchin and Ullrich, 1973).

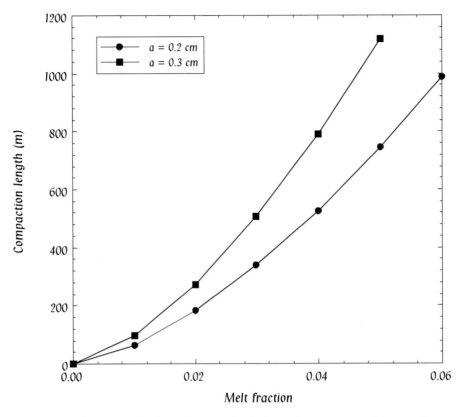

FIGURE 9.12 *Compaction length as a function of grain size (radius = a) and melt fraction (F).*
F = 0.02, 0.04 means 20% and 40% melting, respectively.

CHEMISTRY OF MELTING PROCESSES

Here we consider a few fundamental equations that describe the chemical aspects of melting. During melting, an element may prefer to remain in a particular mineral of the residual rock or it may prefer the melt phase. This selective behavior may be described in terms of partition coefficient or K_d, where

$$K_d = C_i^S/C_i^L$$

and C_i^S, C_i^L are concentrations of any element i in a mineral and coexisting melt, respectively. As an example, consider the $[K_d]_{Sr}^{Ol/L} = 0.005$ for the element strontium (Sr): The partition coefficient indicates that Sr prefers the melt 5,000 times more than olivine. In contrast, Nickel (Ni) prefers to stay in olivine about 12 times as much as melt (i.e., $[K_d]_{Ni}^{OL/L} = 12$). Clearly when K_d for a certain element is 1, it means that the element has no special preference toward either the melt or the mineral. K_d can vary as a function of pressure, temperature, and melt composition. Based on their partitioning behavior, elements are called *incompatible* $(K_d < 1)$ and *compatible* $(K_d > 1)$. An element may be compatible for one mineral but incompatible for another. For example, Sr is a compatible element for plagioclase $(K_d = 2)$, whereas it is incompatible for olivine $(K_d = 0.005)$.

In the case where an element is being partitioned between a melt and a rock consisting of several minerals, the term *bulk distribution coefficient (D)* is used to account for the total partitioning of an element between the bulk solid and melt. This is done by using the partition coefficients of each solid adjusted for their weighted abundances in the bulk rock. For example, for a system in which Element i is being partitioned between a rock composed of the minerals N, Q, and R and a melt, the D is given as:

$$D = X_N \cdot K_d^{N/L} + X_Q \cdot K_d^{Q/L} + X_R^{R/L} K_d^{R/L},$$

where $X_{N, Q, R}$ stand for weight fractions of the mineral phases N, Q, and R, respectively, in the bulk solid, and K_ds are partition coefficients for the element i between these phases and the melt. For example, X_N is the weight (or could be molar) fraction of N in the bulk solid [i.e., $X_N = N/(N + Q + R)$]. $K_d^{N/L}$ is partition coefficient for element i between mineral N and melt L.

Although a number of different melting models are mathematically possible, two end-member type melting processes are generally considered to cover all possible melting types. These two extreme models are the so-called *equilibrium batch melting* (EBM) and *Raleigh fractional melting* (RFM) processes (see Chapter 7). The equations for EBM (Equation H) and RFM (Equation I) are given as follows:

$$C_i^L = C_i^O/\left[F(1 - P) + D_o\right] \qquad \text{[Eq. 9.5]}$$

$$C_i^L = (C_i^O/D_o) \cdot \left[1 - (PF/D_o)\right]^{[(1/P)-1]}, \qquad \text{[Eq. 9.6]}$$

where F is melt fraction and C_i^L and C_i^O are the concentrations of element i in melt and the original rock. P is the bulk distribution coefficient of the melting assemblage (i.e., contributions made by each mineral going into melt). D_o is the bulk distribution coefficient of the starting rock and is generally different from P. As an example, let us calculate the concentration of Sr in the melt formed by 2% melting of a rock composed of 60% olivine $(K_d = 0.005)$, 15% clinopyroxene $(K_d = 0.1)$, and 25% orthopyroxene $(K_d = 0.01)$. The melting mode (i.e., proportion in which the three minerals would be dissolved in the melt; e.g., say, at an isobaric invariant point), is ol_{20} cpx_{50} opx_{30}. We assume that the original source rock had 20 ppm Sr $(= C_i^O)$. First, $D_o = (0.6 \times 0.005) + (0.15 \times 0.1) + (.25 \times 0.01) = 0.021$. $P = (0.2 \times 0.005) + (0.5 \times 0.1) + (0.3 \times 0.01) = 0.054$. Since F = 0.02, and $C_i^O = 20$, C_i^L is calculated to be 501 and 669 ppm for EBM and RFM processes, respectively.

Using an EXCEL® spreadsheet (or other computer programs), one can easily calculate the concentration of Sr in the melt for varying degrees of melting. Figure 9.13 shows a plot of variation of C_{Sr}^L for EBM and RFM processes as a function of F. Because fractional melting rapidly depletes the residue in an incompatible element (which Sr is in this case since its D_o and P are less than 1), RFM-generated melts contain greater amounts of Sr for low degrees of melting. However, at higher degrees of melting, EBM-generated melts must have greater Sr. Also during melting, a mineral may selectively be dissolved into the melt phase relative to the other minerals present in the starting rock, and that mineral may disappear from the residue entirely before all other minerals.

As we noted elsewhere, upper-mantle lherzolite partially melts to produce basalt magma. During the melting process, clinopyroxene selectively dissolves into the melt relative to olivine and orthopyroxene, and thus may completely disappear from the residue. Beneath the global midoceanic ridges, the disappearance of clinopyroxene from the residue is generally believed to be the factor that stops melting in the rising asthenosphere parcel (explained elsewhere).

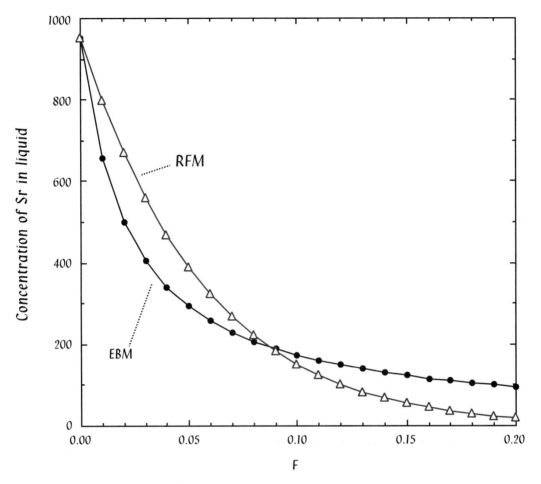

FIGURE 9.13 *The concentration of Sr in partial melt is shown as a function of melt fraction in equilibrium batch melting (EBM) and Raleigh fractional melting (RFM) processes.* The following values were used: Sr concentration in the bulk solid source (lherzolite) = 20 ppm. Sr—partition coefficients: olivine = 0.005, clinopyroxene = 0.1, orthopyroxene = 0.01. Mode of the source rock: $ol_{60}opx_{25}cpx_{15}$. Partial melting mode (i.e., at the pseudoinvariant point): $ol_{20}opx_{30}cpx_{50}$. Note that these values, although not too unrealistic, were arbitrarily chosen for the calculations performed for this figure.

SUMMARY

[1]   Magmas show large variation in terms of their mafic ($MgO + FeO + Fe_2O_3$) and alkali ($Na_2O + K_2O$) contents and can be fundamentally divided into ultramafic, mafic, intermediate, and felsic types. Examples of these magmas are as follows: ultramafic—komatiite; mafic—basaltic magmas; intermediate—andesitic magmas; felsic—rhyolitic magmas.

[2]   In terms of its atomic structure, silicate magma is principally composed of polymerized SiO_4 chains and network modifier cations.

[3]   Density and viscosity of magmas are important properties that control their ascent. Both factors vary as a function of pressure and composition. Viscosity is also controlled by other factors such as crystal content, volatile content, and so on.

[4] Magma formation is controlled by the solidus of the source rock and the ambient temperature at any depth. At a given depth, a rock melts at a lower temperature in presence of volatiles (commonly H_2O and CO_2) than in absence of volatiles. It may form due to decompression or volatile-aided melting.

[5] Magma transportation occurs via porous flow (controlled by interfacial energies between magma and crystal grains in contact) and pipe flow. Porous flow is driven by compaction.

[6] Chemical composition of the primary magma is controlled by partitioning of elements between the solid source and the melt. These factors can be controlled by the melting process, the two extremes of which are equilibrium batch melting and Raleigh fractional melting.

BIBLIOGRAPHY

Agee, C.B. and Walker, D. (1988) Static compression and olivine flotation in ultrabasic silicate liquid. *J. Geophys. Res.* **93**, 3437–3449.

Baker, M. and Stolper, E. (1994) Determining the composition of high-pressure melts using diamond aggregates. *Geochim. Cosmochim. Acta* **58**, 2811–2827.

Hess, P.C. (1989) *Origins of Igneous Rocks*. Harvard University Press (Cambridge, MA.).

Hirose, K. and Kushiro, I. (1993) Partial melting of dry peridotites at high pressures: determination of compositions of melts segregated from peridotite using aggregates of diamond. *Earth Planet. Sci. Lett.* **114**, 477–489.

Kushiro, I. (1980) Viscosity, density, and structure of silicate melts at high pressures, and their petrological applications. In *Physics of Magmatic Processes* (R.B. Hargraves, ed.), 93–120.

Jaques, A.L. and Green, D.H. (1980) Anhydrous melting of peridotite at 0–15 kb pressure and the genesis of tholeiitic basalts. *Contrib. Mineral. Petrol.* **73**, 287–310.

McBirney, A.R. (1993) *Igneous Petrology* (2nd ed.). Jones and Bartlett (Boston, MA.).

McGetchin, T.R. and Ullrich, G.W. (1973) Xenoliths in maars and diatremes with inferences for the Moon, Mars, and Venus. *J. Geophys. Res.* **78**, 1833–1853.

McKenzie, D. (1984) The generation and compaction of partially molten rock. *J. Petrol.* **25**, 743–765.

Ohtani, E. et al. (1995) Melting relations of peridotite and the density crossover in planetary mantles. *Chem. Geol.* **120**, 207–221.

Philpotts, A.R. (1990) *Principles of Igneous and Metamorphic Petrology*. Prentice-Hall (Englewood Cliffs).

Rigden, S.M., Ahrens, T.J. and Stolper, E.M. (1984) Densities of liquid silicates at high pressures. *Science* **226**, 1071–1074.

Sen, G. (1983) A petrologic model for the constitution of the upper mantle and crust of the Koolau shield, Oahu, Hawaii, and Hawaiian magmatism. *Earth Planet. Sci. Lett.* **62**, 215–228.

Spera, F.J. (1980) Aspects of magma transport. In *Physics of Magmatic Processes* (R. Hargraves, ed.). Princeton University Press (Princeton, N.J.), 265–323.

Sparks, S.J. (1992) Magma generation in the earth. In *Understanding Earth: A New Synthesis* (Brown, G.C., Hawkesworth, C.J. and Wilson, R.C.L., eds.). Cambridge University Press (Cambridge, U.K.), 91–114.

Stolper, E.M. and Walker, D. (1980) Melt density and the average composition basalt. *Contrib. Mineral. Petrol.* **74**, 7–12.

Takahashi, E. (1986) Melting of dry peridotite KLB-1 up to 14 GPa: implications on the origin of peridotitic upper mantle. *J. Geophys. Res.* **91**, 9367–9382.

Watson, E.B. (1982) Melt infiltration and magma evolution. *Geology* **10**, 236–240.

Igneous Rock Series: Basalt Magma Evolution

The following topics are covered in this chapter:

Composition and classification of basalt magmas
Igneous rock series
    Primary, parent, primitive, and derivative magmas
Basalt magma evolution and processes of differentiation
    Fractional crystallization
    In situ and convective crystallization
    Flowage differentiation
    Filter pressing
    Wall rock assimilation and magma mixing
    Liquid immiscibility
Basalt magma evolution: Bowen–Fenner debate
Layered igneous intrusions

ABOUT THIS CHAPTER

In the previous chapter, we learned that magmas can vary considerably in chemical composition and hence in their mineralogy. What are the causes for such diversity? Is it possible that all the major magma types are somehow genetically related? In the early part of the twentieth century, scientists began to explore this question. N.L. Bowen, the "father of modern igneous petrology," and some of his notable colleagues of the Geophysical Laboratory of Washington, DC, developed a unique experimental approach to decipher magma's chemical evolution. This approach involved crystallization of magmas in laboratory furnaces and the application of thermodynamic principles, such as the phase rule, to igneous rocks. These pioneering studies and many subsequent others have shown that a number of processes besides crystallization, for example, partial melting, wall rock contamination, magma mixing, and so on, can affect a magma's composition. Bowen believed that rhyolitic and andesitic magmas are derived from a parent basalt magma by a very efficient type of crystallization process called *fractional crystallization*. Hawaiian active lava lakes offered a natural laboratory to study how basalt magmas (of the type that Bowen believed to be the parent magma) can crystallize and evolve through time. Detailed field and laboratory studies of igneous rocks since Bowen's time have made it clear that many different processes affect magma's chemistry and have yielded constraints on what is possible and what is not. The present chapter exposes the student to topics such as how chemical variations in various magma types can be treated and understood using graphical means and in terms of elementary physical chemistry. It also uses examples of natural, deep-seated, fossilized magma chambers, called *layered intrusions*, to explore how the natural systems may actually behave. Basalt magma takes a central role in this chapter because (a) basalts comprise the bulk of the earth's crust, and (b) Bowen proposed that all magmas are derived through fractional crystallization of basalt magma.

CLASSIFICATION OF BASALTIC MAGMAS

Basalts and Basalts

To the early practitioners of geology, basalts were not a very exciting rock type to study. To them, basalt was a black rock that occurs everywhere on Earth but is compositionally rather uniform and therefore boring. As time passed, however, analytical and experimental tools became more sophisticated and accessible to a much larger number of well-trained scientists and students. Vast amounts of high-quality chemical data continue to accumulate at a fierce pace on basalts covering the entire globe. These new data have made it clear that basalts are anything but boringly uniform, and that there are basalts and basalts (i.e., there are distinct types of basaltic magmas that cannot be generated from each other by low-pressure [crustal] differentiation processes). Even within a single volcano, basaltic magma composition can vary significantly over time. On a global scale, chemically distinctive (especially in terms of isotopic ratios of lead, neodymium, and strontium) basaltic provinces appear to exist. Such variations tell us a great deal about the internal workings of the planet—from the small-scale processes within a magma conduit beneath a single volcano to global-scale dynamic processes that drive plate tectonics.

In 1962, H.S. Yoder, Jr. (at the Geophysical Laboratory of Washington, DC) and C.E. Tilley presented a systematic, normative classification of basalts, popularly known as the *basalt tetrahedron* (Figure 10.1), based on their synthesis of experimental and laboratory studies of basaltic rocks. This classification scheme remains popular today. The four apices of the tetrahedron are clinopyroxene (Di; diopside in the simplified basalt system $CaO–MgO–Al_2O_3–Na_2O–SiO_2$), nepheline, olivine (Ol; forsterite in the simple system), and silica (quartz). Plagioclase (Pl) plots in the middle between nepheline (Ne) and silica (because $NaAlSi_3O_8$ [ab] $= NaAlSiO_4$ [ne] $+ SiO_2$). Likewise, hypersthene (Hy) is located halfway between olivine and quartz (since $MgSiO_3 = Mg_2SiO_4 + SiO_2$). The reasons for selecting these minerals as apices of the tetrahedron are simply that (a) together these mineral components generally comprise 90% of the normative composition of a basalt, and (b) all the low-pressure phase relationships of various basaltic magma types can be systematically depicted with reference to various slices through this tetrahedron.

The basalt tetrahedron makes a fundamental distinction between nepheline-normative magmas from those that do not have normative nepheline. Based on the system nepheline-silica (Chapter 7), we know that at 1 atmosphere pressure albite acts as a thermal barrier between nepheline-normative magmas and hypersthene-normative magmas. In the basalt tetrahedron, the triangular plane Ol–Pl–Di is a low-pressure thermal barrier, preventing magmas from one side to go across this plane to the other side by any low-pressure differentiation process. Yoder and Tilley called the plane Ol–Pl–Di *critical plane of silica undersaturation* because it separates silica-deficient nepheline-normative magmas from silica-saturated olivine tholeiites. [Recall that $NaAlSiO_4$ (ne) $+ 2SiO_2 = NaAl Si_3O_8$ (ab)—that is, nepheline has less silica than albite and is therefore silica deficient relative to albite.] The volume Di–Ne–Pl–Ol (i.e., left portion of the tetrahedron in Figure 10.1) is called the *alkali basalt volume* named after the more dominant nepheline-normative magma type (i.e., alkali basalt). A second plane Pl–Hy–Di separates magmas that contain normative quartz (and hypersthene) from those that contain normative olivine (+hypersthene). For an obvious reason, Yoder and Tilley referred to the plane Pl–Hy–Di as *plane of silica saturation*. Note that this plane does

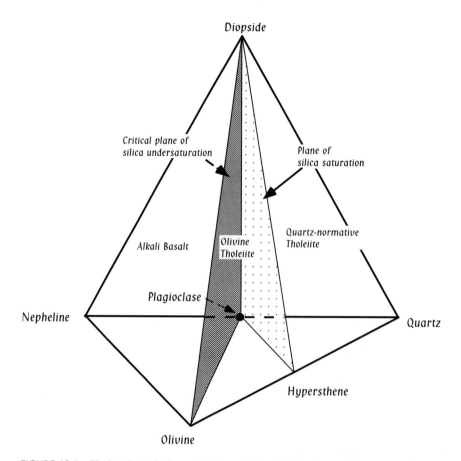

FIGURE 10.1 *The basalt tetrahedron of Yoder and Tilley (1962).* Composition volumes for three main basalt types—alkali basalt, olivine tholeiite, and quartz tholeiite— are shown. The two important planes in this diagram are the diopside-olivine-plagioclase (dark shaded) and diopside-hypersthene-plagioclase (stippled) planes. The diopside-olivine-plagioclase plane, separating the alkali and tholeiite basalt volumes, acts as a low-pressure thermal barrier. Basalt magmas cannot cross this plane during fractional or equilibrium crystallization.

not act as a thermal barrier, and olivine tholeiite magmas can fractionate and evolve to quartz-normative tholeiites (see e.g., the discussion of the system Di–Fo–SiO$_2$ in Chapter 7).

THE CONCEPT OF IGNEOUS ROCK SERIES

In the late 1800s, long before the concept of *plate tectonics* was established, petrologists had recognized the existence of *igneous rock series*—a term that was used to signify the genetic relationship between a group of chemically (and petrographically) different but regionally associated igneous rocks. That a parent magma can evolve to a series of compositionally different daughter magmas by some process of differentiation, such as fractional crystallization, assimilation, and other processes, was realized by Scottish geologists in the 1890s. They recognized the distinction between alkaline series rocks and subalkaline (i.e., tholeiitic) series rocks by the presence of normative feldspathoids in the first series. This key distinction served later as the basis for igneous rock classifications, as attested by Yoder and Tilley's classification. Aside from the alkaline and tholeiitic series, a major global igneous rock series, typically associated with subduction zones, is the calc-alkaline series. The tholeiitic series—typically exemplified by the oceanic crustal basalts, continental flood basalts, and Hawaiian shield volcanoes—exhibits an initial trend of iron enrichment followed by silica and alkali enrichment. Alkaline-series lavas occur in all plate tectonic environments, but are generally much smaller in volume than tholeiitic and calc-alkaline series.

Any discussion of magma evolution must begin with the nature of the starting magma that must undergo differentiation processes and evolve into other magma types. In this context, petrologists often like to distinguish among primary, primitive, and parental magmas. A primary magma is directly generated by melting a source rock. It is doubtful whether any primary magma erupts on the surface of the earth. A primitive magma is generally thought to be a magma that has undergone minimal differentiation. However, it may or may not be primary. A parental magma is generally identified to be the least differentiated magma in an igneous series, which, by differentiation processes, gives rise to a range of derivative magmas. A parental magma is necessarily a primitive magma, but the two terms are not interchangeable since parental magma must refer to a genetically related magma series. Modern practitioners use the term *liquid line of descent* to identify a series of liquids derived from a single parent magma. Identification of primary magmas is an extremely difficult if not impossible task, and scientists generally use element partitioning criteria to constrain their compositions.

A primary magma undergoes differentiation and evolves to derivative liquids in a number of ways. These processes may be very complex in detail. Extensive details of magma differentiation processes cannot be addressed here, and only a brief outline is presented next.

MAGMA DIFFERENTIATION

Fractional Crystallization

As defined in Chapter 7, *fractional crystallization* is an end member process in which tiny fractions of crystals are physically removed from the magma as soon as they form. In general, natural processes probably never reach such extreme conditions, and some amount of equilibrium is maintained for a brief period between crystals and magma. Fractional crystallization may be aided by gravity because crystals that are denser than the magma will settle (crystal settling; Figure 10.2a), obeying Stoke's law if the magma is Newtonian (or other laws if the magma is non-Newtonian). Petrographic and

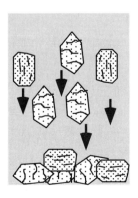

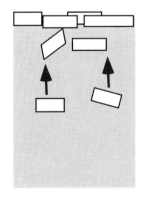

FIGURE 10.2   (a) Schematic diagram showing settling of olivine (with curved fractures) and pyroxene (with cleavage) to the bottom of magma (dark shaded). (b) Flotation of plagioclase crystals in magma.

(a) Crystal settling   *(b) Crystal flotation*

mineral-chemical evidence of settling of olivine crystals, which is usually denser than basalt magma, in thick sills and lava flows is common. In contrast, crystals lighter than the magma will float (crystal flotation; Figure 10.2b). Density of plagioclase is usually close to basaltic magmas, and they may be lighter than slightly Fe-rich differentiated magmas, causing them to float. Lunar highlands (the white areas on the Moon) are composed of a plagioclase-rich rock called anorthosite. It is commonly believed that these anorthosites formed by accumulation of floating plagioclase crystals in a denser magma ocean that once covered the moon.

In Situ Crystallization and Convective Crystallization

Crystallization, without the aid of gravity, can lead to signficant compositional changes in the residual magma. Growth of crystals in situ, and not crystal settling, along walls of magma chamber is believed to be the principal way in which magmas fractionate (Figure 10.3). Detailed examination of several Hawaiian lava lakes (e.g., Kilauea Iki, Alae, Makaopuhi lava lakes) over several years by the U.S. Geological Survey has shown some consistent compositional trends in the residual liquid evolution as a function of crystallization. Because the heat loss is primarily from the roof, these lava lakes quickly quench to form a thin chill zone at the top. With time, a solidification front moves from roof inward. Crystallization is relatively rapid, whereas chemical diffusion (i.e., movement of ions through melt and solid phases) is a very slow process. Thus, the interstitial liquid between crystals cannot maintain chemical equilibrium with the main body of the magma and develop highly differentiated composition.

In large layered intrusions, which are believed to be fossil magma chambers, mineral layering parallel to the roof, bottom, and sidewalls is a constant feature. Layering is the thickest in the bottom–up sequence, and layers of minerals are often repeated (discussed later). The very occurrence of layers of dense minerals parallel to the sides of a layered intrusion makes it clear that gravity (and, by interpolation, crystal settling) could not have played a major role in magma differentiation in such magma chambers because such layers would fall off into the magma. In contrast to early interpretations (principally by L.R. Wager, W.A. Deer and coworkers) of differentiation of layered mafic intrusions by crystal settling, it is generally believed

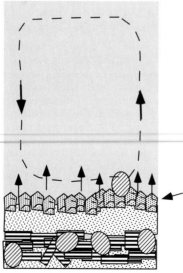

FIGURE 10.3   *Sketch showing in situ growth of crystals on the bottom of a magma chamber.*

Crystals growing in situ on the floor mat of crystals in a convecting magma chamber.

today that some process of in situ crystallization must be the most important process of differentiation in these intrusions.

Side wall or roof cooling and crystallization also leads to development of crystal-laden melt layers close to the roof or sides of the magma chamber. The melt in such layer is generally compositionally distinct from the main magma body because chemical diffusion is too slow to equilibrate them. Also a crystal-rich melt layer is too viscous (as discussed in Chapter 9) to mix easily with the rest of the magma chamber. Thus, such a crystal-laden melt layer may in some cases become gravitationally unstable, sink along the sides, and eventually drop crystals off near the bottom of the magma chamber. Once the crystals are released, the melt portion of the layer may become lighter and rise again. Evidence of such convection currents and crystal precipitation from convective layers has been seen as current bedding/layering in many large, layered mafic intrusions (Wager and Brown, 1968). However, chamber-wide convection is unlikely, and the observed cross-stratification may have been produced by small, intermittently active, local convection cells.

Flowage Differentiation

In some cases, crystals may be concentrated in the center of a moving magma body as the magma flows through a dike or a sill—a process known as *flowage differentiation* (Figure 10.4). This happens when the flow is laminar, and magma develops a velocity gradient across the intrusion. The dispersive force that pushes phenocrysts apart increases from the center to the walls of the intrusion and thus concentrates the crystals in the center of the intrusion—much the same way dead leaves and trash gather at the center of a flowing river.

Filter Pressing

This is a process of compaction that involves squeezing out of melt from a crystal-melt mush. Loading of crystals in a crystal-liquid mush (and, hence, gravity), a stress regime, or flow through constrictions may cause melts to be separated this way. This process is likely to be very important in separating a small melt fraction from its

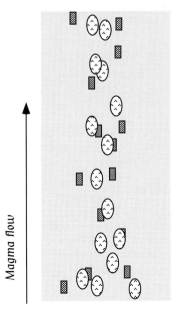

Magma flow

FIGURE 10.4 *Sketch showing flowage differentiation— concentration of crystals in the middle of the flowing magma.*

source region in an area that is undergoing melting (see Chapter 9). It is also important in squeezing out residual liquid from the interstitial spaces in large, layered intrusions.

Assimilation ± Fractional Crystallization

The process of assimilation involves reaction and/or dissolution of wall rock by rising magmas. Such processes are sometimes easily detected with the help of isotopes and trace elements. In some cases, petrographic evidence of reaction textures (Figure 10.5) may be found. Many field examples of assimilation have appeared in the published literature, attesting to the fact that this process does affect magma composition (e.g., McBirney, 1980). For example, xenolith of quartz in alkalic intrusions is clear evidence of partial assimilation of wall rock since we know, based on phase diagrams, that nepheline in such alkalic intrusions cannot co-exist with quartz under equilibrium conditions. On a much larger scale, the general process of assimilation and accompanied fractional crystallization has been proposed by a number of authors to be responsible for the origin of calc-alkaline series magmas in volcanic arcs and for making the subcontinental mantle relatively orthopyroxene rich (e.g., Kelemen, et al. 1998). Long ago, Daly (1933) had proposed that all alkalic magmas form by contamination of subalkaline (tholeiitic) magmas by calcareous sedimentary/metamorphic rocks based on reactions of the following type:

$$CaMg(CO_3)_2 \quad + \quad NaAlSi_3O_8 \quad = \quad NaAlSiO_4 + CaMgSi_2O_6 + 2CO_2.$$

Dolomite in Wall rock	Albite component in magma	nepheline	diopside	gas
			Products	

Although this type of reaction and alkalic liquid production may occur at a local scale, most scientists today do not believe that alkalic magmas commonly originate this way. Rather, the majority of them form by direct melting of the mantle rock.

How much wall rock assimilation can occur will depend on a number of factors—ascent velocity and temperature of the magma, enthalpy difference between magma and wall rock, solidus of the wall rock, physical nature of the contact between magma and wall rock (more fractured the contact more xenoliths are likely to fall off in the magma), and so on. In a vigorously convecting magma chamber, a magma may be able to break off xenoliths of wall rocks more efficiently, expose fresh surfaces on wall rocks, and assimilate them. However, a relatively stagnant body of magma may quickly quench to a thin selvage of chill zone at the contact between wall rock and magma, which may prevent further assimilation of wall rocks.

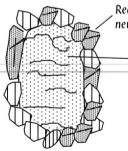

Reaction rim composed of new minerals

Reactant phase

FIGURE 10.5 *Sketch showing reaction rim of minerals between magma and a reactant mineral.*

Reaction texture

ADVANCED READING

The heat released by the magma in assimilation of wall rocks may come from the latent heat of crystallization of its minerals or the magma may be superheated in the sense that its temperature may be above its liquidus temperature. Long ago, Bowen (1928) demonstrated that it is unlikely for a magma to carry significant superheat and that too much superheat in a large mass of magma would be needed to melt a very small mass of rock. In melting a certain mass of wall rock $(= M_r)$, a mass of magma $(= M_m)$ must first heat up the wall rock to its melting point and then provide additional latent heat of fusion (H_f) to initiate melting at a constant temperature. If Cp^r and Cp^m are heat capacities of wall rock and magma, respectively, ΔT_1 = difference in ambient temperature and solidus temperature of the wall rock, ΔT_2 = cooling of the magma, then the heat lost by the magma and heat gained by the wall rock must balance:

$$M_m C_p{}^m \Delta T_2 = M_r(Cp^r \Delta T_1 + H_f)$$
$$M_m/M_r = (Cp^r \Delta T_1 + H_f)/C_p{}^m \Delta T_2 \qquad \text{[Eq. 10.1]}$$

If the wall rock is close to its solidus temperature when the magma encounters it, then:

$$M_m/M_r \cong H_f/C_p{}^m \Delta T_2. \qquad \text{[Eq. 10.2]}$$

As an example, imagine a mass of magma assimilating 1 gm of wall rock whose ambient temperature is 400°C. Assuming that the wall rock's solidus temperature is 1,000°C, the wall rock must first be heated by 600°, and then the magma must provide an additional latent heat of fusion (H_f), around 100 cals, to melt 1 gm of the rock. If we choose a reasonable value for $C_p{}^r$ of 0.30 cals/gm degree for the wall rock, then to raise the temperature of 1 gm of wall rock by 600°C will require 600×0.3 or 180 cals [based on $C_p = (dH/dT)_p$]. Thus, 280 cals (180 cals to heat up the wall rock and 100 cals to melt it) would be needed to melt 1 gm of wall rock. Assuming a reasonable value of 0.25 cals/gm^{-1}K^{-1} for $C_p{}^m$, this heat may be provided by 3.73 gm of magma assuming that it cools by 200° (Equation 10.1). A larger mass of magma would be required if the temperature drop is less than 200° and vice versa. Figure 10.6 shows that the M_m/M_r ratio is >1 (i.e., magma mass will always be significantly greater than the mass assimilated). Also, the hotter wall rocks will require a lower M_m/M_r ratio than cooler wall rocks. A basalt magma at 1,200°C may crystallize entirely if its temperature drops by as much as 300°. Given that upper-crustal rocks are much cooler than deep crustal rocks, assimilation of lower-crustal rocks would require less energy from the assimilant basalt magma than upper-crustal rocks. Also there is a greater probability of a magma crystallizing entirely when it assimilates upper-crustal rocks. In the event that the magma is not carrying superheat, the magma mass M_m would be the same as mass crystallized (or M_c) in the previous equations. Thus, it is possible to calculate the extent of crystallization of a magma during assimilation.

Magma Mixing

Mixing of magma batches in some subterranean chambers or intrusive bodies should be a common process. However, mixing is unlikely to be solely responsible for all geochemical variations in a magma series but is usually accompanied by crystallization processes. A number of hypothetical examples of mixing of silicate liquids in

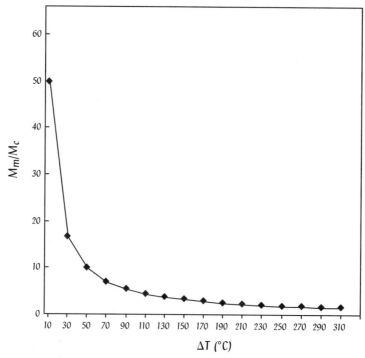

FIGURE 10.6   *Heat loss from magma to wall rock versus amount of magma crystallized during assimilation.* M_m = mass of magma used in the assimilation process. M_c = mass of magma crystallized. Temperature difference between the magma and the wall rock is plotted on X-axis.

binary and ternary systems were considered in Chapter 7, where the effects of such mixing have been portrayed graphically. Most notably, it was pointed out, with the help of a binary solid solution, that textural criteria and the presence of normal and reverse-zoned crystals in the same rock can be used to identify magma mixing. Linear geochemical trends on oxide-oxide plots (e.g., SiO_2 vs. MgO) are also generally attributed to magma mixing (discussed later).

Liquid Immiscibility and Other Processes

Immiscibility between two (or three) compositionally different liquid fractions of a magma has been demonstrated in a number of studies of natural rocks (Figure 10.7). The immiscible liquid droplets maintain a globular structure in the rocks in which they occur. Perhaps the most significant type of immiscibility occurs between silicate and

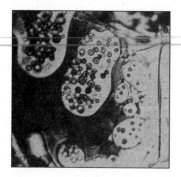

FIGURE 10.7   *Photomicrograph of liquid immiscibility.* (From Philpotts, 1990, *Principles of Igneous and Metamorphic Petrology,* 1st ed., Prentice Hall)

carbonatitic liquids. Ocelli structure, composed of globular masses (a few centimeters thick) of felsic magma in lamprophyre magma (defined in Chapter 8), is commonly attributed to liquid immiscibility. In basaltic magmas, liquid immiscibility between a volumetrically minor sulfide-rich liquid and silicate magma occurs over a wide temperature range. During the late stages of differentiation of a tholeiitic basalt magma, immiscibility commonly occurs between alkali-rich and iron-rich silicate liquids.

There is no doubt that liquid immiscibility occurs in magmas. However, in a volumetric sense, it plays a minor role in magma differentiation. Geochemical and petrographic evidence clearly shows that crystal sorting is the dominant process in magma differentiation.

Other processes of magma differentiation include transfer of volatiles from deep to shallow levels within a magma chamber. Such volatiles may carry some dissolved, light, large ions with them in the process. Volatiles can concentrate in highly differentiated liquids near the roof of a magma chamber. They may also be absorbed from wall rocks by the magma. Often the concentration of volatiles may reach supersaturation and separate (exsolve) from the magma—a process known as *retrograde boiling*. Volatile transfer is of particular significance in the formation of rich ore-bearing pegmatites and hydrothermal veins around large granitoid batholiths.

CHEMICAL SIGNATURE OF DIFFERENTIATION: VARIATION DIAGRAMS

Chemical compositional variations in a genetically related suite of rocks from a geographic area may be plotted in a number of ways to obtain a detailed description of the magmatic differentiation process that may have been responsible for such variations. One such early attempt was made by Alfred Harker (1909), who plotted abundances (weight%) of various oxides against SiO_2 (wt%) and demonstrated that the rock compositions vary in some systematic manner. Such a diagram has since been called a *Harker diagram*, which is only one possible type of diagram among many others; the general term *variation diagram* encompasses all such possibilities of X–Y type compositional plots. One must remember that plots like these only suggest what is a mathematically plausible process in relating rock compositions of a suite, but they are not a definitive way to identify a differentiation process. Ideally, one should use other criteria, such as thermodynamic validity, to test its feasibility.

In terms of oxide–oxide variation diagrams, the modern users commonly prefer the use of the Mg/(Mg + Fe) molar ratio (commonly referred to as Mg#), Mg/Fe ratio, or weight% MgO on the X-axis, instead of SiO_2, against which variations of other oxides are plotted (Figure 10.8). This is partly because the minerals that are likely to fractionate early from a basaltic magma (the most commonly identified parent magma) are ferromagnesian minerals that, on fractionation, leave distinctive signatures in terms of MgO, Mg#, or Mg/Fe ratio of the residual liquid. To explain the utility of MgO variation diagrams, liquid lines of descent (LLD) have been calculated at two pressures (2 and 15 kilobars) for a parent basalt magma with 15.8% MgO, 9.6% CaO, and 9.6% Al_2O_3. Let us first consider the 15 kb LLD: Olivine [$(Mg,Fe)_2SiO_4$] is the first mineral to crystallize (as Bowen noted long ago), and its fractionation (removal from the magma) depletes the magma in MgO and enriches it in CaO and Al_2O_3 (because CaO and Al_2O_3 do not enter olivine and stay in the melt). When the magma composition reaches ~11% MgO, a strongly aluminous clinopyroxene (with significant CaO and Al_2O_3) begins to fractionate along with olivine, depleting the residual liquid in CaO and Al_2O_3 (and, of course, MgO). This

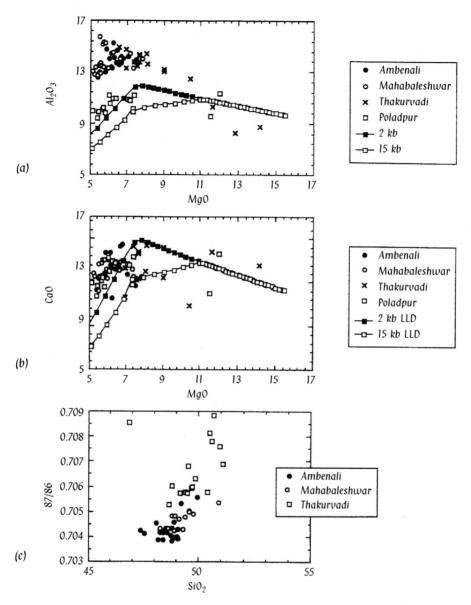

FIGURE 10.8 *Variation diagrams showing Al_2O_3 and CaO variations as a function of MgO in various formations of Deccan Trap basalts (a and b).* (c) Isotopic variation of $^{87}Sr/^{86}Sr$ as a function of SiO_2 content in Deccan lavas.

causes the formation of a kink at ~10% MgO on the LLD trends in both MgO–CaO and MgO–Al_2O_3 diagrams. Finally, a calcic plagioclase (with much higher Al_2O_3 than both clinopyroxene and olivine) appears at ~7.3% MgO and extracts significantly greater amounts of CaO and Al_2O_3 from the melt. This translates to additional kinks on the MgO–CaO and MgO–Al_2O_3 trends at ~7.3% MgO.

Comparison of the 15 kb LLD with 2 kb LLD shows that olivine fractionation is extended to lower MgO values at the lower pressure, such that the melt totally by-passes clinopyroxene fractionation and the residual melts reach higher peak values of Al_2O_3 and CaO. Once plagioclase joins olivine at ~7.3% MgO, however, the melt again starts getting depleted in CaO and Al_2O_3. Pressure therefore clearly affects fractionation trends (further discussed later). To better understand the utility of the calculated trends, we compare them with some well-known lava formations from the Deccan Traps—namely, Ambenali, Poladpur, Bushe, and Mahabaleshwar formations (data set compiled by Sen, 1996). The MgO–Al_2O_3 plot is particularly interesting because it shows that Poladpur basalts are totally different from the others. Our calculated trend at 2 kbar pressure can roughly explain the Poladpur data but not the others. We may conclude that our chosen starting composition (and the model used to calculate the trend) could have been an appropriate parent magma for Poladpur basalts, and that the other formations may have been derived from other parent magmas. We can use other geochemical tracers to explore the validity of our hypothesis; in particular, isotopic ratios (e.g., $^{87}Sr/^{86}Sr$; see appendix) are good to use because they remain constant for a parent–daughter suite of magmas that are only related by crystal fractionation (and not by other processes, such as crustal contamination). Figure 10.8c shows that $^{87}Sr/^{86}Sr$ ratios of each formation are different, and therefore they cannot be related by simple crystal fractionation from a unique parent magma. Much of the variability is due to contamination by the continental crust (very high in $^{87}Sr/^{86}Sr$), which adds variable amounts of $^{87}Sr/^{86}Sr$ to the basalt magma (e.g., Peng and Mahoney, 1995).

Significant progress in chemical analytical techniques has been made over the last decade. As a result, we have seen an explosion of high-quality major, minor, trace element and isotopic (particularly, Nd, Sr, Pb) data on individual rocks and minerals. It appears to be a common practice these days to combine Sr, Nd, and Pb isotopic ratios with elemental or oxide variations to address in detail the melting, crystallization, and assimilation processes (as alluded to in the example used earlier). The choice of diagrams depends entirely on what goal the petrologist is trying to accomplish. It is usually the case that the petrologist makes several different plots before he or she can select the useful ones. Discussion of all attributes of each diagram is well beyond the scope of this book, and the interested reader is further referred to a book by Cox et al. (1979). Here we focus only on oxide–oxide diagrams and discuss some general features of them.

Figure 10.9 shows a series of hypothetical examples of variation diagrams in which %A and %B are plotted, where A and B could be two oxides or two elements. Figure 10.9a shows a single-phase fractionation case where a suite of lavas—namely, 1–2 to 1–6—have been derived from a parent magma by variable extents of fractionation of a mineral "M." Obviously, 1–6 is the most differentiated and 1–2 is the least differentiated lavas of this suite. The Lever rule (Chapter 7) can be used to calculate how much fractionation of M is required to derive each of these lavas from the parent magma 1: For example, %fractionation of M to derive magma 1–6 from parent magma 1 may be given as 100 × [distance M to parent magma 1/distance M to 1–6]. Also, M-fractionation may occur at a single pressure (i.e., isobaric) or over a range of pressures (i.e., polybaric; see prior discussion on Figure 10.8).

Note that fractionation of M from the parent magma 1 to derive magmas 1–2 through 1–6 is not the only possible mechanism by which this suite may be generated. Mixing between magmas, say between 1–2 and 1–6, could also lead to a linear trend like this. Thus, one diagram alone is unlikely to be sufficient to deduce which of the processes (or combination of processes) may have generated such a suite of rocks. Other pieces of evidence, such as petrographic evidence of magma mixing (see Chapter 5), and other chemical criteria may be used to decipher what actually happened.

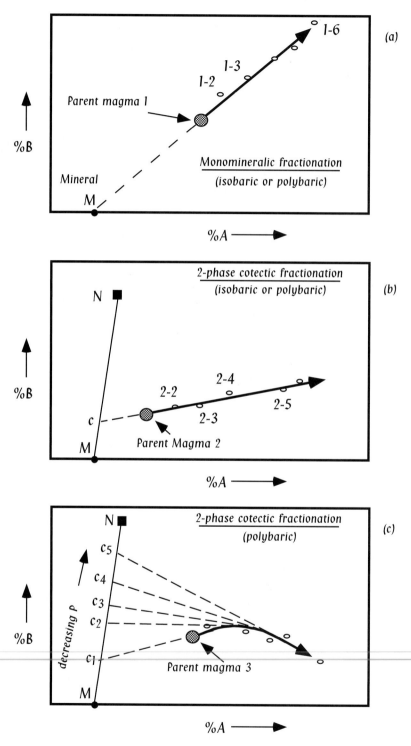

FIGURE 10.9 *Modeling of crystal fractionation processes shown in oxide (A)–oxide (B) plots.* Compositions of hypothetical magmas are shown as little circles. M, N—minerals. Solid arrow—magma fractionation trends. (See text for further details)

Cotectic fractionation involving multiple phases may also generate a linear trend (Figure 10.9b). In Figure 10.9b, fractionation of an isobaric cotectic assemblage of bulk composition "c" [$\sim M_{76}N_{24}$] from a parent magma 2 can generate the suite of lavas 2–2 through 2–5. Once again, the Lever rule may be used to deduce how much fractionation would be necessary to derive one magma from another.

In the case of two-phase fractionation, it is common to assume that the phenocryst phases in the least differentiated rocks of a suite of rocks represent the fractionating phases. In the case where pressure does not significantly change the proportion of M and N in the cotectic assemblage (which is generally unlikely), even polybaric fractionation of M and N may generate this linear trend.

In summary, we state the first rule about linear trends on oxide–oxide plots as follows.

Linear trends represent (a) fractionation (or accumulation) of a single mineral phase lacking significant solid solution (Figure 10.9a), (b) mixing of two magmas (say, mixing between 1–2 and 1–6 in Figure 10.9a), and (c) fractionation of two or more phases in a cotectic proportion at a constant pressure (or could also be polybaric fractionation provided that the cotectic proportion of the two phases is virtually insensitive to pressure; Figure 10.9b).

Proportions and composition of fractionating phases can be deduced from phenocrysts in the lavas or from information on experimental crystallization of lavas in the laboratory. In fact, several computer programs are available these days that can quantify isobaric as well as polybaric fractionation processes and determine the proportions, compositions, and temperature of formation of various fractionating minerals at various stages of magmatic evolution in a suite of igneous rocks (Yang et al., 1997). Because these programs are largely based on data from experimental studies of crystallization, they are generally able to duplicate the experimental data quite well, but all of them are limited in terms of their universal applicability. In particular, most of them cannot deal with Ca-poor pyroxene or amphibole fractionation.

Curvilinear Trends. We now proceed to examine how smoothly curving trends are generated. Smoothly curving trends on a variation diagram may be caused by the separation of a phase(s) with extensive solid solution, by polybaric fractionation, or by complex asimilation-fractional crystallization processes. For example, in Figure 10.9c, the curved trend may be generated by polybaric fractionation of minerals M and N, where decreasing pressure shifts the cotectic proportion from c_1 to c_5. A curved trend should also result if M and N were two end members of a solid solution series, and the fractionating solid phase changes composition from M toward N end member. As an example, consider a suite of lavas where plagioclase fractionation is the only differentiation process. In such a case, a curved trend would appear on a CaO–Na$_2$O diagram because plagioclase is a solid solution between albite (NaAlSi$_3$O$_8$) and anorthite (CaAl$_2$Si$_2$O$_8$) end member. Note, however, that such plagioclase separation will not produce a curved trend but a linear one on a MgO–CaO diagram since plagioclase generally has <0.5% MgO. The point is that one must use multiple plots to deduce the nature of the differentiation process. Mixing of magmas generated by different degrees of polybaric fractionation may result in reversals or wiggles in an otherwise curvilinear trend (discussed later). Curvilinear trends may also be produced by combined assimilation of wall rocks followed by fractional crystallization.

Kinks in Trends. Sharp bends (kinks) in a trend can result from the fractionation of a new mineral phase (illustrated earlier with Figure 10.8). The extent of the slope break generally reveals the culprit phase, and clever application of the Lever rule allows fairly precise determination of the mode of the fractionating assemblage.

ADVANCED READING

Chemical Quantification of Differentiation Processes

Crystallization of magmas may be considered in terms of two extreme conditions: equilibrium crystallization (EC), in which chemical equilibrium between all phases is maintained throughout, and Rayleigh fractional crystallization (RFC), in which crystals are removed from the melt as soon as they form (discussed in Chapter 7). In our illustration, we mostly consider magma evolution in a closed system (i.e., a system, magma chamber in our case, that is closed to any exchange of components with the surrounding materials). We also consider a more plausible open-system phenomenon in which a differentiating magma chamber is periodically replenished with primitive magma, which leads to mixing between differentiated and primitive magma batches accompanied by periodic eruptions.

The equations for EC and RFC are given as Equations 10.1 and 10.2, respectively:

$$C_i^{\text{residual liquid}} = C_i^{\text{parent magma}} / \left[F + \left(D(1 - F) \right) \right] \qquad \text{[Eq. 10.1]}$$

$$C_i^{\text{residual liquid}} = C_i^{\text{parent magma}} F^{(D-1)}, \qquad \text{[Eq. 10.2]}$$

where C_i is the concentration of element i, F is fraction of liquid remaining, and D is bulk distribution coefficient (defined before).

As an example, consider how the abundances of two elements—namely, Sr and Ce—with different degrees of compatibility will change in EC versus RFC. The parent magma is assumed to have 250 ppm Sr and 6 ppm Ce. Using K_ds given in the appendix and assuming crystallization of an assemblage olivine$_{20}$cpx$_{20}$plag$_{60}$, the abundances of Sr and Ce in the residual liquid were calculated using Equations 10.1 and 10.2 as a function of varying F (= liquid fraction remaining) and are plotted in Figure 10.10. As can be detected in the difference between EC and RFC equations, RFC produces rapid depletion of Sr (a compatible element in this case) and enrichment of Ce (an incompatible element) relative to EC.

In contrast to the prior case, with crystal assemblage of constant composition, consider what happens when the fractionating phases change in a closed system magma chamber, in which it only cools and does not lose magma through eruptions nor does it get replenished with fresh magma. We assume that olivine fractionates alone until 60% of the magma has crystallized. Plagioclase then joins as a fractionating phase (fractionating assemblage—Pl$_{80}$Ol$_{20}$). When the remaining liquid reaches 20% of its original mass (i.e., F = 0.2), clinopyroxene appears as an additional fractionating phase. Here the fractionating assemblage is Pl$_{40}$Cpx$_{40}$Ol$_{20}$. Note that the choice of the modes of fractionating assemblages and the F values where they appear are purely arbitrary and not based on experimental petrology. These were chosen to generate sharp changes in residual liquid composition. Because D changes abruptly from 0.005 (= Ol/L K$_d$) to 1.601 (Pl$_{80}$Ol$_{20}$) at F = 0.4, and then to 0.841 (Pl$_{40}$Cpx$_{40}$Ol$_{20}$) at F = 0.2, we see kinks in the trend (Figure 10.11). Such kinks can clearly serve as important indicators of the nature of the fractionating assemblage at any stage of fractionation of a magma in a particular intrusion or a lava sequence.

The prior examples are somewhat idealistic, and in nature it is more likely that the magma chamber is an open system, where a differentiating magma chamber receives periodic influxes of primitive magma and also discharges magma through eruptions to maintain constant volume, the so-called *RTF model*

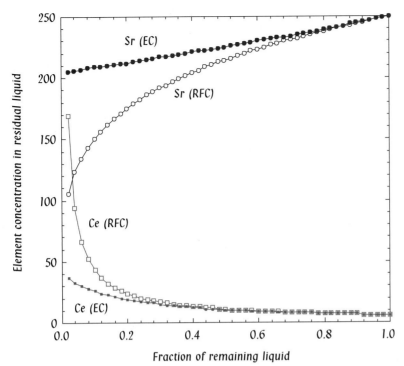

FIGURE 10.10 *Variation in concentration of two incompatible trace elements, strontium (Sr) and cerium (Ce), in the residual liquid as a function of equilibrium crystallization (EC) versus Raleigh fractional crystallization (RFC). Fraction of remaining liquid is plotted on the X-axis, where 1 means zero crystallization (i.e., 100% melt), 0.8 means 80% melt present, and so on.*

(R = replenished, T = tapped, F = fractionating). In this case, concentrations of trace elements in the residual liquid exhibit cyclical variations, as would be expected when new magma is periodically added followed by eruption and fractionation. The net result is that noncompatible and compatible element concentrations increase and decrease, respectively, more rapidly than they would in the closed system RFC with time (Figure 10.12). However, abundances of elements with D ~ 1 do not change. After many cycles of differentiation, replenishment, and eruption, the residual liquid in the magma chamber reaches more or less a steady state, and the steady-state concentration of an element is given by the following equation:

$$C_i^{SS} = C_i^O \cdot [(X + Y)(1 - X)^{D-1}]/[1 - (1 - X - Y)(1 - X)^{D-1}],$$

where C_i^{SS} and C_i^O are concentrations of Element i in the steady-state magma and original parent magma, respectively. X and Y are mass fractions crystallized and discharged (erupted lava) per cycle, and D is bulk distribution coefficient.

During magma's rise through mantle and crust, it is most likely that the magma will react with wall rocks and cool, and cooling leads to fractional crystallization. In such a case of assimilation of wall rock with concomittant fractional crystallization of magma (commonly called *AFC process*), the concentration of Element i changes considerably in the residual liquid owing to the simple mass balance equation:

$$C_i^{\text{residual liquid}} = M_m \cdot C_i^O - M_c \cdot C_i^x + M_{\text{assim}} \cdot C_i^a,$$

where M_m, M_c, and M_{assim} are masses of magma, crystals, and assimilant, respectively. C_i^O, C_i^X, and C_i^a are concentrations of Element i in the parent magma, crystals, and assimilant, respectively.

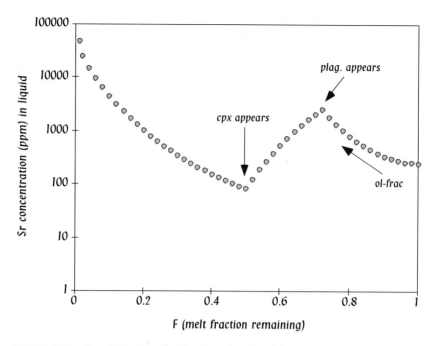

FIGURE 10.11   *Sr variation in residual liquid as a function of changing fractionating assemblages: olivine fractionates alone up to F = 0.7 (i.e., 30% of the liquid crystallized); then an assemblage composed of 80% plagioclase and 20% olivine fractionates up to F = 0.5. Finally, clinopyroxene appears and the fractionating assemblage is $ol_{20}pl_{40}cpx_{40}$. Note that the assemblages were chosen arbitrarily and may not conform to phase equilibrium.*

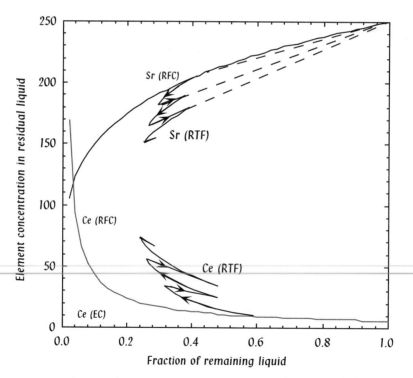

FIGURE 10.12   *Behavior of Sr and Ce in RFC versus RTF process.* (See text for further discussion)

BASALT MAGMA EVOLUTION: AN EARLY CONTROVERSY BETWEEN BOWEN AND FENNER

The subject of petrology was largely treated in a descriptive manner in the 1800s. In the early 1900s, Norman L. Bowen of the Geophysical Laboratory (Carnegie Institution of Washington), who may be thought of as the father of modern igneous petrology, revolutionized this field with his theoretical and experimental phase equilibrium studies. Driven by his desire to explore the origin of magma series and genetic relationships between different magma types of a magma series, Bowen and coworkers set out to perform numerous phase equilibrium experiments on simplified mineral systems that are relevant to crystallization of such magmas. Combining these with theoretical considerations, field associations, and petrography of igneous rocks, Bowen proposed a unifying hypothesis, called *Bowen's reaction principle*, in which he proposed that tholeiite (subalkaline) basalt magmas are *primary magmas* (i.e., they are directly generated from the mantle) and intermediate and felsic magmas are derived by fractional crystallization of basalt magma (Figure 10.13). In his proposal, ferromagnesian minerals form a *discontinuous series* during crystallization of a basaltic magma, in which one mineral crystallizes for a while and then a second mineral appears as the old mineral stops crystallizing further. In Bowen's discontinuous series, a forsteritic olivine appears first in a mafic magma, crystallizes for a while, and then pyroxene forms by reaction between olivine and magma. Bowen proposed that an orthopyroxene forms first and a clinopyroxene later. Eventually amphibole, followed by biotite, replaces clinopyroxene as the crystallizing mineral, and so on. Meanwhile a calcic plagioclase co-crystallizes with olivine. Through progressive crystallization, it becomes progressively albite rich by continuously reacting with the magma. Thus, plagioclase$_{solid solution}$ forms the continuous series of Bowen's reaction series. Note that olivine and pyroxenes individually exhibit

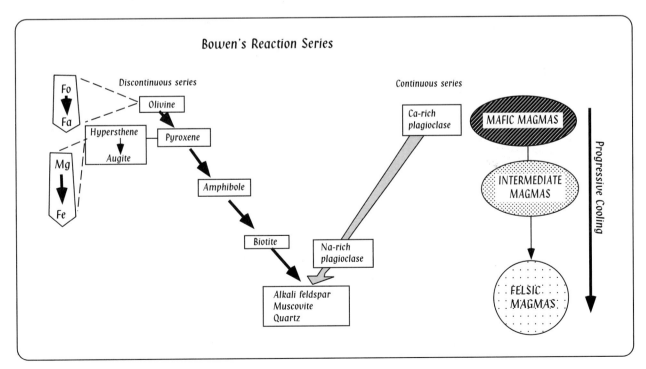

FIGURE 10.13 *Bowen's reaction series.* The two main series are the discontinuous and continuous series. Within the discontinuous series, individual minerals may form continuous series (i.e., solid solution) as illustrated by olivine (Fo–Fa solid solution).

solid solutions, and therefore their compositions also vary continuously during crystallization from Mg-rich toward Fe-rich compositions.

Bowen hypothesized that fractional crystallization of minerals from a basaltic parent magma leads first to an intermediate (andesitic) magma and eventually to a highly differentiated silica-alkali-rich residual liquid, from which mica, alkali feldspar, and quartz (i.e., minerals of granite) crystallize more or less simultaneously. Bowen proposed that all magmas are derived from a parent basalt magma.

Bowen was persuasive in his arguments, and his reaction principle convinced a large number of petrologists. However, based mostly on field relations, some scientists found it hard to accept Bowen's proposal to account for the origin of all granites. Among many problems, two most notable are: (a) the absence of appropriate volumes of associated mafic and intermediate rocks, and (b) field relations in some areas showed transition from large batholiths to veins of partial melts (migmatites) of high-grade metamorphosed sedimentary rocks in deep (eroded and exposed) deformed areas of mountain roots. Subsequent studies have shown that most granites are not produced by fractional crystallization of basalt magma (see Chapter 12).

Returning to Bowen's time, however, it was C.N. Fenner, who was not persuaded by the elegance of Bowen's proposal because he noted that the interstitial materials in coarsely crystallized diabase are mostly Fe-rich granophyre, characterized by their eutectoid or granophyric (wormy) intergrowth between alkali feldspar and quartz and abundance of Fe–Ti oxide minerals (magnetite and ilmenite). Therefore, Fenner argued that fractional crystallization of a (tholeiitic) basalt magma does not lead to the formation of granite, which is poor in iron; instead, an Fe-enriched residual magma is generated. Fenner's contention was later substantiated by the detailed study by L.R. Wager and W.A. Deer of the Skaergaard intrusion (Greenland), which is a mafic intrusion that is considered the best-known example of closed system differentiation of tholeiitic basalt magma. Several other major layered intrusions have been found since then, among which particularly notable is the Bushveld intrusion (South Africa).

LAYERED INTRUSIONS: BASALT MAGMA DIFFERENTIATION

Layered intrusions are enormous, mafic, deep crustal intrusions that are characterized by kilometer-scale layered structure. Table 10.1 summarizes the general features of layered intrusions. In terms of size, they can exceed the dimensions of some small countries put together. For example, the Bushveld intrusion of South Africa occupies an area of about 64,000 km^2. Most of them are pre-Cambrian, although the Skaergaard intrusion of Greenland is tertiary. Simple heat loss calculations suggest that some of these intrusions may have taken as much as 1 million years to cool to the surrounding wall rock (ambient) temperatures. In general, shapes of most layered intrusions are that of a lopolith (Figure 10.14). However, Skaergaard intrusion is funnel-shaped (Figure 10.14). These intrusions generally have a chill zone or fine-grained zone along the contact with wall rocks. Parent magma compositions, as estimated from the composition of chill zone rocks, are most commonly olivine tholeiite to tholeiite. However, alkaline and calc-alkaline layered intrusions do occur as well.

Mineral chemical variations from the least differentiated ultramafic rocks to maximum differentiated granophyric rocks in these layered intrusions suggest multiple magma intrusion and mixing in many cases (e.g., Bushveld intrusion [Wager and

---

TABLE 10.1   *Layered igneous intrusions.*

What they are: Fossil deep crustal magma chambers that were exposed by
tectonic uplift and subsequent erosion of the cap

Size: Bushveld of South Africa (64,000 km^2)
Skaergaard intrusion of Greenland (170 km^2)

Shape: Lopolith or funnel

Parent magma: Olivine tholeiite or tholeiite

Structure and texture: Layering structure and cumulate texture
Layering:
Principal types: Modal layering, phase layering, cryptic layering
Additional curious types: Inch-scale layering, cross-lamination
Attributes: Layering is always parallel to the margins of the
intrusion—from bottom–up, sidewall inward, and roof downward.
Thickest layers occur in the bottom–up sequence. Layering is poorly
developed in the central part of an intrusion.
Origins: In situ crystallization controlled by nucleation versus growth
of crystals, crystal settling, crystal flotation, appearance or disappearance of phases.

Magma evolution
Chemical: Principally iron enrichment followed by alkali enrichment
Mineralogical: Normal cumulus sequence: olivine $\longrightarrow$ opx $\longrightarrow$
plagioclase $\longrightarrow$ augite $\longrightarrow$ pigeonite $\longrightarrow$ magnetite $\longrightarrow$ apatite
Olivine Gap: Olivine disappears and a more Fe-rich olivine reappears late in
the crystallization sequence
Lithological zones (from least to maximum differentiated): dunite and
chromitite, orthopyroxenite, harzburgite, anorthosite, gabbro, ferrogab-
bro, granophyre
Important economic reserve: Important resource for chromite and platinum
group elements

Brown, 1968] and Muskox intrusion, Canada [Irvine, 1980]). Skaergaard intrusion
appears to be an exception in the sense that it seems to have experienced only one
episode of magma intrusion followed by closed system differentiation. Thus, this intrusion shows the extreme development of differentiated liquids, whose nature was
the main issue in the Bowen–Fenner controversy. Figure 10.15 shows that the Skaergaard magma became enriched in iron oxides with fractionation, and only in the very
late stages the residual liquids experienced significant enrichment in alkalis (and silica). It now appears that, in the very late stages, the Skaergaard residual liquid split
into two immiscible silicate liquids—one rich in alkalis and the other rich in iron oxides (De, 1974; McBirney, 1975). Studies of layered intrusions have made it abun-
dantly clear that Fenner was right: *Fractional crystallization of basalt magma does
not lead to the formation of granitic magma of the kind that forms large continental
batholiths but to an iron-rich granophyric liquid.*

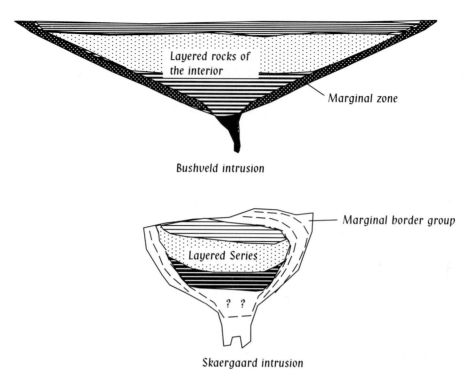

Bushveld intrusion

Skaergaard intrusion

FIGURE 10.14   *Cross-sections of two important layered mafic intrusions—the Bushveld intrusion of South Africa and the Skaergaard intrusion of Greenland.* Note that these diagrams are not to scale.

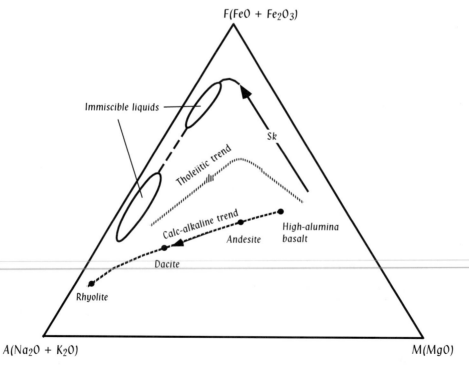

FIGURE 10.15   *AFM diagram showing the various igneous rock series*—(1) the tholeiitic series, illustrated by the Skaergaard trend—"Sk" and the trend shown by commonly erupted tholeiitic lavas and shallow intrusions. (2) The calc-alkaline series and the various rock types that compose this rock series.

FIGURE 10.16   *Field photograph of modal layering in Skaergaard intrusion.*(Courtesy of Gautam Sen)

Layering

Many different varieties of layering occur within a single layered intrusion. In general, the most visible of these is modal layering—the type of layering defined by conspicuous variations in abundances of dark and light minerals on a megascopic scale (Figure 10.16). Cyclical or rhythmic repetition of the layers is fairly common, and hence terms such as *rhythmic layering* and *cyclic layering* have often been used to describe them. A most remarkable variation of modal layering, called *inch-scale layering*, was described by Hess (1960) from the Stillwater intrusion of Montana (United States). This layering is characterized by repetition with amazing regularity of alternate dark- and light-colored layers separated at the scale of inches (Figure 10.17). Aside from these, graded layering is sometimes found, in which crystals are sorted in

FIGURE 10.17   *Hand specimen photograph of fine repetitive layering between chromitite layers and anorthosite layers (Bushveld intrusion).*(Courtesy of Gautam Sen)

size such that individual layers are characterized by a gradual decrease in size of crystals from bottom to top. Cross-stratification structures, similar to those found in classic sedimentary rocks, are also found, suggesting the presence of currents in the magma from which crystals were laid down.

As would be expected, chemical evolution of the magma is best reflected in the changing phase equilibrium and compositions of phases. Discontinuous reaction series, as professed by Bowen, is exemplified by the appearance and disappearance of phases. The "stratigraphy" of a layered intrusion is generally divided into a number of zones on the basis of such appearance and disappearance of mineral phases (Figure 10.18). Lay-

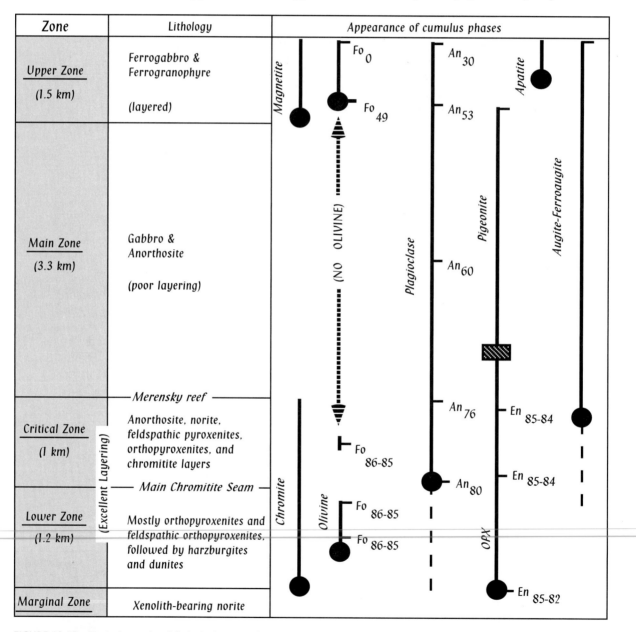

FIGURE 10.18 *Vertical zonation, lithological units, and variation in cumulus phase compositions in Bushveld intrusion.* The dashed lines indicate that the mineral may be present as an intercumulus phase. The Ca-poor pyroxene is orthopyroxene in the lower part of the intrusion and pigeonite (later inverted to orthopyroxene; see Figure Bx. 10.2) in the middle.

ering defined by the appearance or disappearance of phases is referred to as *phase layering*. For example, in the Bushveld intrusion, the appearance of magnetite as a cumulus phase marks the boundary between the Main Zone and Upper Zone.

Slow gradual change in the solid solution composition of an individual phase, which is equivalent to Bowen's continuous reaction series, is called *cryptic layering*. It is called that because this type of layering is not visible in the field and is only discernible after determining compositions of the minerals. As an example of cryptic layering, note how the composition of plagioclase$_{\text{solid solution}}$ changes from An_{85} in the Lower Zone to An_{30} in Upper Zone of the Bushveld intrusion (Figure 10.18).

Cumulus and Cumulate

The term *cumulate texture* has traditionally been used to describe the textures of the rocks in layered intrusions (Figure 10.19). The texture typically consists of layers of well-developed euhedral grains (referred to as *cumulus* grains) with variable proportion of fine-grained interstitial materials that crystallized from an *intercumulus* liquid. Originally these terms were introduced as somewhat of interpretive,

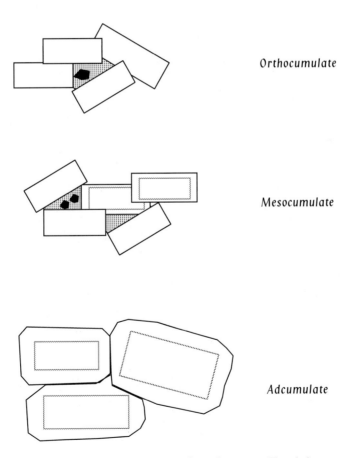

Orthocumulate

Mesocumulate

Adcumulate

FIGURE 10.19   *Three important types of cumulate textures.* The tabular crystals are that of the cumulus phase (plagioclase in this case). Some of these crystals in mesocumulate and adcumulate textures are characterized by overgrowth. The gray and black minerals are intercumulus phases.

10.1 DEEP MAGMA CHAMBER(S) BENEATH HAWAII

This writer has been studying upper-mantle xenoliths in lavas of Oahu since ~1978. In 1990, he accidentally came upon a xenolith in the Dale Jackson Collection of the National Museum of Natural History (Smithsonian Institution) that was unlike anything he had ever seen before. Petrographic examination revealed the presence of distinct layering with olivine and Cr-poor spinel as cumulus phases. What made this rock different from cumulates in any layered intrusions around the world is the presence of intercumulus garnet (Figure Bx 10.1). Experimental petrology indicates that garnet can crystallize from a basalt magma only at very high pressure, and thus garnet's presence suggested the formation of this cumulate in some deep-seated magma chamber, much deeper than any we have known. Comparison with high-pressure phase diagrams suggested that this cumulate formed from a melt at a pressure of ~3 GPa, which is at/near the base of the oceanic lithosphere beneath Oahu. To the author's knowledge, this was (and remains to be) the first report of the operation of cumulus processes in a magma chamber at such great depth.

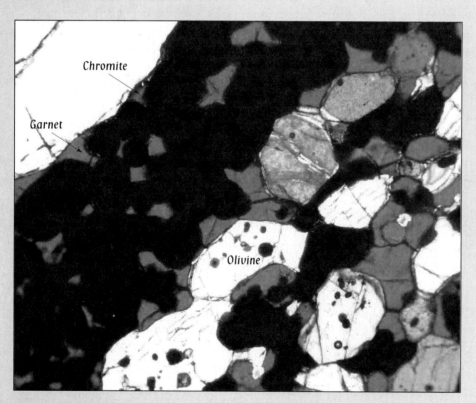

FIGURE BX 10.1 *An unusual cumulate xenolith with layering from the Hawaiian upper mantle.* Cumulus chromite (opaque) and olivine form alternate layers. The occurrence of intercumulus garnet and the other minerals present in this rock indicate a depth of origin at the base of the oceanic lithosphere. (Courtesy of Gautam Sen)

10.2 MERENSKY REEF AND PLATINUM GROUP OF ELEMENTS (PGE)

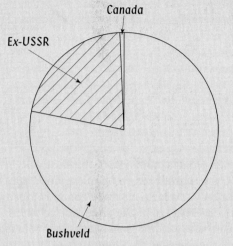

FIGURE BX 10.2   *Global supply of platinum group of elements.*

Merensky Reef of the Bushveld intrusion accounts for about 80% of the world's supply of platinum group of elements (PGE: platinum, palladium, iridium, ruthenium, rhodium, and osmium). The origin of the Merensky Reef continues to be a matter of some debate.

Laterally the thickness of the reef varies from 30 to 90 cm. The average PGE content is about 10 grams per ton, although at places it may be as high as 40 grams per ton. Sixty percent of PGE occurs in solid solution in pyrrhotite, pentlandite, and pyrite; the other 40% occurs as platinoid minerals, such as cooperite (PtS), laurite (RuS_2), Pt–Fe alloys, braggite (Pt,Pd,Ni)S, or their intergrowths (cf. Guilbert and Park, 1988). Origin of the Merensky Reef is a matter of some debate. The hypotheses range from magmatic processes to late-stage hydrothermal concentration of PGE metals. Some sort of liquid immisciblity between PGE-rich sulfide melts and silicate magma likely had occurred at this level of the Bushveld intrusion. It has been suggested that chromite precipitation (forming chromitite layers) may have led to sulfur (and PGE) supersaturation in the melt in the immediate vicinity, which led to immisicible separation of PGE-sulfide melts.

meaning their origin was by some gravity-induced accumulation process. The modern scientist continues to use these terms, although the interpretation concerning their origin has changed considerably. Nowadays it is generally believed that the dominant process by which these rocks form is *in situ* crystallization. Cumulates are generally classified into three types based on the proportion of intercumulus materials: Those with 25% to 50% intercumulus materials are called an *orthocumulates*; cumulates with 7% to 25% intercumulus materials are referred to as *mesocumulates*, and those with <7% intercumulus materials are called *adcumulates*. In mesocumulates and adcumulates, overgrowth on individual cumulus grains is common (Figure 10.19). This overgrowth is believed to take place by continued growth of the cumulus grains by precipitation from the intercumulus liquid. In adcumulate grain boundary, recrystallization is also common. An unusual type of cumulate, called *crescumulate*, also occurs. In it, the crystals are oriented obliquely to the layering. This type of layering is inferred to originate by *in situ* growth of crystals from the base into the magma within the interior of the intrusion. The occurrence of poikilitic inclusions of many smaller crystals in a large (several cms) adcumulus crystal, called an *oikocryst*, is fairly common in layered intrusions. Sometimes they are hard to identify as such because of the grain sizes involved. However, between crossed polars of a petrographic microscope, such oikocrysts appear optically continuous, meaning that the entire grain will go extinct (excluding the inclusions) at the same time. They are clearly identifiable in the field when light reflects off cleavage surfaces on freshly broken surfaces.

Bushveld Intrusion as an Example

Figure 10.18 shows a schematic vertical lithological zonation of the Bushveld intrusion of South Africa. The Marginal Zone rocks are somewhat finer grained than the rocks from the interior, and xenoliths of country rock are often found in this zone (Cameron, 1978). This zone is commonly interpreted as the *chill zone* or where the Bushveld magmas first quenched against the wall rocks. In a sense, Bushveld is not particularly a good example to use here because there is clear evidence that there were multiple episodes of magma intrusion and magma contamination. These episodes of multiple intrusion have been inferred to have occurred when the lower and critical zones were forming. For at least two reasons, the study of Bushveld intrusion is appealing: (a) It is the largest intrusion that seems to have the least to most differentiated rocks, and thus offers an opportunity to explore how magma evolves; and (b) it has important economic mineral reserves of chromite and platinum.

Division of the differentiated rocks into various zones is based on the appearance or disappearance of a phase. For example, the base of the Upper Zone marks the first appearance of magnetite as a cumulus phase. Note that magnetite may be found in rocks from the zones below but only as an intercumulus phase.

The earliest cumulates are layers of chromitites and orthopyroxenites. Dunite layers are virtually absent near the bottom but become more abundant in the middle part of the lower zone (Cameron, 1978). A highly forsteritic olivine appears as a cumulus phase in the Lower Zone but disappears even before it reaches the top of the Lower Zone. The origin of this *olivine gap* can be explained with reference to the system forsterite–fayalite–silica (see "Advanced Reading" below).

ADVANCED READING

Olivine Gap

Crystallization relationships of pyroxene and olivine in layered tholeiitic basalt intrusions are of much interest in petrology. The olivine gap (i.e., olivine disappearance and reappearance of a more Fe-rich olivine in extremely differentiated rocks) in intrusions such as Bushveld and Skaergaard is a result of fractional crystallization. The system forsterite–fayalite–silica may be used to explain this olivine gap. Olivine and Ca-poor pyroxene (orthopyroxene in the forsterite–fayalite–silica system and orthopyroxene + pigeonite in the diopside–forsterite– fayalite–silica system [not shown here]) exhibit reaction relationship. There are three liquidus fields in this system (Figure 10.20): the boundary between olivine and orthpyroxene liquidus fields is partly a reaction curve (1–2) and partly a cotectic (2–3). 3–*e* is a cotectic boundary between olivine and quartz (or some other silica polymorph) liquidus fields. *a–b–c–d–e* is the liquid path of a Magma *a*, whose starting composition is a simplified representation of the Bushveld magma. From *a* to *b*, an olivine with decreasing Fo content will fractionate from the magma. At *b*, the liquid will leave the olivine/orthopyroxene reaction curve and crystallize only orthopyroxene (*b* to *c*). Then from *c* to *d*, orthopyroxene and a silica phase will co-precipitate. At *d*, orthopyroxene will be substituted for by an Fe-rich olivine. Thus, because of early reaction relations between olivine and magma, olivine will disappear and only reappear along a cotectic when the magma has reached appropriate Fe enrichment.

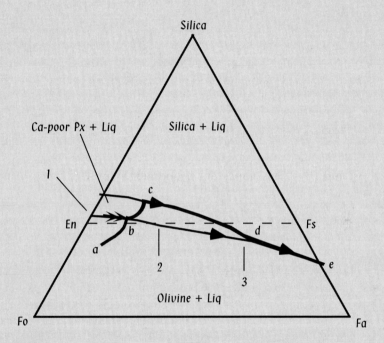

FIGURE 10.20 *The schematic phase diagram forsterite–fayalite–silica is used here to explain how olivine may disappear due to reaction relation with the magma and reappear as a nonreacting precipitate late in a magma's crystallization history.* The light lines are liquidus boundaries and the dark curve (*a–b–c–d–e*) represents the liquid path of our hypothetical magma. Olivine crystallizes from *a* → *b*. At *b*, the reaction *ol + liq = Ca-poor pyroxene* occurs. Because the magma is undergoing fractional crystallization, the liquid moves across the pyroxene liquidus field while precipitating Ca-poor pyroxene only. From *c* to *d*, pyroxene $+SiO_2$ crystallize. When the residual liquid reaches *d*, an Fe-rich olivine (+tridymite—a silica polymorph) reappears in the magma and continues to fractionate until the residual liquid reaches *e*.

Cumulus plagioclase first appears at the base of the Critical Zone. A thick and laterally continuous chromitite layer (called *main chromitite seam*) separates the Critical Zone from the Lower Zone. The Critical Zone rocks are dominantly composed of huge alternate layers of orthopyroxenites and anorthosites with fine, interspersed layers of chromitite.

10.3 PYROXENE CRYSTALLIZATION IN THOLEIITIC INTRUSIONS

Tholeiitic basalt magmas in layered intrusions generally crystallize all three pyroxene types—augite-ferroaugite, orthopyroxene, and pigeonite (Figure Bx. 10.3). With progressive differentiation, all three pyroxene types show considerable Fe enrichment. In the case of Bushveld intrusion, augite + orthopyroxene crystallized early, and pigeonite made its appearance later. In the progressive differentiation sequence, the augite that co-precipitated with the Ca-poor pyroxenes (including orthopyroxene and pigeonite) became strongly Fe rich as its Ca content dropped. At the base of the upper zone, pigeonite became unstable and the magma crystallized a more Fe-rich olivine (and silica polymorph) instead. Figure Bx. 10.3 shows tie lines (dashed), each of which connects two coexisting pyroxenes. These tie lines show that (a) augite-orthopyroxene

co-precipitate first, followed by (b) all three phases, augite-orthopyroxene-pigeonite, crystallize for a short period (note the three-phase triangles), and then (c) ferroaugite crystallizes with pigeonite. During subsolidus cooling of pigeonite, it inverted to an orthopyroxene that has slightly more Ca than the primary orthopyroxenes, which crystallized earlier in the sequence.

The presence of a solvus limits the mutual miscibility between augite-ferroaugite and a Ca-poor pyroxene. Pyroxene phase diagrams (Chapter 5) show that pigeonite may invert to orthopyroxene if it is cooled slowly in the subsolidus region and crosses into the orthopyroxene stability field. Inverted pigeonite crystals are recognized by the development of typical herringbone exsolution (Chapter 5).

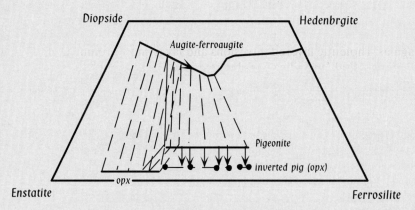

FIGURE BX10.3 *Crystallization of pyroxenes in the Bushveld intrusion.* The dark lines are trends defined by pyroxenes that crystallized from the magma. The dashed lines are tie lines, and each tie line connects compositions of coexisting pyroxenes. The Ca-poor pyroxene that crystallizes first from the Bushveld magma is an orthopyroxene, which became relatively enriched in iron as crystallization proceeded. Eventually pigeonite replaced orthopyroxene as the Ca-poor pyroxene precipitating from the magma. Pigeonite is unstable under subsolidus conditions and inverts to an orthopyroxene. These orthopyroxene (inverted pigeonite) crystals are recognized by their content of herringbone exsolution of augite.

The Main Zone rocks are anorthosites and gabbros and are characterized by the presence of cumulus augite, pigeonite (inverted; see Box 10.3), and plagioclase. These rocks lack good layering. The extremely differentiated rocks of the Upper Zone carry cumulus apatite, magnetite, and very Fe-rich olivine and Ca-rich pyroxene.

SUMMARY

[1] A primary magma is one that is generated directly by melting a source rock. A primitive magma is a magma that has not undergone significant evolutionary change. A parental magma is simply one that is a suitable parent for a more derivative magma. The three definitions may or may not be mutually interchangeable.

[2] The term *differentiation* is used as a general umbrella term that includes all types of processes (crystal sorting, magma mixing, contamination, liquid immiscibility etc.) that may affect magma's composition following its generation. Among various differentiation processes, crystallization-related processes (e.g., in situ crystallization) are considered to be most important.

[3] Variation diagrams are types of diagrams that are used to portray chemical variations in a suite of rocks and to draw inferences about the processes responsible for such variations.

[4] Effects of equilibrium and fractional crystallization processes on magma chemistry can be easily modeled with appropriate equations. Incompatible and compatible elements behave differently in such processes.

[5] Layered intrusions are fossil magma chambers that are characterized by fascinating layered structure ad mineral chemical variations. Their study was important in resolving the debate between Bowen and Fenner on the nature of the remaining melt during fractional crystallization of basalt magmas. Tholeiitic layered intrusions and lavas show prominent iron enrichment. Alkali and silica enrichment occurs in the very late stages of crystallization.

BIBLIOGRAPHY

Bowen, N.L. (1928) *The Evolution of the Igneous Rocks.* Princeton University Press (Princeton, N.J.).

Cox, K.G., Bell, J.D. and Pankhurst, R.J. (1979) *The Interpretation of Igneous Rocks.* George Allen and Unwin (London, U.K.).

Daly, R.A. (1933) *Igneous Rocks and Depths of the Earth.* Hafner press (New York, N.Y.).

De, A. (1974) Silicate liquid immiscibility in the Deccan Traps and its petrogenetix significance. *Geol. Soc. Am. Bull.* **85**, 471–474.

Irvine, T.N. (1980) Rocks whose composition is determined by crystal accumulation and sorting. In: *The Evolution of the Igneous Rocks: Fiftieth Anniversary Perspectives* (H.S. Yoder, Jr., ed.). Princeton University Press (Princeton, N.J.), 245–306.

Kelemen, P., Hart, S.R., and Bernstein, S. (1998) Silica enrichment in the continental upper mantle via melt/rock reaction. *Earth Planet. Sci. Lett.* **164**, 387–406.

McBirney, A.R. (1975) Differentiation of the Skaergaard intrusion. *Nature* **253**, 691–694.

McBirney, A.R. (1980) Effects of assimilation. In: *The Evolution of the Igneous Rocks: Fiftieth Anniversary Perspectives* (H.S. Yoder, Jr., ed.). Princeton University Press (Princeton, N.J.), 307–338.

McBirney, A.R. (1993) *Igneous Petrology* (2nd ed.). Jones and Bartlett (Boston, MA.).

McBirney, A.R. and Nicolas, A. (1997) The Skaergaard layered series; Part II, Magmatic flow and dynamic layering. *J. Petrol.* **58**, 569–580.

Peng, Z.X. and Mahoney, J.J. (1995) Drillhole lavas from the northwestern Deccan Traps, and the evolution of Reunion hot spot mantle. *Earth Planet. Sci. Lett.* **134**, 169–185.

Philpotts, A.R. (1990) *Principles of Igneous and Metamorphic Petrology.* Prentice-Hall (Englewood Cliffs, N.J.).

Ragland, P.C. (1989) *Basic Analytical Petrology.* Oxford University Press (Boston, MA.).

Sen, G. (1990) Cumulate xenolith in Oahu, Hawaii: implications for deep magma chambers and Hawaiian volcanism. *Science* **249**, 1154–1157.

Sen, G. (1996) A simple petrologic model for the generation of Deccan Trap magmas. *Int. Geol. Rev.* **37**, 825–850.

Wager, L.R. and Brown, G.M. (1968) *Layered Igneous Rocks*. Oliver and Boyd (London, U.K.).

Yang, H-J., Kinzler, R. and Grove, T.L. (1997) Experiments and models of anhydrous, basaltic olivine-plagioclase-augite saturated melts from 0.001 to 10 kbar. *Contrib. Mineral. Petrol.* **124**, 1–18.

CHAPTER 11

Origins of Basalt Magmas

The following topics are covered in this chapter:

Basalt magmas
    Nature and origin of basalt magmas in midoceanic ridges
    Hot spot volcanism and basaltic magmas
    Flood basalt volcanism

ABOUT THIS CHAPTER

In the previous chapters, we learned about the physical and chemical nature of magmas, how to name igneous rocks, how they occur, how magmas form and segregate from the source, how magmas may change composition during their transit from source to their sites of emplacement, and how to track such processes using chemical signatures. The present chapter focuses specifically on basalts: where they occur (i.e., their plate tectonic associations), their compositional characteristics, their source rocks, and the pressure-temperature conditions in which they form.

THOLEIITIC VERSUS ALKALI BASALT MAGMAS

The recognition that alkali basalts and tholeiitic basalts are fundamentally different (Chapter 10) makes the dual problems of generation of different types of basalt magmas and the effects of high-pressure differentiation processes on these basalt magma types all the more interesting. Even as early as 1960, it became clear that basalt magmas are generated within the upper mantle. It was only reasonable to ask at that time: (a) Do these different magma types occur in different tectonic regimes? (b) What is/are the appropriate upper-mantle rock(s) that can melt and produce different basalt magma types? (c) What sorts of pressure and temperature conditions are required to produce the different basalt magmas? (d) Do volatiles (and what are they) or some specific elements exert significant control over what type of basalt magma would be generated? (e) Is it possible to generate alkali basalt magma from tholeiite via differentiation processes; if so, then how?

An era of experimental petrologic inquisition thus began to resolve these issues. The experimental approach used by various scientists may be grouped into two types—*naturalists* (i.e., the scientists who perform experiments on natural rocks) and *purists* (i.e., the scientists who prefer to experiment with analog systems with fewer components). Although the first approach allows a semiquantitative understanding of the natural system, the large variance (too many components) does not allow a quantitative evaluation. The advantage of the second approach is that it does allow a quantitative understanding of the phase relationships involved in magma generation and crystallization. However, the disadvantage is that the results must be extrapolated to the natural system for making sense of the natural system, which may be a stretch sometimes. There is clearly merit in both types of approaches. In the following sections, therefore, references are often made to studies on natural rocks as well as simple analog systems.

Partial Melting of Lherzolite: Experimental Generation of Basaltic Magmas

The Source Lherzolite

Lherzolite is generally accepted by scientists to be the principal source rock for basaltic magmas; and eclogite is a minor contributor. As pointed out in Chapter 8, a typical lherzolite is dominantly composed of olivine (Fo_{88-93}), highly magnesian and aluminous ortho- and clinopyroxenes, and usually <5% of an aluminous phase (plagioclase, spinel, or garnet). There are many reasons for believing that lherzolite is the main source for basalt magmas: (a) Lherzolite is the predominant type of mantle xenolith found in strongly nepheline-normative lavas, such as alkali basalt, nephelinite, basanite, and kimberlites, indicating that it is likely an important upper-mantle rock. (b) Lherzolite also occurs in ophiolites (discussed later), which are widely accepted to be slices of the upper mantle obducted on continental crust. (c) Seismic velocity measurements of the upper mantle support the contention that lherzolite probably makes up >90% of the upper mantle. (d) When a lherzolite is melted in the laboratory, composition of small amounts of melts is that of a basalt (further discussed later).

The nature of the aluminous phase provides an important key to the equilibrium pressure of a lherzolite. High-pressure experiments indicate that plagioclase lherzolite is stable at pressures less than ~1GPa, spinel lherzolite is stable at intermediate pressures (approximately 1–3 GPa), and garnet lherzolite is stable at >3 GPa (Figure 11.1; Takahashi and Kushiro, 1983). Among these three rock types, plagioclase lherzolite is ex-

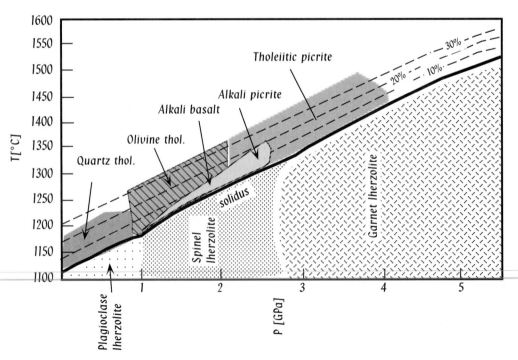

FIGURE 11.1   *This figure shows the (a) volatile-free solidus of lherzolite; (b) types of basaltic magmas generated at different pressure (P), temperature (T), and degree of melting (F); and (c) pressure-temperature stability fields of the three types of lherzolite.* Note that the dashed contours represent %melting. These are based on data from a large number of sources—most notably, Jaques and Green (1980), Takahashi and Kushiro (1983), Hirose and Kushiro (1993), and Baker and Stolper (1995).

tremely rare. Garnet lherzolite (as a xenolith) is more abundant in kimberlites, and spinel lherzolite (also as xenolith) is more common in alkali basalts. The obvious question one may ask at this stage is what do we know about the types of magmas produced from lherzolites as a function of pressure (or depth). The following section explores this topic.

Making Basaltic Magmas in the Laboratory: a Perspective

Pioneering experimental studies in the 1960s by D. H. Green and A.E. Ringwood of the Australian National University on synthetic lherzolitic starting materials (which they called *pyrolite*) showed that basalt magma forms by moderate degrees ($<30\%$) of melting of lherzolite in absence of volatiles such as CO_2 and H_2O. They also showed that magmas generated at low pressure (approximately <0.8 GPa) are quartz-normative tholeiites; with higher pressures, magmas become increasingly olivine-normative (Figure 11.1). In subsequent years, Green and coworkers developed a type of petrogenetic grid for basalts, in which they delineated the P,T fields where various basaltic magmas are generated. Given an appropriate source rock, these authors showed that it is possible to generate alkali basalts by smaller degrees of melting and olivine tholeiites by higher degrees of melting over a P,T range. When $P > \sim 2$ GPa, these authors showed that the magma is more picritic (i.e., very rich in normative olivine and MgO [wt%] $>15\%$).

Subsequent experiments by many scientists have not altered two of the more broad but nonetheless important conclusions reached by Green, Ringwood, and colleagues of the Australian National University: (a) magmas become progressively more MgO rich (i.e., olivine rich) with pressure, and (b) it is possible to generate alkalic and tholeiitic magmas from the same lherzolite (Figure 11.1). Until about 1983, it was generally believed that basalt magma forms by isobaric melting of lherzolite at a pseudoinvariant point (see e.g., the 10-, 15-, and 20-kbar points in Figure 11.5) defined by the confluence of liquidus fields of all the minerals present in the lherzolite, much like isobaric melting at an invariant point (eutectic or peritectic; Chapter 7) in a simple two- or three-component system.

Takahashi and Kushiro (1983) demonstrated that the liquid composition does change as melt% increases at a constant pressure. That is, melting of a lherzolite is not an invariant phenomenon at any pressure. For example, see the dashed curve at 1 GPa in Figure 11.3a (discussed later). In a sense, Figure 11.1 also shows this: At any particular pressure between 1 and 2 GPa, the near-solidus, low % melts are alkali basaltic and higher temperature, higher % melts are olivine tholeiite, all with a lherzolitic residue (i.e., all contain olivine, orthopyroxene, clinopyroxene, and spinel).

In recent years (1992–1993), there has been a significant shift in the philosophy of lherzolite melting experiments due to the recognition that near-fractional melting is the likely process operating in the upper mantle, rather than batch melting (assumed in the pre-1990 experiments), in which large amounts of melts ($\sim 10\%$–15%) equilibrate with the source rock (Chapter 9). As indicated in Chapter 9, McKenzie (1984) pointed out that it is unlikely for basalt magmas to remain in the source region once they reach about 3% of the mass of the original starting lherzolite. This was subsequently verified by Johnson et al. (1990) and Yang et al. (1998) with trace element modeling of oceanic lherzolites. However, the details of their arguments are beyond the scope of this book.

It is clear that a quantitative understanding of the genesis of basalt magmas must rely on high-quality experimental data on the chemical composition of near-solidus (melt% <3) melts as a function of pressure, temperature, and melting extent. It turns out that such near-solidus experiments (melt% <3) are extremely problematic. At the time of this writing, the writer is aware of only two or three groups in the world—most notably, M.B. Baker and E. Stolper (CalTech) and K. Hirose and I. Kushiro (University of Tokyo)—that have had some success. However, even these new-generation experimental data are not free of criticism, and independent verification of the data using thermodynamic means is currently being pursued by some scientists (Asimow et al., 1997; Hirschmann et al., 1998). The next section contains the writer's synthesis of what crucial information scientists have been able to extract from new and old experiments. As the foregoing discussion indicates, this field is currently undergoing a revolution, and there is a real chance that parts of the discussion below will become obsolete within a decade.

Magma Compositions as a Function of Pressure (P) and Degree of Melting (F) in Volatile-Absent Conditions: Recent Experiments

Although generally consistent with the older data, the new experiments on lherzolite melting at relatively low melt fractions show several additional details:

1. SiO_2 abundance in the melt is not strongly affected by the source composition or degree of melting. The only parameter that controls its abundance is pressure (Figure 11.2). SiO_2 content of near-solidus basaltic magma decreases with increasing pressure. Therefore, SiO_2 content of melts is a useful barometer that may be used to determine the depth range over which basalt magmas are generated. However, the absolute abundance of SiO_2 in extremely low melt fractions (<2% melting) has been a matter of some debate. As far as the writer is aware, the verdict is not yet in.

2. Na_2O content of the source lherzolite exerts significant control over the nature of lowest fraction melts generated (Figure 11.2). Na behaves as an incompatible element and prefers melt over the solid residue. Therefore, during partial melting of lherzolite, Na_2O content of the partial melts is the highest at extremely low % melting (i.e., near the solidus). As melting degree increases, the Na_2O content of melt goes down simply due to dilution effect. In fractional melting, the Na_2O decrease in melt is more spectacular than in batch melting because the source gets rapidly depleted in Na_2O as melts are continuously extracted from the source. Although this overall behavior of Na_2O is not disputed, the actual Na_2O contents of the lowest melt fractions at different pressures remain controversial. For example, Figure 11.2 shows that two recent sets of experiments show linear versus logarithmic behavior of Na_2O abundance in melt (based on Sen, 1996).

3. Al_2O_3 content of magmas is the best indicator of the degree of melting (Figure 11.2). It is essentially independent of pressure or the composition of the starting lherzolite (cf. Sen, 1996). In contrast, note that Na_2O content of the partial melts is significantly controlled by bulk composition of the source lherzolite (not shown here).

Figure 11.3 is a type of petrogenetic grid that shows the effect of pressure and degree of melting on normative composition of the partial melts in experiments (mostly recent and some old experimental data). In a general sense, an increase of pressure results in a decrease in normative quartz in the clinopyroxene (cpx)–olivine (ol)–quartz (qz) diagram. Low-degree (<5%) partial melts are: qz-tholeiite at $P < \sim 0.6$ GPa, olivine-tholeiite at $P = 0.6 - 1$ GPa, and alkalic at $P > \sim 1.1$ GPa.

Figure 11.3 illustrates an interesting aspect of melting at a constant pressure: at $P > \sim 1.1$ GPa, the melt composition changes from alkalic to olivine-tholeiitic with increased amount of melting, regardless of the melting process—whether it is batch melting or fractional melting. To illustrate the difference between near-fractional versus batch melting process, we estimate the melt composition paths for both processes

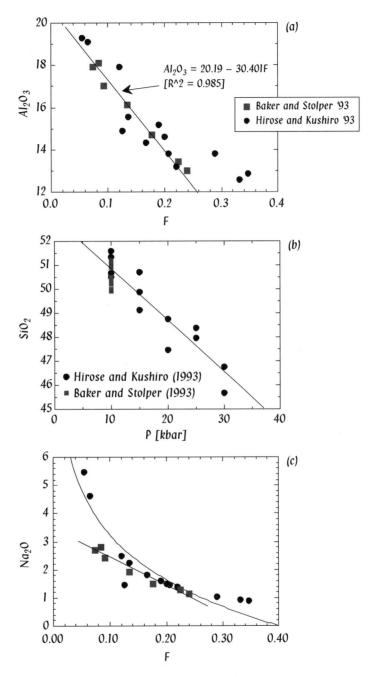

FIGURE 11.2   *Data from two recent sets of experiments using the diamond aggregate technique are plotted here* (from Sen, 1996). (a) Variation of SiO_2 in experimentally generated magmas from lherzolite as a function of pressure. The scatter in SiO_2 content at each pressure represents melt compositions at a range of Fs. It is clear that SiO_2 content of primary basaltic magmas is a sensitive indicator of its depth of generation. (b) Na_2O content of experimental melts as a function of F, showing that it could be a potential indicator of degree of melting. However, the discrepancy between two different data sets at F < 0.1, where fractional melting is operative, suggests that more work needs to be done before Na_2O content can be used as an F indicator. (c) Al_2O_3 content of experimental magmas as a function of F. Note that these data were collected over a broad pressure range. This plot suggests that Al_2O_3 content of primary basaltic magmas is a sensitive indicator of F and is virtually unaffected by pressure.

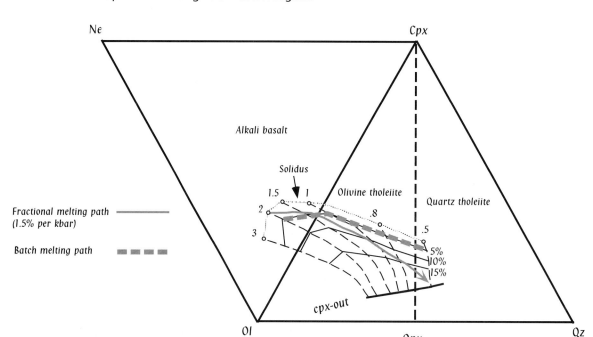

FIGURE 11.3 *Mole percent diagram (petrogenetic grid) relevant to variable percent melting (5% to the point where clinopyroxene disappears from the residue) of lherzolite over a pressure range of 0.5 to 3 GPa (i.e., about 15–90 km depth; pressure shown in bold).* Each light dashed line at a given pressure (say, 2 GPa) represents loci of melt compositions (molar normative) generated by progressive partial melting of a lherzolitic assemblage (ol + opx + cpx + melt) at that pressure (melt% increasing from left to right on each dashed curve). Each light continuous line represents a fixed %melting curve (only three are shown). Also shown is the cpx-out line. A lherzolitic source rock will lose cpx to the melt beyond this line. Sources of data: Takahashi and Kushiro (1983), Hirose and Kushiro (1993), and Baker and Stolper (1995). Polybaric melting paths for batch melting (5% per kb [0.1 Gpa], melt stays with the ascending lherzolite matrix) and fractional melting (1.5% per kbar, melt leaves the source) are shown. Note that it is mainly schematic and does not take into account the changing source composition that must happen as the melt is removed from the source.

for a mantle parcel that is rising and beginning to melt at 2 GPa (Figure 11.3). In the batch melting case, the melt% at each isobar is fixed at 5% (and melt does not leave the source). In the near-fractional melting case, we assume a melting rate of 1.5% per kbar with melt leaving the source. Clearly the difference in the resultant range of melt compositions produced by the two melting processes is remarkable.

Note that the basaltic crust at a midoceanic ridge is made from a mixture of small % melts pooled over a wide depth range (discussed further later). Therefore, the composition of the crust represents an average composition of all such melts, weighted by how much total melt is supplied from what depth. Let us now look back at the examples of fractional versus batch melt composition paths shown in Figure 11.3 and try to understand what the concept of *weighted mean* means. It is possible that the weighted mean melt compositions (= crust) generated by the two melting processes are identical and converge on the cross-over point where the two paths cross. However, the two means may be sufficiently different as to produce very different compositions of the crust.

In the case where magmas have been generated from mantle lherzolite by fractional fusion over a wide volume (as is likely in most plate-tectonic situations) and mixed in some shallow crustal chamber, it may be hard to decipher the depth range of origin using a diagram like Figure 11.3. Alternative procedures using actual oxide analysis of the lavas have been put forth for such purpose by a number of authors, led by McKenzie and Bickle (1984) and Klein and Langmuir (1987). Some additional discussion of this topic follows in a later section.

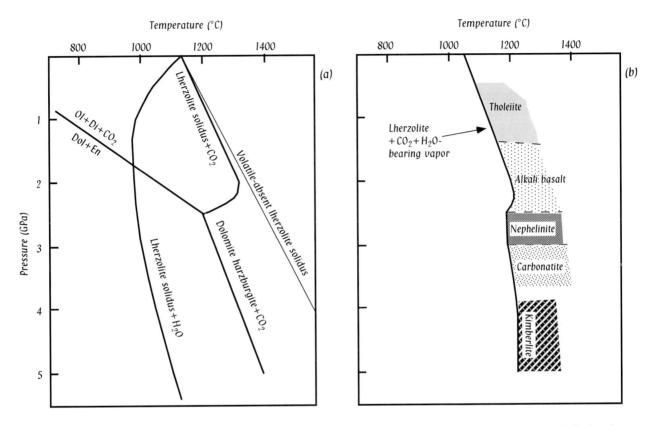

FIGURE 11.4   (a) Relative effects of CO_2 and H_2O on lherzolite solidus in pressure-temperature space are shown. The volatile-free dry solidus is also shown for the purpose of comparison (largely from Wyllie, 1978). (b) Composition of magmas generated from lherzolite in presence of a CO_2 + H_2O-bearing vapor. (Based on the works of Wyllie, Eggler, Mysen, and Boettcher, D.H. Green, and their coworkers)

Melting of Lherzolite in the Presence of Volatiles: A Short Note

The function of CO_2 and H_2O, the two most abundant volatile components in the upper mantle, is to depress the solidus of the mantle lherzolite (Figure 11.4a). H_2O lowers the solidus a little more than CO_2. Their effects on composition of the melts generated near the solidus are also very different. Briefly, high relative abundance of H_2O tends to generate more silicic melts (andesitic to high-alumina basalt magmas of convergent plate margins, discussed later). In contrast, CO_2 tends to make magma compositions more alkalic. The presence of both CO_2 and H_2O can generate alkali basalts at relatively low to moderate pressure ($\sim$1.1–2.5 GPa), nephelinite and carbonatite magmas at moderate pressure ($\sim$2.5–3.5 GPa) and kimberlitic melts at much higher pressure (>4 GPa; Figure 11.4b).

BASALTIC MAGMATISM AT MAJOR TECTONIC ENVIRONMENTS

Basalts occur everywhere—from divergent and convergent plate boundaries to hotspot tracks and major flood basalt provinces. The following is a short review of the characteristics of basalts in all of these environments except convergent boundaries (discussed in a separate chapter) followed by a short discussion of their origin.

Midoceanic Ridge Basalts

Magmatic activity beneath the entire length of the earth's 65,000-km midoceanic ridge system continuously adds new materials to the crust. Because these rocks are dominantly basaltic (in fact, tholeiitic), it has become a common practice to refer to them as midoceanic ridge basalts (MORB). MORB are chemically and isotopically heterogeneous along the entire length of the global ridge system. Klein and Langmuir (1987) and Niu and Batiza (1987) showed that there is correlation between global and local variations in crustal thickness and major element chemical composition of the basaltic rocks (discussed further later).

The general petrologic characteristics of MORB are as follows:

1. Most are olivine and hypersthene-normative tholeiites, and very few are quartz normative (Figure 11.5). The most primitive aphyric or glassy (i.e., not enriched with xenocrysts) MORB contain ~9% MgO. Extremely differentiated iron- and titanium-rich basalts (so-called *Fe–Ti basalts*) occur in some places, most particularly Galapagos rift.

2. Petrographically, these basalts may be glassy to phenocryst bearing to moderately crystallized types. In general, the phenocryst crystallization sequence

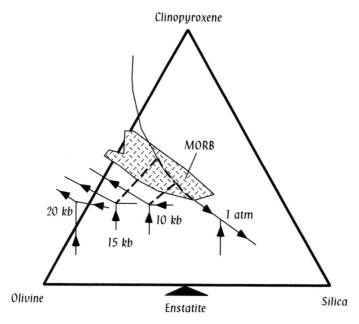

FIGURE 11.5 *Mole percent diagram (based on Stolper, 1980) showing the field of most common MORBs.* Although not shown here, note that nepheline would plot to the left of the cpx–ol–qz triangle (see Figure 11.3 for reference). The isobaric pseudoinvariant points at four different pressures are shown. Note that the number of authors in the post-1980 period have shown that such near-invariant melting does not apply to lherzolite (see Figure 11.3 for comparison). However, this diagram still serves three important purposes: It shows, albeit approximately, (a) how pressure moves the primary melt composition, (b) where common MORB plot in relation to these melt compositions, and (c) at P ~ 10 kb (i.e., 1 GPa) the thermal barrier between alkali basalts and tholeiites no longer exists, and a tholeiite can fractionate olivine + clinopyroxene and become alkalic in composition (see which way the arrow on the 10 kb pseudo-cotectic ol + cpx is pointing). The two dashed paths illustrate how fractionation of olivine from magmas generated at 10 and 15 kb pressures can lead to a large portion of the observed MORB compositions.

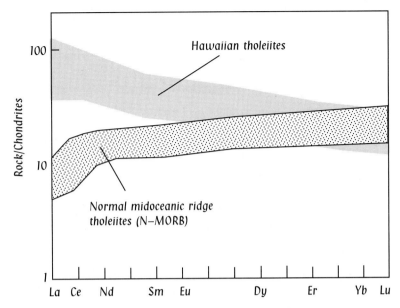

FIGURE 11.6 *Chondrite normalized rare-Earth element compositions of Hawaiian tholeiites are compared with N–MORB.* Calculations involving partitioning of these trace elements between a lherzolitic source and magma (see Chapter 9) would show that an N–MORB source must be depleted in light rare-Earth elements (LREE–La, Ce, Nd) relative to chondritic abundances (chondritic REE pattern would be a horizontal line at 1 because note that all REE patterns are normalized to chondrites). However, Hawaiian tholeiites require partial melting of a different source—LREE-enriched or chondritic. However, it is apparent that a small component contributing to Hawaiian tholeiites may come from the MORB source.

appears to be olivine (most primitive olivine $\sim Fo_{91}$) $\pm$ chrome-spinel $\longrightarrow$ olivine + plagioclase (most primitive $\sim An_{90}$) $\pm$ spinel $\longrightarrow$ plagioclase + augite (e.g., Hess, 1989).

3. Trace element composition of MORB varies greatly, but the vast majority of normal MORB (N–MORB) erupted at normal ridge axes (not influenced by hot spots or plumes) is characterized by LREE-depleted (chondrite-normalized) patterns (Figure 11.6). N–MORB are characterized by depleted $^{143}Nd/^{144}Nd$-isotopic compositions (and Sr, Pb, and He isotopic systems as well) relative to that of the undifferentiated earth (i.e., if the earth did not differentiate into crust, mantle, and core; see explanation in Figure 11.7 caption). These REE and isotopic characteristics of N–MORB are generally interpreted to be due to their derivation from a source upper-mantle layer (= convecting asthenosphere) that has been depleted in LREE and other strongly incompatible elements over the 4.6 billion year (= age of the earth) history of the earth. The extracted elements have been added to the continental crust as continents are very old and enriched in such elements.

Ridge Topography, Spreading Rate, and Magma Chamber

Seismic and gravity data indicate that the crust of the normal ocean floor is about 6 to 8 km thick. In areas where a hot spot coincidentally occurs beneath a ridge axis, such as Iceland, the crust is unusually thick (14–25 km). In older areas of some oceans, where thick basaltic plateaus occur (e.g., Ontong-Java in the Pacific), the crust can be very thick ($\sim$18–25 km). However, it is generally believed that such thick plateaus were created by large plume heads (discussed later), and not by normal midoceanic ridge volcanism.

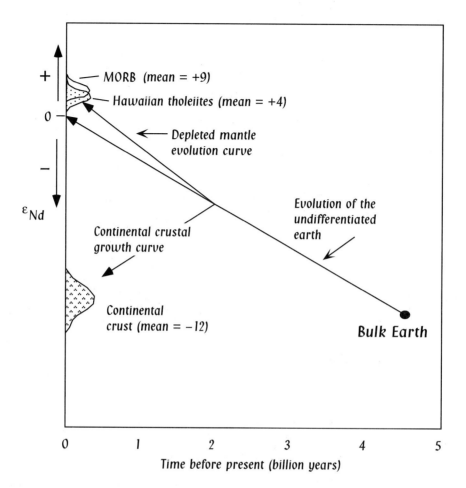

FIGURE 11.7   *This is a schematic Nd-isotopic evolution diagram for the earth's mantle and some crustal rocks* (*details presented in the appendix*). The radiogenic isotope ^{143}Nd forms by the decay of ^{147}Sm. ^{144}Nd is a nonradiogenic isotope. Thus, any portion of the earth formed some 2 billion years ago with some ^{147}Sm/^{144}Nd ratio would have increasingly greater ^{143}Nd/^{144}Nd as time passes. The original ^{147}Sm/^{144}Nd ratio of that portion of the earth will dictate what the present-day ^{143}Nd/^{144}Nd ratio of it should be (see appendix for details). A parameter called ε_{Nd} is plotted on the Y-axis, and time (billions of years) is plotted on X-axis. 0 on the X-axis means present time, and 1 to 5 refer to billions of years before present. The plot *Bulk Earth* refers to the time and composition of the protoearth. Stated simply, the term ε_{Nd} refers to the present-day isotopic ratio ^{143}Nd/^{144}Nd of a rock relative to the same of the undifferentiated earth (i.e., if the earth had not differentiated into crust, mantle, and core). The line marked *evolution of the undifferentiated Earth* shows how the undifferentiated Earth's isotopic composition would have evolved. Thus, ε_{Nd} = 0, where this line intersects the 0 time axis. The continental crust is said to be enriched because the materials that are extracted from the upper mantle to form the continental crust prefer Nd over Sm. Its extraction depletes the upper mantle in terms of Nd. Accordingly, such an upper mantle is said to be *depleted*. Depleted rocks have $+\varepsilon_{Nd}$ values, and enriched rocks have $-\varepsilon_{Nd}$ values. MORB have an average ε_{Nd} of +9, whereas Hawaiian hotspot tholeiites have an average ε_{Nd} of +4. Continental crust separated from the mantle at 2 b.y. ago should have a ε_{Nd} around −12. Whereas MORB appear to come from the depleted asthenosphere that is the residue from continental crust extraction, Hawaiian tholeiites appear to be a mixture of a number of depleted, primitive, and enriched mantle sources.

The topography of the midoceanic ridge varies considerably from place to place and correlates with spreading rate (i.e., magma supply). In slow-spreading (half-spreading rate <3 cm/year) areas of Mid-Atlantic Ridge (MAR), an axial graben or rift valley occurs at the ridge axis bounded on both sides by topographic highs formed by block faulting (Figure 11.8). In fast-spreading (half-spreading rate ~5–6 cm/year) areas (e.g., the East Pacific Rise [EPR]), such axial rift valleys and highs are almost inconspicuous; only a broad, gentle topographic swell is seen (Figure 11.8). Also, the water depth is shallower above a fast-spreading ridge relative to a slow-spreading ridge. The smooth versus rough topography in fast- versus slow-spreading ridges is believed to be related to magma supply rate (discussed in a later section).

Seismic studies indicate the presence of a melt lens (10–100s of m high and 1–2 km wide) beneath the fast-spreading EPR. Such a melt lens has not been found beneath any slow-spreading ridge. This observation suggests that, at slow-spreading ridges, a perennial melt lens does not exist and the magma supply is intermittent. These observations, combined with lithologic characteristics of ophiolite complexes (see Box 11.1) and gabbroic rocks drilled/dredged from the ocean floor, tell us about the nature of the magma chambers beneath fast- versus slow-spreading ridges (Figure 11.8; Sinton and Detrick, 1992). Common to all spreading ridges are a 0.5-km thick volcanic crust composed of basalts with pillow structures (Figure 11.8 and Figure Bx. 11.1). This layer is underlain by a sheeted dike complex, which serves as the conduit for magma delivery to the shallow crust. In the case of fast-spreading ridges, magmatic activity, high heat flow, the presence of a melt lens, and the occurrence of a broad-seismic, low-velocity zone beneath the ridge axis indicates that (a) the melt lens is underlain by a broad region of crystal-liquid mush zone where some amount of interstitial melt is always present between mineral grains, and (b) erupted lavas are processed (i.e., fractionation, mixing) in the melt lens prior to their eruption. It is believed by most recent workers that the mush zone is surrounded by a transition zone, which is a largely crystallized zone with smaller amounts of interstitial melt. The transition zone then grades into surrounding, still hot gabbroic rocks. Similar mush and transition zones occur beneath slow-spreading ridges as well. The lack of a perennial melt lens suggests that the magmas that erupt at the ridge axis do not undergo significant fractionation and mixing in a shallow melt lens and therefore retain a stronger signal from their source mantle chemical characteristics. Such a hypothesis is compatible with the following observations from lavas erupted at the slow-spreading ridges: (a) greater isotopic and trace element diversity, (b) generally less differentiated compositions of the lavas, and (c) the general lack of textural and compositional characteristics indicative of significant magma mixing. Note that in MORB terminology, the melt lens, mush, and transition zones are all considered to be parts of a complex magma chamber, which is never fully molten (cf. Sinton and Detrick, 1992).

Contrasting topography of fast- versus slow-spreading ridges is related to their different styles of magma processing and eruption. The axial rift valley-and-ridge topography of a slow-spreading ridge is controlled by episodes of magma supply (the so-called *waxing* episodes) and of extension not accompanied by magma supply from below (the so-called *waning* episodes). Waxing of the magma supply system gives rise to axial highs, and waning periods lead to magma chamber collapse, block faulting and rotation, and rift valley formation. This type of waxing and waning does not occur beneath a fast-spreading ridge; the broad swell-type topography is perhaps due to magma chamber buoyancy pushing up broad regions beneath the spreading center.

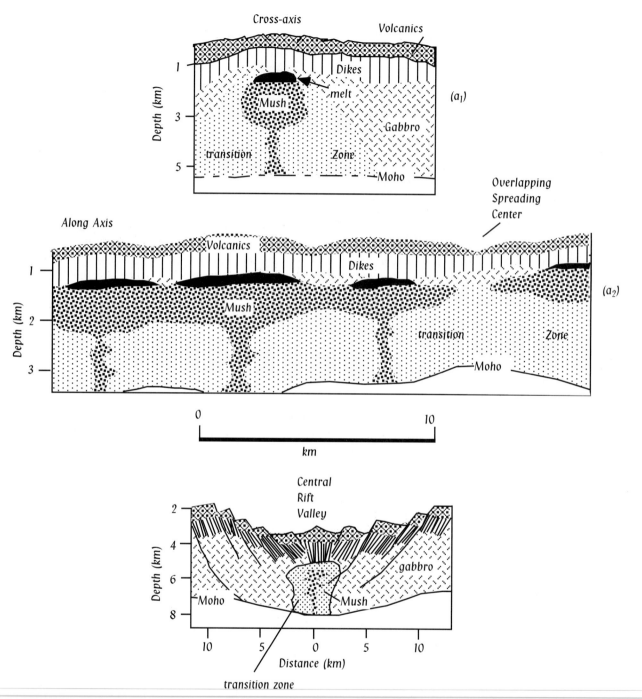

FIGURE 11.8   *These diagrams represent a model of the crust and magma-delivery systems in fast- and slow-spreading midoceanic ridges* (*modified from Sinton and Detrick, 1992*). (a₁) A schematic cross-section of a fast-spreading ridge axis showing the presence of a thin magma lens beneath the ridge axis, which grades into a mush and transition zones containing variable amounts of melts and crystals. The transition zone grades into crystalline gabbro. Lavas erupted at the surface are mainly supplied by the melt lens. (a₂) An along-axis view of the same above showing how the magma lens pinches out beneath discontinuities (such as a transform fault or an overlapping spreading center) that disrupt the along-axis continuity of the ridge axis. (b) Cross-section of a slow-spreading ridge axis. Most notable is the absence of a perennial magma lens. Eruptions on the surface are believed to largely result from magma tapped more or less directly from the upper mantle. The sharply defined ridge-and-valley structure of the slow-spreading ridge is believed to be tied to relatively intermittent magma supply coupled with extension. (From Sinton and Detrick, 1993, published in *Journal Geophysical Research*, American Geophysical Union, Washington, D.C.)

11.1 OPHIOLITES

An ophiolite is a large (several km²), dominantly mafic-ultramafic, fault-bounded, complex that is of oceanic origin but found on land (Figure Bx. 11.1). It is interpreted to be a slice of the oceanic crust and uppermost mantle. The little-understood tectonic process by which an ophiolite is dismantled from the oceanic lithosphere onto a continent is known as *obduction*. A complete ophiolite starts with a highly deformed, discontinuous (maximum thickness of ~12 km) layer of harzburgite or lherzolite layer at the base. This layer is often serpentinized to various extents and is cut across by large (10s of meters to km sized) pods of dunite and by dunite and gabbroic dikes. This deformed ultramafic layer, often called the *tectonite* layer, is generally inferred to be the mantle portion of the ophiolite from which magmas have been extracted. The deformed ultramafic layer is followed above by layers of ultramafic and gabbroic rocks with well-defined cumulate texture (see Chapter 10). By analogy with layered igneous intrusions (Chapter 10), these cumulate-textured layers are interpreted to be the bottom portion of a magma chamber where crystals had accumulated from magma batches. The cumulate layers are followed above by a large section of gabbro that lacks cumulate texture. Because of the absence of any directional features within this gabbro, it is generally referred to as *isotropic gabbro*. A layer composed of parallel, straight, or curved dikes occurs above isotropic gabbro layer and is called the *sheeted dike complex*. Such dikes are believed to be feeder dikes that supplied magma to the overlying crust. The rocks above the sheeted dike complex are pillow basalts (i.e., basalts with a type of bulbous structure that has the appearance of a pillow). Scientists have actually observed the formation of such pillows in actively erupting basalt lavas under water. Basaltic crust forming the ocean floor is also characterized by such pillow structure. Above the pillow basalt layer occurs a thin cherty layer with abundant remains of an organism known as *Radiolaria*. The cumulate layers, isotropic gabbro, sheeted dike complex, pillow basalt layer, and radiolarian chert layer together comprise the entire crustal section of an ophiolite. The maximum thickness of the crustal section of an ophiolite is about 9 km.

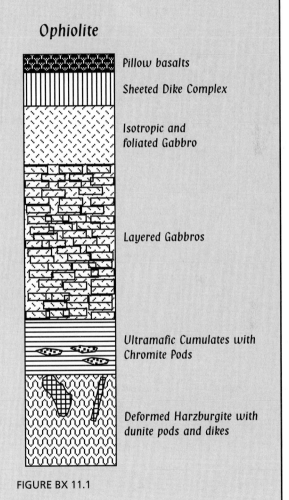

Ophiolite

- Pillow basalts
- Sheeted Dike Complex
- Isotropic and foliated Gabbro
- Layered Gabbros
- Ultramafic Cumulates with Chromite Pods
- Deformed Harzburgite with dunite pods and dikes

FIGURE BX 11.1

Magma Generation and Crustal Thickness

As pointed out in Chapter 10, magma is mostly generated by passive upwelling beneath normal midoceanic ridges, although beneath some ridges a narrow zone of active upwelling also occurs (e.g., Langmuir et al., 1992). In the general case of passive upwelling, the melting regime is believed to have a triangular cross-section (Figure 11.9). It was

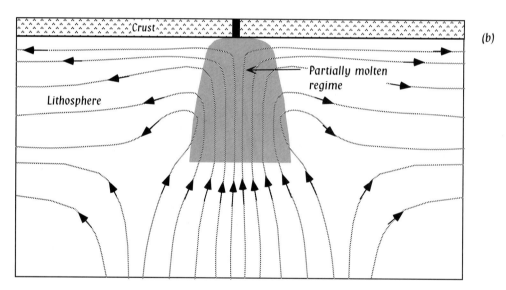

FIGURE 11.9 *Cross-sections of two types of MORB melting regimes:* (a) triangular in case of passive melting due to extension, and (b) pipelike in case of active upwelling (e.g., Plank et al., 1995). Arrows indicate mantle flow lines. Dashed lines represent constant % melting contours. (From Langmuir et al., 1992, *Geophysical Monograph* 71, American Geophysical Union, Washington, D.C.)

also pointed out that the magma-generation process is likely a near-fractional fusion process: As an asthenospheric parcel rises along a flow line, small fractions of melt (<3%) are extracted from it. The depleted residue continues to move along its flow line and becomes a part of the oceanic lithosphere. The magma moves at least initially by porous flow and compaction; it accumulates perhaps in inclined lenslike bodies beneath the lithospheric lid. Eventually magma from these lenses get delivered to the crust via dikes.

    Although this overall picture of magma generation and delivery appears to be accepted by most, it is clear that some important differences must exist in the magma-generation process between ridges that have a thick crust (therefore, greater magma supply rate) and those with a thin crust. What are these differences? Is there any petrologic/geochemical way to understand why such differences occur?

Numerous basaltic samples have been collected and analyzed from the global mid-oceanic ridge system to investigate whether any global correlation exists between basalt chemistry and crustal thickness. Realizing that apples must be compared with apples, Klein and Langmuir (1987) compared rock compositions that have reached a comparable degree of fractionation, as judged by a fixed MgO value of 8 wt%, so that the differences in chemical compositions of the rocks are unlikely to be a product of differentiation processes of a homogenized parent magma but are due to differences in the magma-generation processes. These authors did this by normalizing MORB lava compositions from different geographic areas to a fixed value of 8 wt% of MgO. Two parameters, so-called Na_8 and Fe_8 (which stand for Na_2O and FeO wt%, respectively, corresponding to 8 wt% MgO), although interpolated (and not analyzed) values, appear to correlate well with the crustal thickness on a global scale (Figure 11.10).

R. Batiza (University of Hawaii), C.H. Langmuir (Lamont-Doherty Earth Observatory), and their coworkers also noted that, within an individual locality, the rocks produce a linear correlation that is at a sharp angle to the global correlation vector or global array (Figure 11.10a). Na_8 was found to correlate well with crustal thickness (Figure 11.10b). Because Na_2O abundance in partial melt is largely controlled by the degree of melting (and, to some extent, by the source rock's Na_2O content), Klein and Langmuir proposed that the negative correlation between Na_8 and crustal thickness is simply a result of melt production. In contrast, Fe_8 is reflective of the depth of onset of melting of the mantle peridotite: The greater the Fe_8 value, the greater the depth of onset of melting (Figure 11.10b). For example, the model calculations performed by Langmuir et al. (1992) show that the Na_8–Fe_8 variation in the Reykjanes peninsula (thick crust) can be explained with a depth of onset of melting ~120 km (4 GPa); in the case of SWIR, it would be about 75 km. Klein and Langmuir (1987) and McKenzie and Bickle (1988) pointed out that an initially hotter lherzolite will (a) begin to melt at a deeper level than a cooler lherzolite, as the former would intersect the solidus at a greater depth (see also Chapter 9); and (b) melt over a longer depth range and therefore generate more melt (= more crust). However, a hotter rising mantle parcel may rapidly run out of clinopyroxene and stop melting at a deeper level than one that is somewhat cooler, producing the same (or even less) amount of melt (= crust) as the latter.

As for the source of local variability, several new basalt crystallization experimental studies indicate that polybaric fractionation (and magma mixing) and some source mantle heterogeneity can account for the trends (e.g., Kinzler and Grove, 1992).

The following is a short summary of the hypotheses that have been put forward for the origin of MORB. (a) The cross-section of the melting regime may vary from triangular in passive upwelling to narrow pipelike in active upwelling regimes. (b) The degree and depth range of melting are generally related and together perhaps exert the strongest control on crustal thickness. (c) Based on compositions of primitive MORB projected onto appropriate phase diagrams and numerical modeling of MORB data, it has been proposed that, in general, MORB-parent magmas may be generated over a pressure range of 4 to 0.6 GPa. (d) Magma is generated over a wide region and focused toward a very narrow region of the ridge crest. The efficiency of the melt-focusing process exerts a significant control on crustal thickness because greater melt focusing efficiency results in the delivery of a greater amount of melt to the crust (= more crust).

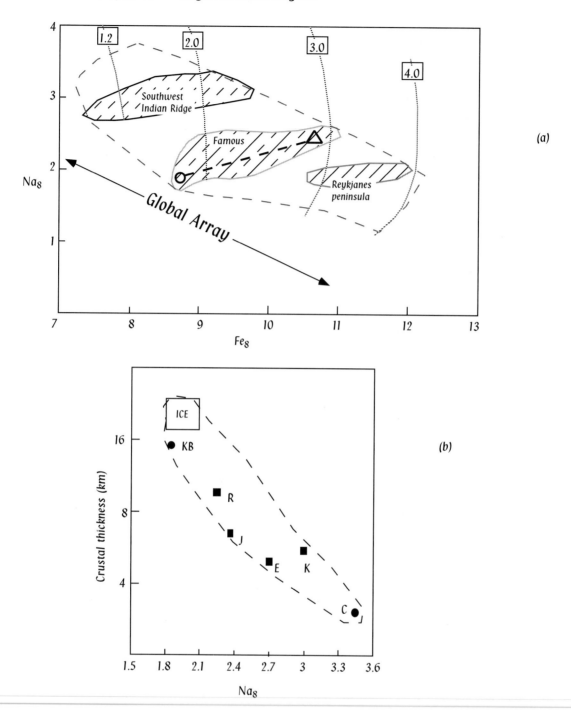

FIGURE 11.10 (a) Na_8 versus average crustal thickness from various midoceanic ridges. ICE = Iceland, KB = Kolbeinsey ridge, R = Reykjanes ridges, J = Juan de Fuca, E = East Pacific Rise, K = Kane Fracture zone, C = Cayman ridge. (From Langmuir et al., 1992, *Geophysical Monograph* 71, American Geophysical Union, Washington, D.C.) (b) Fe_8 versus Na_8 fields of selected ridge locations. Data from global oceanic ridges plot within the dashed field. The dotted lines with numbers marked on them are polybaric fractional melting paths of lherzolite, and each of the numbers indicates the pressure (GPa) of onset of melting in each case. The melting is assumed to end at the base of the crust. (From Klein and Langmuir, 1987, *Journal Geophysical Research*, American Geophysical Union, Washington, D.C.)

BASALTIC MAGMATISM AT HOT SPOTS: THE HAWAIIAN EXAMPLE

Evolution of a Hawaiian Volcano

Hawaiian-Emperor chain is without a doubt the best example of a chain of volcanic islands and seamounts that have been generated by hot-spot volcanism over a migrating lithosphere (see Chapter 1). Each individual volcano evolves through several stages in terms of chemical composition, volume, and eruption frequency as the volcano (and the lithosphere on which it is located) passively migrates over the hot spot (Figure 11.11). In the earliest preshield (Loihi) stage, when the lithosphere has barely approached the hot spot, alkalic lavas formed by very small degrees of volatile-rich melting erupt. This stage is exemplified by the newest Hawaiian volcano, Loihi volcanic seamount. As the volcano gradually moves toward the center of the hot spot, the eruption frequency (magma supply) rapidly increases as the lavas become picritic (with ~15% MgO) to tholeiitic (with $\leq 10\%$ MgO). This is the shield stage, because this is the stage when rapidly erupting, fast-flowing lavas form a giant shield volcano. Shield stage gradually gives way to postshield stage, as the volcano gradually moves away from the center of the hot spot, and the eruption frequency rapidly diminishes and the lavas become alkalic. The volcano eventually stops erupting altogether as it migrates farther away from the hot spot. However, following a ≤ 1 million-year hiatus, during which the shield volcano undergoes heavy erosion, volcanism is renewed (posterosional or rejuvenated stage). During this period, much smaller volumes of strongly alkaline mafic lavas (basanite, nephelinite, alkali basalt) erupt, often carrying mantle xenoliths through small vents scattered along fractures that are scattered across the shield volcano.

There is a large volume of petrological-geochemical data on Hawaiian volcanoes from which the following generalizations may be made: (a) There is a sharp isotopic distinction between the voluminous shield tholeiites (primitive) and the posterosional alkalic lavas (depleted: Figure 11.12). Figure 11.12 shows a comparison between tholeiites that erupted from the Koolau shield volcano and the posterosional Honolulu Volcanics (HV, alkalic) on Oahu in terms of Nd–Sr isotopic system. Koolau field plots close to the Bulk Earth value, whereas HV clearly tapped a depleted source not unlike that tapped by the MORB. (b) Even tholeiitic lavas from individual shield volcanoes have distinctive isotopic characteristics, suggesting that even the plume source that is largely tapped by shield lavas is quite heterogeneous. (c) Although much less voluminous in comparison with shield lavas, posterosional lavas erupt through small vents and fissures over a wider geographic expanse. In fact, such volcanism can occur simultaneously over volcanic islands of very different ages.

Only recently, similarly alkalic flood basalt lavas have been discovered from the ocean floor over the Hawaiian arch surrounding the Hawaiian volcanoes. The major element and isotopic diversity among Hawaiian lavas from different islands is explained in terms of a radially layered plume (Sen et al., 1996; Hauri, 1996). Frey and Rhodes (1993) and Sen et al. (1996) proposed that the Hawaiian plume is more like a series of compositionally heterogeneous blobs, each being composed of a mixture of rocks of different isotopic compositions (Figure 11.13). In the Sen et al. model, the posterosional and Hawaiian arch alkaline lavas are generated by small degrees of melting of the asthenosphere in presence of a H_2O-rich fluid. In this model, the H_2O-rich fluid is supplied to the hot asthenosphere from the outer fluid-rich envelope of a layered plume whose small hot and relatively dry core provides shield-building magmas (Figure 11.13).

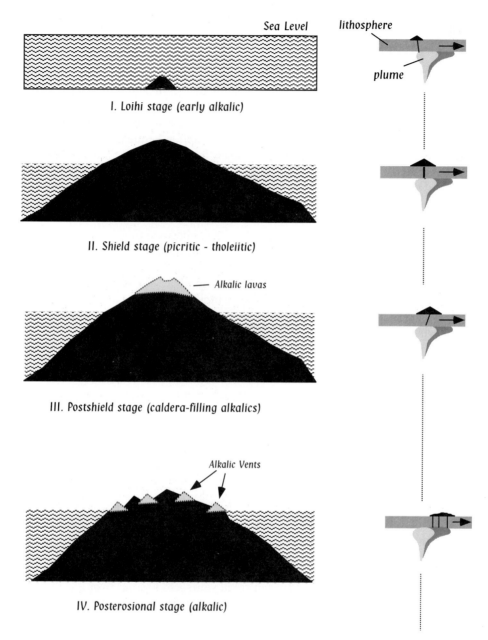

FIGURE 11.11 *Stages of evolution of a Hawaiian volcano are shown in the sequence on the left-hand side.* Location of the volcano relative to the plume corresponding to each stage is shown on the right-hand side.

Two important questions concerned with plume/hot-spot-originated volcanoes are: (a) Where in the mantle is the source of the plume (or hot spot) located? (b) Why does such a hot spot/plume remain relatively fixed over at least 90 million years? No reasonable answer has yet been found for the second question, but there have been some important clues to the answer of the first question. For example, the sharply distinctive rare Earth element characteristics and Nd, Sr, Pb isotopic compositions of N–MORB and Hawaiian shield tholeiites (Figures 11.6, 11.12) require that they tap

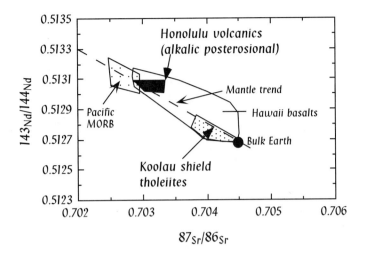

FIGURE 11.12   *Nd–Sr isotopic comparison between Hawaiian basalts and Pacific MORB.* The difference between shield lavas (Koolau) and posterosional lavas (Honolulu) from Oahu are also shown.

Hawaiian Islands

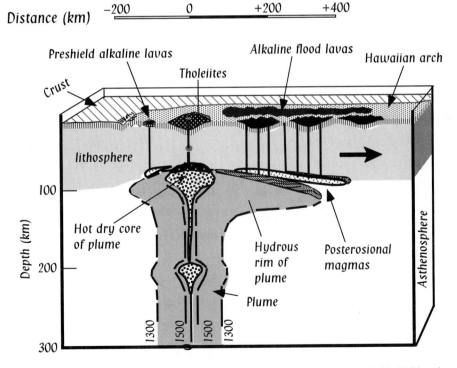

FIGURE 11.13   *A schematic model of the Hawaiian plume and formation of preshield, shield, and posterosional magmas.* The plume is believed to be nonaxisymmetric with a broad hydrous envelope and a dry core composed of blobs from the deep mantle. The dry core supplies the shield lavas and the wet envelope supplies the volatiles to trigger melting (posterosional melts) in the asthenosphere. (Sen et al., 1996)

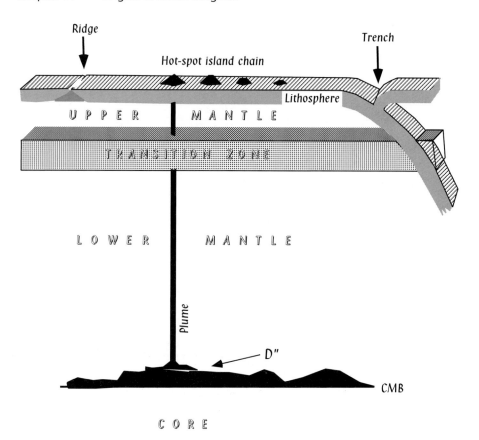

FIGURE 11.14   *A general plate tectonic model depicting the rise of a plume from the D layer occurring at the core-mantle boundary (CMB).*

very different source layers in the mantle. MORB appear to originate in the asthenosphere (isotopically depleted), whereas Hawaiian plume is much less depleted and must be located at some depth below the asthenosphere. Plausible location of the source layer for Hawaiian tholeiites can be the transition zone (TZ), the lower mantle (LM), or the core-mantle boundary (CMB). The writer's own prejudice is that the Hawaiian plume sources are a mixture of TZ and LM components and are triggered by a thermal plume rising from the CMB (Figure 11.14).

LARGE IGNEOUS PROVINCES

From time to time, enormous volumes of dominantly tholeiitic basalts have flooded over continents and ocean basins. They have usually been called by two names: *flood basalt* or *large igneous provinces* (LIP). The most famous LIP is perhaps the Deccan Traps of India (Figure 11.15), whose original volume is estimated to have been ∼1.5 million km³. Its fame is at least partly due to two reasons: (a) its eruption coincided with the cretaceous/tertiary boundary, and (b) the bulk of the volcanism probably occurred in ≤0.5 m.y. The most voluminous of all LIPs is the Ontong-Java plateau in the Pacific Ocean (Figure 11.15; e.g., Coffin and Mahoney, 1997).

There are similarities and dissimilarities between individual provinces that cannot be discussed at length in this book. In general, however, it seems that

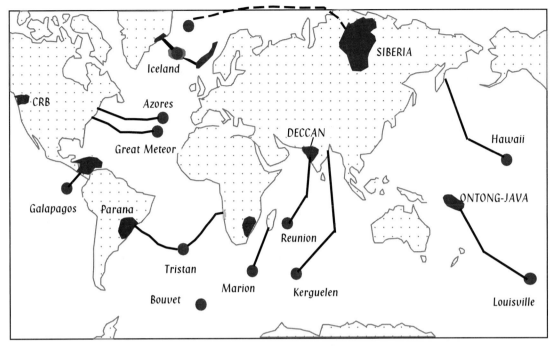

FIGURE 11.15 *LIPs and their corresponding (alleged) hot-spot tracks.* Currently active hot spots are shown as filled circles and the LIPs are shown as shaded irregular areas. (Redrawn after Duncan and Richards, 1989)

1. lavas are predominantly tholeiitic and largely erupted from fissures/dikes and not from well-defined volcanoes;

2. continental LIPs occur mostly on the edge of the continent;

3. many LIPs appear to have erupted over a limited time at phenomenal eruption rates;

4. in many cases, their eruption coincided with a major stratigraphic boundary—for example, Deccan Traps at the K/T boundary and the Siberian Traps (Russia) at the Permian/Triassic boundary;

5. eruption was largely nonexplosive, as indicated by the general rarity of pyroclastics; and

6. many/most of these LIPs are associated with a hot-spot track (much like the Hawaiian hot-spot track of volcanic islands)—for example, the Deccan is connected by a complex hot-spot track believed to have been once linked to the currently active Reunion hot spot in the Arabian sea (Figure 11.14).

Based on Observation 5, it has been widely accepted that an LIP is generated by a large plume head whose tail can remain active for ~90 m.y. (or more) and generate a hot-spot track (Figure 11.16). Observation 2 has led some people to propose that large plume heads that generate continental flood basalts may also be responsible for splitting continents (see Chapter 1; D. Anderson's hypothesis).

A survey of the published literature indicates that little attention has been paid to modeling the P,T conditions of a generation of LIPs. This is perhaps because most have undergone significant shallow-level fractionation, magma mixing, and crustal contamination. Although it is generally accepted that some large plume head must rise

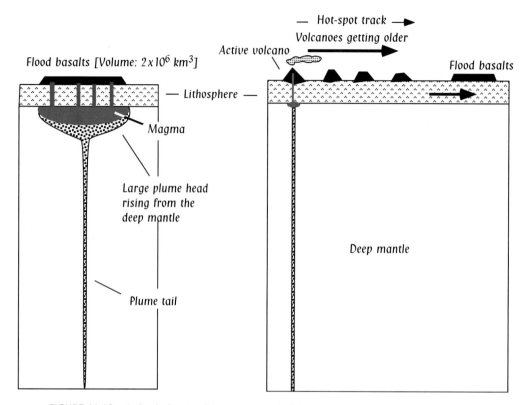

FIGURE 11.16 *A sketch showing the generations of a LIP and a hot-spot track by melting of a large plume head and its narrow tail, respectively.* (After Richards et al., 1989)

from the deep (lower-mantle, transition zone, or core-mantle boundary), where such a plume head must begin (and end) melting, the rate of melt escape from the source remains a subject of speculation. The writer attempted to model such conditions for the Deccan (Sen, 1996) and concluded that primary Deccan magmas were picritic and generated over a pressure range of 2 to 3 GPa.

SUMMARY

[1]   The source rock for basalt magma is primarily lherzolite, the dominant upper-mantle rock. Melting experiments on this rock in the absence of volatiles show that low-degree melts are quartz tholeiite at <0.6 GPa and olivine tholeiite at 0.6–1 GPa. At 1–2 GPa, the near-solidus melts are likely to be *ne*-normative changing to olivine tholeiitic or picritic at higher degrees of melting. At ~2 GPa, the near-solidus melts are picritic.

[2]   Recent laboratory experiments on lherzolite melting show that SiO_2 and Al_2O_3 contents of basalt magma can be good indicators of pressure and degree of melting, respectively. Because magmas are pooled over a range of pressures, it is possible to obtain a mean depth of magma formation and fraction of melting from the aggregate magma composition.

[3]   When a lherzolite is partially melted in the presence of volatiles, CO_2 makes the initial magma composition more alkaline, whereas H_2O makes it more silica rich.

[4]   Ridge topography/morphology is different between slow- versus fast-spreading ridges and is correlated with MORB composition. Global correlations between ridge morphology and MORB chemistry are primarily due to an interplay between depths and degrees of melting.

[5]   MORB magma chambers at fast- versus slow-spreading ridges differ: A thin perennial magma lens is present only in fast-spreading centers. The bulk of the magma chamber in both slow- and fast-spreading ridges is crystalline, with only a few percent interstitial melt present.

[6]   N–MORB are derived from a depleted, convecting, shallow asthenosphere, whereas hot-spot-generated basalts, such as Hawaiian tholeiites, are derived from primitive enriched plumes rising from the transition zone or lower mantle.

[7]   Hawaiian plume is radially zoned with a narrow hot and dry core composed of a series of isotopically heterogeneous blobs and a broad wet envelope. The core produces main shield volcanism, and the wet envelope is responsible for triggering melting in the asthenosphere and generating posterosional volcanics.

[8]   LIPs are voluminous tholeiitic flood basalts that erupted through fissures in the crust. Many of them erupted at major stratigraphic boundaries, and the eruption rates were phenomenal. Most of them can be linked to their individual hot-spot tracks. A currently popular model for their origin is that they are melts that form enormous plume heads whose long tails persist for another ~90 m.y. and produce hot-spot tracks.

BIBLIOGRAPHY

Albarede, F. (1992) How deep do common basaltic magmas form and differentiate? *J. Geophys. Res.* **97**, 10997–11009.

Asimow, P.D., Hirschmann, M.M., and Stolper, E.M. (1997) An analysis of variations in isentropic melt productivity. *Phil. Trans. R. Soc. London, Series A* **355**, 255–281.

Baker, M.B., et al. (1995) Compositions of near-solidus peridotite melts from experiments and thermodynamic calculations. *Nature* **375**, 309–311.

Basaltic Volcanism Study Project (1981) *Basaltic Volcanism on Terrestrial Planets.* Pergamon Press (New York, N.Y.).

Duncan, R.A. and Richards, M.A. (1991) Hot spots, mantle plumes, flood basalts, and true polar wander. *Rev. Geophys.* **29**, 31–50.

Eggins, S.M. (1992) Petrogenesis of Hawaiian tholeiites: 2, aspects of dynamic melt segregation. *Contrib. Mineral. Petrol.* **110**, 398–410.

Falloon, T.J. et al. (1988) Anhydrous partial melting of fertile and depleted peridotite from 2–30 kbar and application to basalt petrogenesis. *J. Petrol.* **29**, 1257–1282.

Fisk, M.R. et al. (1988) Geochemical and experimental study of the genesis of magmas of Reunion island, Indian ocean. *J. Geophys. Res.* **93**, 4933–4950.

Forsyth, D.W. (1992) Geophysical constraints on mantle flow and magma generation beneath mid-ocean ridges. *Geophysical Monograph* **71**, *Am. Geophys. Union*, 1–65.

Frey, F.A. and Rhodes, J.M. (1993) Intershield geochemical differences among Hawaiian volcanos: implications for source compositions, melting processes and magma ascent paths. *Phil. Trans. R. Soc. London, Series A* **342**, 121–136.

Green, D.H. and Ringwood, A.E. (1967) The genesis of basaltic magmas. *Contrib. Mineral. Petrol.* **15**, 103–190.

Hauri, E.H. (1996) Major-element variability in the Hawaiian mantle plume. *Nature* **382**, 415–419.

Hess, P.C. (1989) *Origins of Igneous Rocks.* Harvard University Press (Cambridge, MA.).

Hirose, K. and Kushiro, I. (1993) Partial melting of dry peridotites at high pressures: determination of compositions of melts segregated from peridotite using aggregates of diamond. *Earth Planet. Sci. Lett.* **114**, 477–489.

Hisrchmann, M.M. and Stolper, E.M. (1996) A possible role for garnet pyroxenite in the origin of the "garnet signature" in MORB. *Contrib. Mineral. Petrol.* **124**, 185–208.

Hirschmann, M.M. et al. (1998) Calculations of peridotite partial melting from thermodynamic models of minerals and melts. I. Methods and comparison to experiments. *Journal of Petrology* **39**, 1091–1115.

Johnson, K.T.M., Dick, H.J.B. and Shimizu, N. (1990) Melting in the oceanic upper mantle: an ion microprobe study of diopsides in abyssal peridotites. *J. Geophys. Res.* **95**, 2661–2678.

Kinzler, R.J. (1997) Melting of mantle peridotite *t* pressures approaching the spinel to garnet transition: application to mid-ocean ridge basalt petrogenesis. *J. Geophys. Res.* **102**, 853–874.

Kinzler, R.J. and Grove, T.L. (1992) Primary magmas of mid-ocean ridge basalts, 2, applications. *J. Geophys. Res.* **97**, 6907–6926.

Klein, E.M. and Langmuir, C.H. (1987) Global correlation of ocean ride basalt chemistry with axial depth and crustal thickness. *J. Geophys. Res.* **92**, 8089–8115.

Langmuir, C.H., Klein, E.M. and Plank, T. (1992) Petrological systematics of mid-ocean ridge basalts: constraints on melt generation beneath ocean ridges. *Geophysical Monograph* **71**, *Am. Geophys. Union*, 183–280.

Mahoney, J.J. and Coffin, M.F., eds. (1998) Large igneous provinces: continental, oceanic, and planetary flood volcanism. *Am. Geophys. Union* (Washington, D.C.).

McKenzie, D. (1984) The generation and compaction of partially molten rock, *J. Petrol.* **25**, 743–765.

McKenzie, D. and Bickle, M.J. (1988) The volume and composition of melt generated by the lithosphere, *J. Petrol.* **29,** 625–679.

Nicolas, A. (1997) Seafloor spreading: a viewpoint from ophiolites, *C. R. Acad. Sci. Paris,* **324** 1–8.

Niu, Y. and Batiza, R. (1991) An empirical method for calculating melt compositions produced beneath mid-oceanic ridges: application for axis and off-axis (seamounts) melting. *J. Geophys. Res.* **96**, 21753–21777.

Presnall, D.C. et al. (1979) Generation of mid-ocean ridge tholeiites. *J. Petrol.* **20**, 3–35.

Richards, M.A., Duncan, R.A., and Courtillot, V.E. (1989) Flood basalts and hot-spot tracks: plume heads and tails. *Science* **246**, 103–107.

Sen, G. (1996) A simple petrologic model for the generation of Deccan Trap magmas. *Int. Geol. Rev.* **37**, 825–850.

Sen, G., Macfarlane, A. and Srimal, N. (1996) Significance of hydrous alkaline melts in Hawaiian xenoliths. *Contrib. Mineral. Petrol.* **122**, 415–427.

Sinton, J.M., and Detrick, R. (1992) Mid-ocean ridge magma chambers. *J. Geophys. Res.* **97**, 197–216.

Stolper, E. (1980) A phase diagram for mid-ocean ridge basalts: preliminary results and implications for petrogenesis. *Contrib. Mineral. Petrol.* **74**, 13–28.

Takahashi, E. (1986) Melting of dry peridotite KLB-1 up to 14 GPa: implications on the origin of peridotitic upper mantle. *J. Geophys. Res.* **91**, 9367–9382.

Takahashi, E. and Kushiro, I. (1983) Melting of a dry peridotite at high pressures and basalt magma genesis. *Am. Mineral.* **68**, 859–879.

Takahashi, E., Nakajima, K. and Wright, T.L. (1998) Origin of Columbia River basalts: melting model of a heterogeneous plume head. *Earth Planet. Sci. Lett.* **162**, 63–80.

Walter, M., Sisson, T. and Presnall, D. (1995) A mass proportion method for calculating melting reactions: an application of melting of model upper mantle lherzolite. *Earth Planet. Sci. Lett.* **135**, 77–90.

Yang, H.-J., Sen, G. and Shimizu, N. (1998) Mid-Ocean Ridge Melting: constraints from lithospheric xenoliths at Oahu, Hawaii. *J. Petrol.* **39**, 277–295.

Yoder, Jr., H.S. (1976) *Generation of Basaltic Magma.* National Academy of Sciences (Washington, D.C.).

Origins of Andesitic and Rhyolitic Magmas

The following topics are covered in this chapter:

ABOUT THIS CHAPTER

Tectonic boundaries where two plates converge are characterized by prominent igneous activities. "The Pacific Ring of Fire" or "The Andesite Line," which marks dominantly andesitic volcanism around the rim of the Pacific Ocean, is a classic example of plate convergence magmatism (Figure 12.1). There are three types of convergent plate boundaries: (a) *oceanic–oceanic*, where two different oceanic plates collide and one plate is subducted underneath the other; (b) *oceanic–continental*, where an oceanic plate is subducted beneath a continental plate; and (c) *continental–continental*, where two continental plates converge. When two oceanic plates converge, the result is an *island arc*, which is an arcuate shaped series of volcanic islands. Andesite lavas are typical of island arcs, although basaltic and felsic eruptions also occur. Similarly, a continental arc results from subduction of an oceanic plate beneath a continental plate. In addition to mafic and andesitic volcanism, large granitoid batholiths and rhyolitic lavas typically characterize continental arcs. When two continental plates collide, one rides over the other and granite batholiths form. This chapter's focus is principally on igneous activities in convergent plate boundaries where andesites and granites are common. How these two important magmas form is briefly discussed here.

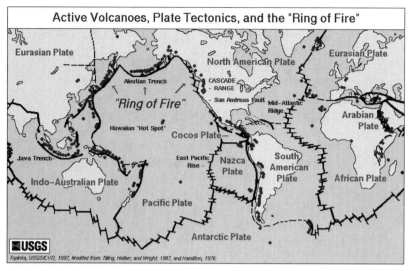

FIGURE 12.1 *Map showing an outline of the Pacific "Ring of Fire."* (From USGS Website, map downloaded from the website)

GENERAL CHARACTERISTICS OF VOLCANIC ARCS: IGNEOUS SERIES AND ERUPTION PATTERNS

A typical schematic cross-section through an island arc is shown in Figure 12.2. The location of the subducting lithosphere (or downgoing slab) is marked by an oceanic trench and dipping zone of active seismicity called the *Wadati–Benioff zone*. Some of

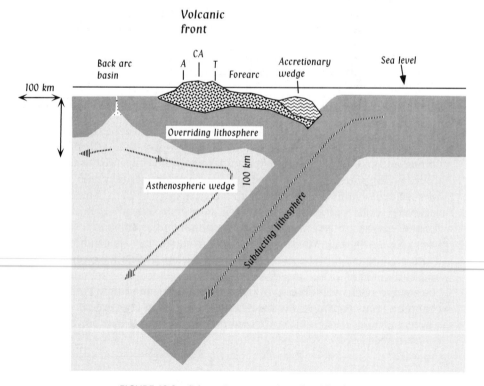

FIGURE 12.2 *Schematic cross-section of an island arc.*

the deepest focus earthquakes (~700 km) come from such zones. The steeper the Wadati–Benioff zone, the narrower the volcanic arc and the closer the arc is to the plate boundary. Generally, the Wadati–Benioff zone dips at ~45°. When it is less than about 25°, there is no volcanism. The overriding lithosphere on which the volcanic front is situated generally lies ~100 km to 200 km above the Wadati–Benioff zone.

A deep trench marks the area where the two oceanic plates come in contact. Some trenches may be sediment starved, whereas other trenches may have an abundance of sediments. Enormous packages of sediments are deformed and thrusted in many continental and island arcs, forming an accretionary wedge. Farther toward the volcanic front occurs a gently dipping area known as the *fore arc*. Igneous activity has been noted in some fore-arc areas. Sea-floor spreading and associated basaltic volcanism often occur in the back-arc basin, which is located on the back side of the volcanic front in an island arc.

When viewed on a map, the volcanic front, composed of a series of volcanoes, forms a gentle geometric arc—for example, the Aleutian arc off Alaska (Figure 12.1). The diverse types of igneous rocks that occur in an island or continental arc may be classified into several igneous series based on their SiO_2 and K_2O contents (Figure 12.3): These are the low-K series, the calc-alkaline series, the high-K series, and the shoshonite series. The low-K series is really the island arc tholeiite series, which is different from the true tholeiites that erupt at midoceanic ridges and other extensional plate boundaries. The main difference is that the island arc tholeiites show mild

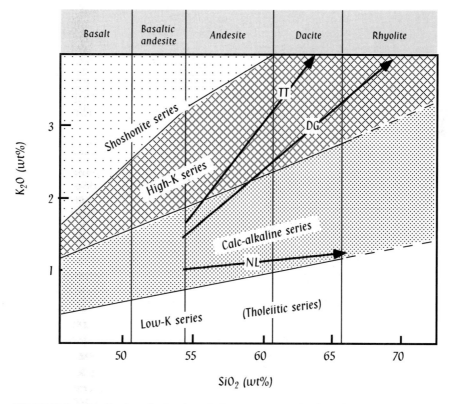

FIGURE 12.3 *Classification of volcanic rocks from convergent plate boundaries.* Fields for four major igneous series—low-K, calc-alkaline, high-K, and shoshonite series are shown. Compositions of lavas from three volcanoes in Chile are plotted: NL–Nevado de Longavi, TT–Tupungatito, and Puy–Puyehue. (Source: Hildreth and Moorbath, 1988; Gerlach et al., 1987.)

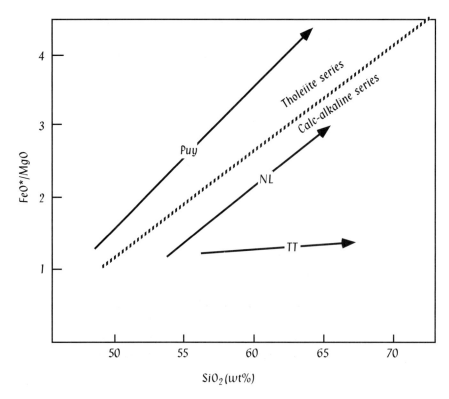

FIGURE 12.4 *The trends for the three Chilean volcanoes of Figure 12.3 are shown together with the boundary between the tholeiitic and calc-alkaline series from Miyashiro (1974).* (Redrawn from Hildreth and Moorbath, 1988.)

Fe enrichment accompanied by SiO_2 enrichment, whereas the true tholeiites show strong Fe enrichment at a fairly constant SiO_2. The high-K and shoshonite series, by virtue of their high alkali contents, are special types of alkaline series. Within each series, the rocks may range from basalt through andesite to rhyolite. Although the definition of igneous series using K_2O and SiO_2 works reasonably well, there are a few exceptions, as shown in Figures 12.3 and 12.4, in which rocks from three specific volcanoes from the Chilean Andes are plotted (Puy–Volcan Puyehue, NL–Nevado de Longavi, TT–Tupungatito; Hildreth and Moorbath, 1988; Gerlach et al., 1987). In terms of the time-tested diagram FeO/MgO versus SiO_2 (which is another way to look at an AFM diagram; see Figure 10.16 in Chapter 10), it is apparent that Puy lavas fall in the tholeiitic series, and yet in the K_2O–SiO_2 diagram its rocks fall within the field for calc-alkaline series. Similarly, in a classical sense, based on FeO/MgO versus SiO_2, TT lavas would be calc-alkaline series. However, in Figure 12.3, these lavas fall in the high-K series. NL lavas do fall within the field of calc-alkaline series in both diagrams. This shows that there is really a continuum among the various series and that no single classification scheme can work fully.

In many arcs, particularly in the case of Japanese islands, the volcanic front is segmented in terms of the types of lavas that erupt from volcanoes aligned parallel to the trench axis: Tholeiitic series lavas erupt from volcanoes located closest to the trench, calc-alkaline series lavas occur in the middle, and shoshonite (alkaline) series lavas erupt farthest away from the trench (Figure 12.2).

The eruption behavior of arc volcanoes is highly variable and related to the igneous series: Tholeiite series volcanoes are generally nonexplosive and erupt basalts and basaltic andesites from small, youthful cones and shield volcanoes. These rocks are more aphyric than calc-alkaline series lavas. Calc-alkaline series volcanoes are generally explosive and commonly erupt pyroclastic materials. These volcanoes are stratovolcanoes and generally dominate mature island arcs. The lavas commonly contain abundant phenocrysts of plagioclase. Late-stage volcanic activity in an island arc is usually of shoshonite type. Shoshonite series volcanoes are minor relative to tholeiitic and calc-alkaline series. Like the calc-alkaline series, they are also explosive. The lavas are predominantly alkali basalt. Aside from these igneous series, a rather unusual type of lava with ~6% MgO and 53% to 55% SiO_2 occurs in the fore-arc region of some island arcs. These lavas are called *boninites*.

PHENOCRYST MINERALOGY

The mineralogy of island arc lavas varies as a function of silica content and series type: Olivine (Fo_{70-85}), augite, and a highly anorthitic plagioclase ($\geq An_{90}$) are common phenocrysts in basalt and basaltic andesites (Table 12.1). A brown hornblende and biotite are common phenocrysts in andesites, dacites, and rhyolites of calc-alkaline and high-K series. These hydrous mineral assemblages suggest high H_2O contents of the lavas. In contrast, the low-K series contains phenocrysts of orthopyroxene or pigeonite. Crystallization of pigeonite requires temperatures that are substantially higher than the temperatures at which hydrous mafic-intermediate magmas may crystallize at a moderate to low pressure. Therefore, the occurrence of orthopyroxene and pigeonite in the low-K series suggests that the low-K series magmas are relatively free of H_2O compared with the other series. Plagioclase, which is a dominant phenocryst phase in calc-alkaline and low-K basalt and andesite, is not nearly as common in high-K and shoshonite series basalts. Pigeonite is a common groundmass mineral in the low-K or tholeiitic series, whereas orthopyroxene is the common groundmass Ca-poor pyroxene in andesites. Sanidine ($\pm$fayalitic olivine) may also occur as a phenocryst phase in rhyolites.

TABLE 12.1   *Phenocrysts in island arc igneous series* (after Hess, 1989).

Rock	Low-K	Calc-alkaline	High-K	Shoshonite
Basalt	Ol + Aug + Pl $\pm$Tmgt $\longrightarrow$ Fo_{70-85} $\longrightarrow$ Max. ~ An_{90} $\longrightarrow$	Ol + Aug + Pl	Ol + Aug $\pm$Pl	Ol + Aug + Oxide $\pm$Pl, Hbl, Bi
Andesite	Pl + Aug + Pig $\pm$Ol, Tmgt An_{50-70} $\longrightarrow$ Some An_{90}	Pl + Aug + Opx $\longrightarrow$ $\pm$Hbl, Tmgt, Bi $\longrightarrow$		Pl + Aug + Oxide + Bi + Hbl
Dacite, Rhyolite	Pl + Aug + Opx +Qz + Tmgt $\pm$Fa, San	Pl + Hbl + Bi + Opx +Qz $\pm$Aug, Fa, San	Pl + Hbl + Bi +San + Qz $\pm$Fa	

SOME TRACE ELEMENT AND ISOTOPIC CHARACTERISTICS

Arc basalts are easily distinguished from N–MORB and hot-spot basalts by their characteristic enrichments in some strongly incompatible elements (cesium and barium) and, particularly, low niobium (Nb) content relative to potassium and lanthanum (Figure 12.5). These differences are generally attributed (but not fully understood) to the different (and complex) source rock characteristics of island arc basalts.

In terms of isotopic ratios of Pb, Nd, and Sr, igneous rocks from island and continental arcs define a rather wide field that overlaps the fields of N–MORB, hot-spot basalt, oceanic sediments, and continental crust (Figures 12.6a, 12.6b). Such great diversity signifies a complex mix of processes by which contributions of isotopic components are made from a variety of sources—asthenosphere (i.e., N–MORB source), mantle plumes (so-called ocean island basalt [OIB] source), oceanic sediments, and the continental crust (in the case of continental arc volcanics). Different arcs can have very different isotopic compositions (Figure 12.6a, 12.6b), indicating that the relative material contributions from these different sources can vary from arc to arc. For example, Figure 12.6b shows magmas erupted at the Marianas and Aleutian arcs receive relatively little isotopic contributions from oceanic sediments. In contrast, volcanoes from the Lesser Antilles clearly show a significant isotopic contribution from the oceanic sediments. The physical means by which such contributions are made (e.g., whether during melting at the source area or through contamination of transient melts by the crust) is often not easy to discern and is still poorly understood.

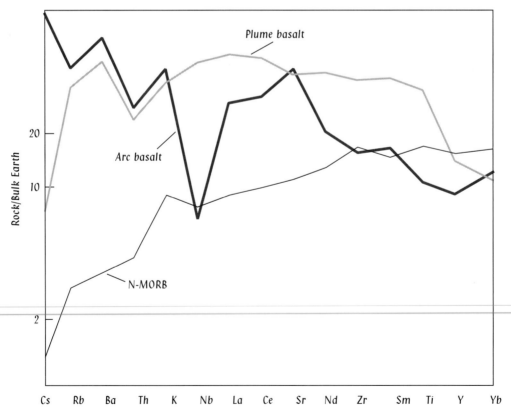

FIGURE 12.5   *A trace-element diagram, more popularly known as a Spider diagram showing abundances of some trace elements (normalized to Bulk Earth values) in N–MORB, plume basalts, and an arc basalt.* (Redrawn from Hickey et al., 1986.)

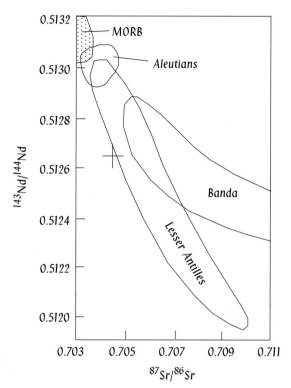

FIGURE 12.6 *Compositions of some island arc basalts are compared with N–MORB, plume basalts, and oceanic sediments in (a) Pb–Pb, and (b) Nd–Sr isotope diagrams.* These diagrams show strong sedimentary influence on the composition of magmas erupted from Lesser Antilles and Banda arcs.. (From Wilson, 1989, *Igneous Petrogenesis,* Unwin Hyman)

(a)

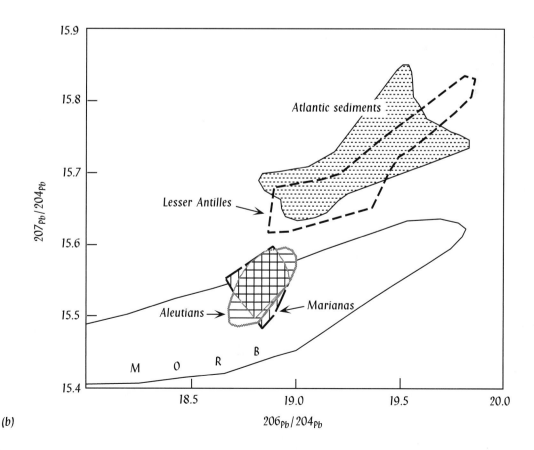

(b)

PHASE RELATIONS AND ORIGIN OF ARC MAGMAS

Evidence for the Role of Water

Evidence shows that water plays an important role in triggering melting beneath a volcanic arc:

1. **Occurrence Constraint**—A somewhat indirect but nonetheless powerful constraint on the origin of arc magmas is the fact that they only occur over subduction zones. Numerous studies of old ocean-floor rocks and ophiolites have shown that H_2O is present in old oceanic crust and mantle as hydrous minerals (e.g., clays, zeolite, chlorite, amphiboles, serpentine, biotite, etc.) formed via hydrothermal metamorphic reactions at or near the spreading center. Plate convergence zones are likely sites where these hydrous minerals may be dragged deep into the upper mantle, where they would eventually break down and release their structurally bound H_2O. Given that H_2O can significantly lower the solidus of a rock (discussed in previous chapters and further later), it is hard to imagine that the H_2O introduced into the upper mantle this way would not play any role in arc magmatism.

2. **Thermal Constraint**—Old lithosphere is cold (relatively speaking), and subduction of cold lithosphere lowers the temperature in the surrounding asthenosphere (Figure 12.7). Whether such a cold lithosphere will melt in the absence of any H_2O or CO_2 can be illustrated with the help of Figure 12.8. As noted in previous chapters, melting occurs when the source rock is hot enough so that its temperature exceeds its solidus temperature at any particu-

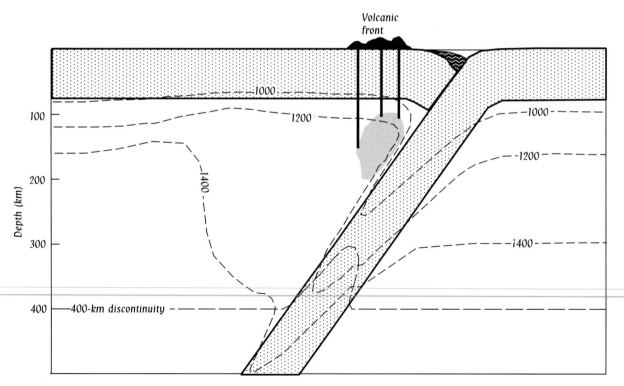

FIGURE 12.7   *A generic thermal structure of the lithosphere in a subducting slab.* (Sources: Toksoz and Hsui, 1978; Tatsumi et al., 1983; Turcotte and Schubert, 1982.) The numbers are in °C. The effect of the exothermic phase change reaction at the 400 km discontinuity on the isotherms is shown. (Based on Turcotte and Schubert, 1982)

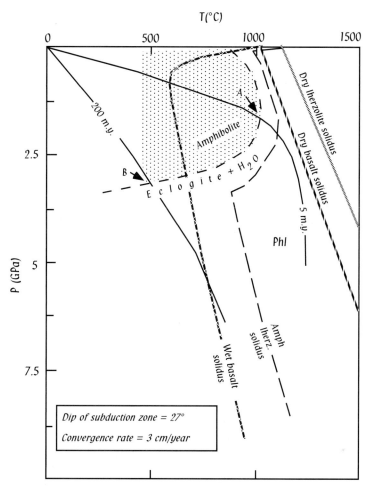

FIGURE 12.8   *Effect of H_2O in lowering the solidi of mantle lherzolite and the subducted crust (wet basalt solidus).* Basaltic crust may be expected to convert to amphibolite, which would eventually break down to eclogite and release H_2O. Two subduction zone geotherms of different ages are shown to illustrate the effect of hot versus cold subducting plates on amphibolite breakdown reaction.

lar pressure. Figure 12.8 shows two calculated geothermal gradients for a hot, 5 million-year-old subducting crust and a 200 million-year-old, cold subducting oceanic crust assuming that the dip of the slab is 27° and the rate at which the subducting slab is moving is 3 cm/year (Peacock, 1987, 1996). Each of these geothermal gradients describes how temperature increases as the subducting crust reaches from shallow to deep levels. Also shown are the dry (i.e., free of H_2O, CO_2, or other volatiles) solidi for basalt and lherzolite, with basalt and lherzolite being the main components of the subducting crust and overlying asthenospheric wedge (or the subducting lithosphere), respectively. The important thing is that the geotherms do not intersect the two dry solidi at any pressure, meaning that the available heat is insufficient to melt the slab if the slab is totally dry. This figure also shows that introduction of H_2O to the subducting basaltic crust can sufficiently lower its solidus (marked as *wet* basalt solidus), via whatever process, so that melting can occur at a shallow versus deep level for the hot versus cold geotherm, respectively.

Based on numerous field studies of ophiolites and metamorphic blocks found in accretionary prisms, it is commonly believed that the subducted basaltic crust undergoes a series of hydration and carbonation reactions with seawater and eventually turns into amphibolite and perhaps some other minor rock types. Figure 12.8 shows the pressure, temperature limits beyond which such an amphibolite crust in the subducting slab would break down into an eclogite composed of garnet and clinopyroxene and give off a H_2O-rich vapor. Note that in the case of the hot subducting crust, amphibole remains stable down to Point A (~1.3 GPa or 40 km; Figure 12.8). Beyond that depth, rather than melting, it breaks down into eclogite and yields a fluid. In summary, this diagram suggests that (a) melting cannot occur in the absence of H_2O in a subduction environment, and (b) it is unlikely that hydrated (amphibolitized) basaltic crust would melt. However, crustal rocks of sedimentary origin (which comprise a small fraction of the oceanic crust) may melt within the descending slab. Furthermore, in the case of a spreading center subduction, the crust would be hot and therefore may undergo partial melting within the subduction zone. Based on arguments such as the one presented earlier, it is commonly accepted that the slab does not undergo significant melting. Instead, the lherzolite in the asthenospheric wedge above the slab melts when its solidus is lowered by introduction of H_2O from the breakdown of hydrous minerals.

3. Chemical Constraint—Basalts from arcs are notoriously high in Al_2O_3 (~17%–18%) compared with basalts from any other tectonic provinces (e.g., N–MORB and hot-spot tholeiites ~12%–14%). Hydrous melting experiments on natural lherzolites and on simple two- and three-component systems have shown that hydrous magmas generated from lherzolite are higher in Al_2O_3 than those produced by dry melting at the same pressure (depth). As an example, let us compare the eutectic melt compositions generated under dry (volatile-free) versus wet (hydrous) conditions in the simple binary system diopside–anorthite (Figure 12.9). Relative to the dry system, the addition of

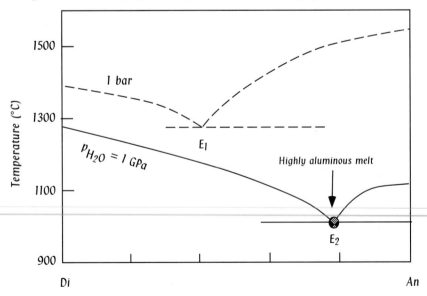

FIGURE 12.9 *The system diopside–anorthite at 1 atmosphere and at $p_{H_2O} = 1$ GPa.* The drastic changes in the temperature and composition of the eutectic point due to the addition of water are remarkable. Note that the hydrous melts generated at the eutectic point E2 at p_{H_2O} of 1 GPa are very high in normative anorthite content (i.e., high in alumina) relative to anhydrous melts that may be generated at 1 atmosphere eutectic point (E1).

H_2O drastically lowers the melting point of anorthite and moves the eutectic point between diopside and anorthite closer to anorthite. The eutectic melt under hydrous conditions is much richer in normative anorthite, and anorthite being the aluminous phase, the melt is naturally a lot more aluminous than the dry eutectic melt.

In passing, it should be noted that there are other ways to generate high-alumina basalts. In fact, plagioclase phenocryst accumulation in an otherwise ordinary mafic magma can make it a high-alumina basalt (e.g., Crawford et al., 1987). This is an equally strong possibility since such basalts are generally strongly porphyritic and the plagioclase crystals are strongly zoned to high-An contents (sometimes as high as An_{95}).

The origin of high-An plagioclase in arc basalts is commonly attributed to their crystallization from a hydrous melt at moderately high pressure. H_2O expands the solid solution loop of the albite–anorthite system and depresses it to lower temperature relative to the dry system (cf. Arculus and Wills, 1980; cited in Sisson and Grove, 1993; Figure 12.10 based on Johannes, 1989). Figure 12.10 shows that the melt X would be in equilibrium with a An_{74} plagioclase in a H_2O-free melt and with An_{93} plagioclase under H_2O-saturated conditions. In an experimental study of crystallization of hydrous versus dry high-alumina basalts, Sisson and Grove (1993) noted that only the hydrous mafic melts crystallized high-calcic plagioclase ($> An_{90}$). Therefore, in short, a wet magma would crystallize a significantly more anorthitic plagioclase than a dry magma.

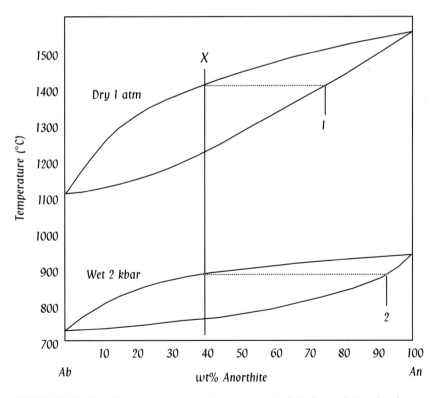

FIGURE 12.10   *The Ab–An loop at atmospheric pressure (volatile free or dry) and under H_2O-saturated conditions at 2 kbar pressure (in the system Ab–An–Qz–H_2O; Johannes, 1989).* Two effects are conspicuous: (a) lowering of melting points in the hydrous system relative to the dry binary system, and (b) crystallization of strongly anorthitic plagioclase in the hydrous system relative to the dry system from a magma with the same Ab–An ratio.

PLAUSIBLE SCENARIOS FOR THE ORIGINS OF ARC MAGMAS

The materials involved in arc magma generation are, broadly, the descending slab, the overlying mantle wedge, and the overriding lithosphere. In detail, the biggest complexity in arc magmatism is introduced by the diverse mineral assemblages and their modes in the subducted crust. Based on isotopic evidence, it is apparent that in some subduction zones (e.g., Lesser Antilles) sediments are being shoved deep into the upper mantle. In other areas, hardly any sediment is being subducted: In these areas, sediments are being scraped off at the trench to form an accretionary wedge. The crustal part of subducted slabs can vary a great deal in chemical composition and in their metamorphic mineralogies. The hydrous minerals that form in the subducted crust have different pressure-temperature stabilities, and therefore they must break down to anhydrous minerals (or less hydrous minerals) at different depths in variously dipping subducting plates of differing ages. Such reactions are very important because they provide the fluids necessary to trigger melting.

Age of the subducting lithosphere is an important factor because a relatively young lithosphere would be hotter than an old lithosphere. Therefore, in the crustal portion of a subducting lithosphere, different dehydration reactions may occur in a young (hot) slab at different depths than those for a old (cold) slab. Dip of the subduction zone, which is related to convergence rate, is clearly important (discussed earlier).

Considering such a great deal of variability in plausible scenarios in arc magma generation, which is also reflected in the great deal of complexities in intra- and interarc volcanism, it is beyond the scope of this book to discuss the many petrogenetic models that exist in the literature. However, some of the more general concepts are discussed next.

Nature of the Primary Magma

An important question is this: *What are the primary magmas in arcs?* It has long been known, with the help of hydrous melting experiments conducted in various laboratories, that under water-saturated conditions a mantle lherzolite or basalt/amphibolite/eclogite generates strongly silicic magmas like a quartz tholeiite to tonalite (e.g., Kushiro, 1969; Rushmer, 1991). For example, Kushiro (1969) showed that in the system diopside–forsterite–silica $\pm$ H_2O, a lherzolite-type source rock, if partially melted at 2 GPa under H_2O-saturated conditions, would generate quartz–tholeiite magma (Figure 12.11). Figure 12.11 shows the location of the isobaric invariant points where the assemblage ol + opx + cpx + melt is stable at (a) 1 atmosphere pressure, (b) 2 GPa with excess H_2O, and (c) 2 GPa dry. Based on our prior discussion of this system in Chapter 7, we know that when the lherzolite is melted at 2 GPa under H_2O-excess conditions, the first melt would form at the invariant point marked 2 GPa (excess H_2O). Because this invariant point lies within the quartz–tholeiite portion (i.e., inside the di–en–silica triangle) of the basalt tetrahedron, this melt would be quartz–tholeiite in terms of its normative composition.

Some high Mg-andesite magmas may indeed be generated this way; however, it is doubtful that the common, more abundant types of andesites are generated as mantle-derived primary magmas. *Most authors believe that basalts are the parent magmas in arcs. These generate the other magma types within a series through crystallization-differentiation and/or contamination by the crust.*

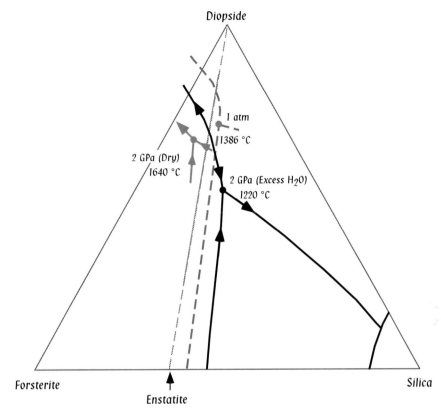

FIGURE 12.11   *Effect of dry versus wet pressure on liquidus phase relations in system diopside–forsterite–silica (wt% diagram).* The 1 atm triple-point liq + di + en + fo occurs at 1,386°C. This temperature increases to 1,640°C at a pressure of 2 GPa (anhydrous conditions) as the liquid at the invariant point becomes enriched in forsterite component. In contrast, H_2O lowers the temperature of the same triple point to 1,220°C at a total pressure of 2 GPa, and the liquid composition becomes more silicic.

Origins of the Different Series

Noting that H_2O would stabilize magnetite early in a basaltic liquid, Kennedy (1955) first suggested that the calc-alkaline magma series may evolve from a hydrous basaltic magma by early precipitation of magnetite, which would extract the iron from the magma and leave the residual melt enriched in silica and alkalis. Kennedy's proposal may be understood with the help of the schematic high-pressure (speculative) phase diagram forsterite–magnetite–SiO_2 (Figure 12.12a). We assume that Fo and En do not have a reaction relationship at this pressure, and consider the case of fractionation of the olivine-normative magma *m* at this pressure. Forsterite fractionates first from *m*, followed by forsterite and magnetite. During this last stage, the magma is enriched in SiO_2 and depleted in Mg and Fe. In contrast, a dry basaltic magma would continue to become iron enriched with differentiation because magnetite would not fractionate early (Figure 12.12b).

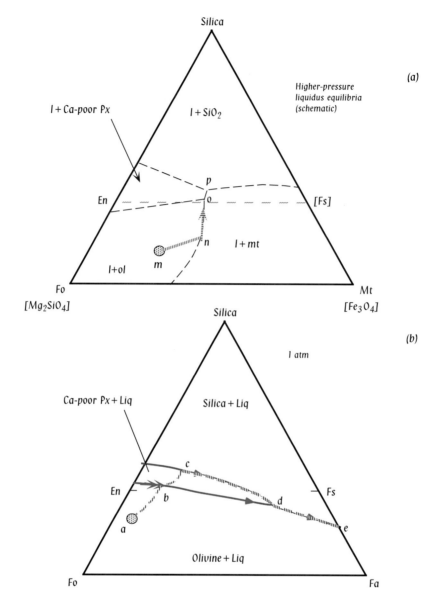

FIGURE 12.12 (a) A purely schematic representation of the liquidus phase relations on the join Fo–Mt (magnetite) -SiO$_2$ in the system MgO–FeO–Fe$_3$O$_4$–SiO$_2$ at some (schematic) high pressure >1 GPa. Although Fo–En boundary is cotectic much of the way, a small portion of it shows reaction relation (it is for the student to determine which portion). A hypothetical magma m fractionates the following minerals in the following sequence: ol → ol + mt → ol + mt + opx → mt + opx → mt + opx + silica. The liquid path is $m → n → o → p$. The net result is silica enrichment of the residual liquid. (b) The system Fo–Fa–SiO$_2$ at 1 atmosphere. The liquidus boundary between Ca-poor pyroxene$_{ss}$ and olivine$_{ss}$ fields is somewhat complicated in the sense that the magnesian part of it is a reaction boundary, whereas the Fe-rich part is a cotectic one. A typical liquid path followed during fractional crystallization of a tholeiitic magma (represented by a) is shown: $a → b → c → d → e$. It is apparent that the residual liquid gets progressively enriched in FeO, which is typical of tholeiites. Note that at high pressures (P > 0.6 GPa) the pyroxene–olivine liquidus boundary moves across the En–Fs line so that it is no longer a reaction curve. In such a case, the liquid does not move across the Ca-poor pyroxene field but along the ol/px liquidus boundary. At any rate, the result is still one of Fe-enrichment in the residual liquid.

We have already considered the generation of calc-alkaline versus tholeiitic series magmas during a discussion of the Bowen–Fenner controversy in an earlier chapter. Grove and Baker (1984) reexamined the phase relationships and melt compositions in tholeiitic and calc-alkaline series and showed that the tholeiitic series (or low-K series) may be derived by the fractionation of olivine, plagioclase, and augite from basaltic parents at relatively low pressure, with plagioclase dominating the fractionating assemblage. On the other hand, calc-alkaline series can be generated by fractionation of olivine + plagioclase + augite in nearly equal mass proportion from high-alumina basalts at moderate pressure and water-undersaturated conditions (Figure 12.13). Extended fractionation with additional participation of orthopyroxene, magnetite, and amphibole at ~0.2 GPa can lead to the formation of andesites and dacites (Sisson and Grove, 1993).

Tatsumi et al. (1983) performed experiments on three very primitive magnesian basalts of the tholeiite, calc-alkaline, and alkali basalt series from the Japanese

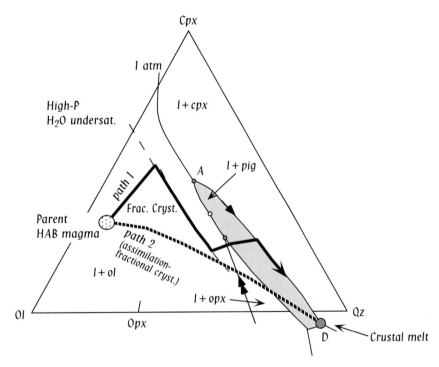

FIGURE 12.13   *A scheme proposed by Grove and Baker (1984) to generate tholeiitic versus calc-alkaline trends with reference to the phase equilibria as projected from plagioclase onto the plane cpx–ol–qz in the pseudoquaternary system pl–cpx–ol–qz (wt%).* The 1 atmosphere liquidus boundaries are shown as thin, continuous lines, and a high-pressure, H_2O-undersaturated, pseudocotectic ol+aug+liq is shown as a dashed curve. Grove and Baker pointed out the high-pressure (H_2O-undersaturated) shift of the plagioclase-saturated augite + ol + liq boundary (dashed curve) from the cpx-apex of the triangle. This allows a basaltic magma to become silica (and alkali, although not shown here) enriched without any Fe-enrichment (i.e., generate a calc-alkaline trend) while crystallizing olivine, plagioclase, and augite in roughly equal proportions—*path 1*). Grove and Baker also pointed out that assimilation of minor volumes of silicic melts (derived by melting of the continental crust) and accompanying fractional crystallization can also generate the calc-alkaline trend (*path 2*). On the other hand, the same basalt magma may undergo significant ol + pl-fractionation before reaching the low-pressure (as represented by the 1 atmosphere phase relations) plagioclase-saturated aug + ol + liq pseudocotectic boundary. This path would result in significant initial Fe-enrichment, which is a typical tholeiitic trend.

islands (Figure 12.14). Their goal was to determine how such magmas may have actually formed in the upper mantle beneath the Japanese volcanic front. The Tatsumi et al. approach was to determine the pressure-temperature and water-content conditions at which these primary magmas would be saturated with the upper-mantle minerals. They found that the tholeiite is saturated with a harzburgite (olivine + orthopyroxene) assemblage at ~1.1 GPa, 1,320°C; and calc-alkaline and alkali basalt were found to be saturated with a lherzolitic assemblage under variable water contents at ~1.7 GPa, 1,325°C and ~2.3 GPa, 1,320°C, respectively. The calc-alkaline and alkali basalt magmas were found to contain 1.5 and 3 wt% water, respectively. Based

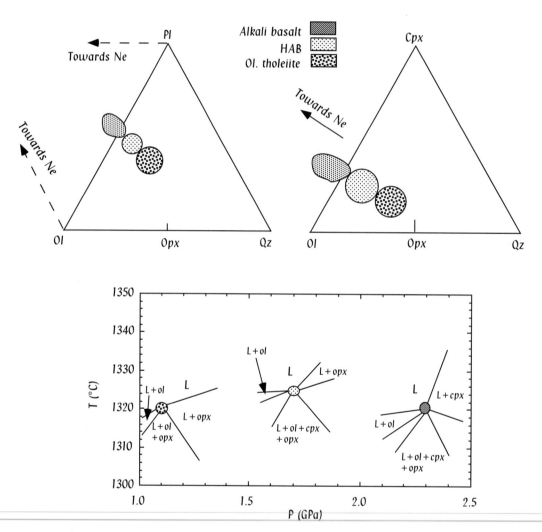

FIGURE 12.14 *The Tatsumi et al. (1983) experiments on high-Mg alkali basalt, high-alumina basalt (HAB), and olivine tholeiite from the Japanese arc are summarized in this figure.* The top two diagrams show where these basalt types plot in two projections of the basalt tetrahedron. The bottom diagram shows P,T points, where the three basalt types were found to be multiply saturated (i.e., the experimental melts were saturated with olivine and two pyroxenes—that is, mantle residue). Tatsumi et al. noted that the dissolved H_2O contents of each of these multiply saturated melts were different (see text for details).

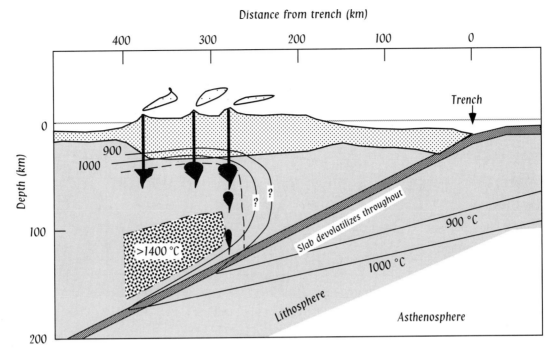

FIGURE 12.15 *The Tatsumi et al. (1983) model for the generation of magmas beneath the Japanese arc based primarily on their experiments (see the previous figure).* Tholeiites are generated closest to the subduction zone, HABs in the intermediate area, and alkali basalt farthest away from the trench.

on these and other observations, Tatsumi et al. presented a model for the origin of Japanese volcanic series (Figure 12.15), in which the parental magmas of the three magma series are generated over the depth range of ~30km–70 km. However, based on what we know about melt formation and transport in the mantle, it is more likely that these parental magmas are aggregates of smaller batches of magma generated over a longer depth range. Therefore, the experimentally determined pressures (depths) of multiphase saturation points of the three magmas simply represent the depths where they last equilibrated with a mantle rock.

Because H_2O-bearing fluids are somehow involved in the melt-generation process beneath arcs, it is important to understand how and where these fluids are released into the subduction environment. Figure 12.16 shows the stability fields of some of the hydrous minerals that are likely important in this fluid-generation process. Also shown are the geothermal gradients in hot versus normal subduction zones because these curves indicate what hydrous phase break down at what depth. The observation that, on average, volcanic fronts are located about 100 km to 200 km above the Wadati–Benioff zone suggests that hydrous fluids involved in magma generation should be released at a minimum pressure of ~3 GPa. This would make amphibole, talc, dense hydrous magnesium silicates (DHMS), MgMgAl-pumpellyite (MMA Pump), and Mg-chlorite as likely candidate phases to break down and generate the fluids needed for magma generation (Figure 12.16). In the case of subduction of a young, hot lithosphere, amphibole breakdown is likely to play an important role in fluid generation.

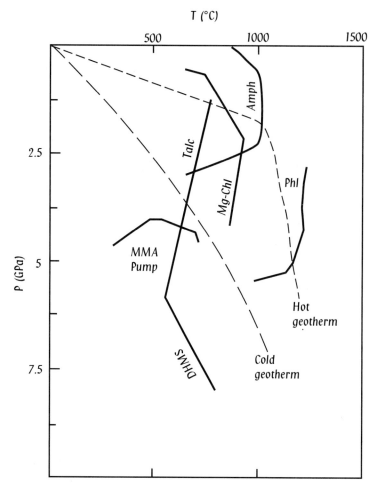

FIGURE 12.16 *This figure shows the stability limits of several hydrous minerals that are potentially significant in arc magma production because their breakdown would produce the fluid needed to lower the solidus temperature of the asthenospheric wedge (figure adapted from Staudigel and King, 1992; Earth Planet Sci. Lett.).* MMA Pump–MgMgAl Pumpellyite. Amph–Amphibole. DHMS–Dense hydrous magnesian silicate. Mg-Chl–Mg-chlorite. Phl–Phlogopite. Note that MgMgAl pumpellyite is stable on the high-pressure side of the curve shown. In all other cases, the hydrous phases are stable on the lower temperature side of each of the curves. Dashed curves—geothermal gradients at two different subduction zones—one cold and the other hot. (After Peacock, 1991, see Chapter 15 for comparison).

A General Model for Island Arcs

A highly schematic model (modified after Tatsumi et al., 1983; Gust and Perfit, 1987) may be used to explain how "average" arc calc-alkaline series magmas (if there is such a thing) may be generated in a subduction zone environment (Figure 12.17). In Tatsumi's model of Japanese island arc, high-alumina basaltic magmas are considered to be primary magmas. These magmas are generated in the asthenospheric wedge by H_2O-induced partial melting. The H_2O-rich fluid involved in such magma production is released from the subducted crust as a result of the breakdown of hydrous phases. These magmas pool and ascend by buoyancy toward the overriding lithosphere. As it reaches the Moho, the magma loses its buoyancy-driven desire to ascend

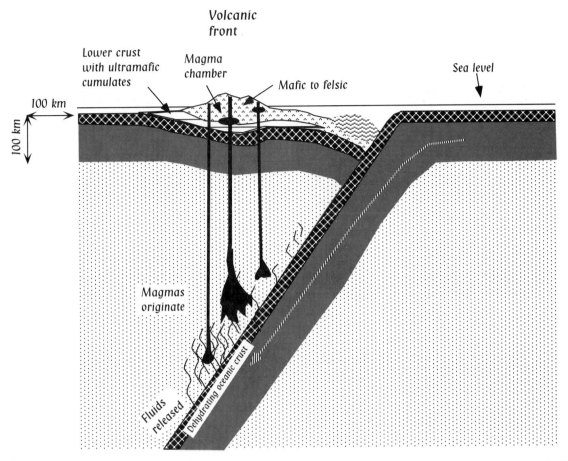

FIGURE 12.17   *A schematic, highly generalized model for the origins of the three main magma series (alkali basalt, tholeiites, and calc-alkaline) in an island arc.* Fluids are generated at different depths from the breakdown of different hydrous minerals in the crustal portion of the downgoing slab. These fluids depress the solidus of the asthenospheric wedge materials and trigger melting beneath the volcanic front. Parental magmas of alkali basalts are produced at the greatest depth, whereas high-Al_2O_3 basalts and tholeiitic basalt magmas (or, more likely, their parent magmas) are generated at intermediate and shallow depths, respectively. Greater melt production leads to thicker crust in the middle of the front. Here, high-Al_2O_3 basalt magmas pond within the crust or at the Moho and undergo fractionation and episodic mixing with fresh batches of magma supplied from deeper zones. Such fractionation and mixing generate the calc-alkaline series. Closer to the fore arc, the crust is thinner and the tholeiite magmas ascend to shallow crustal magma chambers. In such chambers, the magmas undergo fractionation and magma mixing to produce low-K series lavas. Alkali basalts rise rapidly and hardly undergo any fractionation.

and ponds in a magma chamber. Here it differentiates to andesitic and, in extreme cases, to more felsic differentiates by crystal fractionation. These magmas, being lighter than the surrounding ultramafic and gabbroic cumulates, ascend and erupt. Many additional complications may be added to this model. For example, hornblende and magnetite, two likely participants in the fractionating assemblage, would also be present in the crust of the volcanic front. Deep diapirs of partially molten material may rise from much deeper levels along the interface between the subducting slab and the overlying asthenosphere, and materially contribute to the magmas pooled at a shallower depth.

Variations of this scheme occur in many other published petrogenetic models presented for the origin of island arc magmas. For example, in the Gust and Perfit model, the high-Al_2O_3 basalt magma is not considered to be primary, but is derived from a high-MgO mafic magma that is generated by partial melting above a subduction zone. The high-MgO magma ascends through the asthenospheric wedge and ponds at the Moho, where it fractionates to a high-Al_2O_3 basalt. This high-Al_2O_3 basalt magma ascends through the crust and further fractionates at low pressures to evolved andesite, dacite, and rhyolite magmas.

Continental arcs are more complicated due to the presence of a thick crust of varied lithologies and ages. Also such rocks melt and generate highly silicic (rhyolitic) magmas when they are heated by basaltic magmas rising from below or by lowering their solidi with the introduction of volatiles released from the descending slab (discussed later). This phenomenon is substantiated by the relative volumes of silicic magmas in island arcs versus continental arcs: Continental arcs have a dominant silicic (rhyolitic/granitoid) component, whereas island arcs are characterized by a much greater abundance of mafic and intermediate igneous rocks.

GRANITOIDS

It is a common practice to use the broad term *granitic* or *granitoid* (instead of *granite*) to refer to the rocks that compose giant plutonic intrusions made dominantly of mixtures of quartz, alkali feldspar, and plagioclase feldspar, with minor amounts of other minerals. In reality, however, based on the IUGS classification, the rock names may vary from tonalite, quartz diorite, granodiorite, to quartz monzonite, and granite (see Chapter 8).

Although the IUGS classification of granitoids is used throughout this book, two other classifications have also been commonly used in the literature, one of which is based on major element chemical composition (Shand, 1947; cited in Clarke, 1992), and the other is more of a genetic classification based on rock chemistry (Chappell and White, 1974, and their subsequent papers). The Chappell–White work is presented later with respect to granitoid genesis. Shand's classification uses the concept of *alumina saturation* and divides granitoids into three types: peraluminous, metaluminous, and peralkaline (Table 12.2). The alumina saturation concept is based on whether the magma has more Al_2O_3 than what is required to make feldspar, and thus the major classes are based on Al_2O_3 content vis-à-vis alkali contents. Table 12.2 lists average major element and isotopic composition of the three classes (from Clarke, 1992). Clarke (1992) significantly updated Shand's classification and showed that granitoids, as defined in the IUGS nomenclature (and used in this book), belonging to the three classes are also characterized by their distinctive minor minerals (i.e., minerals other than quartz and feldspar; Table 12.2). Table 12.2 shows that Shand's classes also correlate with plate tectonic environments.

The connection between granitoids and their plate tectonic environments has been investigated by a number of workers. As an example, note that Maniar and Piccoli (1989) subdivided granitoids according to their tectonic affiliation (Figure 12.18). Figure 12.18 shows that granitoids from island arcs and continental arcs are more granodioritic to tonalitic. However, granitoids from continent–continent collision zones are proper granites.

Considerable compositional variability is exhibited by batholiths from a particular tectonic province. For example, modal variations in three important granitoid batholiths from western North America are shown in Figure 12.19 (Hyndman, 1985). Seventy-five percent of the Coast Range batholiths of British Columbia have a high proportion of plagioclase compared with the Sierra Nevada and the Idaho batholiths.

TABLE 12.2 *Shand's classification of granitoid rocks (after Clarke, 1992).*

	QAP $60\% > Quartz > 20\%$ *Alkali feldspar/[Alkali feldspar + Plagioclase] = 0–1*		
	Peraluminous	Metaluminous	Peralkaline
Major oxides	$Al_2O_3 > CaO + Na_2O + K_2O$	$CaO + Na_2O + K_2O > Al_2O_3 > Na_2O + K_2O$	$Al_2O_3 < Na_2O + K_2O$
Characteristic minerals	Al_2SiO_5-polymorphs, cordierite, garnet, topaz, tourmaline, corundum	Opx, Cpx, amphibole	Fa-rich olivine, alkali amphiboles, aegirine
Other common minerals	Biotite, muscovite	Biotite, minor muscovite	Minor biotite
$^{87}Sr/^{86}Sr$ isotopic comp.	0.705–0.720	0.703–0.708	0.703–0.712
Plate tectonic association	Continent-continent collision	Continental and Island arcs	Post-tectonic extension
Average Compositions			
SiO_2	71.45	67.43	74.01
Al_2O_3	14.76	14.67	11.59
FeO^*	2.49	4.13	3.08
MgO	0.78	1.64	0.55
CaO	2.01	3.53	0.48
Na_2O	3.72	3.72	4.33
K_2O	3.52	3.20	5.09

353

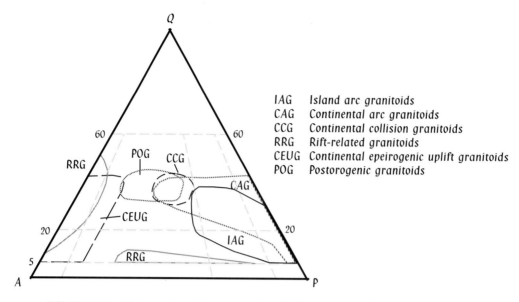

IAG Island arc granitoids
CAG Continental arc granitoids
CCG Continental collision granitoids
RRG Rift-related granitoids
CEUG Continental epeirogenic uplift granitoids
POG Postorogenic granitoids

FIGURE 12.18 *Plate tectonic environments and granitoid compositions.* (Redrawn from Maniar and Piccoli, 1989)

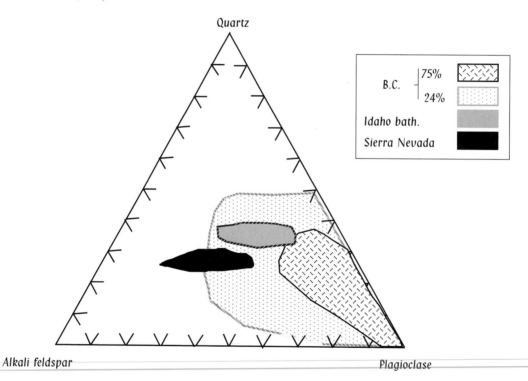

FIGURE 12.19 *Modal variations in three North American batholiths.* (Based on Hyndman, 1985)

An-content of plagioclase, proportion of mafic minerals (mostly hornblende and biotite), and proportion of hornblende to biotite increase with the increasing proportion of plagioclase. Aside from these major minerals, many other minor minerals occur in variable abundances, including pyroxene, muscovite, zircon, sphene, magnetite, ilmenite, and epidote.

Although granitoid (and their volcanic equivalent, rhyolite) rocks occur in a variety of plate-tectonic environments, at midoceanic ridges, island arcs, continental arcs, and hot-spot areas, batholiths that occur in continental arcs and rhyolitic eruptions in continental extension areas are the most important in terms of their volumes. Therefore, in the following discussion, the focus remains on continental arc plutons and rhyolites in continental extension areas.

Granitoid Plutons

Granitoid plutons cover areas of $\sim$ 100s of km^2 and are commonly found in the cores of eroded mountain belts that develop at continental arcs and continent–continent collisional zones. The actual exposed surface area of a granitoid pluton is related to the extent of erosion and not so much to the volume of the pluton. Himalayan batholiths serve as an example of granite plutons in a continent–continent collision zone (Figure 12.20). Some of the more notable examples of plutons in continental arcs are the coastal batholiths of British Columbia, Peru and Chile, the Sierra Nevada batholith of California, and the Idaho batholith of Montana. The shapes (map view) of granitoid plutons can vary from circular to elliptical to almost linear (cf. Clarke, 1992). Their three-dimensional structure is often difficult to determine. However, geophysical studies and analogue model experiments indicate that they vary from dome, tabular, and mushroom to inverted tear-drop shapes (Clarke, 1992; Pitcher, 1993).

Rarely a window into the deep structure beneath a batholith is provided by xenoliths, as is the case for the Sierra Nevada Batholith (SNB) of California (Ducea and Saleeby, 1988). Based on the types of xenoliths and their pressure-temperature conditions of equilibration, Ducea and Saleeby proposed a vertical structure through the root of the SNB as it was some 10 to 100 million years ago (Figure 12.21). Figure 12.21 suggests that the upper to middle crustal section of the SNB is largely composed of granitoids with more mafic-rich granitoids coming in at intermediate to deeper levels (about 40 km). In the lower part of the batholith occurs eclogitic rocks (composed of garnet + clinopyroxene), which are considered to be early fractionates of the granitoid-parent magmas as well as partially melted residues. Spinel lherzolites,

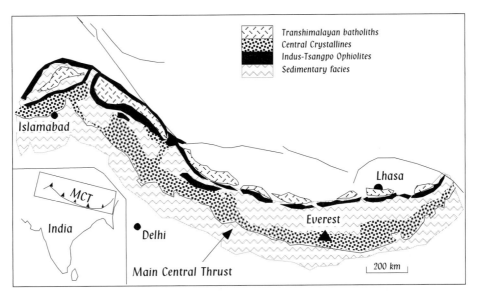

FIGURE 12.20   *Granitoid batholiths of the Himalayas.*

Depth (kms)

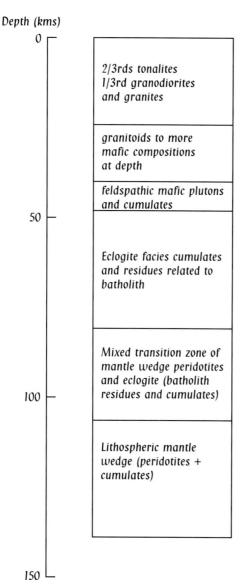

FIGURE 12.21 *Xenolith-based vertical section through the Sierra Nevada Batholith.* (From Spear, 1993, *Metamorphic Phase Equilibria and Pressure-Temperature-Time Paths*, Mineralogical Society of America, Washington D.C.)

representing the upper mantle, occur at about 75 to 100 km depth. Interestingly, there is no clearly defined Moho beneath the SNB, but a broad transition between the crust and mantle occurs at ~75 km to 100 km.

Granitoid plutons are not generally formed by a single-phase intrusion, but rather by multiple intrusions that take place very slowly over a few tens of millions of years. The magma composition with each phase of intrusion changes as well. The coastal plutons of Peru consist of a chain of 800 plutons that were gradually emplaced over a period between 110 and 300 million years ago. The Tuolumne Intrusive Series, a relatively small pluton that occurs within the SNB, had evidently seen at least four distinct episodes of magma emplacement (Figure 12.22; Bateman and Chappell, 1979). A hornblende + biotite-bearing quartz diorite occurs at the peripheral region and was the first batch of magma to be emplaced. The intermediate region is composed of granodiorite, and a porphyritic granite occurs at the core of the pluton. Bateman and

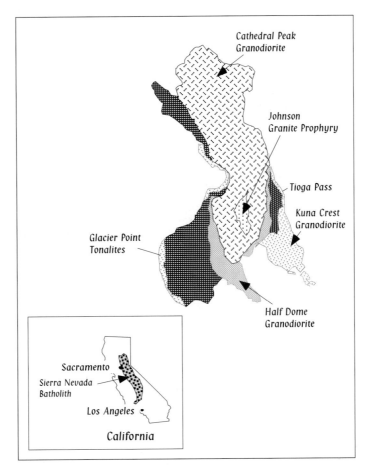

FIGURE 12.22   *A map showing distribution of the various lithologic types of granitoids intruded at different times within the Toulumne Volcanic Series* (Sierra Nevada batholith of California).

Chappell (1979) showed that these distinct episodes of magma intrusion occurred in the core region of the pluton while the pluton was solidifying. Intrusion of these magma batches resulted in episodic breaching of the solid margins of the intrusion by the fresh magma, which then intruded the surrounding wall rocks.

Buddington (1959) recognized three classes of plutons based on their field relations and inferred depths of emplacement. In his model, the shallowest depth of batholith emplacement, 0 km to 5 km, is known as *epizone*, the deepest level (> ~10 km) is the *catazone*, and the intermediate depth range of ~5 km to 15 km is *mesozone* (Figure 12.23). Epizonal batholiths are relatively small and characterized by (a) sharp, usually discordant contacts with wall rocks; (b) intense hydrothermal alteration (and ore mineralization) of the wall rocks; and (c) generally weak contact metamorphism of wall rocks. Catazonal batholiths are generally large and surrounded by high-grade metamorphic rocks. Their contacts with the wall rocks are generally diffuse, and migmatite veins are common. Migmatites are generally believed to be low degrees of melt that are sweated out of the wall rocks. Mesozonal batholiths have intermediate characteristics, whose contacts with wall rocks may vary from sharp to gradational. Wall rocks often show metamorphism of a previously weakly metamorphosed rock.

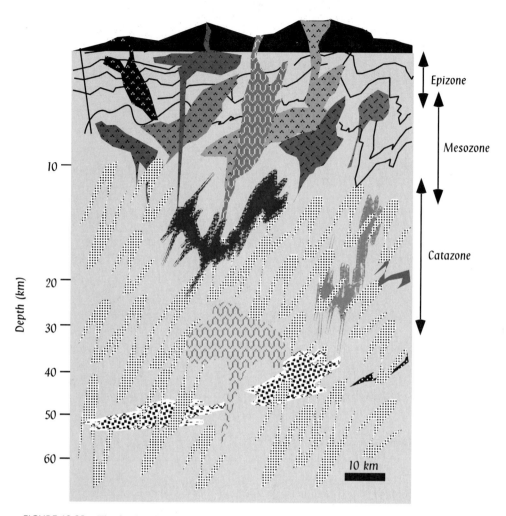

FIGURE 12.23 *The depths of emplacement of batholith types based on Buddington's classification.* (From Best, 1982, *Igneous and Metamorphic Petrology*, W.H. Freeman)

Feldspars in Granitoids

Granitoids are often divided into hypersolvus and subsolvus granitoids on the basis of whether they contain a single feldspar (say, an alkali feldspar: hypersolvus granitoid) or two feldspars (alkali feldspar and plagioclase: subsolvus granitoid). At low pressure, the binary system albite–orthoclase is characterized by a minimum and the presence of an immiscibility gap in the subsolidus region (see Figure 7.17 and the accompanying discussion in Chapter 7). As a result of this phase relationship, granitoid plutons emplaced at shallow levels in the crust would be expected to contain a single feldspar$_{ss}$, and such granitoids are called *hypersolvus granitoids*. Upon very slow cooling, the single feldspar$_{ss}$ crystals break down into two different feldspar$_{ss}$, one of which comes out as an exsolved phase (see Chapter 7). Depending on the composition of the feldspar$_{ss}$ that crystallized from the liquid, the exsolved phase may be Or rich and the host Ab rich or vice versa. In a single crystal, when the host phase is Or rich, it is called a *perthite*; when the host is Ab rich, it is called an *antiperthite*. Cooling kinetics dictate the nature of the resultant texture exhibited by such feldspar$_{ss}$ grains: The slower the cooling rate, the larger the exsolved blebs. Various textural terms such as *macroperthite, microperthite*, and so on are used to describe the relative size of the exsolved phase.

At higher pressures and at water-saturated conditions, the solvus (or immiscibility gap) touches the solidus in the Ab–Or join, creating an eutectic point between two feldspar$_{ss}$ phases—one rich in the Or component and the other in the Ab component (see Figure 7.18 and accompanying discussion in Chapter 7). Under such conditions, granitoid magma at the eutectic crystallizes two feldspar$_{solid\ solution}$ (i.e., subsolvus granitoid). Therefore, the presence of two feldspars in a granitoid causes a deep crustal emplacement. Because the composition of feldspar$_{ss}$ is sensitive to temperature changes, its Ab/(Or + Ab) or Or/(Ab + Or) ratios may be used as a thermometer to estimate the temperature to which the granitoid cooled.

The Granite System

The granite system is a ternary system whose three components are quartz, albite, and K feldspar, which are the three principal constituents of all granitoid or granitoid rocks. In this system at low pressures, a ternary minimum (m) occurs on the cotectic between quartz and alkali feldspar$_{ss}$ liquidus fields (Figure 12.24). This point is not a eutectic because, in such a case, three solid phases and a liquid should coexist in a three-component system, whereas in the Ab–Or–Qz system at low pressure, only a single alkali feldspar$_{ss}$ and quartz coexist with a liquid right at the minimum. Also in a ternary eutectic system, the last liquid must always reach the eutectic point. In the Ab–Or–Qz system, many liquid paths may be constructed that, under equilibrium conditions, will completely crystallize long before reaching the minimum. For example, the liquid X (Figure 12.24) follows a curved path (because of solid solution behavior of the crystallizing feldspar$_{ss}$)

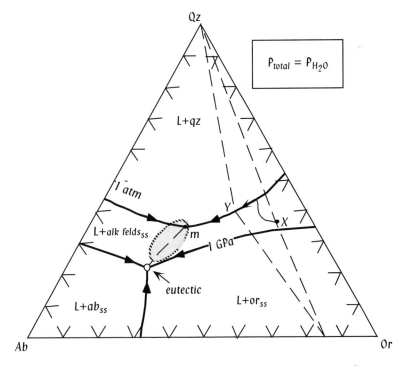

FIGURE 12.24   *The granite system at 1 atmosphere and at 1 GPa (H_2O saturated). m* represents a ternary minimum at 1 atm. Most common batholithic granitoids plot in the trough defined by the minima at low pressures and the high-pressure ternary eutectic. (From Best, 1982, *Igneous and Metamorphic Petrology*, W.H. Freeman)

until it reaches the 1 atmosphere cotectic. The liquid then travels down the cotectic to Y while crystallizing quartz and an alkali feldspar$_{ss}$. The last three-phase triangle connecting Y, quartz, and an alkali feldspar$_{ss}$ is shown as a dashed triangle in Figure 12.24. The last liquid will crystallize at Y. A granite formed on the cotectic will be a hypersolvus granite because of the presence of a single feldspar$_{ss}$ phase.

At higher pressures (e.g., 1 GPa in Figure 12.24) in the water-saturated system Ab–Or–Qz, the alkali feldspar$_{ss}$ liquidus field splits up into two different feldspar$_{ss}$ liquidus fields, creating a ternary eutectic point where quartz and two feldspar$_{ss}$ phases crystallize together from a liquid. Thus, a granite formed at such an eutectic point is a subsolvus granite. The conversion of the ternary minimum to an eutectic point probably happens at about 0.4 GPa (Hess, 1989).

Figure 12.24 is of great importance in petrology because, prior to its existence, there were two fundamentally distinct schools of thought on the origin of granites (or granitoids): One of these invoked an igneous origin and the other supported their formation by metasomatic replacement of wall rocks (called the *granitization hypothesis*). O.F. Tuttle and N.L. Bowen (1958), the originators of this phase diagram, pointed out that granites plot along a polybaric trough (gray shaded in Figure 12.24) that is formed by the positions of the minima at low pressures and eutectic point at higher pressures. This was an important piece of evidence to support Bowen's original hypothesis that granitoid liquids form by fractional crystallization of basaltic magmas because such liquids should be expected to plot at the lowest temperature region of the diagram. However, Tuttle and Bowen noted that the invariant melts formed by melting of sedimentary rocks containing a large proportion of quartz and feldspars also plot in this trough. Therefore, either way one looks at it, an igneous origin of granites is certain. This diagram thus killed the granitization hypothesis.

ORIGINS OF ARC GRANITOIDS

Source Rock and Granitoid Types

As indicated in the prior section, Tuttle and Bowen's experiment reduced the problem of the origin of large granitoid batholiths in continental arcs to one that is igneous in nature. Bowen's own preference was to derive granitoid liquids via fractional crystallization of basalt magma. In contrast, his colleague, Tuttle, apparently favored the origin of granite magma by direct melting of sedimentary (or metamorphosed sedimentary) rocks.

There is no doubt that some volumetrically minor granitoid bodies found on the ocean floor or in Iceland were produced by fractional crystallization of basaltic magma. However, most modern petrologists do not believe in such an origin of batholithic granitoids because, quite simply, the required volumes of cumulates of intermediate and mafic composition that may have fractionated to form such batholiths are not found. Also many scientists have cited a large number of field examples of silicic melts caught in action as they were trying to escape their high-grade metamorphic crustal source rocks. Therefore, it appears that at least some arc granitoids are generated by melting of crustal rocks. Then the question becomes one of what source rock or rock types are melting and under what P,T, and p_{H_2O} conditions.

The plausible source rocks could be: (a) middle or lower crust, particularly sediments (or metasediments) that accumulate to form great thickness in arcs and mafic (or intermediate) metamorphic rocks (granulites and amphibolites), (b) subducted crust (amphibolite or eclogitic material), (c) old (≥ 3 billion years) subcontinental

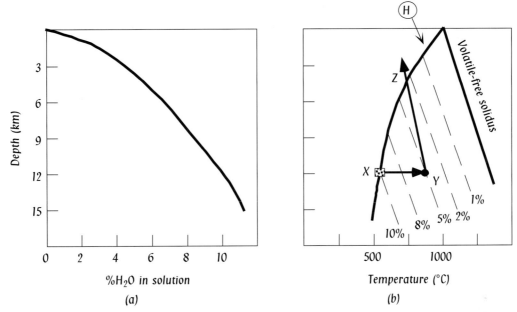

FIGURE 12.25   (a) Dissolved H$_2$O content of granitoid magma as a function of depth. (b) Dry and wet solidi of a granitoid with schematic melting and ascent paths (discussed in the text). (After McBirney, 1993)

mantle that may have been reprocessed by basaltic and andesitic magma passing through it, and (d) pyroxenites or amphibole-rich veins in the mantle wedge. Whatever the source rock, some involvement of water in the melting process is almost guaranteed because (a) hydrous minerals commonly occur in major granitoids, and (b) temperatures in the melting region (generally lower crust) are not high enough for the dry solidus to be reached. However, it is unlikely that melting occurs in a water-saturated environment because the slope of the water-saturated solidus is such that most magmas would solidify long before reaching the upper-crustal levels (Figure 12.25, and elaborated further later). Yet in many areas, rhyolites have erupted that can be genetically linked to batholiths in the same general area.

An understanding of ascent and eruption of granitoid magmas requires a look at the controls imposed by H$_2$O solubility in such magmas as a function of pressure and the solidus of granitoid magmas. Figure 12.25a shows that the solubility of H$_2$O in granitoid magmas rapidly increases with depth so that at about 15 kms the solubility is about 11 wt%. Such variation has significance in the eruptibility of granitoid magmas, as illustrated with Figure 12.25b. This figure shows the H$_2$O-saturated (H) solidus and volatile-free (dry) solidus of a granitoid magma. Also shown are dashed lines with constant H$_2$O contents in the magma. Consider a case where a granitoid rock is heated up at 12 km under H$_2$O-saturated condition. This rock begins to melt when the rock is heated to its wet solidus temperature (Point X; slightly higher than 500°C). If this partially molten rock is heated up further, its dissolved water content will decrease: At ~800°C, the H$_2$O content will be about 5% (Point Y). Now imagine that the magma leaves the source region and ascends toward shallower depths along an adiabat Y → Z. Note that, as the magma ascends, it has to exsolve some H$_2$O because it can hold less H$_2$O in solution at shallower depths. Most important, if this magma ascends slowly, by the time it reaches about 4 km (Point Z), it would have crossed the wet solidus (i.e., it must entirely solidify). However, the magmas that contain little dissolved water have more of a chance to stay molten because their ascent paths would stay above the volatile-free solidus. The implication of this diagram is that rhyolite lavas may have less dissolved H_2O then granitoid plutons.

$^{87}Sr/^{86}Sr$ and $^{143}Nd/^{144}Nd$ isotopic systems, when combined with other stable isotopic systems (e.g., oxygen), can provide strong constraints on the nature of the source rock and/or contamination. This is a vast area of study, and a detailed discussion is beyond the scope of this book. Instead only one radiogenic isotopic system, as applied to batholithic granitoids, is discussed herein. In general, $^{87}Sr/^{86}Sr$ values of ~ 0.700 to 0.7030 may be thought of as depleted mantlelike values from which N–MORBs are derived. A Hawaiian-type plume is believed to be characterized by $^{87}Sr/^{86}Sr$ values of ~ 0.7035 to 0.7045. However, old continental crust is characterized by very high $^{87}Sr/^{86}Sr$ values. For example, a 2.5-billion-year old crust can have a $^{87}Sr/^{86}Sr$ value of ~ 0.718, and a 1-billion-year-old crust can have a $^{87}Sr/^{86}Sr$ value of 0.710. Therefore, a granitoid with a mantlelike value is likely to be generated by partial melting of a basaltic rock that was derived from a depleted oceanic-type mantle. A granitoid rock characterized by a very high Sr-isotopic ratio may be produced by melting an old crustal source or, alternatively, it may have still been generated from a source with mantlelike values but was subsequently contaminated by a very old continental crust.

Based on Sr-isotopic ratios and other geochemical criteria, Chappell and White (1974) distinguished between two geochemical classes of granitoids: the I (for igneous) and S (for sedimentary) type granitoids (Table 12.3). *I type* stands for an origin by partial melting of metamorphosed igneous (e.g., eclogite, granulite, amphibolite) rocks. *S type* stands for an origin by partial melting of sedimentary or metasedimentary rocks. S-type granitoids have high $^{87}Sr/^{86}Sr$ ratios and relatively high Al_2O_3 content. In contrast, I-type granitoids have low $^{87}Sr/^{86}Sr$ ratios and relatively high Na_2O contents. In general, I-type granitoids include most diorites, quartz diorites, and tonalites. S-type granitoids include granodiorites and granites. A rather distinctive feature of the two types of granitoids is the nature of restites they each carry. Restites are believed to be residues of the partially molten source rock, and these often occur as lenticular enclaves within the batholith. Restites in I-type granitoids include amphibole, pyroxene, and plagioclase, whereas those in S-type granitoids are biotite, garnet, sillimanite, cordierite, and plagioclase. It is apparent that S-type granitoids are equivalent to the peraluminous granitoids of Shand; whereas I-type is equivalent to metaluminous type.

TABLE 12.3   *Comparison of I- and S-type granites.*

	I-Type Granitoid	S-Type Granitoid
$[^{87}Sr/^{86}Sr]_{initial}$	0.700–0.703	≥ 0.710
Al_2O_3	Moderate	High
Na_2O	High	Relatively low
Rock types	Diorite	Granodiorite
	Quartz diorite	Granite
	Tonalite	
Restite types	Amphibole	Biotite
	Pyroxene	Garnet
	Plagioclase	Sillimanite
		Cordierite
		Plagioclase

Although several other types of granitoids have been proposed in the literature, the so-called A-type granitoids cannot be ignored because of their volume and distinctiveness from I and S types. A-type granitoids typically intrude late in an arc environment or in continental extensional regimes (e.g., Rio Grande rift in North America). These include all or most hypersolvus granites. They are relatively rich in alkalis. Fluorite and biotite are commonly present. These magmas are believed to originate by melting of a source that has already been depleted of an I-type granitoid magma.

Subsequent authors, including Chappell and White, have discovered that there can be many exceptions to the major chemical types just described. Nonetheless, as far as source rocks are concerned, the prior discussion suggests that there may be old and young or mixed sources of metasedimentary and metaigneous rocks.

Origins of Granitoids

This is a vast area of research, and it is extremely difficult to summarize the large volume of literature that currently exists on this topic. The interested reader is well advised to consult the books by Pitcher (1993), Clarke (1992), and Johannes and Holtz (1996) and excellent reviews by Patiño Douce (1996, 1999). What follows is in some ways a sweeping coverage of ideas on the origins of granitoid magmas. In passing, students are reminded that the rocks which are sources and residues of melting related to granitoid melt production are generally all metamorphic rocks. Therefore, it is useful for students to look at Chapter 15 (or ask an instructor).

As with any melting process, there are two things that must be known: the source rock that provides the magma and the conditions of melting. Through earlier discussions, we know that the potential source rocks could include a variety of sedimentary, metasedimentary, and metaigneous rocks, such as graywacke, sandstone, mudrock, amphibolite, granulite, amphibolite, and eclogite (or garnet clinopyroxenite). The conditions of melting include temperature, H_2O content of the source rock, and depth of melting. The heat needed for melting can be delivered by two processes: thickening of the crust and/or ponding of basalt magmas at the base of the Moho. The crustal thickening occurs at continent–continent collision sites due to overthrusting of one continent over another, and the underthrusted crust is heated by the regional geothermal gradient over time. It is also possible that the heat is supplied by hot, basaltic, or andesitic magmas ponding at the base of the crust. Heating of the thickened continental crust beneath continental arcs by mafic-intermediate magmas is probably the only plausible way to generate melting in the lower continental crust.

It is unlikely that free aqueous fluids exist in the continental crust except for the upper few kilometers and where low-grade, H_2O-rich, metamorphic rocks are rapidly underthrust below relatively dry, higher-grade metamorphic rocks (Patiño Douce, 1999). Therefore, granitoid magma formation probably happens in an environment where free aqueous fluid is either too small or absent. The melt-forming reaction is likely a dehydration reaction(s). The three important hydrous minerals that are of importance in this context are muscovite, biotite, and amphibole. Dehydration melting of these minerals in a range of appropriate continental crustal rock compositions is therefore of interest to us as far as origin of granitoid magmas is concerned. These dehydration melting reactions are all incongruent (see Chapter 7) of the type:

$$A = B + C + \text{hydrous melt,}$$

where A is a hydrous phase and B and C may be some anhydrous phases. Generally, in these melt-formation reactions, the hydrous minerals break down producing refractory residues (high in MgO, CaO, FeO) and felsic melts. For example, note the following amphibolite-melting reaction that produces a granitoid melt and an eclogite residue:

$$\text{Hornblende + Plagioclase = Garnet + Clinopyroxene + Melt.}$$

There are significant differences in composition of the melts generated from mica-rich sources (e.g., metamorphosed shale, graywacke) versus those that are produced from amphibole-rich sources (Patiño Douce, 1999): (a) The melts from mica-rich sources tend to have distinctively lower MgO + FeO than those from amphibole-rich sources, (b) the melts from mica-rich sources have distinctly higher Na_2O + K_2O but lower CaO than amphibole-rich sources, and (c) the compositions of melts generated from metamorphosed mafic melts are also more aluminous than those from amphibole-rich sources.

Aside from bulk composition and H_2O content of the source rock, pressure and degree of melting can also affect the composition of the melt generated from a given source rock. For example, in their melting study of four amphibolites (metamorphosed basalts), Rapp and Watson (1995) observed that from 0.8 to 1.6 GPa, $\sim 5\%$ melting produces a high-K_2O granitic melt, whereas the melts produced by 20% to 40% melting have compositions of tonalite, granodiorite, quartz diorite, and diorite.

Based on the previous discussion, one may come to a broad general conclusion that plagioclase-rich granitoids (i.e., tonalite, quartz diorites, etc.) form by partial melting of metamorphosed rocks of mafic-intermediate bulk compositions (basalts–andesites), whereas peraluminous granitoids are derived by melting of mica-rich aluminous source rocks. However, scientists who have extensively mapped granitoid plutons through the years have often proposed a hybrid (mixed) origin of many of the granitoid magmas (e.g., Reid and Hamilton, 1987). In the field, these authors often noted large modal (and therefore chemical) variation occurring within individual batholiths due to mixing between variably disintegrated mafic/intermediate enclaves (could be melts or restites) and the host felsic intrusive. Such a hybrid origin in many batholiths is also supported by chemical and isotopic mixing arrays.

In the earlier context of the hybrid origin of some granitoids, mention may be made of an interesting observation by Patiño Douce (1999) while evaluating all available experimental data on partial melting of mica-rich source rocks relevant to granitoid origins. He compared the SiO_2 contents of experimental melts with those of calc-alkaline granitoids (CAGS), peraluminous S-type granitoids (PSGS), metaluminous A-type granitoids (MAGS), peraluminous leucogranites (PLGS, as found in the Himalayan continent–continent collision zone), and some other types of granitoids (an explanation of this nomenclature may be found in Patiño Douce, 1999). Patiño Douce noted that the experimental melts have $SiO_2 \geq 70$ wt%, whereas all granitoid types except the PLGS have SiO_2 ranging to lower values (Figure 12.26). He concluded that (a) only the PLGS are true crustal melts, and (b) the other groups must have different origins, including a mixed origin (whether melt–melt mixing or crystal-rich melt–melt mixing).

Figure 12.27 shows a general phase diagram relevant to the origin of granitoid melts based on the work done by Rapp and Watson (1995). It is mostly relevant to the melting of meta-basalt (because that is what they used as source compositions). Two geotherms, a hot one (possibly related to heating by ponded basalt magma) and a cooler one (possibly thickened crust without heating by basalt magmas), are also shown. It shows that, under hot conditions, dehydration melting begins at A (about 10 kbar or ~ 30 kms); in the cooler case, melting begins at B. Depending on the depth, the residue that would be left behind would be garnet amphibolite followed by eclogite and garnet granulite (lower crust). (The student may wish to figure out what the crustal litho-

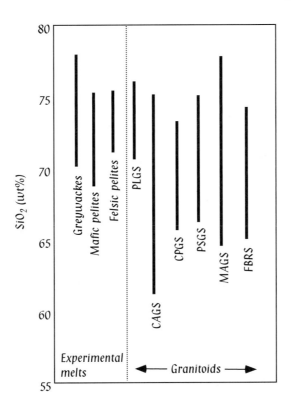

FIGURE 12.26 *Comparison of SiO₂ in experimentally produced melts and natural granitoids.* (From McBirney, 1993, 2nd ed., *Igneous Petrology*, Jones and Bartlett)

logic sequence would be for the two geotherms. Compare them to Figure 12.22. Deduce what sort of geothermal conditions would produce the Sierra Nevada batholith.)

On the basis of the above discussion of granitoids it should be apparent to the students that granitoid batholiths are complex and no unique hypothesis can explain the origin of all granitoid types. Although the source rocks for granitoid (or their parent) magmas can vary a great deal, it is unlikely that direct melting of peridotite (wet or dry) can generate the voluminous batholithic granitoid magmas. Plagiogranites (with plagioclase, alkali feldspar, quartz, amphibole) found in ophiolites and at mid-oceanic ridges may form by differentiation of basaltic magmas produced in the mantle.

Batholithic granitoids vary from granodioritic to tonalitic to granitic compositions. However, intermediate compositions are predominant. Their parent magmas may be produced by many different mechanisms—melting of crustal rocks, differentiation of high-alumina basalt magmas, assimilation of silicic crustal rocks by mafic magmas, and mixing of magmas. Magma production mechanism is probably one of dehydration breakdown melting and not water-saturated melting. Peraluminous granitoid-parent magmas with $^{87}Sr/^{86}Sr > 0.710$ and $\delta^{18}O > 7$ are probably generated (a) by the melting of silicic-aluminous metasedimentary continental crustal rocks, or (b) by assimilation of such rocks (or melts derived from them) by mafic magmas. (Briefly, $\delta^{18}O$ value for mantle rocks is generally less than 7. Values of greater than 8 generally represent the continental crust.) Metaluminous granitoid-parent magmas are probably produced by 20–60% melting of metaigneous rocks. Small degrees of melting of such rocks generate parent magmas of perlkaline granitoids. The origin of granitoids with mantle-like $^{87}Sr/^{86}Sr$ (<0.706) and $\delta^{18}O$ (<6.8) is particularly interesting. A straightforward interpretation would be that they formed by differentiation of mafic magma generated in the mantle or by partial melting of mafic material, which may be crystallized, deeply ponded, magmas. The granitoid magmas so formed do not assimilate old continental crustal materials.

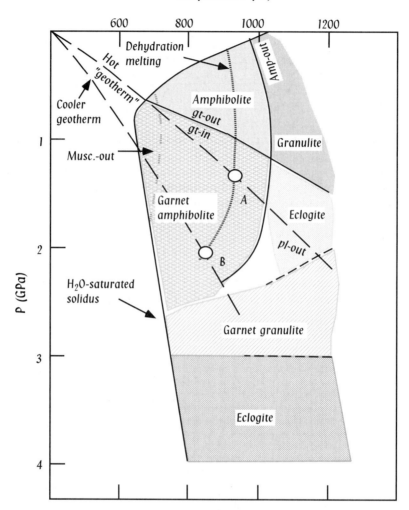

FIGURE 12.27   *A generic phase diagram applicable to the formation of granitoid melts and their residues (partly based on Rapp and Watson, 1995).* The two geotherms shown are entirely schematic and only used to illustrate how magma may be generated at two contrasting conditions, as well as how that affects the residues that constitute the lower continental crust (and mantle).

This discussion would remain incomplete without mentioning the Peninsular Ranges Batholith of southern California (USA) and Baja California (Mexico), which has been examined in great geochemical detail by many scientists (e.g., DePaolo 1981; Gromet and Silver 1987; Ague and Brimhall 1988; Silver and Chappell 1988). This batholithic complex runs for about 1000 km, approximately parallel to the coast, and is about 100 km wide. It is spatially zoned from west to east with respect to the ages of the rocks and their isotopic, major and trace element compositions (Figure 12.28). The plutons of the western zone are 120–105 million years old, are compositionally diverse (from ultramafic to quartz monzonite), and were emplaced in low-K mafic volcanic rocks. Their $\delta^{18}O$ values (5.5–6.5) and $^{87}Sr/^{86}Sr$ ratios (0.703–0.705) are substantially lower than the plutons of the eastern zone ($\delta^{18}O$ = 8.5–12; $^{87}Sr/^{86}Sr$ = >0.706). The

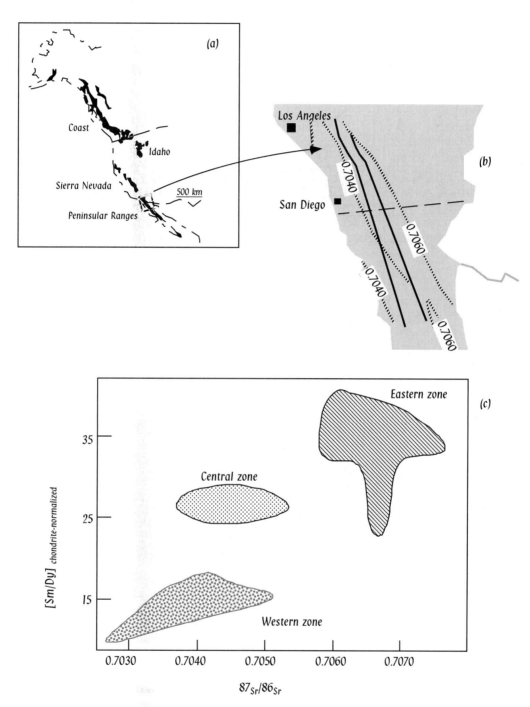

FIGURE 12.28   (a) Map showing the location of major Mesozoic cordilleran batholiths in western North America. (b) Close-up of the Peninsular Ranges batholith. Dashed lines–lines of constant $^{87}Sr/^{86}Sr$ composition. Thick continuous lines–rare earth element based zones. (c) Covariation of Sr-isotopic ratio and the [Sm/Dy] ratio. The three zones are clearly distinctive. (After Gromet and Silver, 1987)

plutons of the eastern zone are dominantly tonalitic and granodioritic, and show an age progression of 105 to 80 million years from west to east. There is a concomittant increase from west to east in SiO_2, alkalis, $^{87}Sr/^{86}Sr$, $\delta^{18}O$, REE and several other compositional parameters in the plutons of the eastern zone. Because the crust is considerably thicker to the east, one interpretation is that the mantle-generated magmas have assimilated more crust in the easterly direction. The progressively greater trace element abundance from west to east apparently also requires a gradual increase in the depth of magma generation. The western zone plutons have been suggested to be an old offshore island arc that has been sutured by the batholiths to the main continent.

RHYOLITES AND BIMODAL (MAFIC AND SILICIC) VOLCANISM IN CONTINENTAL EXTENSION AREAS

Rhyolites are volcanic equivalents of granitoid rocks and are most commonly found in continental arcs and continental extension zones. They can be glassy (a variety known as *obsidian*), porphyritic with common phenocrysts of sanidine (a high-temperature polymorph of orthoclase), and quartz or pyroclastic rocks composed of ash (called tuff) or larger fragments. Rhyolites from arcs and continental extensional rift zones typically have >70% SiO_2, but highly variable Al_2O_3 contents and Sr-isotopic ratios.

In extensional regimes, such as the Rio Grande rift in western United States, volcanism is bimodal in the sense that basaltic and rhyolitic lavas erupt in these areas and intermediate composition lavas are typically lacking (Figure 12.29). This absence of intermediate composition lavas is referred to as the *Daly Gap*, named after R.A. Daly, who first noted it in the early part of the 20th century. Such gap is also found in continental areas and is easily explained by a model that calls for basaltic magmas as the provider of the heat needed to generate rhyolite magmas by melting of the crust (Figure 12.30: Hildreth, 1981). In this model, basalt magma generated in the upper mantle ascend to the Moho and ponds there because it reaches a state of neutral buoyancy (i.e., it is no longer buoyant enough to rise through the crust). These basaltic magma bodies cool and crystallize gabbros and anorthosites. The crystallization process releases latent heat to the crust, and additional heat is provided by the basalt magma's ambient (adiabatic) heat. This heating results in melting of the crust. These crustal melts are rhyolitic in composition.

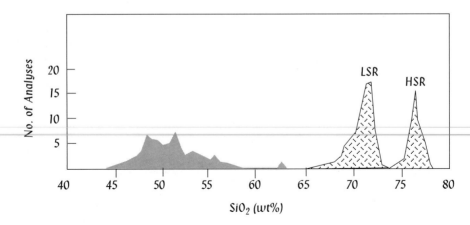

FIGURE 12.29   *Silica variation in four areas of bimodal volcanism in Arizona.* (Source: Moyer and Nealey, 1981)

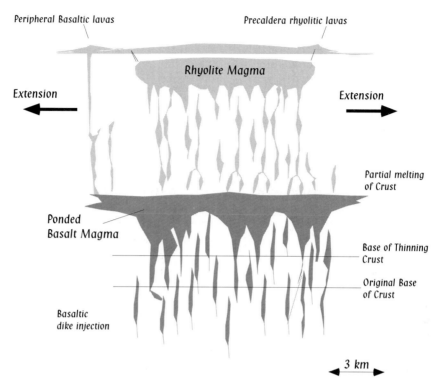

FIGURE 12.30   *A schematic model for the origin of bimodal volcanism in areas of continental extension.* (After Hildreth, 1981; discussed in the text)

SUMMARY

[1]   Plate margin volcanism is characteristically andesitic in nature, although high-alumina basalts, tholeiitic basalts, alkaline magmas (shoshonite series lavas), and some other minor lava types also erupt.

[2]   The volcanic front generally lies 100 km to 200 km above the Wadati-Benioff zone. The volcanic front is often composed of subparallel belts of volcanoes that show a progressive change from tholeiitic to calc-alkaline to shoshonitic series further away from the subduction zone.

[3] H_2O plays a major role in the magma generation process. Magmas are generated within the asthenospheric wedge by dehydration melting. The H_2O in the source region is supplied by the breakdown of hydrous minerals of the subducted crust.

[4]   It is commonly accepted that andesitic magmas are not generated from the mantle but are products of differentiation of basalt magmas.

[5]   Granitoids are broadly of three types—peraluminous, metaluminous, and peralkaline. They occur as batholiths in plate convergence sites and in continental extension areas.

[6]   Granitoid magmas can have a variety of source rocks and origins: only the peraluminous leucogranites, such as the continent-continent collisional granites are likely crustal melts. Tonalitic and other plagioclase-rich granitoids may form by higher pressure melting of amphibolites. Mixing of magmas, crustal assimilation, magma differentiation and possibly other mechanisms contribute significantly to the origin of peraluminous granitoids.

BIBLIOGRAPHY

Ague, J.J. and Brimhall, G.H. (1988) Regional variations in bulk chemictry, mineralogy, snd the compositions of mafic and accessory minerals in the batholiths of California, *Geol. Soc. Am. Bull.* **100**, 891–911.

Barker, D.S. (1983) *Igneous Rocks.* Prentice-Hall (Englewood Cliffs, N.J.).

Bateman, P.C. and Chappell, B.W. (1979) Crystallization, fractionation, and solidification of the Tuolumne Intrusive series, Yosemite national park, California, *Geol. Soc. Am. Bull.* **90**, 465–482.

Bose, K. and Ganguly, J. (1995) Experimental and theoretical studies of the stabilities of talc, antigorite and phase A at high pressures with applications to subduction processes, *Earth Planet. Sci. Lett.* **136**, 109–121.

Buddington, A.F. (1959) Granite emplacement with special reference to North America, *Geol. Soc. Am. Bull.* **70**, 671–748.

Chappell, B.W. and White, A.J.R. (1974) Two contrasting magma types, *Pacific Geol.* **8**, 173–174.

Clarke, D.B. (1992) *Granitoid Rocks.* Chapman and Hall (New York, N.Y.).

DePaolo, D.J. (1981) A neodymium and strontium isotopic study of the Mesozoic calc-alkaline granitic batholiths of the Sierra Nevada and Peninsular Ranges, California, *J. Geophys. Res.* **86,** 0470–10488.

Ducea, M. and Saleeby, J.B. (1998) A case for delamination of the deep batholithic crust beneath the Sierra Nevada, California, *Internat. Geol. Rev.* **40**, 78–93.

Gerlach, D.C. et al. (1987) Recent volcanism in the Puyehue—Cordon Caulle region, southern Andes, Chile (36°S), *Contrib. Mineral. Petrol.* **29**, 333–382.

Gromet, L.P. and Silver, L.T. (1987) REE variations across the Peninsular Ranges Batholith: implications for batholithic petrogenesis and crustal growth in magmatic arcs, *J. Petrol.* **28**, 75–125.

Grove, T.L. and Baker, M.B. (1984) Phase equilibrium controls on the tholeiitic versus calc-alkaline differentiation trends, *J. Geophys. Res.* **89**, 3253–3274.

Gust, D. A. and Perfit, M.R. (1987) Phase relations of a high-Mg basalt from the Aleutian island arc: implications for primary irland arc basalts and high-Al basalts. *Contrib. Mineral. Petrol.* **97**, 7–18.

Hess, P.C. (1989) *Origins of Igneous Rocks.* Harvard University Press (Cambridge, MA.).

Hickey, R.L. et al. (1986) Multiple sources for basaltic arc rocks from the Southern Volcanic Zone of the Andes: trace element and isotopic evidence for contributions from subducted oceanic crust, mantle and continental crust, *J. Geophys. Res.* **91**, 5963–5983.

Hildreth, W. (1981) Gradients in silicic magma chambers: implications for lithospheric magmatism, *J. Geophys. Res.* **86**, 10153–10192.

Hildreth, W. and Moorbath, S. (1988) Crustal contributions to arc magmatism in the Andes of Central Chile, *Contrib. Mineral. Petrol.* **98**, 455–489.

Hyndman, D.W. (1985) *Petrology of Igneous and Metamorphic Rocks* (2nd ed.). McGraw-Hill (New York, N.Y.).

Johannes , W. (1989) Melting of plagioclase-quartz assemblages at 2 kbar water pressure, *Contrib. Mineral. Petrol.* **103**, 270–276.

Johannes, W. and Holtz, F. (1996) *Petrogenesis and Experimental Petrology of Granitic Rocks.* Springer-Verlag (New York, N.Y.).

Kushiro, I. (1969) The system forsterite-diopside-silica with and without water at high pressures, *Am. J. Sci.* **267A**, 269–294.

Le Fort, P.E. (1981) Manaslu granite: a collision signature of the Himalaya, *J. Geophys. Res.* **86**, 10545–10568.

McBirney, A.R. (1993) *Igneous Petrology* (2nd ed.). Jones and Bartlett (Boston, MA.).

Patiño-Douce, A.E. (1996) Effects of pressure and H_2O contents on the compositions of primary crustal melts. *Trans. R. Soc. Edinburgh, Earth Sciences,* **87**, 11–21.

Patiño-Douce, A.E. (1999) What do experiments tell us about the relative contributions of crust and mantle to the origin of granitic magmas? In A. Castro et al. (eds), Understanding granites: integrating new and classical techniques. *Geol. Soc. London Special Publ.* (in press).

Pitcher, W.S. (1993) *The Nature and Origin of Granite.* Blackie (London, U.K.).

Reid, J.B. and Hamilton, M.A. (1987) Origin of Sierra Nevada granite: evidence from small scale composite dikes, *Contrib. Mineral. Petrol.* **96**, 441–454.

Rapp, R.P. and Watson, E.B. (1995) Dehydration melting of metabasalt at 8–32 kbar: implications for continental growth and crust-mantle recycling, *J. Petrol.* **36**, 891–931.

Rushmer, T. (1991) Partial melting of two amphibolites: contrasting experimental results under fluid-absent conditions, *Contrib. Mineral. Petrol.* **107**, 41–59.

Silver, L.T. and Chappell, B.W. (1988) The Peninsular Ranges batholith: an insight into the evolution of the Cordilleran batholiths of southwestern North America. *Trans. R. Soc. Edinburgh* **79**, 105–121.

Sisson, T.W. and Grove, T.L. (1993) Experimental investigations of th role of H_2O in calc-alkaline differentiation and subduction zone magmatism, *Contrib. Mineral. Petrol.* **113**, 143–166.

Stern, R.J. (1998) A subduction primer for instructors of introductory-geology courses and authors of introductory-geology textbooks, *J. Geosci. Edu.* **46**, 221–228.

Tatsumi, Y. et al. (1983) Generation of arc basalt magmasand thermal structure of the mantle wedge in subduction zones, *J. Geophys. Res.* **88**, 5815–5825.

Tuttle, O.F. and Bowen, N.L. (1958) Origin of granite in the light of experimental studies in the system $NaAlSi_3O8$– $KAlSi_3O_8$–SiO_2–H_2O, *Geol. Soc. Am. Mem.* **74**, 153.

Wilson, M. (1989) *Igneous Petrogenesis.* Unwin Hyman (Boston, MA.).

CHAPTER 13

Sediments

The following topics are covered in this chapter:

Origin of sediments
 Mechanical, chemical, and biogenic processes of derivation
Sediment transport processes
 Mechanical, dissolved, colloidal, and biogenic
Sediment depositional processes
 Mechanical, chemical, and biogenic
Postdepositional changes
 Cementation and diagenesis

ABOUT THIS CHAPTER

This chapter is about sediments from which sedimentary rocks are derived. Sediments and sedimentary rocks together cover about 66% of the exposed crust. Sediments include all loose (unlithified) solid materials that accumulate at or near the surface of the earth. These particles are either derived from preexisting rocks from weathering processes, direct chemical precipitation (e.g., from seawater), or organisms' skeletal remains. Transport processes are diverse and depend on the environment of origin and various factors on the earth's surface such as climate and tectonic forces. Sediments can be deposited in a variety of subaerial and subaqueous environments (Figure 13.1). Following deposition, sediments are ultimately converted to sedimentary rocks by a transformation process known as *diagenesis*, with lithification (i.e., transformation into rock) as one of the most obvious changes. (It is generally believed that temperature during diagenesis does not exceed a temperature of $\sim 200°C$, beyond which it is a metamorphic regime.)

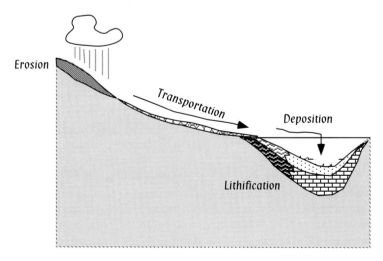

FIGURE 13.1 *A schematic representation of sediment derivation, transportation, and deposition.* Following deposition, the sediment is lithified into a sedimentary rock. (From McLane, 1995, *Sedimentology*, Oxford University Press)

SEDIMENT PRODUCTION: WEATHERING PROCESSES

Prolonged exposure of rocks to the atmosphere causes them to gradually break down into smaller pieces either by mechanical action (mechanical weathering) and/or chemical reactions (chemical weathering). Both are irreversible processes. Chemical weathering produces new minerals such as clays. The rate of weathering processes varies depending on the nature of the source rock and the environment in which it is taking place (discussed further later). The term *erosion* is generally used to include mobilization processes that move the particles generated by weathering from the source area to a different site. Transportation of sediments by glaciers, rivers, winds, and so on are all examples of different types of erosional processes.

How important is weathering? Weathering breaks down rocks and produces the residual particles (called *regolith*) that may accumulate as sediments that eventually become sedimentary rocks. Sediments (soil included) play the most important role for the development of a vegetative cover on the earth, and thus in the production of oxygen in the atmosphere and in the carbon and nitrogen cycles, all of which are vital for sustaining life. Weathering is a slow process that, together with erosional processes, has the power to bring down the highest mountains in the world to nearly sea level. In the long range, by affecting mountains in such a way, weathering indirectly acts as the earth's thermostat because mountains are an important factor in controlling climatic patterns on Earth.

HOW WEATHERING TAKES PLACE

Chemical weathering occurs because many rocks are unstable at the surface—a situation accelerated by rain water, which, particularly with CO_2 of the atmosphere and from plants, launches a fierce attack on rocks to destabilize them further and eventually disintegrate them completely. Mechanical weathering (also called *physical weathering*) processes reduce the rock into smaller fragments without altering its chemical composition. Biological processes can also lead to weathering of rocks, but their effects are less significant as compared with mechanical and chemical weathering processes. Mechanical and chemical weathering generally operate together because the former process increases the surface area in the rocks through cracks where reactive solutions can then penetrate and chemically alter the rocks more efficiently.

Examples of Mechanical Weathering

Mechanical weathering can occur in a number of ways. For example, in colder climates, water may seep into cracks in rocks and freeze. Ice has a 9% greater volume than water, and therefore when the water inside a crack turns into ice, it expands and breaks the rock apart. This process is called *frost wedging*. Many freezing and thawing cycles like that can turn a large boulder of rock into a pile of rubble. Frost wedging is also responsible for damaging freeways and water pipes of houses in cold regions.

Another example of mechanical weathering is *exfoliation*, which occurs when a plutonic rock is exposed at the surface via erosion and removal of overlying rocks and develops parallel sheets of curved fractures along which the rock eventually peels off like an onion skin. When the pressure on the rock from the overburden is removed, the rock naturally attempts to expand in volume, which then results in the development of concentric fractures. This type of weathering is also called *spheroidal weathering* because it generally results in spheroidal boulders.

Examples of Chemical Weathering

Chemical weathering processes are highly variable; they include oxidation, hydrolysis, dissolution, and other types of chemical reactions. Oxidation is a type of chemical reaction in which oxygen combines with a metal and makes a metallic oxide. Rusting of iron is an example of oxidation in which iron reacts with oxygen and forms hematite:

$$4Fe + 3O_2 = 2Fe_2O_3$$

Hydrolysis requires the presence of water. It involves a chemical reaction in which H^+ and $(OH)^-$ ions of water molecules replace various cations in a mineral, thereby transforming it into a different mineral. Hydrolysis is responsible for converting feldspars, a dominant mineral in the earth's crust, into clay minerals. One such reaction is shown in the following example, in which the white clay, called *kaolinite* (used in making pottery and china), is formed at the expense of K feldspars by hydrolysis:

$$2KAlSi_3O_8 + 2H^+ + 9H_2O = Al_2SiO_5(OH)_4 + 4H_4SiO_4 + 2K^{2+}$$

K feldspar   kaolinite (clay)   silicic acid   K ion
(in solution)

Thus, a granite composed dominantly of quartz and feldspar will eventually decay into loose residues as feldspars are all altered into clays and other dissolved ions are removed by meteoric water.

Formation of acids by reaction between rain water and atmospheric carbon dioxide, sulfur, or nitrogen gases are particularly effective in accelerating weathering. The following reaction shows how carbonic acid (in acid rain) is formed:

$$CO_2 + H_2O = H_2CO_3 = H^+ \cdot CO_3H^-$$

An interesting example of weathering by carbonic acid may be observed by dropping a fingernail into a bottle of carbonated drink. Watch how long it takes for the fingernail to get totally dissolved.

Factors That Affect the Rates of Weathering Processes

1. *Nature of rocks and minerals.* Rock composition plays an important part in weathering because the mineralogy of some rocks are more amenable than others to certain types of breakdown mechanisms. Goldich presented a flow chart that indicates how susceptible common rock-forming silicate minerals of the crust are to chemical weathering (Figure 13.2). The sequence in which minerals are arranged is essentially a reversed order of Bowen's reaction series, with quartz being the most stable mineral (i.e., least susceptible to weathering) and olivine and calcic plagioclase being the least stable (i.e., most susceptible to weathering). This concept is generally correct, but one should bear in mind that the stability of a mineral depends on dissolution rates that vary from one environment to the other. For example, although quartz may be extremely stable in most sedimentary environments, it may rapidly dissolve in strongly alkaline waters because silica is unstable at pH $\geq$ 9.

   During chemical weathering, feldspar alters to clay minerals, commonly smectite or illite, silica is released in solution, and the alkali ions are carried away in dissolution. A granite, for instance, may get completely altered to sediments rich in clay and quartz. Similarly, olivine alters to serpentine and iddingsite. An ultramafic rock such as a dunite, composed largely of olivine, may become serpentinite. Thus, the source rock exerts the primary control on the composition of the sediments.

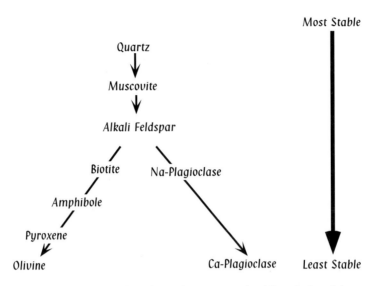

FIGURE 13.2   *Goldich's flow chart indicating general stability of minerals in sedimentary environments:* A mineral lower in the series is less stable than a mineral higher in the series. Quartz is the most stable mineral, whereas olivine is least stable.

2. *Climate.* A rock usually weathers much faster in wet tropical climates than in dry arid areas. This is because the chemical and biological reactions associated with weathering are faster at higher temperature, and the bipolar structure of a water molecule makes it an effective agent for chemical weathering (see Box 13.1). In cold or arid areas, physical weathering may be active, but chemical weathering is minimal. For instance, centuries-old tombstones made of the same types of rock in cold regions may be cracked, but the inscriptions are generally readable. In contrast, in humid areas, tombstones of similar ages would be smoothed and worn down by chemical weathering to the extent that the inscriptions would be illegible or totally erased.

3. *Structural weaknesses.* Rocks with well-developed joints or fractures are more amenable to weathering as they offer greater exposed surface area to the weathering agent. Thus, slate, a fine-grained, low-grade, metamorphic rock with extremely well-developed rock cleavage, will break off along those planes of weakness during weathering. Under the same conditions, a massive granite with little or no fractures may suffer minimal weathering compared with the slate.

4. *Topography.* The topography of the source rock area (generally termed *provenance*) dictates to a great extent the effectiveness of weathering and transportation processes. For example, in extremely rough terrains, where the topography is composed of steep hills and deep valleys, mass movement processes such as landslides or rockfalls are very effective in uncovering subsurface rocks and exposing them to weathering agents.

5. *Soil cover.* Regolith and soil cover may create an environment that is rich in reactive chemical solutions. In addition, the vegetative cover and associated bacteria and burrowing organisms further sustain an environment that is highly conducive to weathering. Thus, a rock with a soil cover may weather faster than a bare rock. In regions of steep topography, soil cannot remain stable on the slopes and is easily removed by erosional processes, and thus chemical weathering is slower.

13.1 POLAR NATURE OF WATER MOLECULE

Water is the most effective agent of weathering. Its global effectiveness may be attributed to three fundamental factors: (a) It is the most abundant naturally occurring compound. Therefore, rocks all over the globe are easily accessible to water. (b) It occurs in all three physical states—solid, liquid, and vapor. Thus, from hot to extremely cold areas on Earth, water makes its way through the rocks in one state or another. In cold regions, frost becomes important as a weathering agent. In very hot regions, water and steam are important. (c) Water is a highly reactive liquid. This is because of the polar nature of its molecule (Figure Bx 13.1). In a water molecule, two hydrogen anions are attached to an oxygen cation by sharing their single electrons with the electrons in the outer shell of oxygen anion (known as a hydrogen bond). The two hydrogen ions occur on the same side of the oxygen ion and make a 104° angle between them. Because of this uneven distribution of hydrogen ions, there is greater concentraion of protons (+) at the pole around which the hydrogen ions occur; at the other pole of the water molecule, a weak negative charge develops. Thus, the water molecule is polar with a weak positive and a negative charge at the opposite poles. Such polar structure allows water to break down ionic solids and dis-

solve them. Also during solidification, water molecules arrange themselves into lattices, which force them to stay farther apart than they would in the liquid state—that is why water expands in volume when it turns into ice. Frost wedging, as we saw earlier, can be a powerful weathering process in cold climates.

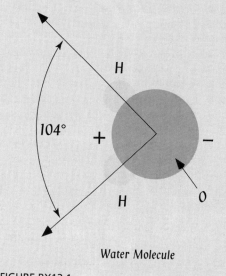

Water Molecule

FIGURE BX13.1

6. *Time.* The longer a rock is exposed to the atmosphere, the more altered it becomes. A case in point is the fact that the highest peaks of older mountains (e.g., the Appalachian mountains) are much lower than the peaks of younger mountains (e.g., the Rockies and the Himalayas). Using sophisticated dating and experimental techniques, geologists have successfully measured weathering and erosion rates of mountains of all ages. Similarly, by measuring the rate at which rivers carry loads of sediment from mountain peaks to lower grounds, scientists have been able to provide some estimate of the rates at which these mountains are eroding. On average, it seems that beveled-down continents are eroding at rates of about 0.03 mm per year. Mountains may erode at a much faster rate (0.2 mm per year).

Products of Weathering

Weathering produces essentially two kinds of products: dissolved ions and solid particles. Dissolved ions include Na^+, K^+, Ca^{2+}, and Cl^-. Solid particles include fragments of the bedrock, stable minerals like quartz, or new minerals produced by weathering (e.g., clay minerals). The general term *regolith* is used to describe the sediment that accumulates on rocks as a by-product of weathering processes.

Soil is simply a type of regolith that has been processed by water and near-surface chemical biological reactions. Soil forms in situ at the highest rate in wet humid regions under topographic conditions where erosion may be minimal. The names *residual soil* versus *transported soil* are given to those that remain *in situ* versus those that have been transported from elsewhere. Under most suitable wet tropical climatic conditions, thick accumulations of residual soil can take about 400 to 500 years to form.

The color, composition, and texture of a soil cover may change from the surface down to its contact with the bedrock. Vertical sections (called *soil profile*; Figure 13.3) through soil covers are often best exposed in road cuts. In humid areas

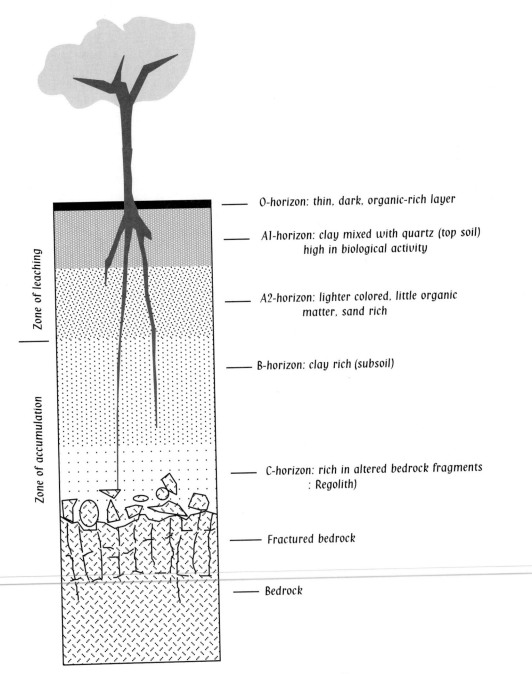

O-horizon: thin, dark, organic-rich layer

A1-horizon: clay mixed with quartz (top soil) high in biological activity

A2-horizon: lighter colored, little organic matter, sand rich

B-horizon: clay rich (subsoil)

C-horizon: rich in altered bedrock fragments : Regolith)

Fractured bedrock

Bedrock

Zone of leaching

Zone of accumulation

FIGURE 13.3 *A well-developed soil profile.*

13.2 PALEOSOLS AND PAST ATMOSPHERES

Soils dating back 3.5 billion years are sometimes preserved, and such old soils are called *paleosol*. Because climate and vegetation are closely tied to the chemical weathering processes that produce soils, paleosols are often extremely useful as indicators of past climates. For example, the discovery that 2 billion-year-old paleosols on granitic rocks were oxidized but paleosols of same age on basalt were not led scientists to conclude that, although oxygen was present in the atmosphere, it was not enough to oxidize the iron-rich soil derived from basalt. By 850 million years ago, red (oxidized) paleosols became abundant, indicating the abundance of oxygen in the atmosphere. Scientists have used more sophisticated techniques involving the use of C- and O-isotope composition of paleosols to determine the past climatic balances of oxygen and carbon dioxide. Such studies suggest that the earth underwent a global warming due to a high abundance of carbon dioxide in the atmosphere from 570 to 410

million years ago. This *greenhouse effect* gradually gave way to extreme cooling and ice ages some 270 million years ago. Figure Bx 13.2 shows a summary diagram (from Rye and Holland, 1998) based on a comparative study of many paleosols throughout the world. One important feature in this figure is the big jump in partial pressure of oxygen (pO_2) that appears to have occurred ~2.2 million years ago.

Paleosols are also useful indicators of past locations of continents and can be used in determination of past atmospheric flow patterns. For example, we know that laterite typically forms under relatively uniform maritime conditions on the windward sides of continents under warm, humid conditions. Based on the occurrence of 80 million-year-old laterites in various continents and reconstructing geographic positions of such continents, some scientists have been able to reconstruct global wind flow patterns some 80 million years ago.

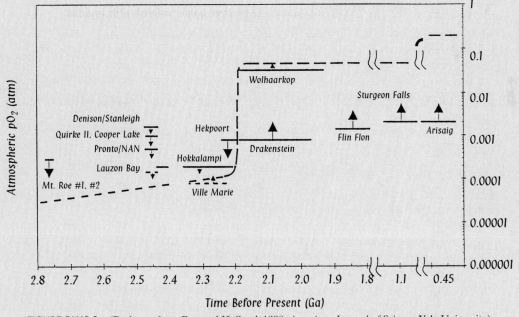

FIGURE BX13.2   (Redrawn from Rye and Holland, 1998, *American Journal of Science*, Yale University)

with thick soil cover, five distinctive soil zones, called *soil horizons*, can be identified. These horizons develop through time depending on local climatic and physiographic conditions. At the top is the O horizon, which is essentially the organic-rich top soil where bacteria, algae, insects, and dead leaves are abundant. Below the O horizon is the A horizon, which is composed of inorganic residues and humus (a black, carbon-rich organic substance). The A horizon downgrades into a lighter colored B horizon. Its lighter color owes to the general absence of dark organic matter. Water percolating downward through this zone also carries away ions and finer particles into the lower zones. B, A, and O horizons together are called the *zone of leaching* because groundwater leaches out ions and inorganic materials from these zones and transports them to lower horizons, where clay particles removed from the overlying horizons are deposited. This horizon is also enriched in iron and aluminum. B horizon is underlain by the C horizon, which is composed of partially altered and fragmented bedrock. Part of the B horizon and all of the C horizon represent the zone of accumulation because all the leached materials from the overlying horizons accumulate at these levels.

Transportation and Deposition

The modes in which transportation takes place may be mechanical (e.g., by a moving fluid), chemical (i.e., as dissolved ions), or biochemical.

Mechanical transportation includes processes by which loosened solid particles (called *detritus* or *clastic particles* or *clastics*) are physically picked up and transported by water, wind, or glaciers. It is by far the most important process in terrestrial environments, generally accounting for more than 90% of the sediments transported by major river systems from continents to continental shelves. Gravity-induced mass wasting processes, such as landslides, rockfalls, submarine landslides (referred to as *olistostromal flows*), and so on (Figure 13.4), are also important mechanical transportation processes.

Transportation often induces two kinds of textural changes in the nature of the sediments depending on the dynamics of the transport mechanism, travel distance, nature of the source rocks and minerals, and initial grain size. The first change involves grain size reduction and rounding of grains: Continuous knocking and grinding action due to collision of grains results in diminution of grain size and rounding of sharp edges and corners of detrital fragments. The second type of change is sorting, whereby the kinetic conditions of the transporting medium sort out grains according to size, shape, and composition (density), such that when these sediments are deposited they may be distributed according to these characteristics.

Figure 13.5 shows the relationship between the nature (i.e., texture and compositon) of transported detrital sediments and energy of the transporting medium. The term *maturity* is used to describe the sediments in terms of their sorting characteristics, roundness of grains, and percentage of clays present in the matrix. Long transportation of detritus and/or a dynamic transporting medium causes clays to be quickly removed from the larger framework grains, and the framework grains become well rounded and sorted through constant knocking and grinding processes. As a result, the sediments that have been transported in a dynamic environment (such as the beaches) are well sorted with rounded grains with very little intergranular clay. Terms such as *immature* and *supermature* describe the degree of maturity of the sediment, where a supermature detritus is a well-sorted assemblage of well-rounded detrital

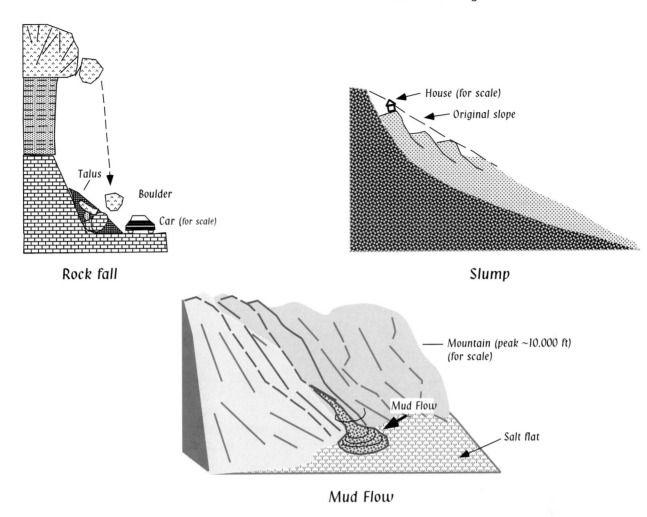

FIGURE 13.4   *Examples of mass wasting processes.*

grains (i.e., framework) with very little clay in the matrix, and an immature detritus is composed of poorly sorted, angular framework grains and an abundance of clays.

Because quartz is generally the most stable product of all weathering processes, a sand deposit composed of well-rounded quartz grains of nearly similar sizes (i.e., well sorted) points to its long transportation history during which the grains have gone through enough attrition and abrasion. In general, sand bars at the mouths of rivers carrying sediments over 1,000s of kilometers are typically characterized by such well-rounded, well-sorted quartz sands. Thus, the physical nature of a clastic sedimentary deposit can provide clues to its transportation history.

In chemical transportation, the transporting medium carries the sediment as a dissolved ion. A familiar example of this is the transportation of Na^+ and Cl^- (i.e., ions that form common salt) in seawater. Chemical transportation and depositional processes are affected by two factors: alkalinity (pH) and oxidation/reduction (or Eh). In nature, pH (see Box 13.3) generally varies between 4 and 9, with 7 = neutral, <7 = acidic, and >7 alkaline. Ocean water normally has a pH between 8.1 and 8.3,

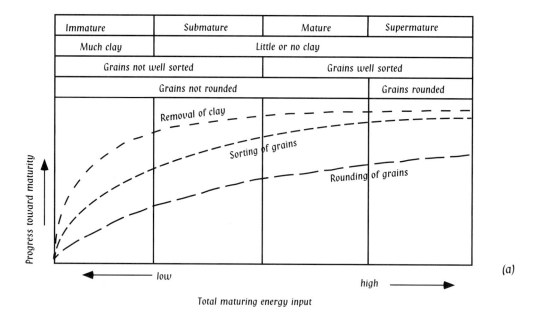

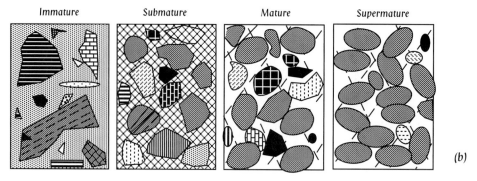

FIGURE 13.5 *The upper panel shows the relationship among mineralogy, sorting, and other textural features of sediments and the kinetic energy of the transportation process.* The lower panel is a schematic graphic representation of the nature of textures in sediments with varying degrees of maturity.

whereas water in desert playas (which are small lakes) has a pH ≥ 9. Rivers and streams in humid areas normally have a pH between 5 and 6.5.

Eh (see Box 13.3) is a measure of the oxidizing or reducing ability of a solution. Positive values of Eh mean oxidizing environment and negative values mean reducing conditions. Eh–pH variations in the natural system are shown in Figure 13.6a. When this diagram is used with the Eh–pH fence diagram (Figure 13.6b) of Krumbein and Garrels (1956), it becomes clear whether a specific type of material is dissolved as an ionic solution or precipitated as a chemical precipitate. Let us consider the following example to see how this works. Near the bottom of coastal estuaries or lakes, the water is strongly reduced with Eh < -0.2 and pH between ~ 7 and 8. The fence diagram indicates that in such an environment pyrite and organic matter are stable and are deposited. In fact, mud that accumulates in such environments is black (organic rich) and contains pyrite crystals.

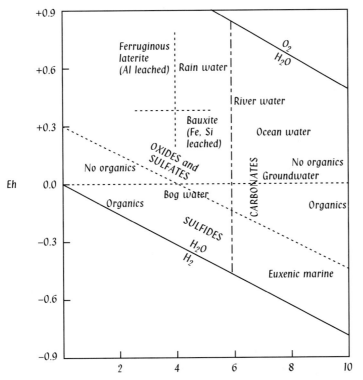

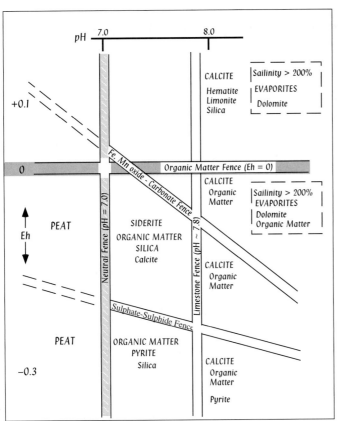

FIGURE 13.6   (a) Eh–pH conditions of natural environments. (b) Eh–pH fence diagram showing stabilities of minerals in various sedimentary environments.

13.3 EH–pH ELABORATED

pH:

pH can be explicitly understood in terms of the reaction:

$$OH^- + H^+ = H_2O$$

In a neutral solution, by definition, the concentrations of both H^+ and OH^- are equal to 10^{-7} moles/liter (M). The ionic product $[H^+] \cdot [OH^-] = 10^{-14}$ is constant at room temperature (25°C) for all aqueous solutions, be it neutral, acidic, or alkaline. A 1N solution (N = *normal*, meaning equivalents of acid or base per liter of solution) of a strong acid has $1M$ H^+, a neutral solution has 10^{-7} M of H^+, and a strongly alkaline 1N solution has 10^{-14} H^+. Negative logarithms (base 10) of the neutral solution is 7, whereas that of the alkaline solution is 14, and so on. Therefore, the formal definition of pH is that it is the negative logarithm of the hydrogen ion concentration in a solution.

Eh:

Oxidation number or Eh represents the tendency of an element to oxidize or reduce in an aqueous environment and is expressed in terms of volts of electricity. The reduction of H^+ to H_2 is assigned to have a standard Eh value of 0 volts. Eh increases in oxidation and decreases in reduction. For example, when a copper plate and a zinc plate are submerged in a copper sulfate solution and the plates are connected by a wire, the zinc plate gradually dissolves and new copper is precipitated onto the copper plate. This coupled process may be symbolized as:

$$Zn \longrightarrow Zn^{2+} + 2e^- \text{ and } Cu^{2+} + 2e^- \longrightarrow Cu,$$
$$\text{(where } e^- \text{ refers to 1 electron)}$$

Eh is measured as the emf or potential difference in volts for these reactions and can be measured from the wire.

Deposition of sediments occurs when the conditions are appropriate such that dissolved ions and transported particles can no longer be transported. Like transportation processes, deposition processes may be of three kinds: mechanical, chemical/inorganic, and biological (organic). Mechanical deposition of detritus is an important process. It generally happens when the velocity of the transporting medium (river, wind, glaciers) drops and its kinetic energy is no longer able to induce movement of the particles. Globally, the most voluminous sites of modern clastic sediment accumulation are at the mouths of about five rivers, and the thickest sediment accumulation is at the mouths of the Ganga and Brahmaputra rivers (India) in the Bay of Bengal (Figure 13.7). These rivers have been transporting vast quantities of sediments derived from the Himalayas.

In chemical or inorganic deposition, dissolved ions are no longer able to stay in solution due to supersaturation caused by evaporation, temperature, or some other changes (say Eh and/or pH) and are deposited as some mineral phase. For example, evaporation of water in a playa (i.e., desert pond or lake) results in supersaturation and deposition of salt (evaporite) on the playa floor, which is an inorganic deposition process.

Biochemical or organic depositional processes include those in which certain organisms (e.g., foraminifera, corals, molluscs, etc. in the marine environment) secrete skeletons from dissolved ions; when they die, their skeletal debris may accumulate on the bottom of oceans or lakes. Such deposits may be called *biogenic accumulates*.

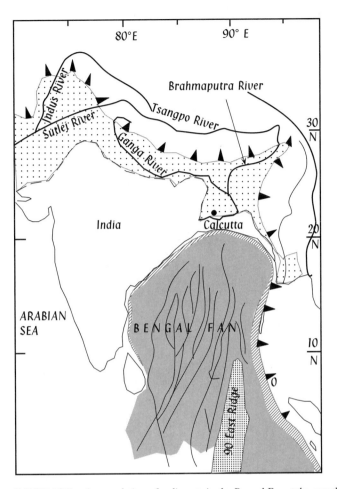

FIGURE 13.7   *Accumulation of sediments in the Bengal Fan at the mouth of the Ganga and Brahmaputra rivers in the Bay of Bengal.* (Redrawn with permission from Miall, 1990)

NATURE OF ACCUMULATED SEDIMENTS

Composition

Sediments collecting at a depositional site may include clastic particles, inorganically or biogenically precipitated minerals, organically produced materials, and pyroclastic materials from volcanic eruptions. Clastic particles are easy to recognize as they include fragments of preexisting rocks or mineral grains. Pyroclastic materials include volcanic ashes, blocks, or bombs that accumulate at a depositional site. However, when such materials are reworked (i.e., transported from the original site of deposition), they become part of the clastic category. An example of chemically (inorganically) precipitated minerals is evaporite minerals (such as gyspsum, halite, etc.) that form via evaporation of, say, waters in a desert playa.

Clastic or detrital sediments are classified according to their size (Table 13.1). Although the term *sand* is commonly (but erroneously) used to mean quartz grains, note that in geology the name *sand* is based on a specific size. Similarly, the use of the term *clay* also refers to a particular size of sediments in engineering and causes a problem in geology because clays are minerals that are types of sheet silicates (as we saw in Chapter 5).

TABLE 13.1 *Classification of sediments based on size.*

Particle Diameter (mm)	Name
>256	Boulder
256–64	Cobble
64–2	Pebble
2–1/16	Sand
1/16–1/256	Silt
1/256–1/4096	Clay

We noted earlier that the texture and composition of a deposit of detrital sediments offer clues to its transportation and deposition histories. For example, rapid transportation followed by quick deposition would result in a body of sediment that contains a poorly sorted mixture of all sorts of grain sizes and shapes. Sedimentary deposits of such character are in fact found in glacial environments and landslide deposits. At the other extreme, long transportation followed by deposition of detritus by a river may result in a sedimentary deposit composed of mature sediments.

Sedimentary Structures

Internal Structures

Sedimentary deposits may develop structures that are indicative of the dynamics of transportation and deposition processes, as well as biological and physical (mechanical) chemical conditions within the depositional site. *Layering* or *stratification* is a first-order depositional structure common to almost all sediments. *Bedding* refers to a type of layering in which individual layers are more than 1 cm thick; if each layer is less than 1 cm thick, it is called *lamination*. Using a $\sqrt{10}$ scale (i.e., each smaller unit is $\sqrt{10}$ thinner than the next bigger unit), Ingram (1954) further classified beds and laminae into several types, as shown in Table 13.2.

Regardless of the thickness, layers are discrete units separated by individual planes. Bedding planes (i.e., planes between adjacent beds) may represent short changes in sedimentation rate, composition, or texture in an otherwise continuous sequence of sediments that produce the layers. Practically each unit is produced by a single episode of sedimentation. As Table 13.2 shows, the thickness of beds varies greatly and depends on sedimentation rate, duration of the depositional episode, and subsequent compaction. Rapid sedimentation produces thick beds, and compaction makes them thinner.

Bedding (or lamination) may be of several types, of which the fundamental ones are parallel bedding (lamination), cross-bedding (lamination), and graded bedding (lamination; Figure 13.8). In parallel bedding, bedding planes are parallel to each other. Parallel beds may be deposited from a fluid in a lower flow regime or in the absence of currents. The term *varve* is used to describe a type of deposit with contrasting laminated couplets. For example, in glacial varves, light layers are silt, which

TABLE 13.2 *Ingram's (1954) $\sqrt{10}$ scale of nomenclature of beds/laminae.*

Size of Individual Unit	Name
≥ 100 cm	Very thick bed
31.6–100 cm	Thick bed
10–31.6 cm	Medium bed
3.2–10 cm	Thin bed
1–3.2 cm	Very thin bed
0.3–1 cm	Thick lamina
<0.3 cm	Thin lamina

are derived from the melting of the glacier during summer months; the dark layers are organic-rich winter layers. Thus, the relative thickness of the two laminae in such couplets accumulated in a glacial lake is an indicator of the seasonal cycle: Long summer and short winter would result in thin, dark laminae and thick, light laminae and vice versa.

In cross-bedding (or current bedding), layers within an individual bed are inclined in the down-current direction. Cross-bedding is a characteristic feature of small grains of loose sediments deposited by a moving fluid. The slope and thickness of the inclined layers are controlled by the grain size of the sediments and the velocity of the current: In a fast-flowing regime, coarse sediments are deposited at a steeper angle.

Graded bedding is a general term used for any layer in which the grain size decreases progressively from the bottom to the top of a single unit. A graded bed (or laminae) is a subaqueous deposit formed when a moving mix of detrital grains of various sizes carried by a density or turbidity current (these terms are explained later)

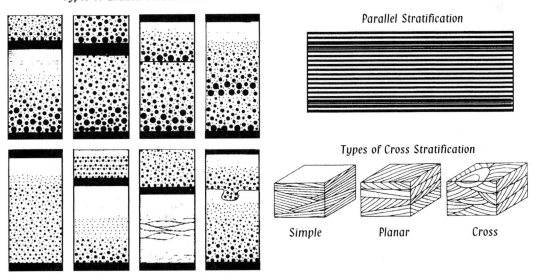

FIGURE 13.8   *Stratification.* (Redrawn with permission from Miall, 1990, 2nd ed.)

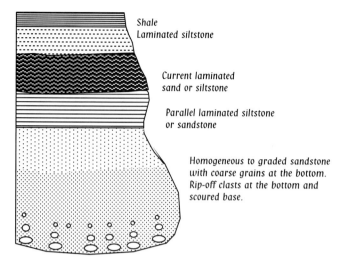

Shale
Laminated siltstone

Current laminated
sand or siltstone

Parallel laminated siltstone
or sandstone

Homogeneous to graded sandstone
with coarse grains at the bottom.
Rip-off clasts at the bottom and
scoured base.

FIGURE 13.9 *Bouma sequence.*

loses momentum and the particles settle differentially by mass (coarse grains first) on the bottom obeying Stoke's law (see Chapter 9). Graded beds are often repeated in a vertical sequence, with each representing a different turbiditic event (see the next section on turbidity currents).

As noted earlier, continental shelf-slope region at the mouth of a major river system is a prime depositional site for the detritus transported by such a system. Such sites undergo significant subsidence as more and more sediments accumulate and apply load pressure on the lithosphere. Sometimes an excessive supply of sediments (which may be due to heavy rainfall or excessive melting of the mountain glaciers that feed the river) or an earthquake may trigger subaqueous landslides (i.e., the slope fails). In such a case, a fast current, called a *turbidity current*, driven by the water + detritus mixture moves down the continental slope. When such a current comes to rest, the sediments deposited from it may form a five-layer sequence called the *Bouma sequence* (Figure 13.9).

At the bottom of the Bouma sequence is a graded bed that has the coarsest particles at its base; it may also contain clasts ripped off the bottom. Within this unit, grains become finer upward and grade into a subunit of parallel laminated silt or sand. Above this unit is a cross-laminated silt or sand unit, which is successively followed upward by parallel laminated silt and clay. The Bouma sequence is typically found in submarine fans, which are enormous, fan-shaped submarine deposits off continental shelves and slopes. A good example is the Bengal fan, which occurs in the Bay of Bengal off the southern coast of eastern Indian and Bangladesh (Figure 13.7).

In carbonate sediments, the following internal structures may be found: stromatolites and reefs. Stromatolites are much larger, internally laminated structures that can have a variety of forms—domal, conical, or irregular mounds (Figure 13.10). These features developed due to trapping of sediments by algae growing as individual laminae. Reefs are domal to elongate structures built in situ by corals or other organisms. Bioturbation is a process caused by organisms burrowing into an unconsolidated sediment. Burrowing organisms, like crabs, shrimps, and so on often dig tubular structures (called *burrows*; Figure 13.11) into the soft sediments, which may be so extensive that they may completely disrupt the original depositional layering. The term *bioturbated structure* is used to describe such disruption of primary depositional structure by burrowing.

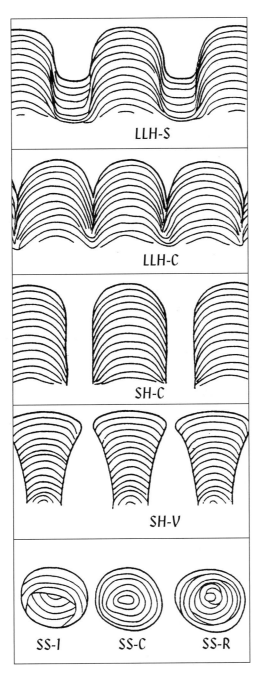

FIGURE 13.10   *Stromatolites.* (Reprinted with permission from McLane, 1995, *Sedimentology,* Oxford University Press)

Concretions (Figure 13.12) are masses of mineral matter with spherical to sub-spherical or irregular shapes that are commonly characterized by a composition different from the surrounding rock. Fossils of plants and organic remains often make up the nucleus of concretions. Colored rings called *liesegang rings* develop different bands of colors due to oxidation and reduction, and ion diffusion through a porous rock.

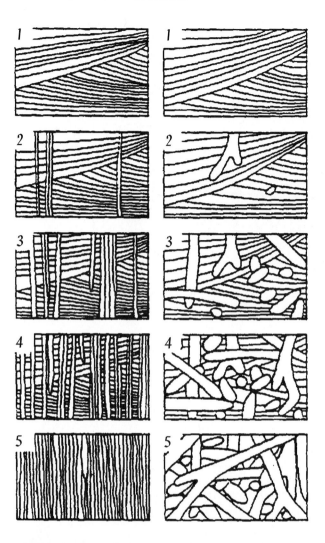

FIGURE 13.11 *Burrowing structures in soft sediments.* (Reprinted with permission from McLane, 1995, *Sedimentology*, Oxford University Press)

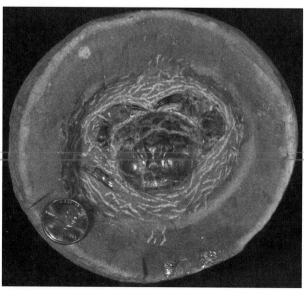

FIGURE 13.12 *A concretion with a fossil crab in its core.* (Courtesy of Gautam Sen)

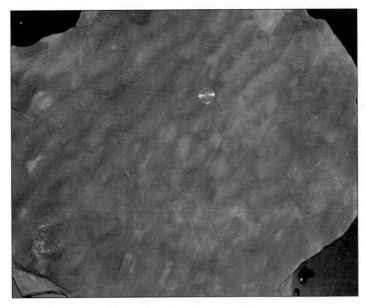

FIGURE 13.13   *Ripple marks.* (Courtesy of Gautam Sen)

Surface Structures

Ripple marks are centimeter-scale undulations on the surface of a sedimentary deposit (Figure 13.13). They may be formed by oscillating water waves and are called *oscillation ripple marks,* or by unidirectional currents and are called *current ripple marks*. In highly deformed, folded sedimentary rocks, the pointed crests of oscillation ripple marks on bedding planes may help in deciphering the tops of beds. Mudcracks or dessication cracks (Figure 13.14) are polygonal cracks developed in fine-grained sediments (clay to silt). They form due to drying and shrinking from the surface. They can also be helpful as a top-determining criterion in highly deformed sedimentary rocks.

FIGURE 13.14   *Mudcracks.* (Courtesy of Gautam Sen)

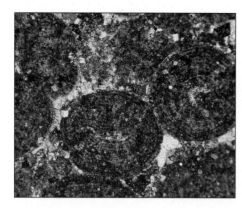

FIGURE 13.15 *Photomicrograph of ooids.* (Courtesy of Gautam Sen)

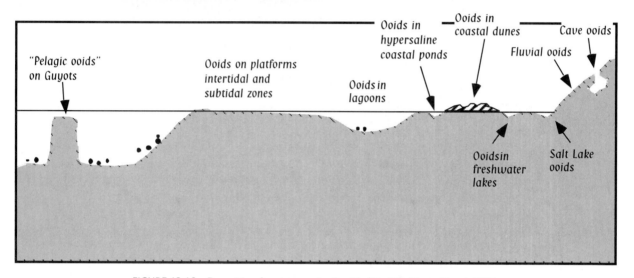

FIGURE 13.16 *Depositional environments of ooids.* (Modified from Flugel, 1982)

Oolites are small (<2 mm) egg-shaped grains with internal, onion-like, concentric laminae (Figure 13.15). These typically form in shallow water and involve rolling of individual grains accompanied by chemical precipitation of calcium carbonates. Ooids can be deposited in a variety of environments (Figure 13.16). Other similarly rounded structures include fecal pellets, which are essentially mud excreted by burrowing organisms.

ENVIRONMENTS OF DEPOSITION

An important reason for studying modern sedimentary environments and their deposits is to be able to use that knowledge to reconstruct the paleo-environmental history recorded in sedimentary rocks now exposed on land. Such knowledge is vital in (a) understanding the tectonic evolution of continents and oceans; (b) deciphering global climate and air/ocean circulation patterns through time; and (c) exploring economic deposits (oil, coal, uranium, etc.) that occur in sediments deposited in particular environments.

Environments in which sediments are deposited may be grouped into three kinds: continental, marine, and transitional (Figure 13.17). Continental environments include river, desert, glacial, and lacustrine (i.e., lake) systems on land. Transitional environments

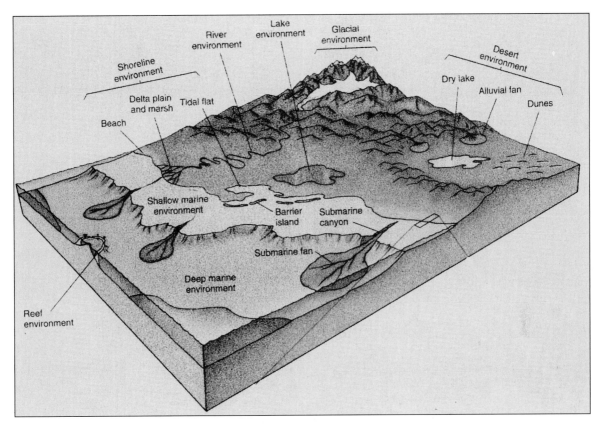

FIGURE 13.17   *A schematic diagram showing various continental, transitional, and marine environments of deposition.* (From Davidson et al., 1997, 1st ed., *Exploring Earth: An Introduction to Physical Geology*, Prentice Hall)

include deltas, off-shore barrier islands, lagoons, and so on, along the coastal regions where marine environments come in contact with continental environments. Transitional environments are of particular importance as two important natural resources, oil and coal, come from sedimentary rocks that are most frequently deposited in such environments. Marine environments include continental shelf, slope and rise, and abyssal plains.

Marine Environments

Table 13.3 gives a general summary of the various marine depositional environments and the nature of the deposits associated with them. In marine environments, the controlling factors, such as temperature, salinity, wave activity, and sunlight availability, are subject to a great deal of variability, which is reflected in the types of sediments that characterize various marine environments. On the continental shelf, the water depth is generally less than 200 meters, and this area receives both continentally derived sediments as well as carbonate biogenic particles generated in the ocean. A large variety of animals and plants generally thrive in these environments. In warm, equatorial regions, coral reefs commonly grow on continental shelves.

In the open-ocean basins, both terrigenous (i.e., continentally derived) clay and skeletal remains of carbonate-secreting organisms (foraminifera, pteropods, and coccolithophores) and silica-secreting organisms (diatoms and radiolaria) dominate the sediments in the hemipelagic environment, whereas only skeletal remains dominate in the pelagic environment. The calcite compensation depth (CCD) is the depth ($\sim 3,500$–$4,500$ m depending on latitude and the ocean) below which the ocean water is undersaturated with calcite because of dissolved CO_2 (Figure 13.18). Therefore,

TABLE 13.3 *Sedimentary environments and sedimentary deposits.*

Type	Subtype	Site	Sediment Composition and Structure
Continental	*Aeolian*	Desert	Wind-blown deposit made entirely of Frosted sand with dune cross-bedding. Sand grains may be frosted indicative of long transport by wind. Bahamites are wind-transported oolitic limestones
	Loess	Adjacent to major rivers and draining glaciers	Wind-blown angular silt, rock flour produced by glacial grinding of bedrock and later blown away by wind
	Fluvial	Alluvial Fans formed at fault mountain base	Wedge-shaped deposit in cross-section Very poorly sorted mix of mud and sizes up to boulders
		Meandering river	Point bar deposit of cross-bedded sand; thinly laminated mud in flood plain
		Braided river	Braided pattern in a plan view. Generally gravel bars, lenticular or sheet-like
	Lacustrine	Lakes	Lake margins have sandy beaches and deltas; centers are black mud, often parallel laminated
	Glacial	Alpine glacial valleys	*Till* (very poorly sorted mix of boulders to clays)
Transitional	*Deltaic*	Mouth of a river	Complex subenvironments and sedimentary *facies* Coal and oil deposits are often associated with Deltaic environments
	Peritidal	Coastal swamps	Mangrove swamps and tidal flats. Cross-beds, oolites, bioturbation. Well rounded quartz sand
	Barrier complexes		Elongated cross-bedded sand deposits with lagoon (mud deposits) on one side and open ocean on the other
Marine	*Continental Shelf/Slope*		Turbidites, lime mud, coral or algal reefs
	Abyssal Plain		Carbonate and or siliceous ooze, continentally derived fine detritus extraterrestrial fine debris

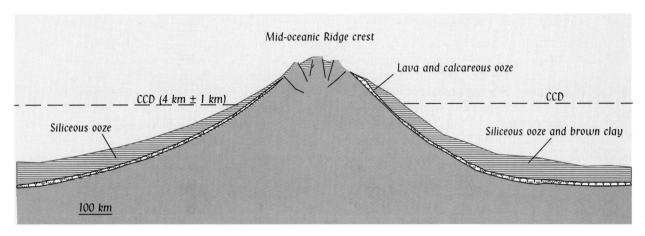

FIGURE 13.18 *The calcite compensation depth (CCD).* (Redrawn with permission from Blatt and Tracy, 1995, 2nd ed.)

calcareous tests of all organisms are totally dissolved below the CCD. Similarly, the aragonite compensation depth (ACD) occurs at a shallower depth (~2,000 m). Aragonite is used as a secreting material by the pteropods, which are pelagic molluscs.

Note that, although foraminifera occur in all latitudes, the siliceous organisms such as the diatoms and radiolaria only thrive in upwelling areas of the oceans. Siliceous tests are preserved at all depths in the deep radiolarian or diatomaceous oozes (which are sediments on abyssal plain rich in radiolaria and diatom, respectively). They become particularly prevalent below the CCD due to dissolution of calcareous skeletons. Siliceous oozes also predominate in high-latitude upwelling regions near the poles, where $CaCO_3$ productivity is at a minimum (Figure 13.19).

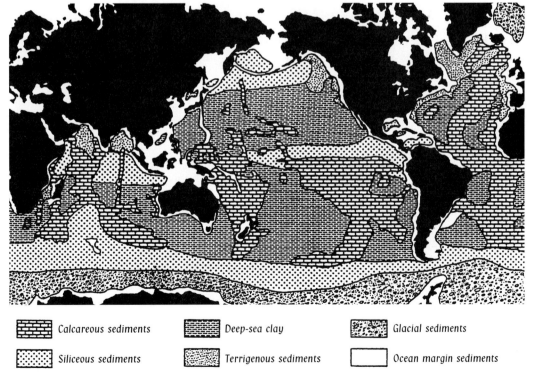

Calcareous sediments	Deep-sea clay	Glacial sediments
Siliceous sediments	Terrigenous sediments	Ocean margin sediments

FIGURE 13.19 *Distribution of siliceous and carbonate sediments on the ocean floor.* (Reprinted with permission from Raymond, 1995, *Petrology*, William C. Brown)

Deltaic Environment

The physiography of an average delta may be divided into three parts: the delta plain (i.e., the flat, landward, platform area), delta front (i.e., the steeper sloping frontal area that dips into the prodelta region), and prodelta (i.e., the deeper water, flatter area in front of the delta). The prodelta is composed of fine (suspension load) fluvial sediments. Delta front is composed of coarser fluvial sediments. The delta plain is composed of braided stream channel (cross-bedded sand) and interchannel (laminated silt and clay) deposits (Figure 13.20). Delta plain, delta front, and prodelta deposits are often equated to topset, foreset, and bottomset beds. Subsidence, due to sediment compaction and/or sediment loading, is a key factor in the sediment thickness and other depositional characteristics in large deltas. Deltas composed dominantly of mud can go through rapid subsidence because of great compaction.

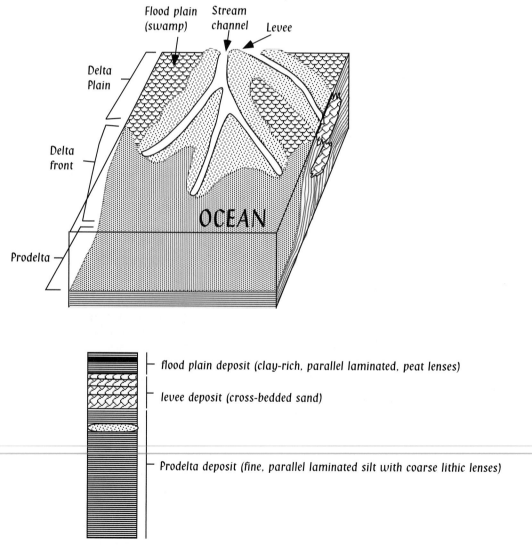

FIGURE 13.20   *Parts of a delta showing typical deltaic deposits.*

Estuarine Environments

An estuary occurs at the mouth of a river where it receives both fluvial and marine sediments. Thus, estuarine sediments develop mixed characteristics depending on whether it is tide-dominated or wave-dominated (Figure 13.21; after Dalrymple et al., 1992). Estuaries are typically associated with marine transgressions, which turn them into a delta if sediment supply is abundant. Similarly, a wave-dominated estuary, bar, barrier island, or spit develops, being composed of sand and gravel. Such a sand bar may stop ocean waves from reaching the shore, thus creating a lagoon in between the barrier island and delta. The lagoon may become a hypersaline marsh with stagnant bottom waters from which mud may be deposited.

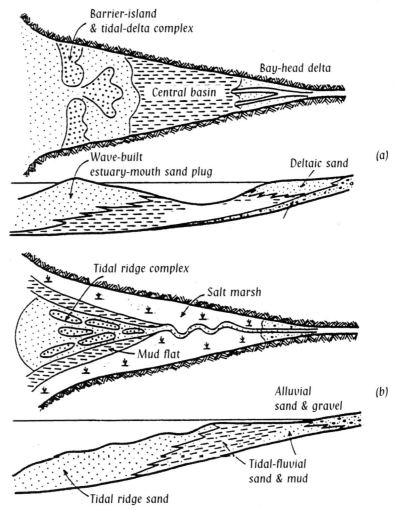

FIGURE 13.21 *Estuarine environments and their deposits:* (a) wave-influenced estuary. (b) tide-influenced estuary. The top panel is a plan view, and the bottom panel is a cross-section. (Reprinted with permission from McLane, 1995, *Sedimentology*, Oxford University Press)

Terrestrial Environments

Terrestrial environments encompass diverse types, from deserts to streams/rivers to glacial. Discussion of the characteristics of each type of terrestrial environment is beyond the scope of this book, and only overall observations are made here.

Mass movements in desertic environments are typically dominated by wind and supplemented by occasional flash flooding. Alluvial fans are enormous fan-shaped deposits (Figure 13.22) commonly found in fault-bounded regions, such as the Death valley region in the western United States. Episodic flash flooding causes enormous debris and mud flows that are responsible for rapid sedimentation in such environments. The sedimentary deposit is characteristically a poorly sorted mix of boulders to clay-sized particles. Sand dunes of several different types, salt flats, playas, and so on are all characteristics of a desert.

The environment of streams/rivers is referred to as a *fluvial environment*. Streams/rivers can have a wide variety of channel patterns—straight, meandering, or braided (Figure 13.23). A typical cross-section of a mature braided stream (Figure 13.24)

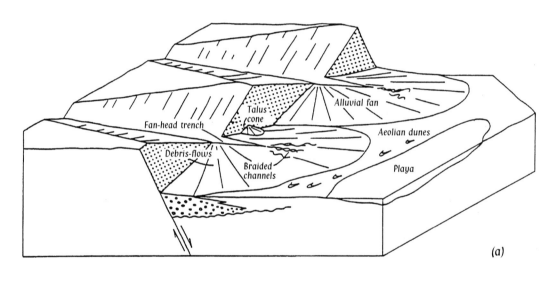

(a)

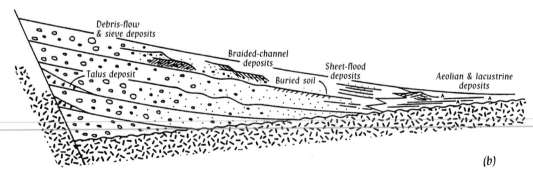

(b)

FIGURE 13.22 (a) Alluvial fans. (b) Cross-section through an alluvial fan. (Reprinted with permission from McLane, 1995, *Sedimentology*, Oxford University Press)

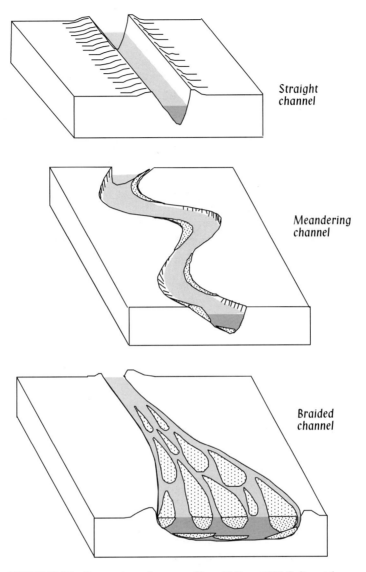

FIGURE 13.23   *Stream channel patterns.* (From McLane, 1995, *Sedimentology*, Oxford University Press)

shows that the water in a stream is restricted to the channel, which is bounded on both banks by ridges made of sand and gravel called *natural levees.* Occasional flooding may occur when water breaks the levees and fills the surrounding flood plains. Certain flood plains are essentially marshy land with mud and organic materials (peat) deposited from flood waters. In meandering streams, sand deposition takes place on the convex curve of the channel (point bar) where water velocity is lower, whereas the concave curve gets eroded because of faster flowing water on that side of the river (Figure 13.25). Through geologic time, erosion of the outer, concave, channel-wall and simultaneous deposition on the convex inner wall causes the stream channel to migrate laterally.

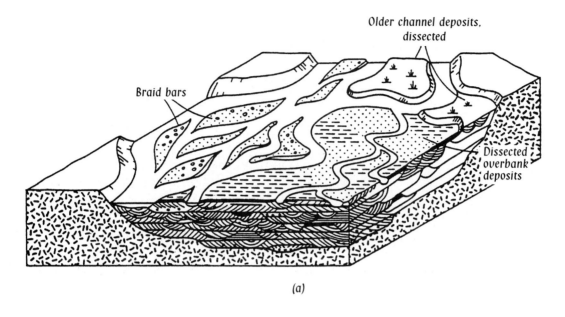

(a)

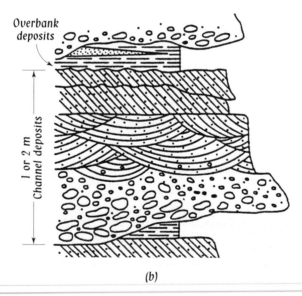

(b)

FIGURE 13.24 *Sedimentary deposits in braided streams.* (Reprinted with permission from McLane, 1995, *Sedimentology*, Oxford University Press)

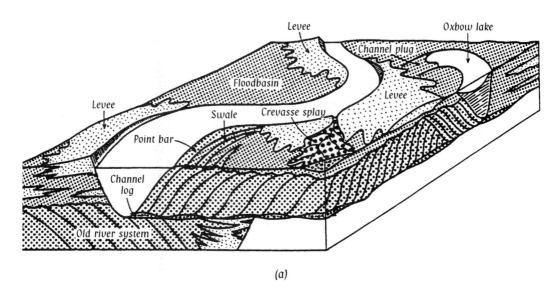

(a)

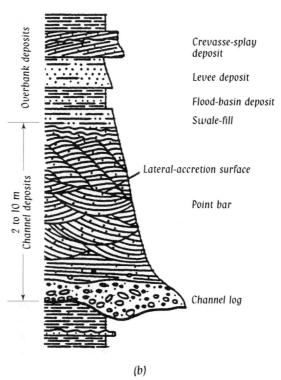

(b)

FIGURE 13.25 *Sedimentary deposits in meandering streams.* (Reprinted with permission from McLane, 1995, *Sedimentology*, Oxford University Press)

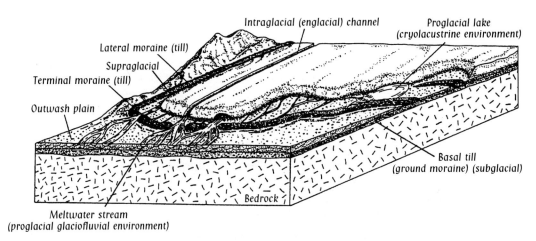

FIGURE 13.26   *Glacial environment.* (Reprinted with permission from Raymond, 1995, *Petrology*, William C. Brown)

The environment of a glacier (Figure 13.26) is of considerable interest because a glacier is a very effective agent capable of carrying house-sized boulders over great distances. *Till* is a typical sedimentary deposit from glaciers, composed of an unsorted mix of mostly angular gravel to silt-sized sediments. Glacial meltwater streams generally form braided networks that lead to the accumulation of cross-bedded channel sand and gravel. The coarser grains in such a deposit are often striated due to the grinding action of the glacier against bedrock. Another typical glacial meltwater deposit is *varve* (described in an earlier section). Glacial grinding of bedrock produces rock powder of silt- to clay-sized particles known as *rockflour*. Rockflour deposited by glacial meltwater streams in the front of a glacier may be picked up by strong winds and deposited en masse downstream as a *loess*, which is a structureless, light-colored deposit composed dominantly of silt.

LITHIFICATION OR DIAGENESIS

Deposition of sediments is followed by lithification, which is the process of conversion of sediments into a sedimentary rock. In detail, this process includes: (a) compaction, (b) dissolution, (c) recrystallization, (d) cementation, and (e) replacement. Although diagenesis occurs mainly at depth, in some cases it has been observed to operate at the surface (e.g., beachrocks). Three stages of diagenesis are commonly recognized: (a) *eogenesis*, a geologically brief period of diagenesis at near-surface conditions; (b) *mesogenesis*, a long-term diagenesis during deep burial; and (c) *teleogenesis*, a late-stage diagenesis of an already lithified sedimentary body.

The nature of the pore fluids is an important factor in diagenesis, which is caused by chemical reactions such as dissolution of old minerals and precipitation of new minerals. These fluids may be primarily of three types: (a) modified meteoric waters that seep through the pore spaces, (b) compaction waters that migrate upward or laterally from sites of greatest compaction, and (c) thermobaric waters, which are structurally bound waters released from clay minerals in the deeper parts of a sedimentary basin. Meteoric waters are most important in diagenesis because they are constantly replenished in the site of diagenesis.

Compaction, Pressure Solution, and Recrystallization

Accumulation of large amounts of sediments in a depositional basin leads to the compaction of the lower layers due the load (lithostatic) pressure of the overburden. Such compaction reorganizes the framework and squeezes out the pore fluid, resulting in a net reduction of primary porosity [Figure 13.27; porosity = (100 × volumes of pores) ÷ total volume (i.e., grains + pores)].

The extent to which compaction can reduce porosity (and thus volume) of a sedimentary body depends on:

1. texture of the sediment—the greater the clay percentage, the more the pore water, and thus more compacted it will be;

2. sorting—for instance, well-sorted quartz sand will compact much less than poorly sorted glacial till;

3. shape—well-rounded grains will compact less than irregular grains; and

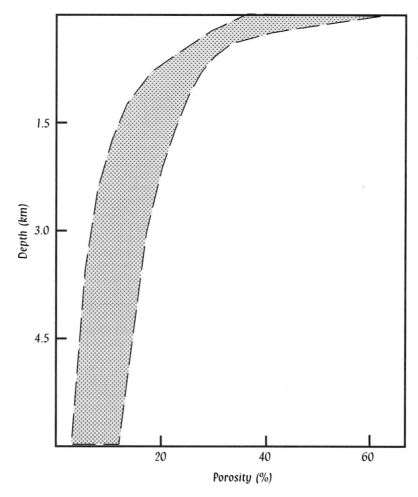

FIGURE 13.27   *Reduction of primary porosity as a function of burial depth in shales and sandstones (Dzevanshir et al., 1986).*

4. other accompanying processes (e.g., cementation) that simultaneously precipitate cement or matrix in the pore spaces. In this context, reference may be made to the development of pressure solution as a result of intense pressure along contacts between grains. The solution, containing dissolved ions of the minerals being squeezed, migrates to a more favorable site where it precipitates a mineral (cement) in pore spaces, thereby reducing porosity. Grains affected by pressure solution would recrystallize jointly, and the only clue as to their originally being different grains is usually the presence of a dusty suture-type boundary that may be visible under the microscope.

An easy experiment involving two ice cubes helps illustrate the recrystallization process: If two ice cubes are squeezed against each other for about 30 seconds and then released, the two cubes will recrystallize along their contact and become one. This effect is also seen in picnic-size bags of ice: If a bag is picked up from the bottom of the ice machine, the ice cubes would have to be broken apart.

Cementation

The process by which new minerals precipitate from migrating solutions in the pore spaces and bind the larger framework grains into a rock is called *cementation*. Composition of cements and the effectiveness of the cementation process depend on solute concentration, Eh, pH, and temperature of the pore fluid. Cements are generally made of a select variety of carbonates (dolomite, calcite, aragonite, and Mg calcite), silicates (chalcedony), and iron oxides (hematite, goethite). Several generations of dissolution (thus, development of secondary porosity) and cementation may occur through time within a sedimentary rock.

Replacement

New minerals may precipitate from a fluid migrating through sediments/sedimentary rock and replace existing coarser mineral grains. Often the replacing mineral maintains the form of the originally deposited mineral. In such a case, the replacing mineral is said to be a *pseudomorph* after the original mineral.

SUMMARY

Derivation, transportation, and depositional processes of sediments may be mechanical, chemical, or biogenic. Postdepositional change that transforms sediment to a sedimentary rock is called *diagenesis*. Mineralogy and texture of sediments reflect the compositions of the original source rocks and sedimentary processes. Depositional environments can be fundamentally grouped into continental, oceanic, and transitional types. Sediment composition and structure are tied to the conditions of deposition in these various environments. Two first-order postdepositional changes to sediments are porosity reduction (due to burial) and cementation. Cementation and all chemical sedimentation processes are largely controlled by Eh/pH conditions.

BIBLIOGRAPHY

Boggs, S., Jr. (1992) *Petrology of Sedimentary Rocks.* McMillan (New York).

Dzevanshir, R.D., Buryakovskiy, L.A., and Chilingarian, G.V. (1986) Simple quantitative evaluation of porosity of argillaceous sediments at various depths of burial, *Sed. Geology* **46**, 169–175.

Flugel, E. (1982), ed. *Microfacies Analysis of Limestones.* Springer-Verlag (New York).

Garrels, R.M. and Christ, C.L. (1965) *Solutions, Minerals, and Equilibria.* Freeman (San Francisco).

Hibbard, M.J. (1995) *Petrography to Petrogenesis.* Prentice-Hall (Englewood Cliffs).

McClean, M. (1995) *Sedimentology.* Oxford University Press (New York).

Miall, A.D. (1990) *Principles of Sedimentary Basin Analysis* (2nd ed.). Springer (New York).

Plummer, C.C., McGeary, D., and Carlson, D.H. (1997) *Physical Geology* (8th ed.) WCB McGraw-Hill (New York).

Rye, R., and Holland, H.D. (1998) Paleosol and the evolution of atmospheric oxygen: A critical review, *Am J. Sci.* **298**, 668.

CHAPTER 14

Sedimentary Rocks

The following topics are covered in this chapter:

Classification of sedimentary rocks
> Dominant types of sedimentary rocks

Texture and structure of sedimentary rocks

Description of sedimentary rocks
> Conglomerates
> Sandstones
> Mudrocks
> Carbonates
> Others—evaporites, cherts, and iron-rich sedimentary rocks

ABOUT THIS CHAPTER

In the previous chapter, we considered how sediments, the precursor to sedimentary rocks, form and ultimately become lithified. Sedimentary rocks are important to the geologist for two principal reasons: (a) They host oil, natural gas, coal, uranium, and many other economically significant deposits, and (b) their composition, structure, texture, and fossils provide clues to paleogeography, paleoclimate, and past life on Earth. It is clearly a vast topic and not every aspect, however important, can be covered here in sufficient detail. Instead, in this chapter, we consider largely the broad petrogenetic aspects of some major types of sedimentary rocks.

CLASSIFICATION OF SEDIMENTARY ROCKS

Sediments are primarily composed of three types of materials (Folk, 1959): (a) terrigenous/siliciclastic materials, which are derived through weathering of rocks—these materials may also be called *siliciclastics* because they are predominantly siliceous (quartz-rich) clastic sediments transported by detrital processes; (b) orthochemical (or authigenic) materials, which are chemical or biogenic precipitates that are generated in situ—these would include all evaporites and cements; and (c) allochemical materials, which are mechanically broken fragments of original chemical or biogenic materials—these may or may not be derived from within the depositional site.

Sedimentary rocks may be classified on the basis of modal contents of the three types of materials listed earlier (Figure 14.1). Rocks that are composed of >50% terrigenous materials are called *terrigenous rocks*. Sandstones, mudstones, and conglomerates are all examples of this class of rocks. Allochemical rocks must have >10% of allochemical materials and <10% terrigenous materials. The adjective *impure* is added if the terrigenous materials exceed 10% of the volume of the rock. Orthochemical rocks contain >90% orthochemical materials.

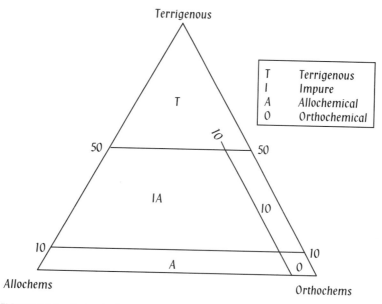

FIGURE 14.1   *A simple classification of sedimentary rocks.* (From McLane, 1995, *Sedimentology*, Oxford University Press)

TERRIGENOUS ROCKS

Terrigenous rocks dominate the sedimentary rocks of the continental crust. They show a wide range of compositions and texture. Terrigenous rocks typically consist of clasts (framework), fine-grained detritus (matrix), or chemically precipitated cement that bind the clasts together (Figure 14.2). Table 14.1 shows the classification of sediments according to the size of their clasts using the Wentworth scale.

FIGURE 14.2   (a) Photomicrograph of clay-rich matrix in an immature sandstone. (b) Photomicrograph of cement in a different immature sandstone. [Magnification in these and all subsequent photomicrographs is 80×, unless otherwise mentioned]. (Courtesy of Gautam Sen)

TABLE 14.1 *Classification of terrigenous rocks.*

Clast Size	Sediment	Sedimentary Rock
>2 mm	Gravel	Breccia (if angular), conglomerate (if rounded), or diamictite (if mixed)
0.063–2 mm	Sand	Sandstone
0.004–0.063 mm	Silt	Shale or mudrock
<0.004 mm	Clay	Claystone or mudrock

Breccias, Conglomerates, and Diamictites

This class of rocks have >25% *gravel*, which is a mixture of particles of all sizes. If the framework grains (e.g., pebbles) are rounded, then the rock is called a *conglomerate*; if they are angular, the rock is called *breccia*. *Diamictite* refers to unsorted, matrix-supported siliciclastics that are generally deposited as glacial tills, icerafted, landslides, mud flows, and so on. Conglomerates are composed of pebbles that have been rounded during transportation. Certain conglomerate horizons serve as excellent indicators of an erosional unconformity, which is a discontinuous surface that represents gaps in the depositional history of the sedimentary basin where the conglomerate occurs. Conglomerates may form in many ways, but the most common are actually depositional (e.g., fluvial gravel) in origin.

Breccias may have a number of origins: They may form from talus at the bottom of steep slopes or cliff (talus breccia), along fault planes during fault movement (fault breccia), from the collapse of reefs due to repeated pounding by ocean waves (reef talus), or from meteorite impact (impact breccia; Figure 14.3).

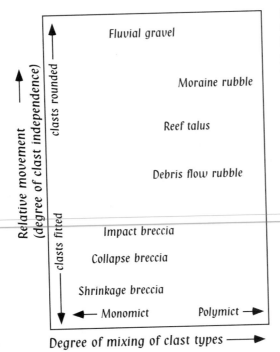

FIGURE 14.3 *Breccia types and their origins.* (Redraw with permission from McLane, 1995, *Sedimentology*, Oxford University Press)

In describing a breccia or conglomerate, it is important to recognize whether the clasts are of a single-lithic type (termed a *monomict* breccia or conglomerate) or multiple types (a *polymict* breccia or conglomerate). Also the clasts in a conglomerate may all have some sort of a crude orientation, in which the conglomerate would be called an *organized conglomerate*; if not, it would be called a *disorganized conglomerate*. The framework clasts may also be graded in some cases. Each of these characteristics is helpful in unraveling the geologic history of a conglomerate.

Sandstones

Sandstones are an abundant and important class of sedimentary rocks. The three principal framework components of sandstones are quartz, feldspar (which may become partially or entirely altered to clay minerals), and lithic (rock) fragments. The minerals are bound together by a fine-grained matrix or chemically precipitated cement. The matrix may be composed of clay, silt, and/or iron oxide. Carbonate and silica cements are common. In some cases, glauconite, a chemically precipitated, green, claylike mineral, and iron oxide form the cement.

Quartz grains may be single crystals or polycrystalline aggregates, which are easily observable under the microscope (Figure 14.4). The nature of quartz clasts may offer clues as to the type of the source rock from which they are derived. For example, polycrystalline quartz grains are generally derived from metamorphic rocks. Another important (but obscure) example is the texture of quartz grains deformed by meteorite impacts: These are typically characterized by three sets of twin planes that cut across each other at acute angles (Figure 14.5). Such shocked quartz grains have been found in the cretaceous/tertiary boundary layers from around the world. They have been used as evidence in favor of a meteorite impact that ushered in a series of events culminating in a major mass extinction event.

The abundance, variability, shape, and size of lithic fragments offer clues to the provenance (i.e., source area), transportation history, and tectonic settings. The maturity of the clasts indicates the kinetics of the transportation process (see Chapter 13). Extremely long transportation removes all of the clay particles and makes a deposit of well-rounded sand grains. A poorly sorted mixture of mud and angular sand grains would thus be clearly texturally immature. In a mineralogical sense, mature rocks may be expected to be essentially composed of quartz because this is the most stable mineral in all sedimentary processes.

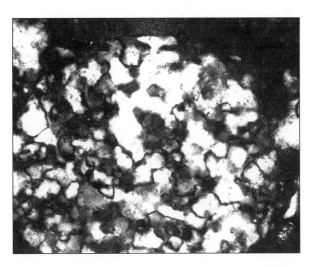

FIGURE 14.4 *Photomicrograph showing a polycrystalline quartz grain.* (Courtesy of Gautam Sen)

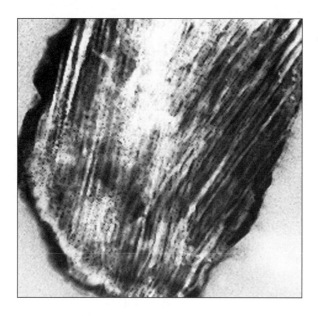

FIGURE 14.5 *Photomicrograph of a shocked quartz grain.* (Courtesy of Gautam Sen)

Feldspar composition and its degree and nature of clay alteration often provide important genetic clues. For example, in arid climatic regions, detrital feldspar may not alter at all at near-surface conditions nor during diagesis. The percentage of plagioclase over K feldspar can be a source indicator: A sandstone whose feldspar is dominantly a K feldspar likely has a granitic source.

Several different schemes to classify sandstones exist in the published literature. These are generally based on the character of the clasts. A convenient and widely used classification is that by Dott (1964; Figure 14.6), which is based on the clast

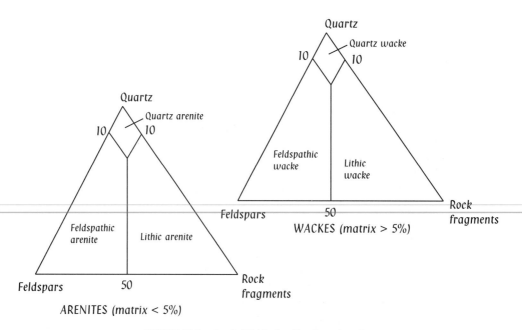

FIGURE 14.6 *Dott's (1964) classification of sandstones.*

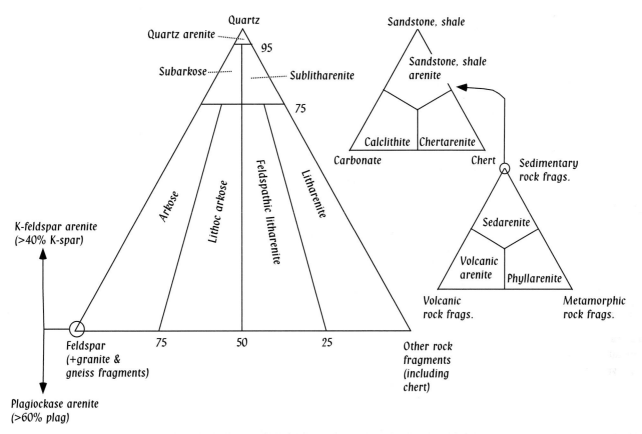

FIGURE 14.7   *Folk's (1974) classification of sandstones.*

composition as well as matrix percentage. In Dott's classification, a first-order division is recognized between arenites and wackes based on the proportion of matrix present. Each of these groups is further classified according to their modes with the use of a triangular diagram whose apices are quartz, feldspar (partially or wholly altered to clay), and lithic fragments. Thus, based on these diagrams, a sandstone with <5% matrix and 50% quartz, 40% feldspars, and ~10% rock fragments should be called a *feldspathic arenite*. If the same rock had >5% matrix, it would be called a *feldspathic wacke*.

Another popular classification is that by R.L. Folk (1974; Figure 14.7). Folk's classification makes a fundamental subdivision between rocks with >15% matrix, which he called *graywackes*, and those with <15% matrix. The latter group is further subdivided into various *arkoses* and *arenites*, with the general sense that arkoses are feldspar rich and arenites are feldspar poor.

Although Dott or Folk may not have intended this classification to have any tectonic significance, Dickinson and Suczec (1979) found that the proportions of quartz, feldspar, and rock fragments can be a good indicator of plate tectonic regimes (Figure 14.8). For example, sandstones rich in K feldspar and lithic fragments are considered to form in continental arc settings. However, later authors pointed out various problems with this approach, citing complications due to multiple provenance, transportation, and diagenesis processes, as well as climatic influence (cited in Raymond, 1995).

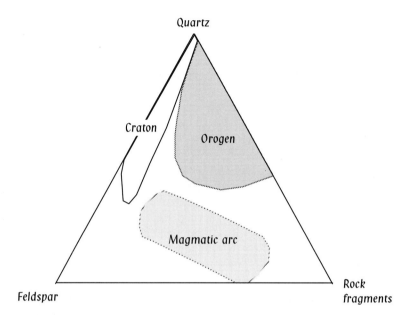

FIGURE 14.8 *Sandstone composition and plate tectonic environments.* (Dickinson and Suczec, 1979)

In general, clastic fragments in sandstones are indicative of provenance; and their textures are indicative of transportation and depositional histories. Their structures (e.g., cross-bedding), nature of fossils, and association with other lithic types (e.g., peat, evaporite, etc.) must be considered together to decipher the depositional environment and paleogeography. To understand how this may be accomplished, let us consider an example in which a geologist maps the lithological units of an area and reconstructs a cross-section such as in Figure 14.9. These units are composed of a coarse conglomerate at the bottom followed successively above by a parallel bedded sandstone, a thick sequence of cross-bedded sandstone, thin layers of mudstone with peat, and massive lenses of medium-grained sandstone. Based on what we learned about depositional environments in Chapter 13, this sequence can develop in a fluvial environment (meandering stream).

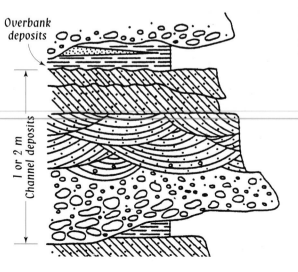

FIGURE 14.9 *A geological cross-section.* (Reprinted with permission from McLane, 1995, *Sedimentology*, Oxford University Press)

Mudrocks

Mudrocks are composed of silt-sized, angular, quartz grains and clay minerals in variable proportions. The four most common clay minerals in mudrocks are illite, smectite, mixed illite-smectite, and kaolinite. Because the grain size is so fine, it is impossible to discern the proportion of quartz to clay with a pocket lens. Blatt et al. (1980) outlined an alternative method of roughly estimating silt to clay proportion based on whether the rock tastes gritty (more quartz than clay) or slimy (more clay than quartz) when nibbled on. *Mudrock* is the general name given to all rocks that consist of silt and clay particles. The term *shale* is used to denote a mudrock with excellent fine lamination, which often causes splitting along parallel planes—a property referred to as *fissility*. Fissility is a result of parallel stacking of the sheets of clay minerals. Fissility may not develop in *siltstone* or *claystone*, which are defined on the basis of whether silt or clay, respectively, dominate the mode. Although mudrocks are generally associated with aqueous environments, there are two classes that have different modes of origin: *Bentonite* is a special type of mudrock that has its origin as a volcanic ash deposit; and *loess* is essentially wind-blown rockflour derived by powdering of bedrock by a glacier.

As a group, mudrocks are a very important class of sedimentary rocks, comprising 60% of the stratigraphic record. Because mudrocks occur in a wide range of depositional environments and are distributed throughout much of the geologic column (from pre-Cambrian to present), they have proved to be very useful in unraveling the earth's paleogeography and environmental conditions. Mudrocks may form as floodplain deposits, alluvial fan deposits, in transitional environments, and in the marine environments.

The color of mudrocks can be an indicator of the depositional environment. For example, black shales, which occur at different levels throughout the phanerozoic rock record, are generally believed to have formed in dysoxic/anoxic, stagnant waters of lagoonal environments. They commonly contain pyrite as an indicator of anaerobic-reducing conditions that exist in the basin. Approximately 150 to 85 million-year-old black shales occur extensively as laterally and vertically discontinuous bodies in the Atlantic Ocean floor, suggesting that, during this period, there were large areas of the Atlantic Ocean floor that suffered large-scale anoxia. Such anoxia was likely caused by the lack of open circulation within the early Atlantic. Paleozoic black shales are well exposed in the northern Appalachians (Figure 14.10).

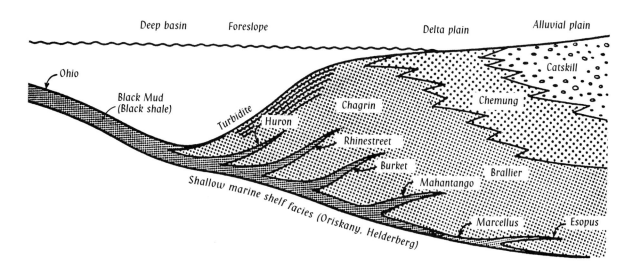

FIGURE 14.10   *Environment of deposition of the Paleozoic black shales from the northern Appalachian mountains.* (Reprinted with permission from McLane, 1995, *Sedimentology*, Oxford University Press)

Black shales owe their color to their content of minor amounts of organic materials, which remain undecomposed by virtue of lack of oxygen availability. Commonly, a mudrock may be gray-black or red-yellow-green. The green color is due to the presence of green clay minerals, chlorite and illite. Red to yellow colors may be due to other minerals such as hematite (red) and limonite (yellow).

Mud readily undergoes compaction following deposition and burial. The initial porosity of water-saturated mud is about 80%, but it can be reduced to almost zero in the first 8 km of burial. During such a process, hydrocarbons that may have originally formed in the pore spaces may be squeezed out to a more suitable rock (e.g., a sandstone, which is less compressible and whose open pore spaces can serve as a reservoir for oil by virtue of its greater porosity). Clay minerals are particularly sensitive to burial pressure and temperature causing their diagenesis. Smectite and mixed-layer (illite-smectite) clays are dominant in muds and young mudstones, but are not nearly as abundant in older and deeply buried clastic rocks. During diagenesis, smectite alters to illite or chlorite. The percentage of illite in mixed-layer clays gradually increases with increasing degree of diagenesis. With further increases in pressure and temperature, illite coarsens and becomes chlorite/mica. In summary, clay minerals are particularly useful in recording the diagenesis of mudrocks.

ALLOCHEMICAL ROCKS

Carbonate Rocks

Carbonate rocks contain >50% carbonate minerals and comprise ~10% of all sedimentary rocks. Carbonate rocks, their depositional environments, and diagenesis of carbonate rocks are subjects of great interest to the petroleum exploration industry because many oil deposits occur in carbonate rocks. The two minerals that are the dominant constituents of carbonate rocks are calcite and dolomite, and carbonate sedimentary rocks are broadly grouped into limestones and dolostones based on whether they are dominantly composed of calcite or dolomite, respectively.

The common calcite crystallizes in the rhombohedral class and is sometimes referred to as *low-Mg calcite*. It forms the shells of most planktonic organisms and has little or no magnesium in solid solution (<8 to 10 mol% $MgCO_3$). Low-Mg calcite can also form during diagnesis. Chemically precipitated low-Mg calcite is not common in limestones. In contrast to the low-Mg calcite of planktic organisms, calcite of benthic organisms can have as much as 30% Mg and is commonly referred to as *Mg calcite* or *high-Mg calcite*. High-Mg calcite can also form as a cement-forming, chemically precipitated mineral. Aragonite is a polymorph of calcite that can occur in abundance in shallow marine deposits. Aragonite crystals are typically needlelike; they form the shells of organisms or may be chemically precipitated. Aragonite converts to calcite or is dissolved in interstitial fluids during diagenesis, and thus aragonite is relatively rare in pre-Pleistocene rocks. Wilkinson (1979) noted that skeleton composition of marine organisms as a whole has changed through geologic history: Paleozoic organisms were composed of high- and low-Mg calcite, Mesozoic organisms were made of high-Mg calcite, and Cenozoic organisms are dominantly composed of aragonite.

Primary precipitates of dolomite form only in exceptional cases in present-day environments, and yet dolostones are an abundant component throughout the stratigraphic record. This suggests that, in such dolostones, dolomite likely formed by diagenetic processes.

Carbonate sediments can form in a wide variety of environments, but they are most common in tropical shallow marine waters and lakes. As discussed in Chapter 13, in shallow marine environments, calcite and aragonite form the skeletons of many organisms. When these organisms die, their skeletal debris may accumulate on the ocean bottom and form a carbonate deposit. However, below the carbonate compensation depth (CCD; see Chapter 13), all carbonates are dissolved.

Carbonate rocks in general are composed of allochemical particles (or allochems) and a matrix or cement. Allochems are grains that form the framework in limestones and dolomitic limestones. They can be of several types—fossils/skeletals, ooids, peloids, and intraclasts. It is generally possible for the petrographer to identify a few fossils, but a more detailed identification must be left to a professional paleontologist.

Ooids are nearly spherical, sand-sized grains with radial and concentric internal structure (Figure 14.11). Modern carbonate ooids are composed of aragonite needles organized in a radial fashion and tangentially in concentric spherical layers around a nucleus composed of any preexisting solids in the environment of deposition, such as a quartz grain, peloid, or some skeletal material. Ancient ooids often formed during diagenesis. Oolitic limestones are commonly cross-bedded, indicating their deposition from currents after precipitation in water. Oolitic limestones exposed on the Bahama islands are called *Bahamites* and are actually eolianite or wind-blown deposits. In such a case, the ooids became exposed under dry conditions and were reworked by winds to form oolitic dunes that later became lithified.

Based on laboratory experiments, Davies et al. (1978) concluded that ooids growing in calm waters develop radial internal structure and tangential lamination in agitated waters. In a study of the ooids in hypersaline waters of Baffin Bay (Texas Gulf coast), Land et al. (1979) noted that tangential aragonite crystals grow in agitated conditions and radial structure, composed of high-Mg calcite, grow in calmer waters.

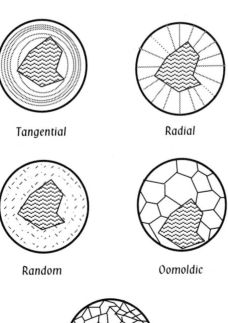

Tangential

Radial

Random

Oomoldic

Pseudospar

FIGURE 14.11   *Ooids types.*

Peloids are the most abundant type of allochems and originate from fecal pellets. Peloids are often crushed during burial to form the matrix of lime mud. However, many may survive but deform under pressure. Intraclasts, another common type of allochem, are fragments of previously formed limestones or partially lithified carbonate sediments.

The matrix or cement of carbonate rocks is generally composed of calcite or aragonite. The cement may also originate by chemical precipitation from pore waters. The matrix may originate as finely crushed fecal pellets, intraclasts, or skeletal debris. When originally deposited, the grain size of the fine matrix is ~1 to 5μm and is called *micrite*. Diagenesis leads to recrystallization and coarsening of grain size, whereby micrite turns into *spar*, which is a textural term to denote coarse, blocky carbonate crystals (Figure 14.12). Note that if a micrite were originally composed of aragonite, diagenesis would convert the micrite into spar of calcite or dolomite.

Carbonate rocks are classified in several different ways. One that is popular with the oil industry is Dunham's (1962) classification (Table 14.2). A first-order distinction is made between rocks that are mud supported versus those that are grain (allochem) supported. In the former, mud separates the allochems; in the latter, grains touch each other. Limestones lacking mud are called *grainstones*. Limestones made of rigid, organically bound, frameworks of skeletons are called *boundstones* and are normally formed in reefs. When mud is < ~25% to 30%, the grains touch each other and mud fills the pore spaces. Such a rock is called a *packstone*. Mudstone and wackestone are both mud supported, but the difference is that the latter has >10% allochems and mudstone has <10% allochems. It is a common practice to use a forename based on a distinctive character of the rock. For example, a grainstone composed almost entirely of ooids would be named an *oolitic grainstone* or *ooidal grainstone*.

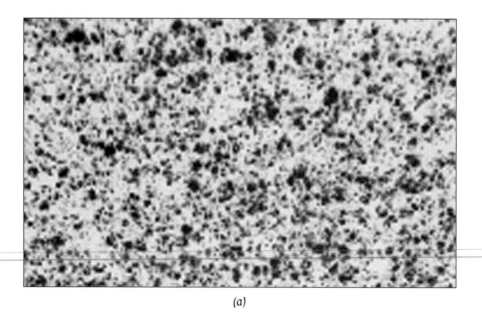

(a)

FIGURE 14.12   (a) Photomicrographs showing micritic cement. (b) Photomicrograph showing coarse dolomitic cement between ooids. (Courtesy of Gautam Sen)

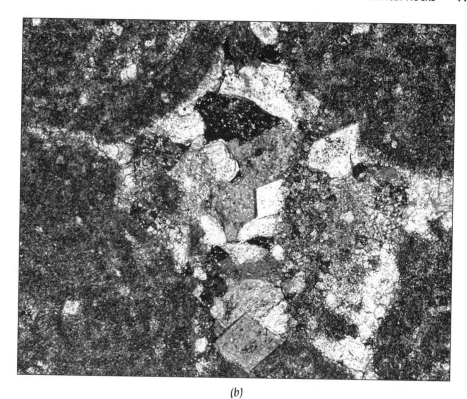

(b)

FIGURE 14.12   (continued)

TABLE 14.2   *Classification of carbonate rocks.*

Components not organically bound during deposition					Organically bound sediments
Carbonate mud present			Mud absent		
Mud supported		Grain supported			
<10% allochems	>10% allochems	Grain supported			
Mudstone	Wackestone	Packstone	Grainstone		Boundstone

Dolomitization

Dolomite may form in one of two ways: as a primary precipitate (Equation 14.1) or as a diagenetic replacement of calcite or aragonite (Equation 14.2):

$$Ca^{2+}[\text{soln.}] + Mg^{2+}[\text{soln.}] + 2(CO_3)^{2-}[\text{soln.}] = CaMg(CO_3)_2[\text{dol. cryst.}] \quad [\text{Eq. 14.1}]$$

$$2CaCO_3[\text{calc/arag.}] + Mg^{2+}[\text{soln.}] = CaMg(CO_3)_2[\text{dol. cryst.}]$$
$$+ Ca^{2+}[\text{soln.}] \quad [\text{Eq. 14.2}]$$

As discussed earlier, dolomite has not been found to precipitate in any present-day *normal* environments, nor has it been successfully made in the laboratory under normal conditions. Thus, those dolostones containing structurally well-ordered dolomites that are common in the geologic record must have formed by diagenesis.

In terms of physical evidence, there are many field examples of dolomite replacing limestone or evaporite and microscopic replacement of calcite by dolomite rhombs. Therefore, the majority of scientists believe that dolomite in stratigraphic sequences is entirely produced through diagnetic replacement of preexisting carbonate deposits—a process known as *dolomitization*. An overall correlation seems to exist between the abundance of dolostones and periods of high stands of sea level through geologic time (Figure 14.13). Also dolostones are commonly associated with limestones and sometimes with evaporites (principally gypsum and anhydrite) or pseudomorphs of gypsum. These features have led scientists to propose different models of origin of dolostones via diagenetic replacement of calcium carbonate by dolomite, which are discussed next.

In the evaporative reflux model (Figure 14.14), the two required conditions are (a) stagnant pools of ocean water in supratidal regions, and (b) carbonate deposits underneath such pools (Patterson and Kinsman, 1982). In this model, evaporation of hypersaline waters in these pools leads to supersaturation and precipitation of gypsum ($CaSO_4 \cdot 2H_2O$), an evaporite that depletes the water in Ca^{2+} but enriches it in Mg^{2+} and increases its density. The brine sinks into the substratum because it is heavier than the accumulated layers of gypsum. The Mg-rich brine dissolves the calcite or high-Mg calcite as it seeps into the underlying carbonate substratum, thus converting the limestone into a dolostone. This model of dolomitization requires enormous masses of evaporite (or at least pseudomorph or other structures reminiscent of previous existence of evaporite minerals, such as gypsum or anhydrite) to be associated with dolostones. In reality, however, many dolostone occurrences are not associated with evaporite (or former evaporite) beds.

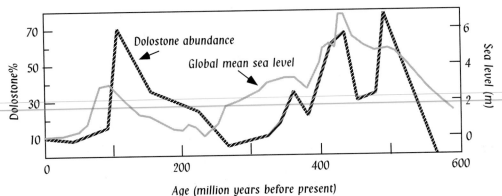

FIGURE 14.13 *Correlation of dolomite abundance and sea-level stand through time.* (After Blatt and Tracy, 1995, 2nd ed., *Petrology*, W.H. Freeman)

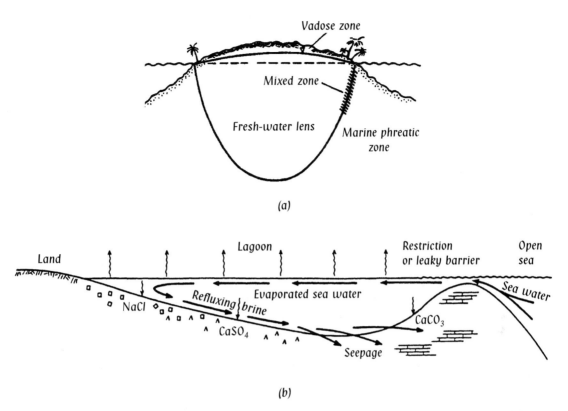

FIGURE 14.14 *Models of dolomitization.* (a) Dorag (mixing zone) model. (b) Evaporative reflux model. (Reprinted with permission from McLane, 1995, *Sedimentology*, Oxford University Press)

A second model of dolomitization involves mixing between marine waters and fresh waters and is known as *dorag dolomitization* (Badiozamani, 1973). In the subsurface of coastal areas, the lighter fresh water floats above denser saline (marine) waters in the pore spaces of the bedrock/regolith. The contact surface between the two pore waters crops out near the coastline and dips landward. In a detailed study of the formation of dolostone in Pliestocene reefs of Jamaica, Land (1973) noted that the dolostone replaces limestone along such a dipping zone marking mixing between marine and fresh pore waters. This works as follows: The curvature of the solubility curves of calcite is such that when the two different waters, both saturated with calcite, are mixed, the mixed water is likely to be undersaturated in calcite. Therefore, it dissolves calcite of the limestone (Figure 14.15). At the same time, it is supersaturated with dolomite and therefore precipitates dolomite. Whereas a wider range of mixing between seawater and fresh water is permitted for ordered dolomite precipitation (Figure 14.15a), the composition of the mixed water is rather limited (~30%–40% seawater and the rest meteoric [fresh] water) for the case of precipitation of disordered dolomites. Given this fact, it is truly surprising why ordered dolomite does not precipitate in natural environments and must be due to some kinetically inhibiting factor(s).

Some authors believe that mixing between fresh water and seawater is unnecessary, and dolomite may precipitate from normal seawater as long as enough unmodified seawater is continuously flushed through a limestone or limemud because such flushing would provide a constant source of Mg^{2+} and remove Ca^{2+}. Carballo et al. (1987) proposed such a model for dolomitization of limemud at Sugarloaf Key, Florida, where they noted flushing by normal seawater through limemud has been

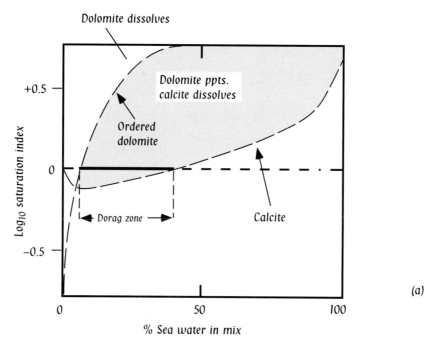

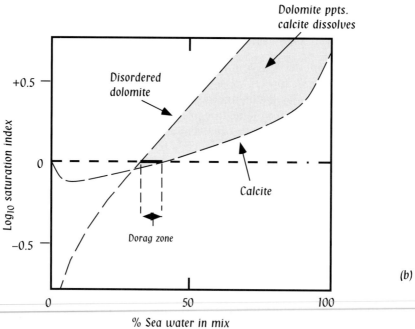

FIGURE 14.15 *Calcite and dolomite saturation conditions.* The shaded area marks the area where dolomite precipitates and calcite dissolves. (a) Conditions for ordered dolomite precipitation. (b) Conditions for disordered dolomite formation. (After Boggs, 1992, *Petrology of Sedimentary Rocks,* Mcmillan and Co.)

happening through tidal rise and fall of sea level. As a result, the pore fluids are constantly replenished in Mg^{2+} while Ca^{2+} is being removed, forming dolomite in the process. This process has been called *tidal pumping*.

The models discussed previously are all concerned with shallow dolomitization, which most authors believe to be the most common mode of dolomitization. It seems plausible that such early dolomitization can happen in a wide variety of environments. Late dolomitization during deep burial is considered by other authors to be an important process. In such a process, the source of Mg in the dolomitizing fluids is often a problem. Some authors believe the source of Mg^{2+} to be from the breakdown of clay minerals (smectite). Others believe that seawater can continuously percolate through to deep levels, obeying Darcy's law of fluid movement, and replace calcite with dolomite. The dolomite problem is far from being over. However, it is beyond the scope of this text to discuss the merits and demerits of every published model.

In passing, it should be noted that dolomite may dissolve in fluids under certain conditions and calcite may be precipitated in its place. Such a process is referred to as *dedolomitization*. In such a case, it is generally possible to recognize a calcite pseudomorph after dolomite rhomb. Figure 14.15 shows that dedolomitization will likely occur if dolomite is exposed to open marine or restricted hypersaline (lagoonal) environments.

ORTHOCHEMICAL ROCKS

Evaporites

Evaporites are rocks composed of minerals that precipitate directly from saline waters as a result of evaporation. The three most common classes of evaporite minerals are sulfates, halides, and carbonates. Carbonates have already been discussed in the preceding paragraphs. Among the sulfates and halide minerals, the most abundant evaporite minerals are gypsum ($CaSO_4 \cdot 2H_2O$), anhydrite ($CaSO_4$), and halite or rock salt (NaCl). In general, evaporites are found in two types of environments: Continental deposits generally formed in desert playas (ephemeral lakes) and warm shallow marine environment (salinas or sabkhas; Figure 14.16). In both cases, the rate of

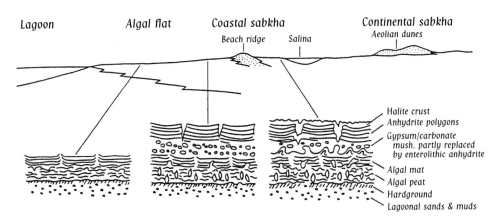

FIGURE 14.16   *Warm, shallow marine (sabkha) environment.* (Reprinted with permission from McLane, 1995, *Sedimentology*, Oxford University Press)

evaporation is much greater than the rate of influx of water. In the shallow marine, salinas type environment, deposition of evaporite occurs in a restricted basin largely isolated from open sea, such that the basin receives only episodic influxes of seawater interrupting strong, long-term evaporation. In this type of environment, evaporites are often associated with limestones and dolostones. In desert mudflats and playas, evaporite deposits are often associated with poorly sorted siliciclastic talus deposits.

Field occurrence of evaporite deposits and laboratory experiments indicate a definite sequence in which different evaporite minerals precipitate during evaporation of seawater (Figure 14.17). Carbonates appear first when the water volume is reduced by half. The next precipitate is gypsum, when the original water volume is reduced to ~20%, followed by halite. Sylvite (KCl) is last, as water volume is further reduced to less than 2%. Anhydrite either forms as a primary precipitate or by dewatering of gypsum during diagenesis.

Lamination, composed of alternate layers of white gypsum/anhydrite and gray calcite or aragonite ± organic carbon, is common in evaporite deposits. Mudcracks, slump structures, and ripple marks may also be found. Many other diagenetic structures are also found in evaporites.

Evaporite deposits have been found in sedimentary sequences dating from the late Cambrian to the Miocene in various parts of the world. Two notable examples are the >2 km-thick evaporites of the Messinian (Miocene) of the Mediterranean basin and the >600 m-thick Upper Silurian evaporites of the Michigan basin. Interpretive models for the origin of such gigantic evaporite deposits have all been based on modern analogs, and are thus not all satisfactory. One proposal for producing such ancient deposits of enormous thickness is a subsiding basin with occasional influx of seawater.

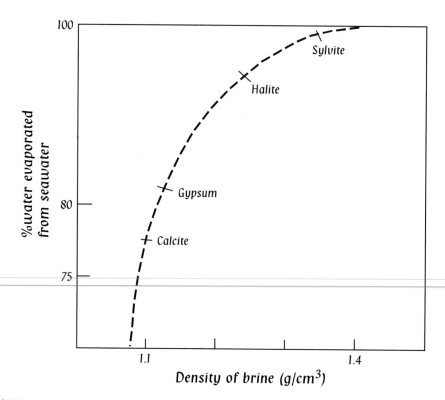

FIGURE 14.17   *Sequence of crystallization of evaporite minerals as a function of evaporation and density of brine.* (Redrawn with permission from Hillard, 1995)

Chert

The name *chert* (*flint*) is used to broadly include all siliceous, generally biogenic sedimentary rocks. They comprise <1% of all sedimentary rocks, but are found in stratigraphic sequences of all ages and from different facies, including the deep oceanic (pelagic) environments. Many different morphological and mineralogical types of cherts have been recognized. The silica minerals in cherts include chalcedonic quartz, opal A (amorphous silica), and opal CT (disordered cristobalite and tridymite). Many different types of cherts have been recognized in geologic record: Jasper is a red-colored chert often associated with pre-Cambrian banded iron formations. Novacullite is a particular fine-grained, dense variety of chert found in the mid-Paleozoic of Arkansas. Porcelanite is a type of chert that takes its name from porcelain because it fractures like porcelain.

It appears that a great number of cherts are biogenic in origin and form by remobilization of silica from the accumulation of skeletal debris of siliceous Protista. Seawater is seriously undersaturated in SiO_2, and therefore it is unlikely that SiO_2 can precipitate directly from seawater unless the pH is locally increased by some special situation to >10. In a sense, it is surprising that radiolaria and diatoms can survive in seawater, which is so undersaturated in silica. Protection against dissolution may be provided when they are alive by an organic coating that prevents physical contact with seawater. On their death, the organic coating decomposes, and the siliceous skeletal matter gets dissolved in seawater unless their deposition is faster than the rate of dissolution and productivity is high.

Opal CT and quartz in ancient cherts probably all formed by diagenetic alteration of opal A. The origin of bedded cherts or ribbon cherts, typically characterized by rhythmic bedding, is somewhat difficult to determine. These are common in old sedimentary rocks, where they appear to be associated with shallow marine deposits.

Iron-Rich Sedimentary Rocks: Iron Stones and Banded Iron Formations

All sedimentary rocks contain some amount of iron, but those containing >15% iron are generally referred to as *iron-rich* sedimentary rocks. Volumetrically, these rocks are really a minor component of the earth's sedimentary rocks; however, they are extremely important for a number of reasons. Among others, perhaps their economic significance is the most important because 90% of the world's iron ore comes from pre-Cambrian iron-rich sedimentary rocks.

Significantly, global deposits of iron-rich sedimentary rocks formed at three distinct time intervals—pre-Cambrian/Archean (3400–2900 m.y.), Early Proterozoic (2500–2000 m.y.), and Jurassic-Cretaceous (200–65 m.y.). These deposits can be broadly grouped into two types: *iron stones* and *iron formations* (also *banded iron formations* [BIFs]). The more abundant and economically important of these two are the iron formations. These are pre-Cambrian in age and alternately banded with jasper. Individual deposits are enormously thick (50–600 m), widespread (hundreds of km), and predominantly composed of magnetite and hematite. In contrast, iron stones are Phanerozoic in age, not as thick (few tens of m), and poorly banded or nonbanded. They are typically oolitic, associated with carbonates, and appear to have been deposited in shallow marine or lacustrine conditions. The iron mineral is goethite or limonite forming the oolitic coatings on a nucleus of quartz, carbonate, or other allochems.

The origin of BIFs is an important and largely unresolved problem. Most authors believe that the source of iron was continental crust. Lepp and Goldich (1964) first suggested that the oxygen level in pre-Cambrian atmosphere was extremely low, allowing transportation of dissolved iron in Fe^{2+} state in rivers to shallow seas. Many authors believe that it is here that somehow oxygen-rich environment was locally created by bacterial and/or photosynthetic algal activities, so that iron oxide was precipitated. There are many variations to these postulations, including groundwater transport and/or colloidal transport of iron. However, none of these explanations has been entirely satisfactory. Also these models cannot explain the easily observed differences between BIFs and iron stones.

14.1 HOW TO DESCRIBE SEDIMENTARY ROCKS

Color, mineralogical composition, texture, and megascopic (field scale) structures are all important parameters in deciphering the provenance, transportation, and depositional histories of sedimentary rocks. Accordingly, these elements must be present, if appropriate, in the routine description of a rock. The description must also be sufficiently detailed such that anyone can have a fairly good idea of what the rock may look like without actually seeing it.

The following is a list of items that should be included in the description of a rock:

Field appearance (i.e., megascopic/hand specimen characteristics)

1. *Lithology.* Thickness, color, mineral/organic constituents, grain size (range as well as mean), roundness, shape, and sorting of clasts.

2. *Structure.* Large-scale structures, such as the nature of bedding (e.g., cross-bedding, how thick are foreset beds, the slope angle of foreset beds, etc.), and small-scale structures (e.g., mudcracks) must be described in as much detail as possible.

3. *Associated lithic types and their spatial relationships.* Similar descriptions for all other rock types present and how they are related to

the sedimentary bed of interest should be made.

Microscopic characteristics

The description must start out by identifying the overall textural elements (e.g., matrix-supported or clast-supported, porosity, fossiliferous) and separate the nature of allochemicals from matrix or cement.

Allochemicals

1. *Modal composition (i.e., relative abundance of minerals, rock fragments, fossils, ooids, peloids, etc.)*

2. *Grain size distribution (range, mode, etc.)*

3. *Grain shapes and roundness (spherical, well rounded, angular, etc.)*

4. *Sorting (well sorted, medium sorted, poorly sorted)*

Cement/Matrix

Compositions and generations of cement and matrix and the physical relationships. (It may be possible to identify different generations of cement in which one may replace [cut across] another and so on. Cathodoluminiscence and SEM [back-scattered electron] imaging of cements often provide strong clues to zoning and replacement features in cement.)

BIBLIOGRAPHY

Boggs, S., Jr. (1992) *Petrology of Sedimentary Rocks.* Mcmillan (New York).

Dzevanshir, R.D., Buryakovskiy, L.A., and Chilingarian, G.V. (1986) Simple quantitative evaluation of porosity of argillaceous sediments at various depths of burial, *Sed. Geology* **46**, 169–175.

Flugel, E. (1982), ed. *Microfacies Analysis of Limestones.* Springer-Verlag (New York).

Garrels, R.M. and Christ, C.L. (1965) *Solutions, Minerals, and Equilibria.* Freeman (San Francisco).

Hibbard, M.J. (1995) *Petrography to Petrogenesis.* Prentice-Hall (Englewood Cliffs).

McClean, M. (1995) *Sedimentology.* Oxford University Press (New York).

Miall, A.D. (1990) *Principles of Sedimentary Basin Analysis* (2nd ed.). Springer (New York).

Plummer, C.C., McGeary, D., and Carlson, D.H. (1997) *Physical Geology* (8th ed.) WCB McGraw-Hill (New York).

Rye, R., and Holland, H.D. (1998) Paleosol and the evolution of atmospheric oxygen: A critical review, *Am J. Sci.* **298**, 668.

Metamorphism and Metamorphic Rocks

The following topics are covered in this chapter:

Metamorphism
Definition and metamorphic environments (types of metamorphism)
Agents—P, T, and fluid
Metamorphic textures
Nomenclature of metamorphic rocks

ABOUT THIS CHAPTER

In the previous chapters, we learned about igneous and sedimentary rocks. This chapter introduces the reader to some general concepts concerning the process of metamorphism, the nature of metamorphic rocks, pressure-temperature conditions in which metamorphism may occur, various types of metamorphism, and some useful concepts that allow the reader a first-order impression of the P–T conditions of various metamorphic rocks' formation.

METAMORPHISM

Metamorphism is the sum of all processes that involve mineralogical, textural, and chemical transformation of an igneous, sedimentary, or metamorphic rock into a different rock, called a *metamorphic rock*. Metamorphic reactions occur largely in solid state, with only limited changes in chemical composition of the original rock or protolith. They occur in response to changes in environmental conditions (e.g., changes in pressure, temperature and fluids migrating through them), related to such diverse geological phenomena as mountain building to emplacement of igneous intrusions. Geological time is a significant factor in metamorphism. It is generally believed that large, regional-scale metamorphic processes (i.e., orogenic metamorphism) operate over 10 to 50 million years (Bucher and Frey, 1994).

The boundary between sedimentary diagenesis and the lower limit of metamorphism is commonly placed at $\sim 150 \pm 50°C$, which is roughly the temperature at which certain metamorphic minerals, including glaucophane, lawsonite, prehnite, pumpellyite, or stilpnomelane, begin to form in the rock undergoing increasing intensity of metamorphism. The high temperature limit of metamorphism is constrained by melting of the rock (Figure 15.1). In Figure 15.1, the solidi of a wet (hydrous; $p_{H_2O} = p_{total}$) and a dry (volatile-free) granite are shown. Because crustal rocks are generally expected to form granitic melts on melting, the two solidi simply show the minimum temperatures of melting under wet versus dry conditions, at which metamorphism gives way to igneous phenomena. It should be noted that, in reality, intermediate and mafic composition rocks also exist in abundance in the crust, and their solidi occur at relatively higher

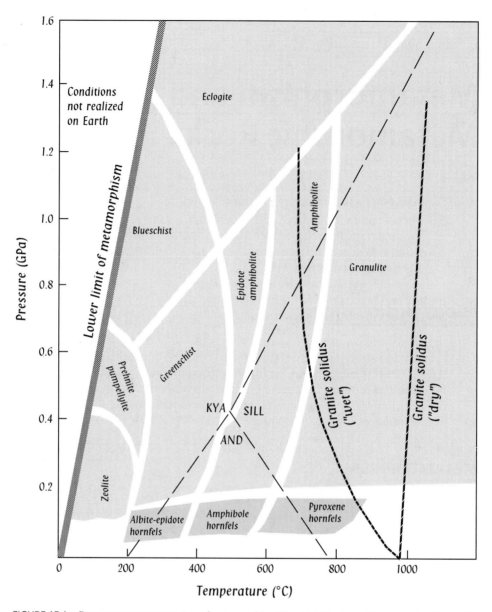

FIGURE 15.1 *Pressure-temperature regime of metamorphism.* The Al_2SiO_5 phase diagram and metamorphic facies (discussed in Chapter 16) are shown for reference. (After Spear, 1993, *Metamorphic Phase Equilibria and Pressure-Temperature-Time Paths*, Mineralogical Society of America, Washington D.C.)

temperatures. Therefore, the upper temperature limit of metamorphism (i.e., melting) will vary from place to place depending on the bulk composition of the protolith and the abundance and composition of the fluid species. In terms of pressure limits, metamorphism can occur at fairly shallow levels in the crust to as deep as the entire depth of the mantle. In general, most authors study metamorphosed crustal rocks that formed at pressures less than 1.6 GPa.

Until recently, there was not much knowledge available about the types of metamorphosed rocks of crustal origin (excluding peridotitic compositions) that may exist deep within the earth. New experimental studies at high pressure and recent discoveries of metamorphic rocks with minerals like coesite and microdiamonds (altered) from a number of localities around the world (such as Dabie Shan, eastern

China; Coleman and Wang, 1995) have led to the recognition of a new class of metamorphism: Metamorphism occurring at pressures of greater than 2.8 GPa is referred to as *ultrahigh-pressure metamorphism* (Coleman and Wang, 1995).

AGENTS OF METAMORPHISM

Temperature

The heat required for metamorphism may be supplied by (a) the heat released by igneous intrusions to the surrounding country rocks, (b) the geothermal gradient, (c) radioactive decay, (d) latent heat of crystallization given off by a crystallizing magma, (e) exothermic metamorphic reactions, and (f) shear heating (Spear, 1993). Among them, magmatic heating and heating by the geothermal gradient are by far the two most important mechanisms. Large (100s of km³) sections of continental crust buried deeply beneath fold-thrust belts over a subduction zone are likely to be metamorphosed due to pressure and heat provided by the geothermal gradient as well as by granitic batholiths or other deep-seated igneous intrusions. At the other extreme, on a much smaller scale, contact metamorphism takes place: Shallow crustal wall rocks around an igneous intrusion may be metamorphosed principally by the heat transferred from the intrusion to the wall rocks.

Pressure

The pressure within the earth's lithosphere is fundamentally of two kinds (Figure 15.2). (a) *Confining pressure* (lithostatic pressure or load pressure), which is the pressure applied on a deeply buried rock by surrounding rocks, is like hydrostatic pressure (which is the pressure applied by water on an immersed object) in the sense that

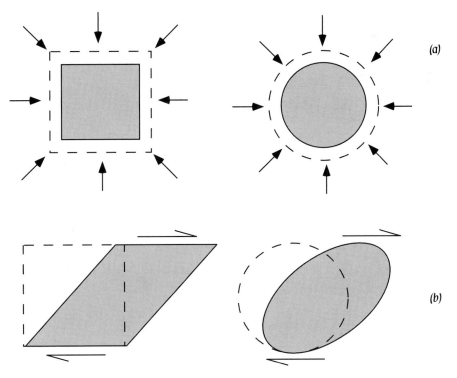

FIGURE 15.2   *Types of pressure responsible for metamorphism.* (a) Confining pressure (also called hydrostatic pressure or lithostatic pressure) and its effect on a cube and sphere (shown in two dimensions). (b) Deviatoric stress and its effect on a cube and a sphere (in two dimensions).

15.1 HEATING OF WALL ROCKS BY MAGMA

In the event of an igneous intrusion, the wall rocks surrounding the intrusion are heated by the intrusion. Such heat transfer may result in nucleation and growth of new minerals or coarsening of mineral grains (or both) in the wall rocks, which is metamorphism. The mechanism by which heat is transferred to the wall rocks is largely conduction, although advective (i.e., one-way, irreversible) transport by fluids percolating through the wall rocks is also likely to be important.

Here we examine cases where igneous intrusions of different dimensions heat up the surrounding wall rocks and determine how long it takes for the wall rocks to heat up via conductive transport as a function of time and distance from the intrusion. Consider the case of transfer of heat by conduction from an intrusion along a specific direction x into the wall rock. Fourier's law indicates that the rate of heat flow across a unit area (i.e., $dQ/[dt\,A]$, where Q is heat, A is area, and t is time), known as the *heat flux* (J_x), is proportional to the temperature difference (dT) between the transmitter (magma) and the receiver (wall rock) and inversely proportional to the distance (dx) between them:

$$J_x \propto [dT/dx]$$

or

$$J_x = k[dT/dx], \quad \text{[Eq. Box 15.1]}$$

where k is a proportionality constant known as *thermal conductivity*. The unit for thermal conductivity is $J\,s^{-1}\,m^{-1}°C^{-1}$ (joules per sec per meter per degree Celsius). The unit for heat flux is mWm^{-2} (milliwatts per square meter; 1 watt $= 1\,Js^{-1}$). Fourier's law indicates that the heat flux from a deep-seated pluton into the surrounding rocks occurs much more slowly than from a shallow-level intrusion (with same initial emplacement temperature) to the surrounding wall rocks because dT (which is the temperature difference between magma and wall rock) will be greater at shallow levels than at deep levels.

As noted in Chapter 7, the heat capacity of a rock at a constant pressure, C_p, is the amount of heat necessary to raise its temperature by 1°. Assuming that this rock has a density of ρ, the amount of heat required to raise the temperature of unit volume of the rock by 1° is $C_p\rho$. Thermal diffusivity, κ, is an important property of any material: It tells us the rate at which the temperature of that material may increase and is given as:

$$\kappa = k/[C_p\rho]. \quad \text{[Eq. Box 15.2]}$$

magnitude of this pressure is equal from all sides. The two examples—a square and a circle—in Figure 15.2a exemplify the first-order effect of confining pressure: It mainly reduces the volume and increases the density of the rock. Confining pressure is important in deep, hot, crustal roots of mountains in orogenic belts. (b) *Deviatoric stress* (shear stress, directed pressure) is one in which the pressure applied is not uniformly distributed, but is strongly directional. Figure 15.2b shows two examples in which the pressure, stress, is being applied only on the top and bottom. In each case, the top and bottom are being dragged in opposite directions so that both the square and circle deform such that they assume more elongated geometric forms. How a rock responds to deviatoric stress depends on confining pressure, temperature, and the strain rate: At low-confining pressure, a rock may fracture due to brittle deformation, whereas the same rock may flow (ductile behavior) at very high-confining pressure. Strain rate is the rate of deformation of a mineral grain (i.e., change of the mineral's geometrical attributes, such as shape and volume in response to pressure applied on it). As an example, consider the rock peridotite. At low crustal pressures and relatively cooler conditons, peridotite behaves as a brittle material. The same rock flows (i.e., exhibits ductile behavior) in the asthenosphere due to high temperature and a very low strain rate.

κ is roughly equal to 10^{-6} m^{2}s^{-1} for most magmas and rocks.

In the one-dimensional case of heat flow from an intrusion into the wall rock, the rate of temperature rise (i.e., dT/dt) is given as follows (Fourier's equation):

$$dT/dt = \kappa(\partial^2 T/\partial x^2). \qquad \text{[Eq. Box 15.3]}$$

The temperature T at a distance x within the wall rock away from an intrusion can be determined by a solution to the previous equation. A simple solution was given by Carslaw and Jaegar (1959):

$$[T - T_o]/[T_s - T_o] = 1/2\big[\mathrm{erf}[(h - x)/(2\sqrt{(\kappa t)})]$$

$$+ \mathrm{erf}[(h + x)/(2\sqrt{(\kappa t)})]\big], \qquad \text{[Eq. Box 15.4]}$$

where T_o is the initial magma temperature, T_s is the solidus temperature of the magma, and h is the half thickness of the intrusion (in meters). erf is something called an *error function* and may be found in tables of some textbooks (such as Table 4–5 in Turcotte and Schubert [1982]). Equation Box 15.4 suggests that the initial thermal perturbation due to the emplacement of an igneous intrusion decays exponentially with time.

A reasonable, straightforward estimate of how long it may take for an igneous intrusion to cool to ambient temperature can be obtained from the equation:

$$t = h^2/\kappa. \qquad \text{[Eq. Box 15.5]}$$

Using Equation Box 15.5 and assuming a $\kappa = 1 \times 10^{-6}$ m^2/sec, we calculate that a 0.2 km-thick ($h = 100$ m) dike will take $\sim$317 years to cool back to the ambient temperature, whereas a 20-km thick ($h = 10$ km) pluton will take 3.17 m.y. to do the same. Therefore all other factors being approximately equal, a thick pluton will be more likely to have an extensive metamorphic aureole (i.e., zone of metamorphism surrounding the pluton) compared with a narrow dike.

Spear (1993) pointed out that Fourier's law of heat conduction is similar to Fick's law of chemical diffusion: $\partial C/dt = D(\partial^2 C/\partial x^2)$, where $\partial C/\partial t$ is the change of concentration of a chemical species as a function of time, and D is chemical diffusivity (equivalent to thermal diffusivity). Although κ is around 10^{-6} m^{2}s^{-1} (10^{-2} cm^{2}s^{-1}), D is orders of magnitude smaller (e.g., 10^{-16} to 10^{-20} cm^{2}s^{-1} for cation diffusion in garnet at 500–700°C; Spear, 1993). What this comparison means is that thermal equilibrium will be achieved much faster than chemical equilibrium in a volume of rock undergoing metamorphism.

Aside from confining pressure and deviatoric stress, shock waves may be sent through a rock due to the impact of a meteorite. Such shock pressure may induce metamorphism in the impacted rock. Finding shock metamorphic minerals at/near the earth's surface is sometimes the only way to positively identify the site of an ancient meteorite impact because weathering, erosion, and sedimentation erase much of the tell-tale topographic signatures of an impact crater in most cases.

Fluids

The role of fluids in metamorphism is of great interest to metamorphic petrologists. Although metamorphism is largely a solid-state process, highly reactive intergranular fluids commonly participate in metamorphic reactions as a reactant or product, and/or as a facilitator that accelerates reactions between solid reactants. Without any involvement of fluids, metamorphic reactions would be tremendously slow because diffusion of chemical components are orders of magnitude slower in solids than in fluids. The intergranular fluid is generally a mixture of H_2O and CO_2 and may simply be groundwater heated by geothermal gradient or by an igneous intrusion, or it may be derived directly from the igneous intrusion.

Most petrologists distinguish between *isochemical metamorphism* (which is commonly accepted as metamorphism) and *metasomatism* depending on the extent to which fluids alter the bulk chemical composition of the original rock: If the bulk chemical composition is significantly modified, the process is called *metasomatism*; if not, it is called *isochemical metamorphism*. Some common, but limited, bulk chemical changes of the protolith occur—mainly by hydration or dehydration and change in Na and K contents.

METAMORPHIC CHANGES

Metamorphism results in the following changes in the mineralogy and texture of rocks:

1. *Recrystallization.* Changes in the grain size in response to temperature change and/or deviatoric stress is the simplest type of change. *Coarsening* of grains is an example of recrystallization. Grain size reduction (*pulverization*) due to shearing along fault zones is another.

2. *Neomineralization.* Growth of new minerals is the most common type of metamorphic change. For example, in Chapter 7 we noted that when a kyanite-bearing metamorphic rock is heated to very high temperature, pressure, the kyanite will transfrom into sillimanite. Thus, formation of the new mineral sillimanite is due to a a polymorphic transformation reaction. There are other ways for new minerals to form, for example, by hydration (e.g., formation of serpentine at the expense of olivine) or dehydration reactions (e.g., formation of K-feldspar at the expense of mica).

3. *Development of oriented fabric.* Metamorphism often results in the development of oriented fabric, such as foliation (a perasive [i.e., present throughout the rock] planar structure defined by parallel structural planes in the rock, defined below) and lineation (pervasive alignment of linear elements [e.g., prismatic mineral grains] in a rock).

4. *Metasomatism (discussed earlier).*

TYPES OF METAMORPHISM

Metamorphic rocks can form in a variety of geological environments—from midoceanic ridges to island arcs, continental arcs, and orogenic belts (i.e., fold-thrust belts) to continent–continent collision zones (Figure 15.3). P–T conditions are very different in each of these environments. Beneath the midoceanic ridges, hydrothermal solutions (hot water solutions rich in dissolved chemical species) percolate into the crustal rocks and trigger metamorphic reactions. The water may have an igneous origin or it may simply be ocean water filling up cracks in the crust and heated by shallow magma bodies. In this type of environment, pressure has virtually no role in initiating metamorphic reactions, and therefore the P–T path taken by the rocks undergoing metamorphism is one of very low pressure but low-to-high temperatures (Figure 15.1). In contrast, in subduction zone environments, and in areas where rapid overthrusting of continental slices is a characteristic feature, pressure generated by the convergence of two plates is the dominant control and temperature is a relatively minor factor. Therefore, metamorphic rocks produced in such an environment are characterized by a high P–low T path (Figure 15.1). Orogenic belts are located along active continental margins hundreds to thousands of kilometers long and are characterized by extreme deformation and recrystallization. In this type of environment, both pressure and temperature increase (Figure 15.1).

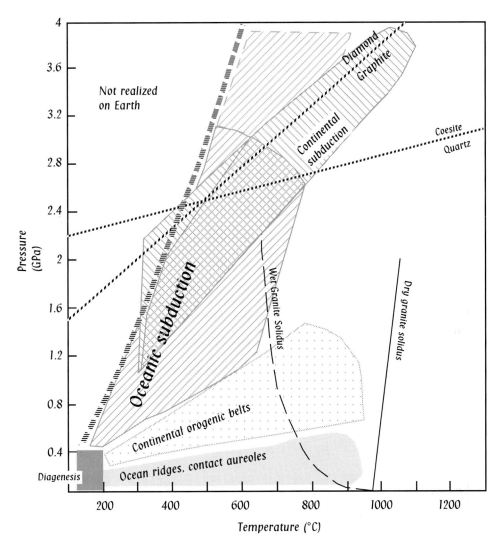

FIGURE 15.3   *P–T conditions in various plate tectonic regimes.* (After Spear, 1993, *Metamorphic Phase Equilibria and Pressure-Temperature-Time Paths*, Mineralogical Society of America, Washington D.C.)

Many different classes of metamorphism have been recognized. These classifications are based on the principal agent of metamorphism (i.e., P and/or T), geologic setting (i.e., local or regional), or plate tectonic environment. Excellent accounts of classification schemes of metamorphism may be found in textbooks by Spear (1993) and Bucher and Frey (1994). The reader is strongly advised to peruse these books for this purpose. This book focuses on the principal characteristics of the various types of metamorphism.

Following Bucher and Frey (1994), a first-order distinction is made between the types of metamorphism that are of local extent (less than a few km) and those that are of regional extent (cover 100s to 1000s of sq km; Figure 15.4). Further classification is based on the geological affinity. Note that, as with any classification, the boundaries between the divisions are gradational.

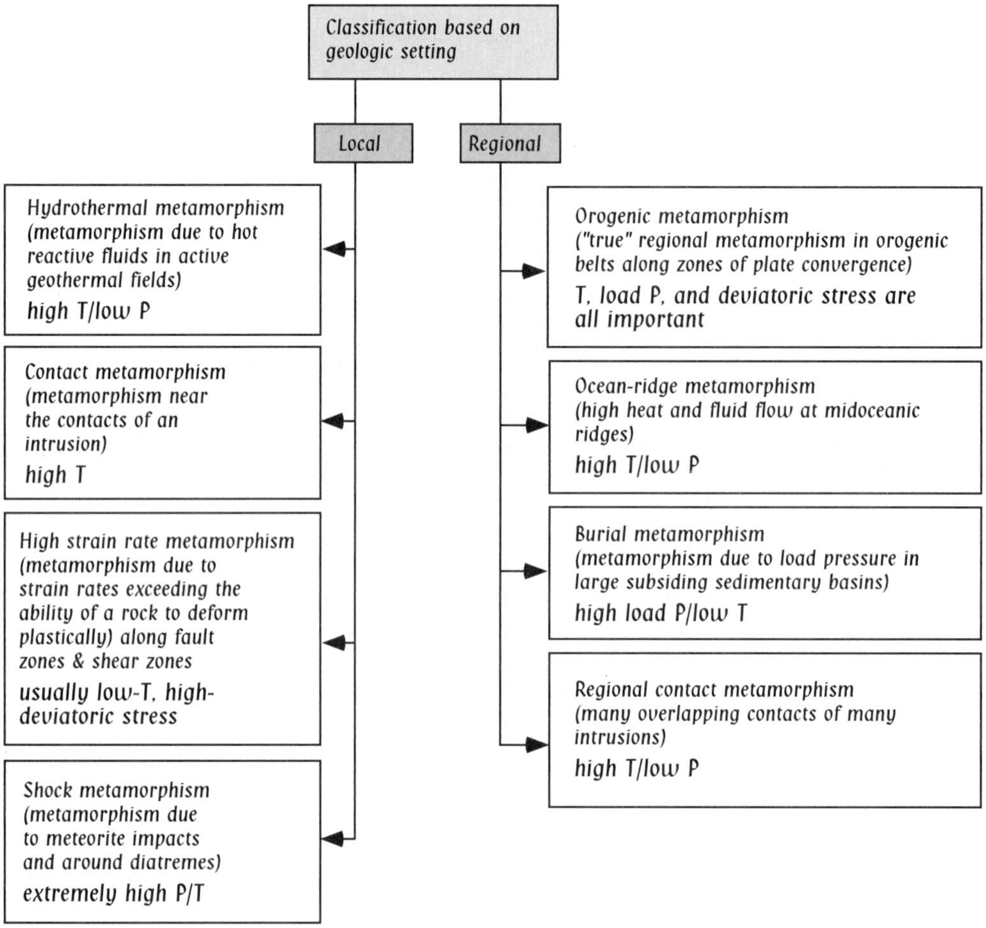

```
                        ┌─────────────────────┐
                        │ Classification based on │
                        │ geologic setting        │
                        └─────────────────────┘
                                   │
                     ┌─────────────┴─────────────┐
                 ┌───────┐                   ┌──────────┐
                 │ Local │                   │ Regional │
                 └───────┘                   └──────────┘
```

Hydrothermal metamorphism (metamorphism due to hot reactive fluids in active geothermal fields) **high T/low P**	**Orogenic metamorphism** ("true" regional metamorphism in orogenic belts along zones of plate convergence) **T, load P, and deviatoric stress are all important**
Contact metamorphism (metamorphism near the contacts of an intrusion) **high T**	**Ocean-ridge metamorphism** (high heat and fluid flow at midoceanic ridges) **high T/low P**
High strain rate metamorphism (metamorphism due to strain rates exceeding the ability of a rock to deform plastically) along fault zones & shear zones **usually low-T, high-deviatoric stress**	**Burial metamorphism** (metamorphism due to load pressure in large subsiding sedimentary basins) **high load P/low T**
Shock metamorphism (metamorphism due to meteorite impacts and around diatremes) **extremely high P/T**	**Regional contact metamorphism** (many overlapping contacts of many intrusions) **high T/low P**

FIGURE 15.4   *Types of metamorphism.*

REGIONAL TYPES

Regional Contact Metamorphism

Although contact metamorphism is a localized phenomenon, in orogenic areas, many intrusions may be emplaced fairly close to each other in space and time. As a result, their contact metamorphic aureoles may overlap and form a broad region of metamorphism. In these areas, the conditions are mainly low pressure and high temperature.

Burial Metamorphism

This type of metamorphism is typical of sedimentary basins where deeply buried sediments undergo compaction and metamorphism due to load pressure of the overlying sediments. In this case, temperature increases with pressure and closely follows the geothermal gradient. In burial metamorphism, the boundary between diagenesis and the beginning of metamorphism is not at all distinct. An excellent example of burial metamorphism comes from southern New Zealand (Boles and Coombs, 1977).

Ocean-Ridge Metamorphism

Basalts, gabbros, and ultramafic rocks are often exposed along fracture zones that cut across midoceanic ridges. Dredged and drilled samples of fracture zone rocks frequently show the presence of three types of metamorphic rocks: serpentinite, amphibolite, and greenstone. Serpentinites are metamorphosed ultramafic rocks, whereas greenstones and amphibolites are metamorphosed basalts and gabbros. This type of metamorphism results from high heat flow along the ridge axis and hot circulating fluids running through pore spaces and fractures in the rocks; it was called *ocean-ridge metamorphism* by Miyashiro et al. (1971).

Orogenic Metamorphism

This type of metamorphism is typical of orogenic belts along active continental margins (i.e., margins where subduction is active: continent–continent collision and continent–ocean collisional areas) and island arcs. The term *regional metamorphism*, often used by many authors, is generally synonymous with orogenic metamorphism. Sometimes the term *dynamo-thermal metamorphism* is also used to mean the same thing because both pressure and temperature are important factors in this type of metamorphism. Orogenic metamorphic rocks form the most extensive belts among all types of metamorphic rocks. Most of the research on metamorphic rocks—present and past—has been focused on them because they provide valuable information on burial, uplift, and erosional histories of orogenic belts.

LOCAL TYPES

Contact Metamorphism

This type of metamorphism occurs in the wall rocks of shallow crustal intrusions and is mainly caused by the heat provided by the intrusion. Pore water in the wall rocks and/or fluids released by magma (intrusion) often play a role in the metamorphism of the wall rocks. The metamorphosed contact zone is referred to as the *contact aureole*. Pressure is limited to a maximum of 0.3 GPa in contact metamorphism. Contact aureoles are generally thin, ranging up to only a few km. The thickness of the aureole depends on the thickness of the intrusive and the temperature difference between a magma and wall rock. Because the temperature difference between a magma and wall rock is likely to be the highest at shallow crustal levels, contact aureoles are generally most spectacular around shallow-level intrusions.

Contact metamorphic rocks typically lack directional textures, such as foliation (a penetrative planar texture) and thus develop a hornfelsic texture (a fine-grained texture lacking any directional features). Note that these textural terms are further illustrated in a later section. *Skarn* is a type of contact metamorphic rock formed due to reaction between limestone and fluids released by the igneous intrusion. Skarns are composed of calc-silicate minerals (such as wollastonite, diopside, grossular) and often serve as loci for ore deposits. Other contact metamorphic rock types include marble, hornfels, and so on.

Shock Metamorphism

This type of metamorphism results from shock waves created by the impact of a meteorite. Typical minerals produced by meteorite impact are high-pressure polymorphs of crustal minerals such as coesite, stishovite, and, rarely, microdiamonds.

High-Strain Metamorphism

In areas along fault boundaries or shear zones, strain rates (and deviatoric stress) are very high, exceeding the ability of the rocks to deform plastically. In such an environment, rocks undergo crushing and grinding (cataclasis) without much temperature change. Such mechanical action pulverizes the rock into a fine-grained rock with fine foliation (called a *cataclasite*). Another typical rock that develops under such conditions is a *fault breccia*, which is a type of cataclasite with visible angular fragments.

In shear zones, however, ductile flow occurs, which results in bending of grains, internal slip within grains, and some recrystallization (i.e., grain growth). In this case, the rock that develops is called a *mylonite*. Best (1982) provided an excellent explanation of the differences between fault zones and shear zones and illustrated how different textures may develop in high-strain metamorphic rocks. The reader may wish to consult his book.

Hydrothermal Metamorphism

This type of metamorphism typically occurs in geothermally active areas where hot reactive fluids react with the wall rocks and form new minerals. Hydrothermal metamorphism may be studied by drilling into the rocks in such geothermal areas. The source of the fluids may be igneous intrusions or groundwater heated by deep-seated magmatic intrusions or hot ground water heated by the geothermal gradient.

METAMORPHIC TEXTURES AND THEIR ORIGINS

Many types of textures develop during metamorphism depending on the nature of the protolith and the relative intensities of various metamorphic agents (Figure 15.5). The terms *blast* and *blastic* are used to describe metamorphic minerals and texture, respectively. The terms *idioblastic, subidioblastic*, and *xenoblastic* are used to describe grain shapes in metamorphic rocks; these are equivalents of *euhedral, subhedral*, and *anhedral*, respectively, which were used in earlier chapters on igneous rocks. Similarly, the terms *porphyroblast* and *porphyroblastic texture* are equivalents of *phenocryst* and *porphyritic* texture of igneous rocks, and *poikiloblastic texture* is the metamorphic equivalent of poikilitic texture. Note that the prefix *blasto-* is used to refer to the original texture (i.e., relict texture in the metamorphic rock) of the protolith. For example, blastoporphyritic texture in a metamorphic rock means the original porphyritic texture of the igneous protolith that may be seen in the metamorphic rock.

Subsolidus or fluid-aided reaction between mineral grains in a metamorphic rock often results in the development of a corona or rim of grains of the product phases around the unstable reactant phase. Such corona structures provide important clues to the pressure-temperature paths followed by the rock. Decomposition of a homogeneous solid solution into two or multiple phases during subsolidus cooling results in exsolution structures in metamorphic minerals. It is particularly common in alkali feldspars that have reequilibrated at lower temperatures. Exsolution structures are most common in metaigneous rocks, especially granulites.

In terms of overall texture of the rock on a microscopic to hand-specimen scale, metamorphic textures can be fundamentally divided into two classes based on whether foliation or some other preferred orientation (e.g., lineation) is present. The term *foliation* is a form of penetrative texture and refers to pervasive (i.e., penetrative) parallel to subparallel planar structure present in a metamorphic rock. These planes are usually defined by subparallel alignment of platy or prismatic minerals like micas and amphiboles, which have a strong preference to grow along the direction of least compressive stress (Figure 15.6). Slaty cleavage, schistosity, and gneissose texture are

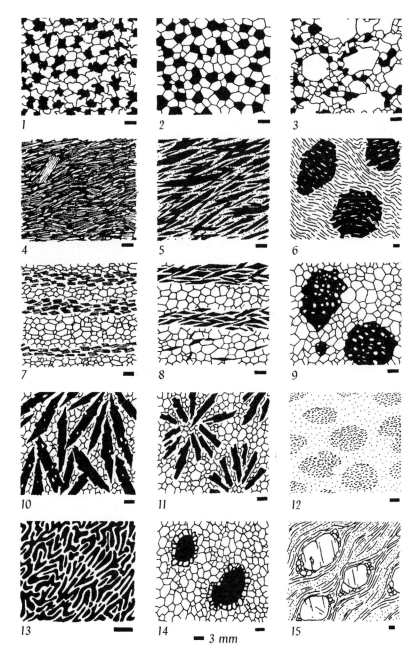

FIGURE 15.5 *Common metamorphic textures.* (a) Sketches of several common textural features of metamorphic rocks (scale: same as in Figure 15.5b). (b) Principal metamorphic textural types. (Reprinted with permission from Bard, 1986)
A. Granoblastic types   1. Isogranular texture (equal size xenomorphic/xenoblastic crystals).
2. Mosaic texture (with more triple junctions).   3. Heterogranular texture (different grain sizes).
B. Rock textures dependent on the habit of constituent minerals   4. Lepidoblastic texture (foliation defined by mica-type platy minerals).   5. Nematoblastic texture (foliation defined by needle-like or prismatic (e.g., amphiboles) minerals).   6. Porphyroblastic texture.   10. Sheaf texture.   11. Rosette texture.   12. Vermicular texture
C. Combinations of A and B   7, 8, 9 are combinations of textural types *A* and *B*.
D. Rocks with spheroidal type minerals or groups of minerals   12. Nodular texture.   14. Reaction corona (explained in Chapter 7).   15. Augen (lens-like) texture.
E. Tectonite textures   (see Figure 15.6d). (From Bard, 1980, *Microtextures of Igneous and Metamorphic Rocks*, Kluwer)

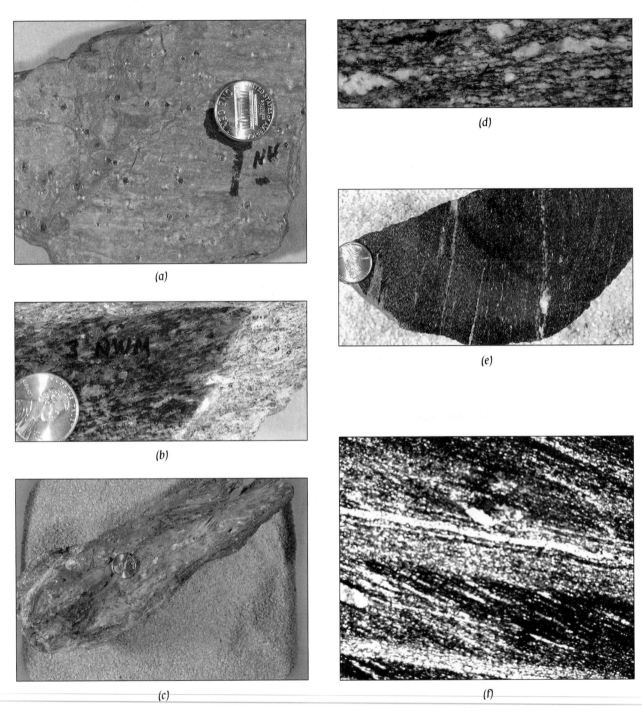

FIGURE 15.6 *Photographs of hand specimens and photomicrographs of some metamorphic rocks.*
Hand specimens (a) Garnet phyllite. Garnet crystals are the pophyroblasts protruding from the surface.
Well developed cleavage is visible in the rock. (b) Mica schist. (c) Kyanite schist. (d) Flaser texture. (e)
Gneissose texture. Photomicrographs (80× magnified). (f) Cleavage in a phyllite.
(g) Andalusite porphyroblast in a slate. (h) Foliation in a mica schist. (i) Amphibolite. (j) Sillimanite
grains in a sillimanite gneiss. (Courtesy of Gautam Sen)

(g)

(j)

(h)

(i)

all different types of foliation. *Slaty cleavage* is typically developed in a low-grade (i.e., low P, T) metamorphic, fine-grained rock called *slate*. It is characterized by fine subparallel-to-parallel sheets (i.e., rock cleavage). Individual minerals in the matrix are extremely fine grained in a slate. Schistosity is typically characterized by subparallel alignment of sheets of mica. The individual layers in a schist are relatively easy to peel off by hand. Gneissose texture refers to alternate bands of dark- (frequently amphibole or biotite) and light-colored (quartz and feldspar) minerals.

   Lineation is a penetrative linear structure in a metamorphic rock. Lineation in a rock may be defined by the alignment of needlelike or prismatic minerals (such as sillimanite, amphiboles, pyroxenes, etc.), intersection of two cross-cutting sets of foliation, alignment of fold axes in a folded set of foliation planes, and intersection of foliation and bedding (in the case of a metasedimentary rock in which bedding is preserved through metamorphism; Figure 15.5).

The term *granoblastic texture* may be used to describe the texture of a metamorphic rock that lacks any preferred orientation and contains equant grains of one or more minerals. Triple-point junctions formed by the merger of three polygonal grains at a point are common. Contact metamorphic rocks typically develop a texture known as *hornfelsic texture*, in which the grains are randomly oriented and well-developed polygonal aggregates are absent. Rocks possessing this texture are called *hornfels*. Hornfelses are called *pyroxene hornfels*, *amphibole hornfels*, or *albite-epidote hornfels* depending on the type of minerals that characterize them. Note that porphyroblastic (or poikiloblastic) texture, which is defined by the presence of porphyroblasts (or poikiloblasts), can occur in any rock regardless of whether it has a preferred orientation.

A special class of textures is exhibited by high strain metamorphic rocks developed along fault boundaries and shear zones (discussed before). *Cataclastic texture* refers to fine grained texture with strong foliation developed in fault zone rocks. *Porphyroclasts* are like porphyroblasts except that their grain boundaries show evidence of cataclasis and granulation. If the presence of porphyroclasts is the single most distinctive feature of a rock then its overall texture would be best described as *porphyroclastic texture* (Figure 15.5).

Processes and Textures

Table 15.1 provides a summary of the common types of textural changes that may be generated as a result of metamorphic conditions. Coarsening of grains is a common change that happens due to a process commonly known as *Ostwald ripening* in material sciences. As we have learned in a previous chapter, all systems work toward minimizing their Gibbs free energy. In the case of a metamorphic rock, a new mineral assemblage can develop at the expense of preexisting minerals because of its lower free energies. However, even when a mineral assemblage is stable because of its free energies, individual grains of each mineral may continue to grow to minimize grain boundary energies (same as interfacial energies in Chapter 9) resulting from disordered distribution of atoms along grain boundaries. Interfacial energy is high for small grains and for irregularly shaped grains. If Ostwald ripening were allowed to go to completion in a rock, then all grains of each mineral phase should fuse to form only one grain of each phase in a metamorphic rock. However, that does not happen because diffusion of components from one grain to another is extremely slow. Instead, different stable grain geometries or textures develop depending on the mineral phases that constitute the rock. These geometries minimize the interfacial energies.

Shapes of mineral grains are often controlled by their surface energy anisotropy (SEA). Minerals with a natural tendency to grow equant grains have very low SEA, and those that prefer growth along specific crystallographic directions have high SEA. This degree of SEA is shown in Table 15.2. The SEA series is not to be confused with the crystalloblastic series, which shows a hierarchy of minerals that naturally tend to be idioblastic to those that have a tendency to be xenoblastic (Table 15.3). A comparison of Tables 15.2 and 15.3 shows that, although some minerals (e.g. sillimanite) have high SEA and sit high on the crystalloblastic series, others, particularly garnet, have low SEA and yet are high on the crystalloblastic series because they have a tendency to form well-formed crystals.

Metamorphic rocks such as a quartzite (>90% quartz), which are dominantly composed of a single phase with relatively low SEA, tend to develop soap-bubble or honeycomb type texture with well-developed triple-point junctions and ~120° interfacial angles

TABLE 15.1 *Summary of common textural changes.*

Nature	Description	Causes
Coarsening	Increase of grain size, triple-point junctions	Minimization of contribution from grain boundary energies
Lineation	Pervasive development of linear alignment of mineral grains	Growth of elongated minerals along a direction of least compressive stress; or by intersection of two sets of foliation planes
Foliation	Pervasive development of parallel to subparallel set of planes, defined by alignment of micas, amphiboles, etc.	Coarsening and alignment of micaceous and other minerals parallel to the least compressive stress
Slaty Cleavage	A type of foliation, or rock cleavage, shown by slate, a low-grade metamorphic rock	
Schistosity	Another type of foliation, defined by close, parallel-subparallel alignment of micaceous minerals. Rocks with such texture are called *schists*.	
Gneissosity	A third type of foliation, defined by alternate, parallel bands of colored minerals (e.g., hornblende) and colorless minerals (e.g., feldspar). A rock that possesses such texture is called a *gneiss*.	
Cataclasis or *Granulation*	Development of finer grains at the expense of partially "ground down" coarser crystals (porphyroclasts)	Brittle deformation at shallow depths
Mylonitic	Commonly lenticular or eye-shaped coarse grains surrounded by fine grained materials	Brittle-to-ductile (grinding) deformation generates fine grains and rounding of coarse ones
Growth of Porphyroblasts	Coarse crystals (called porphyroblasts) set in a smaller grains-metamorphic equivalent of igneous phenocrysts	Inherent tendency of some minerals to grow faster than others, coupled with fewer nuclei of the coarse mineral per square area
Exsolution	Exsolved blebs, rods, or lamellae of one mineral in another. Common in pyroxenes	Limit of solid miscibility of one mineral phase in another under subsolidus conditions
Corona	Reaction rims of newly grown grains of one or more minerals around another, unstable, mineral grain	Reaction breakdown of one or minerals and growth of new minerals

437

TABLE 15.2 *Surface energy anisotropy in minerals.*

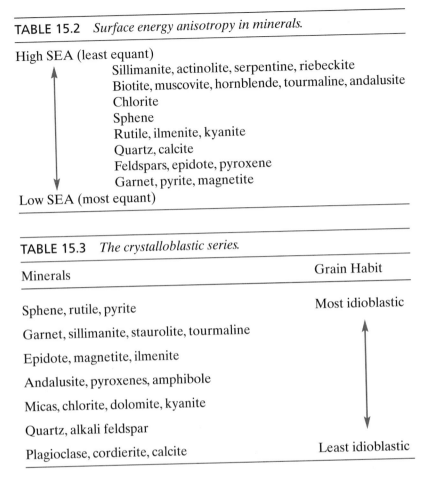

High SEA (least equant)

Sillimanite, actinolite, serpentine, riebeckite
Biotite, muscovite, hornblende, tourmaline, andalusite
Chlorite
Sphene
Rutile, ilmenite, kyanite
Quartz, calcite
Feldspars, epidote, pyroxene
Garnet, pyrite, magnetite

Low SEA (most equant)

TABLE 15.3 *The crystalloblastic series.*

Minerals	Grain Habit
Sphene, rutile, pyrite	Most idioblastic
Garnet, sillimanite, staurolite, tourmaline	
Epidote, magnetite, ilmenite	
Andalusite, pyroxenes, amphibole	
Micas, chlorite, dolomite, kyanite	
Quartz, alkali feldspar	
Plagioclase, cordierite, calcite	Least idioblastic

(Figure 15.7) because that is the stable configuration for a low-SEA phase in which interfacial energies are minimized. Prismatic minerals like amphiboles, pyroxenes, sillimanite, and kyanite have a natural tendency to grow along the *c*-crystallographic direction. Thus, in a two-phase rock, such as a sillimanite-bearing quartzite, elongated or fibrous sillimanite crystals may disrupt the smooth triple-point junctions between quartz grains,

FIGURE 15.7 *"Soap bubble" texture with prominent triple junctions.*

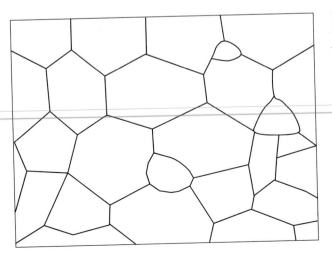

resulting in a somewhat lesser number of interfacial angles of ∼120°. Rocks with a higher abundance of grains of high-SEA phase tend to develop decussate-type texture (Figure 15.7), although polygonal aggregates still dominate the texture.

Porphyroblasts in metamorphic rocks may grow over an extended period or may grow late during a metamorphic event. A number of factors contribute to the growth of porphyroblasts, particularly relative growth rate and relative abundance. In an earlier discussion of nucleation, we noted that for a nucleus to become a crystal it must exceed a certain critical radius. The critical radius is larger for some minerals, such as garnet, and thus these minerals have a natural tendency to form large crystals or porphyroblasts. Also in a rock containing a number of different minerals, some of the minerals may grow faster than others and thus form porphyroblasts. In particular, when such a mineral forms relatively few nuclei, some of the nuclei may grow to large sizes while absorbing the rest of them. Porphyroblasts grow by pushing aside grains surrounding them or simply including them as they grow (Figure 15.8). The evidence

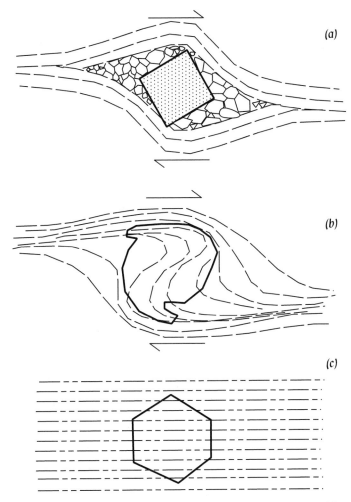

FIGURE 15.8 *Porphyroblasts in a mica schist.* (a) Resistant prophyroblast pushing aside surrounding groundmass while rotating and growing under shear stress. Note the coarser groundmass grains growing in the strain shadow areas. (b) Rotating porphyroblast that has incorporated S-shaped inclusions that are continuous with the surrounding groundmass foliation. (c) Porphyroblasts growing without pushing aside surrounding groundmass.

of shear stress and resulting rotation of porphyroblasts is often preserved as subparallel S-shaped groups of inclusions in porphyroblasts. In a foliated rock, foliation may wrap around a porphyroblast, and slightly coarser crystals in the matrix may form in the so-called *pressure shadows* in front of and behind the porphyroblast. Porphyroblasts are sometimes compositionally zoned because of very slow diffusion rates of components through solids, and therefore they record a long history of the pressure, temperature, tectonic, and compositional environment in which they grew. Study of porphyroblasts is therefore an important endeavor for metamorphic geologists.

Foliation may develop in a number of ways (Figure 15.9): (a) maximum crystal growth of minerals along the direction of the least compressive stress or of maximum extension, (b) pressure solution, and (c) re-orientation of glide planes in individual minerals. The development of foliation by the first mechanism has been confirmed by

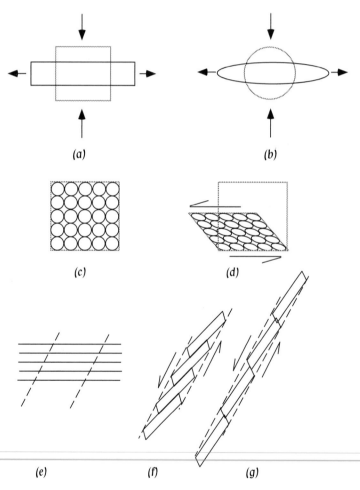

FIGURE 15.9 *Some schemes of development of foliation.* (a) and (b) Growth of crystals in a direction perpendicular to the maximum compressive stress (arrows pointing toward the grain). Pressure solution causes dissolution of the grain surfaces facing the maximum compressive stress directions; and reprecipitation along the surfaces where the compressive stress is minimum causes grain growth. Thus, square and spherical grains will become elongated. (c) and (d) Flattening of grains due to shear stress. c and d show before and after scenarios. (e), (f), (g) Foliation due to transposition of original glide planes in a mineral. e shows the original glide planes. f and g represent progressive deformation and transposition.

numerous studies of deformed and flattened ooids and fossils elongated along the foliation planes in a slate. Ooids (and other features) serve as a passive strain marker because their originally spherical shapes are modified by deformation. By noting the extent of shape modification, one can calculate the amount of strain. Strongly anisotropic (in a crystallographic sense) minerals, such as the micas, grow faster in specific crystallographic directions. In mica schists, for example, mica forms blades and flakes that grow in the direction of least compressive stress and define its schistosity.

Pressure solution of quartz grains can result in preferred orientation defined by elongated quartz grains in a foliated quartzite. Basically, in an environment of unequal stress distribution, the faces of a quartz crystal oriented perpendicular to the maximum compressive stress will dissolve, and the faces aligned along the least compressive stress will grow by reprecipitation.

Rotation of glide planes in some mineral grains may also result in an overall alignment of the grains in the direction of foliation. Olivine and quartz are well-known examples of this process.

Finally, deviatoric stress and accompanying brittle to ductile deformation can produce a series of textures such as mylonitic and cataclastic textures. In a fault zone environment, a layered rock consisting of bands of hard, brittle minerals in a matrix of softer, clay-rich materials can develop lenses of the brittle mineral (often called *Boudins*)—the rounding off the rough edges of such lenses being a result of rotation and grinding during translation and rotation imposed on the brittle band by shear stress. Such textures are often described as *flaser texture* or *augen texture*.

COMMON METAMORPHIC ROCK NAMES

In contrast to the nomenclature of igneous rocks, naming metamorphic rocks is a relatively simple effort and is generally based on modal mineralogy and texture (or structure at a mesoscopic scale). A metamorphic rock name usually has a root name and one or more prefixes. The root name may indicate dominant mineralogy (e.g., *amphibolite*—rock composed of amphibole and plagioclase) and/or structure/texture (e.g., *slate*—a fine-grained, low-grade metamorphic rock with distinctive slaty cleavage; greenschist—a chlorite [therefore the green color]-bearing rock with schistose texture). The prefix(es) may include some specific textural/structural or mineralogical attribute (e.g., garnet amphibolite, pyroxene granulite, etc.). Sometimes the name may have a built-in reference to the protolith from which the metamorphic was derived (e.g., a *metabasalt* is a metamorphosed rock that was originally a basalt). Sometimes the prefixes *ortho-* and *para-* are used to identify the protolith: *Ortho-* means that the rock was derived from an igneous rock, whereas *para-* means that the rock had a sedimentary protolith (e.g., orthogneiss, paragneiss, etc.).

The following is a summary list of some commonly used metamorphic rock names (somewhat modified from Bucher and Frey, 1994):

> *Gneiss* A metamorphic rock with gneissose texture. One can add a prefix to describe the rock more adequately (e.g., sillimanite gneiss, garnet-biotite gneiss).
>
> *Schist* A rock that has schistose texture, such as a mica schist.
>
> *Slate* A fine-grained, low-grade metamorphic rock with slaty cleavage.
>
> *Phyllite* Another fine-grained, low-grade metamorphic rock with strong foliation in which the grains are slightly coarser and impart a shininess to the cleavage surface.

Blueschist A bluish-colored schist whose blue color is due to the abundance of the bluish, pleochroic, amphibole crossite. This rock is typical of the high P/low T environment of a subduction zone. It is often erroneously equated to *glaucophane schist*, whose amphibole is in fact glaucophane and not crossite.

Greenschist A green-colored rock with schistose texture. The green color is due to the green-colored minerals like chlorite, actinolite, epidote, and pumpellyite.

Greenstone Also a fine-grained, green-colored rock, but lacks any foliation.

Hornfels It is a fine-grained rock that lacks any foliation and generally retains relict features (minerals, texture) of the original protolith. A hornfels is produced via contact metamorphism.

Amphibolite A rock dominated by amphibole and plagioclase. It is a metamorphosed basalt (or diabase or gabbro).

Eclogite A high-pressure metamorphic rock with dark green color (due to omphacitic clinopyroxene) and roundish brown patches (due to brown garnet). It is composed essentially of garnet and an omphacitic clinopyroxene. This rock occurs at outcrop scale and is believed to be an important component of subducted lithosphere. Eclogites have also been known to occur as xenoliths in kimberlites.

Granulite It is typical of granulite facies (discussed in the next chapter) and displays characteristic granulitic texture with well-formed triple junctions. It generally contains an ortho- and clinopyroxenes and plagioclase or garnet and has a granoblastic texture. The lower continental crust is believed to be dominantly granulitic.

Serpentinite A rock composed almost entirely of serpentine group of minerals. Serpentinites have an ultramafic protolith.

Marble A metamorphosed limestone or dolostone composed of calcite and/or dolomite.

Calc-silicate rock As the name suggests, this rock has Ca-silicate minerals, including epidote, zoisite, diopside, wollastonite, anorthite, scapolite, Ca-amphibole, and Ca-garnet (e.g., grossular).

Skarn A contact metasomatised rock that exhibits compositional bands formed due to interaction between fluids released by an igneous intrusion and a calcareous wall rock. The minerals typically are: wollastonite, forsterite, grossular, zoisite, anorthite, and so on.

Quartzite A quartz-dominated rock (>90% quartz) with a distinct metamorphic texture.

Migmatite These are complex stringers or lenslike rocks with felsic and mafic parts found near batholiths in exposed, deep to moderately deep crustal sections of orogenic belts. They represent partial melts (felsic) and their mobilized residues (mafic).

SUMMARY

Metamorphism includes all solid-state processes that transform one rock into another. The two dominant changes are in mineralogy and texture. Bulk chemical composition changes only little, mostly due to hydration or dehydration reactions. Extensive chemical change of the protolith, a process known as *metasomatism*, is not volumetrically important. The three agents of metamorphism are pressure (both load pressure and deviatoric stress), temperature (sources: magmatic heat, geothermal gradient, shear heating, crustal thickening), and intergranular fluid (sources: heated ground water, magmatic water). The two most important textural changes are the development of foliation and the growth of porphyroblasts. Two fundamental classes of metamorphism, local versus regional, are recognized. Metamorphic rock names are relatively simple and based on texture or mineralogy.

BIBLIOGRAPHY

Bard, J.P. (1986) *Microtextures of Igneous and Metamorphic Rocks*. Kluwer (Boston).

Best, M.G. (1982). *Igneous and Metamorphic Petrology*. W.H. Freeman (San Francisco).

Carslaw, H.S. and Jaeger, J.C. (1959) *Conduction of Heat in Solids* (2nd ed.). Oxford University Press (Oxford, U.K.).

Coleman, R.G. and Wang, X. (1995) *Ultrahigh-pressure Metamorphism*. Cambridge University Press (New York).

Spear, F.S. (1993) Metamorphic phase equilibria. Pressure-Temperature-Time paths. Monograph. *Mineralogical Society of America* (Washington, D.C.).

Turcotte, D.L., and Schubert, G. (1982) *Geodynamics: Applications of Continuum Physics to Geological Problems*. J. Wiley and Sons (New York).

Miyashiro, A. (1973) *Metamorphism and Metamorphic Belts*. J. Wiley and Sons (New York).

Metamorphic Facies, Reactions, and P–T–t Paths

The following topics are covered in this chapter:

Relating metamorphism to P–T conditions: Concepts of zones, grades, facies, and petrogenetic grid
Graphical representation
Metamorphism of mafic and pelitic rocks
Ultra-high-pressure metamorphism
Geothermobarometry, reaction kinetics, and P–T–t paths
Seismometamorphism

ABOUT THIS CHAPTER

In the previous chapter, we learned that metamorphism occurs because the mineral assemblage of the protolith becomes unstable in a new environment (defined by P–T fluid). Chemical reactions lead to the breakdown of old minerals (reactants) and development of new ones (products). Both the reactant minerals (often preserved as relicts) and new minerals serve as recorders of the pressure-temperature conditions to which the rock was subjected. Recent developments in theory and analytical techniques have allowed the metamorphic petrologist to decipher the times at which the metamorphosed rocks of a particular region were subjected to different P–T conditions. In other words, the evolutionary history of such a region in terms of pressure–temperature–time (P–T–t) has been worked out. This chapter examines the complexities involved in studying metamorphic reactions and exposes the reader to early developed concepts of metamorphic zones, isograds, and facies, as well as to more modern concepts of thermobarometry (i.e., pressure-temperature determination using mineral composition) and P–T–t paths. Perhaps the most exciting new development in metamorphic petrology has been the study of metamorphic rocks that have been recycled from the crust to the deep upper mantle (and maybe the transition zone) and back to the surface. A brief presentation on these rocks is made in the last part of this chapter.

MINERALOGICAL CHANGES: ZONES, ISOGRADS, AND FACIES

A study in 1893 by George Barrow of Dalradian rocks from the Scottish Highlands (Figure 16.1) was the first field-based documentation of mineralogical changes as a function of metamorphic intensity in what was originally a mudrock (commonly called a *pelite* in metamorphic petrology). Barrow and many later workers have shown that distinct zones, the boundaries of which are marked by the appearance/disappearance of a specific mineral, called an *index mineral*, can be mapped on the outcrop not only in the Scottish Highlands, but throughout the world. These zones and their typical

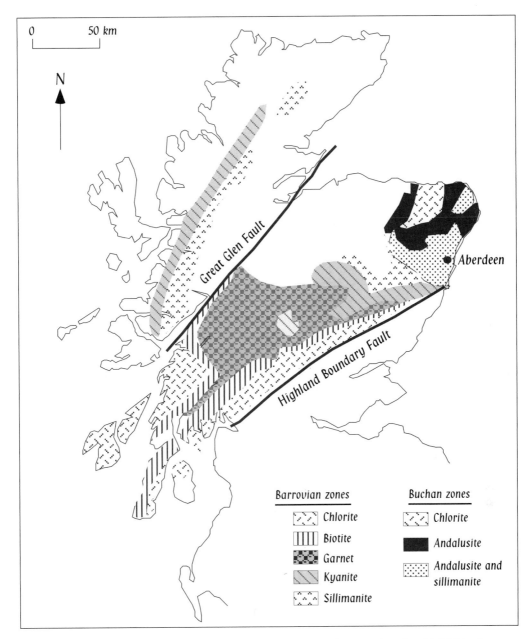

FIGURE 16.1   *Map showing Barrovian and Buchan metamorphic zones from Scottish Highlands.*

minerals in an original mudrock are as follows (in order of increasing metamorphic intensity or grade):

1. ***Chlorite Zone:*** *Chlorite* + Muscovite + Quartz + Albite
2. ***Biotite Zone:*** *Biotite* + Chlorite + Muscovite + Albite + Quartz
3. ***Garnet Zone:*** *Almandine* + Quartz + Biotite + Muscovite + Albite-rich Plagioclase
4. ***Staurolite Zone:*** *Staurolite* + Almandine + Quartz + Muscovite + Biotite + Plagioclase

5. ***Kyanite Zone:*** *Kyanite* + Almandine + Quartz + Muscovite + Biotite + Plagioclase

6. ***Sillimanite Zone:*** *Sillimanite* + Almandine + Muscovite + Biotite + Quartz + Plagioclase + K feldspar

Barrow's interpretation that the metamorphic grade or intensity increased from the chlorite to the sillimanite zones was based on his observation that grain size also increased. These zones became known as *Barrovian Zones*. Barrovian Zones are now recognized as an intermediate P–T metamorphism (orogenic metamorphism; Figure 16.2). C.E. Tilley (cited in Spear, 1993) later extended Barrow's study and

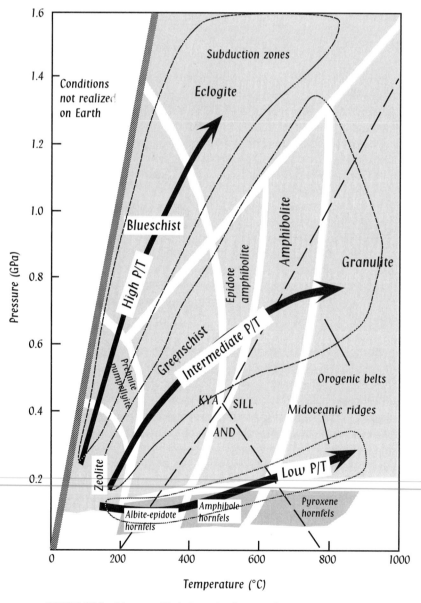

FIGURE 16.2 *Metamorphic facies series, facies, and tectonic associations.*

introduced the concept of an *isograd*. An isograd is a contour on a geological map (which is really a surface in three dimensions) that marks the first appearance or disappearance of an index mineral.

Following Barrow's work, many other studies have found similar as well as different types of metamorphic zonation in rocks of pelitic composition throughout the world. Not too far from Barrow's study area, in the Buchan region of eastern Dalradian, a very different sequence of metamorphic zones occurs in a pelitic protolith: Here the index mineral sequence goes from *staurolite* $\longrightarrow$ *cordierite* $\longrightarrow$ *andalusite* $\longrightarrow$ *sillimanite*. These zones, called *Buchan Zones*, represent much lower pressure conditions than the Barrovian Zones (low P–T: Figure 16.2).

Although mapping of zones is a useful practice because of its simplicity and aid in developing a sense of overall P–T conditions for the field geologist, many problems may occur in the interpretation stage. For example, one can get cross-cutting isograds, as has been shown by D.M. Carmichael (1969) from the Whetstone Lake area of Ontario. Carmichael and many other studies have shown that the bulk composition (including fluid composition) exerts an important control on the types of mineral reactions that can occur in a particular protolith, and thus affect the single-index, mineral-based isograds. Many modern authors use two index minerals instead of one in mapping out zones.

Bulk composition is particularly important because it dictates what minerals may form at any given P–T and fluid composition. Broadly speaking, in an area of regional metamorphism, different starting rocks would end up becoming very different metamorphic rocks at a similar metamorphic intensity because of their bulk compositional differences. The following is an example of how very different rock types may result from different protoliths under very similar P–T conditions:

Protolith		*Metamorphic Rock*
Sandstone	$\longrightarrow$	Quartzite
Limestone	$\longrightarrow$	Marble
Basalt	$\longrightarrow$	Amphibolite
Granite	$\longrightarrow$	Granitic gneiss
Shale	$\longrightarrow$	Sillimanite gneiss
Peridotite	$\longrightarrow$	Olivine-tremolite schist

An obviously important factor in the study of metamorphism is to decipher what reactions occurred because only then is there any hope of figuring out the P–T history of a series of metamorphic rocks in any given geographic area. This is not always easy, and one must consider many different observations. For example, one may find reaction corona type structures (in such a structure, the product minerals form a reaction rim around the reactant mineral). The reactants may also be preserved as relict inclusions in the product minerals.

It is important to recall that metamorphic reactions do not instantaneously go to completion, but happen over geological time. Although metamorphism is driven by chemical equilibrium, how far a reaction would progress is determined by kinetic processes, such as diffusion of ions from one mineral to the other. Thus, the stages of metamorphic reaction progress may be preserved in the form of mineral zoning, in which the core of a mineral grain may retain the reactant composition while its rim may be in chemical equilibrium with the surrounding ground mass minerals. For any given protolith composition, the modal proportions of the product minerals increase as the reactant mineral proportions decrease with the pogression of a reaction. Thus, the metamorphic geologist may be able to determine what reaction has occurred in a particular metamorphic terrain from changing modal proportions of minerals.

Facies

Recognition of problems that are inherent in the isograd concept encouraged metamorphic petrologists to think of other ways to represent variations in metamorphic intensities, which ultimately led to the development of the concept of *metamorphic facies*. The foundation of this concept was laid by V.M. Goldschmidt (1911) in an outstanding study of a wide range of sedimentary rocks, including mudrocks, sandstones, and limestones, which were all metamorphosed by a pluton. He noticed that, despite the wide variation in bulk composition of the starting, unmetamorphosed parent rocks, these rocks developed metamorphic assemblages whose mineralogies were relatively simple; each lithology consists of four or five of the following minerals— quartz, K feldspar, plagioclase, cordierite, wollastonite, diopside, hypersthene, and grossular garnet. He also noted that for a particular bulk composition the mineral assemblage is the same. Thus, a mudrock has quartz-K feldspar-andalusite-cordierite, and the equivalent calcareous metamorphosed rock has a diopside-grossular-calcite assemblage (cited in Turner, 1968). More or less at the same time, in a study of some hornfelses from Finland, P. Eskola (1915) made a similar observation: that metamorphic mineral assemblages are simple in terms of number of minerals and are specific to a certain bulk composition. More important, Eskola noted that the mineral assemblages in the Finlandian example were very different from those in Norway even though the chemical compositions of the rocks from the two areas were more or less identical. Eskola concluded that these differences must be due to different P–T conditions of metamorphism at the two places. Thus was born the concept of metamorphic facies, which was perhaps best defined by Fyfe and Turner (cited in Turner, 1968) as: "a set of mineral assemblages, repeatedly associated in space and time, such that there is a constant and therefore predictable relation between mineral composition and chemical composition."

It is important to recognize that a metamorphic facies does not represent a single rock type, but includes a wide range of rock types that form under the same (or very similar) P–T and fluid composition conditions. Although Eskola recognized that rocks belonging to different facies must have formed at different P–T conditions, appropriate experimental data that are available today did not exist then, therefore he was unable to quantify the P–T differences between different metamorphic facies in terms of absolute ranges in P–T space. Subsequent studies by many workers have led to the development of a general facies diagram (Figure 16.2; after Spear, 1993). Note that the names given to the various facies are based on mineral assemblages that would develop in a mafic bulk composition at various P–T conditions, and the minerals that define each of these facies are as follows (Spear, 1993; Table 2–1):

Zeolite Facies: zeolites
Prehnite-Pumpellyite Facies: prehnite and pumpellyite
Blueschist Facies: glaucophane + lawsonite or epidote (+albite ± chlorite)
Greenschist Facies: chlorite + albite + epidote (or zoisite) ± actinolite
Epidote-Amphibolite Facies: plagioclase (albite-oligoclase) + hornblende + epidote ± garnet
Amphibolite Facies: plagioclase (oligoclase-andesine) + hornblende ± garnet
Granulite Facies: orthopyroxene (+ clinopyroxene + plagioclase ± hornblende ± garnet)
Eclogite Facies: omphacitic pyroxene + garnet

The boundary between any two facies is *gradational*. These boundaries are generally based on discontinuous reactions (discussed in an earlier chapter and in a later section) in which a new mineral assemblage appears or another one disappears.

Continuous reactions, involving solid solutions, and exchange reactions, involving exchange of components between different minerals, occur within each facies and across facies boundaries (these are also discussed further in a later section). Because many of the reactions that mark the facies boundaries involve loss or gain of volatile species, their temperature can vary somewhat depending on the pressure.

It is important to note that although rocks undergo metamorphism over an increasing (or *prograde*) as well as decreasing (or *retrograde;* during the exhumation period) set of P–T conditions, the assignment of a metamorphic rock to a particular facies is almost always based on the *peak metamorphic conditions* it reached. For example, before a basalt becomes an amphibolite, it must go through several other facies conditions (such as the greenschist facies), and it must undergo retrograde P–T conditions as well during its exhumation to the earth's surface. However, the retrograde conditions are generally incapable of obliterating the peak P–T conditions reached by the rock, and hence the rock gets assigned to the amphibolite facies although it may be apparent that it has gone through a lot more complex metamorphic history.

Metamorphic Facies Series and Plate Tectonics

A. Miyashiro (1961) noticed the consistent differences between the Barrovian versus Buchan type sequences in his studies of Japanese metamorphic belts. In fact, he recognized three types of sequences or barric types (because the main difference between them is pressure) or metamorphic facies series:

1. zeolite → prehnite → pumpellyite → blueschist → eclogite (high P–T or Sanbagawa type)
2. greenschist → epidote-amphibolite → amphibolite → granulite (intermediate P–T, Barrovian type: shows kyanite → sillimanite conversion)
3. greenschist → amphibolite → granulite (low P–T, Buchan or Abukuma type: shows andalusite → sillimanite conversion)

Even before the concept of *plate tectonics* was presented, Miyashiro had recognized the existence of subparallel belts of high-P–T and low-P–T metamorphic rocks in the Japanese islands parallel to the Trench zone, called *paired metamorphic belts* (Figure 16.3). The low-P–T belt, composed of andalusite-sillimanite bearing facies assemblages, occurs to the northwest of a major tectonic discontinuity (Median Tectonic Line), and the high-P–T belt occurs to the southeast of it. The high-P–T belt goes from zeolite facies rocks to blueschist/greenschist facies rocks (+ some amphibolite) from southeast to northwest. Miyashiro noted the existence of similar high-P–T and low-P–T paired belts throughout the Mesozoic-Cenozoic rocks of the entire Pacific rim.

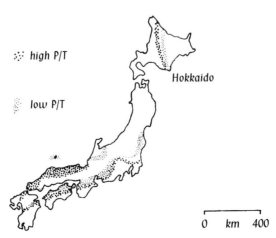

:: high P/T

:: low P/T

Hokkaido

0   km   400

FIGURE 16.3 *Paired metamorphic belts of the Japanese islands.* (Reprinted with permission from Best, 1982, *Igneous and Metamorphic Petrology*, W.H. Freeman)

Paired metamorphic belts have since been found in many other parts of the world, and their origin is clearly related to plate tectonics (Miyashiro, 1973; W.G. Ernst, 1971, 1976; Figure 16.4). In the case of Japan, the high-P–T belt marks the po-larity of the subduction zone in that the higher-grade rocks are further to the north-west (which is the direction in which the subducting plate is moving). The low-P–T belt is an ancient island arc that has been thrust against the high-P–T belt. Such thrust-ing is common in subduction zones. The schematic diagram in Figure 16.4 shows that the collision and subduction of the cold lithospheric plate create a situation where high-P–T facies series rocks are bound to be juxtaposed against a low-P–T series.

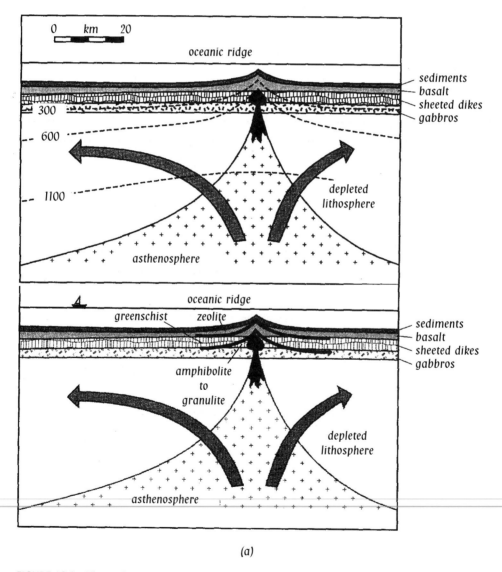

(a)

FIGURE 16.4 *These schematic diagrams show the close relationship between plate tectonics and metamorphism (From Ernst 1976).* (a) Ocean ridge metamorphism. The upper panel shows the nature of the crust at a mid-oceanic ridge and the isotherms (dashed with numbers) in the lithosphere. The lower panel shows the location of various metamorphic facies. (b) Metamorphism associated with a subduction zone environment. The upper panel shows a typical cross-section of an island arc and the isotherms. The lower panel shows the location of various metamorphic facies.

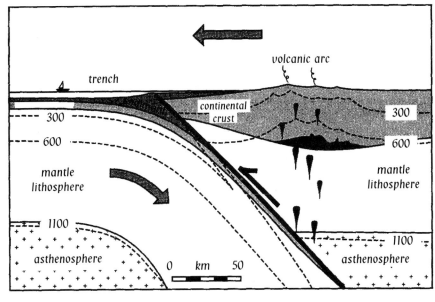

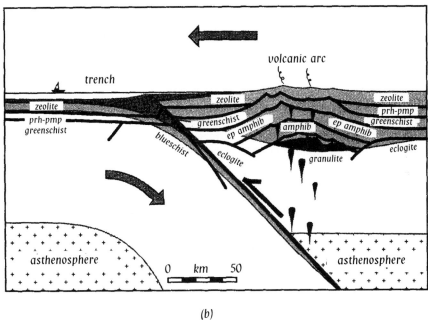

(b)

FIGURE 16.4   *(continued)*

At a *mid-oceanic ridge*, the isotherms are closest to each other above the active zone of magmatism (Figure 16.4). The gap between them increases as a function of distance away from the ridge axis (therefore, age of the lithosphere). Rocks dredged from oceanic fracture zones, where the middle and lower crustal rocks are often exposed, indicate that the oceanic crust probably does not get metamorphosed beyond the amphibolite facies conditions under most circumstances. Note also that in order for the igneous crust to turn into metamorphic rocks, water must have access to great depths via fractures. This probably occurs only in the vicinity of fracture zones and rift zones. Therefore, extensive metamorphism, beyond the low P/T metamorphism of the outermost 1 km or so, probably does not occur in a mid-oceanic ridge situation.

METAMORPHIC REACTION EQUILIBRIA, GRAPHICAL REPRESENTATION, AND PETROGENETIC GRID

Reactions

The realization that different mineral assemblages can form in two different rock types, although they both belong to the same facies, led to the obvious need for the study of many different reaction equilibria involving minerals and volatiles. In a basic sense, one can classify metamorphic reactions into four fundamental types for the convenience of treatment:

1. **Discontinuous reaction.** In this type of reaction, a phase breaks down and/or another appears at a univariant curve. A discontinuous reaction may be of two types: (a) *polymorphic* (e.g., kyanite = sillimanite), and (b) *net-transfer* (e.g., the components of reactant minerals are transferred to the product mineral[s]). A good example is:

$$\text{jadeite } [NaAlSi_2O_6] + \text{quartz } [SiO_2] = \text{albite } [NaAlSi_3O_8].$$

2. **Continuous reaction.** In this type of reaction, the product and reactant minerals do not disappear, but continuously change composition by exchanging components as P–T are varied. A good example of this is the following exchange reaction between garnet and biotite (Figure 16.5):

$$\text{Mg } [gar] + Fe^{2+} [bio] = Fe^{2+} [gar] + \text{Mg } [bio].$$

Figure 16.5 resembles the types of solid-solution diagrams encountered in Chapter 7; the only difference here is that we are not dealing with liquidus and solidus, but both curves represent compositional variation in coexisting garnet and biotite as a function of temperature. For the bulk composition chosen as an example, garnet and biotite exchange Mg and Fe^{2+} and both become progressively more magnesian with increasing temperature (i.e., prograde metamorphism) as the proportion of garnet in the rock also increases.

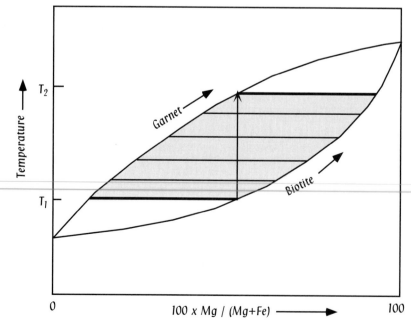

FIGURE 16.5 *Continuous reaction between garnet and biotite.*

The sensitivity of this exchange reaction to temperature makes it a very good geothermometer that can be used to determine the temperature of equilibration of metamorphosed rocks that contain the two phases.

3. Exsolution/solvus reaction. In this type of reaction, a single phase breaks down into two or more phases via exsolution (or decomposition) during retrograde cooling of a rock. An example of this is the following reaction:

$$Al_2O_3\text{-rich orthopyroxene} = Al_2O_3\text{-poor opx} + \text{garnet}$$
$$\text{[High Temp.]} \qquad\qquad \text{[Low Temp.]}$$

4. Reaction in which volatiles (primarily CO_2 and H_2O) are involved. Examples involving volatiles follow:

$$CaCO_3 \text{ [calcite]} + SiO_2 \text{ [quartz]} = CaSiO_3 \text{ [wollastonite]} + CO_2 \text{ [gas]}\Uparrow$$
$$(\textit{decarbonation reaction})$$

$$\text{staurolite} + \text{chlorite} + \text{muscovite} = \text{biotite} + \text{kyanite} + \text{quartz} + H_2O\Uparrow$$
$$(\textit{dehydration reaction})$$

$$\text{margarite} + 2\,\text{quartz} + \text{calcite} = 2\,\text{anorthite} + 1\,CO_2\Uparrow + 1\,H_2O\Uparrow$$
$$(\textit{mixed volatile reaction})$$

In Chapter 7, thermodynamic methods have been presented that allow calculation of univariant curves in P–T space. The methods employ calculation of the slope of a given reaction using the Clapeyron equation (i.e., $dP/dT = \Delta S/\Delta V$, where ΔS and ΔV are entropy and volume changes, respectively). Because volatile-free solids are not very compressible (i.e., ΔV is constant), discontinuous and exchange reactions involving solids (volatile free) are generally linear in P–T space. However, volatiles are very compressible (i.e., ΔV varies as a function of P and T, and hence volatile-bearing reactions always possess a curvature in P–T space).

Graphical Representation

Visual display of compositions and equilibrium assemblages is generally preferable over writing lengthy metamorphic reactions. Therefore, the goal of early to modern practitioners of metamorphic petrology has been to develop graphical means to illustrate what mineral assemblages would be stable at a given P–T regime in a particular protolith. Binary and ternary diagrams, similar in treatment to those we saw examples of in earlier chapters (7–10), were originally developed by Eskola (1920) to include the bulk composition effect on metamorphic reactions. Because natural rocks are composed of multiple components, their bulk compositions, mineral compositions, and mineral (±volatile) reactions occurring within a given rock need to be treated so that they may be plotted on a given ternary (in reality, they are all pseudoternary) diagram either as projections (i.e., when they are projected from an end-member mineral component) or condensations (i.e., when the components are combined in some way and projected from some exchange component—that is, say, solid solution rather than an end member).

In reality, any plot of a natural rock in a pseudoternary diagram is a condensed plot. For example, consider plotting a protolith consisting of nine or more major and minor chemical components [SiO_2, TiO_2, Al_2O_3, Fe_2O_3, Cr_2O_3, FeO, MnO, MgO, CaO, Na_2O, K_2O ($\pm CO_2$, H_2O, OH etc.)] on an Al_2O_3-CaO-FeO (ACF) diagram. Eskola suggested that, in so doing, one must combine the components in this way:

$$A = Al_2O_3 + Fe_2O_3 - (K_2O + Na_2O + CaO)$$
$$C = CaO$$
$$F = FeO + MnO + MgO$$

Clearly any composition plotted in a pseudoternary ACF diagram this way is a condensed plot (Figure 16.6a). Note that SiO_2, the most abundant component of all silicate minerals and rocks, is not part of the combination process. What this means is that all compositions plotted on an ACF diagram are being projected from SiO_2, and thus the underlying assumption is that the plotted rocks have excess SiO_2 (either

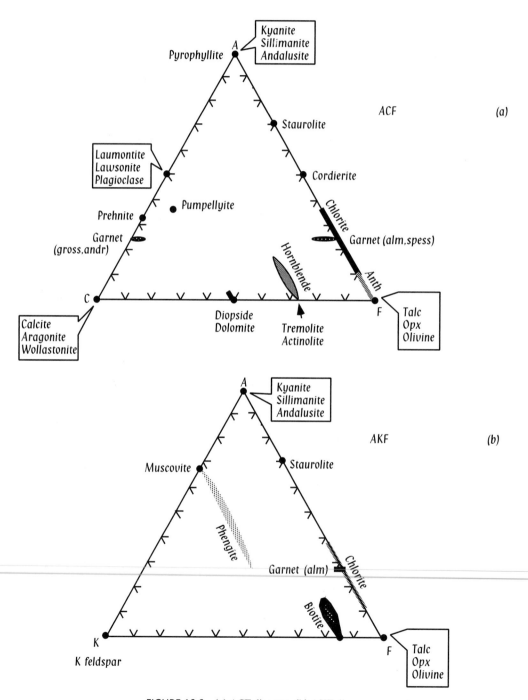

FIGURE 16.6 (a) ACF diagram. (b) AKF diagram.

as free quartz or other SiO_2 polymorphs). Note also that solid solutions are shown as extended fields in this diagram.

The ACF diagram was initially proved to be of value in portraying changing facies assemblages in mafic (i.e., basaltic) protoliths (Figure 16.6a). Unfortunately, many important reactions occur that involve minerals that are (Fe, Mg) solid solutions. The Fe, Mg variation in such minerals cannot be shown in an ACF diagram. One faces the same problem in using AKF diagrams (Figure 16.6b). Later studies pioneered by J. B. Thompson (1982) showed that any reactions in mafic and pelitic rocks are best represented in an AFM diagram (discussed later).

It is instructive to see how we can explore changes in mineral assemblages that may develop in mafic rocks using the ACF diagram (Figure 16.7). Because some mineral compositions plot as points and others plot as elongated areas (due to solid solution), each of the ACF triangles shown in Figure 16.7 contains three-phase triangles (shown as lightly shaded areas; recall from Chapter 7 that the apices of a three-phase triangle represent the three minerals in equilibrium) and two-phase regions (tie lines shown; each tie line connects two coexisting minerals). The field of basaltic rocks and three specific basalt compositions are shown in Figure 16.7 to illustrate how different mineral assemblages may develop in each of these bulk compositions:

Protolith 1 (a basalt)

Facies	*Assemblage*
Zeolite facies	heulandite (a zeolite) + chlorite + calcite
Prehnite-Pumpellyite facies	pumpellyite + actinolite + chlorite
Blueschist facies	lawsonite + glaucophane + aragonite
Greenschist facies	epidote + chlorite + actinolite
Amphibolite facies	plagioclase + hornblende + garnet
Granulite facies	clinopyroxene + garnet + plagioclase

(The reader may wish to determine what mineral assemblages will form in the other two bulk compositions.)

The AFM diagram is the most powerful in depicting reaction equilibria and facies assemblages because it is the only one that depicts the Fe $\Leftrightarrow$ Mg exchange reactions between ferromagnesian minerals. It also depicts the thermodynamically valid phase relations very closely (Thompson, 1957). It is important to recognize that the mineral and rock compositions plotting in an AFM diagram (mole fraction) are actually located within the tetrahedron AKFM (and H_2O) and are projected from muscovite (and quartz) onto the plane AFM that is extended beyond the base of the AKFM tetrahedron (Figure 16.8a). Naturally, the assumption is that the rock must contain modal muscovite and quartz, which is valid for low and medium grades. At high grades, muscovite breaks down, K feldspar forms in a metapelite, and the projection is made from K feldspar instead of muscovite (Figure 16.8b). Because most of the minerals plot on the AFM plane, the change from muscovite to K feldspar as a projection point does not affect their relative position in the AFM diagram, with the exception of biotite, which plots inside the tetrahedron AKFM (its position is shifted slightly higher; Figure 16.8b).

The recalculation of a multicomponent rock into AFM components is done the following way:

$$A = Al_2O_3 - 3K_2O$$
$$F = FeO$$
$$M = MgO$$

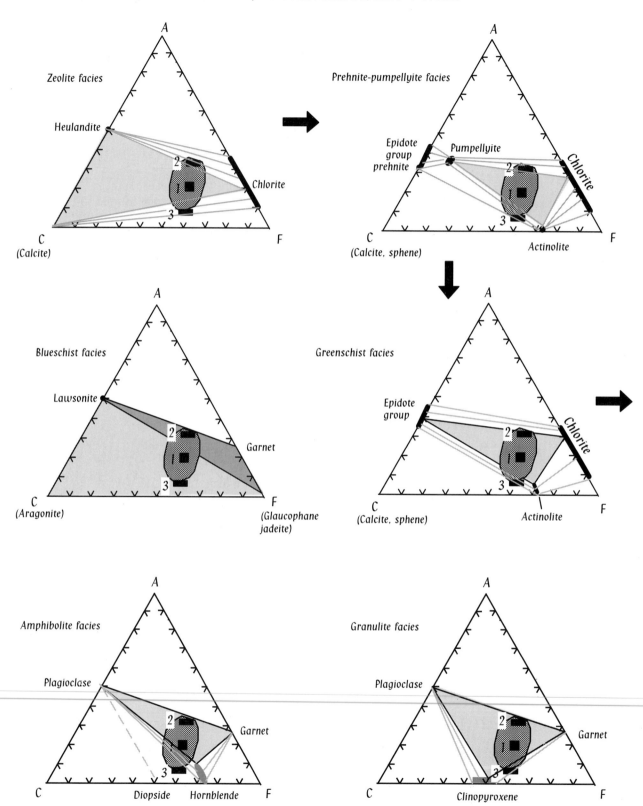

FIGURE 16.7 *Metamorphism of mafic rocks portrayed in a series of ACF diagrams representing various facies (discussed in text).* The arrows indicate prograde metamorphism. (Based on Figure 11.15 in Best, 1982, *Igneous and Metamorphic Petrology*, W.H. Freeman)

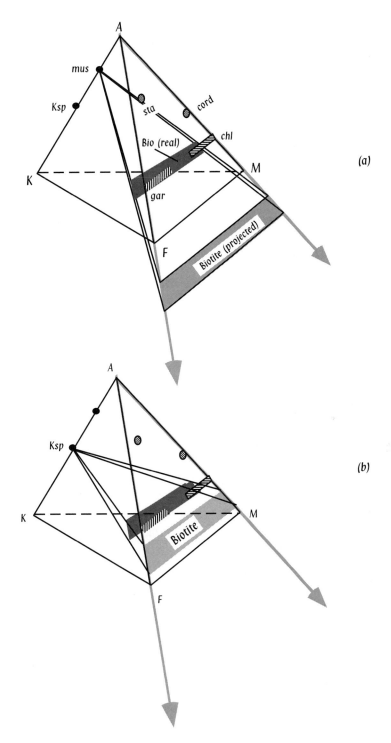

FIGURE 16.8 *The AKFM tetrahedron.* The minerals lying within the tetrahedron are projected from (a) muscovite, and (b) K feldspar. (c) AFM diagram (projected from muscovite).

(continues on next page)

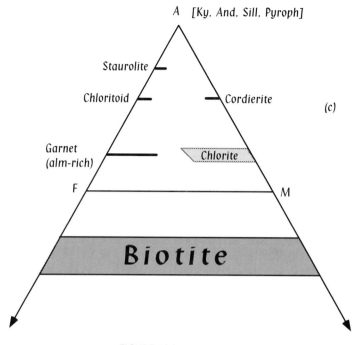

FIGURE 16.8   *(continued)*

Because these are mole percent plots, one must first convert the weight per-cent analyses of rocks and minerals by the appropriate moleculer weights and then add A (i.e., $Al_2O_3 - 3K_2O$) + F (FeO) + M (MgO) and normalize each compo-nent so that the total is 100%. In other words, in the plane AFM, the calculated com-ponents of A, F, and M in a rock or a mineral should be as follows (note that the oxides below are mole fractions):

$$A = 100 \times [Al_2O_3 - 3 K_2O]/[(Al_2O_3 - 3 K_2O) + FeO + MgO]$$
$$F = 100 \times FeO/[(Al_2O_3 - 3 K_2O) + FeO + MgO]$$
$$M = 100 \times MgO/[(Al_2O_3 - 3 K_2O) + FeO + MgO]$$

Note that the expression $Al_2O_3 - 3 K_2O$ derives from the formula of mus-covite, which has three times as much Al_2O_3 as K_2O. K-free minerals would plot in-side the AFM plane, and therefore their plotting is free of complications. However, in the case of biotite $[K_2O \cdot 6(FeO, MgO) \cdot Al_2O_3]$, its composition must be plotted from muscovite onto the AFM plane because its solid solution composition lies with-in the AKFM tetrahedron. Note that the A component in biotite should be (based on the prior formula): $100 \times [1 - 3]/[(1 - 3) + 6]$ or $100 \times (-0.5) = -50$. The neg-ative A number suggests that biotite must plot outside of the AFM triangle at a dis-tance of -50% away from the FM line (Figure 16.8).

If compositions are to be plotted from K feldspar, the A component must be re-defined (= $Al_2O_3 - K_2O$ based on K-feldspar formula, $K_2O \cdot Al_2O_3 \cdot 6SiO_2$). Thus, biotite now has the following A component (in terms of mole percent): $100 \times [1 - 1]$ $/[(1 - 1) + 6] = 0$. That is, the projected biotite composition would plot right on the FM line. In reality, however, because of solid solutions, biotite compositional field plots within the AFM triangle (Figure 16.9j,k).

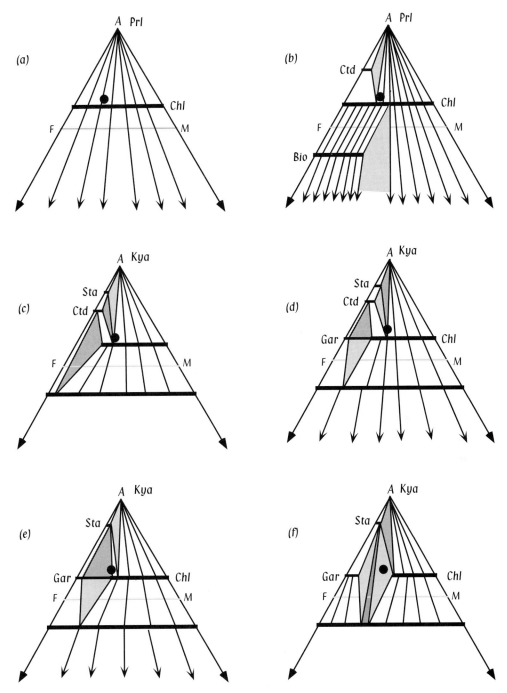

FIGURE 16.9   *A sequence of AFM diagrams (a–k) representing prograde metamorphism of pelitic rocks.* The dot represents a selected bulk composition that is discussed in the text. Note that *a–i* are projected from muscovite and the others from K feldspar. (From Blatt and Tracy, 1995, 2nd ed., *Petrology*, W.H. Freeman)

(continues on next page)

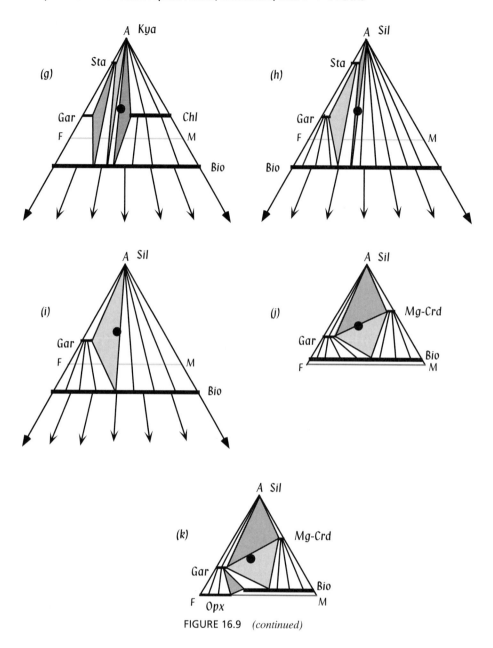

FIGURE 16.9   *(continued)*

AFM Diagram, Reactions, and Petrogenetic Grid: Metamorphism of a Pelitic Protolith as an Example

Figure 16.9 shows how the AFM diagram allows us to graphically portray metamorphic assemblages and reactions in a protolith of pelitic composition. In the example shown, the protolith is undergoing progressive intermediate P–T-type (or regional) metamorphism, and Figures 16.10a–16.10k represent various P–T points on such a trajectory. These diagrams are based on Spear's (1993) Figures 10–5 and 10–6, and Bucher and Frey's (1994) Figure 7.4. In the lowest grade of metamorphism, the assemblages pyrophyllite + chlorite + muscovite or chlorite + K feldspar + muscovite may be stable in high-Al versus low-Al pelites. Note that the arrows in Figure 16.9 point toward K feldspar.

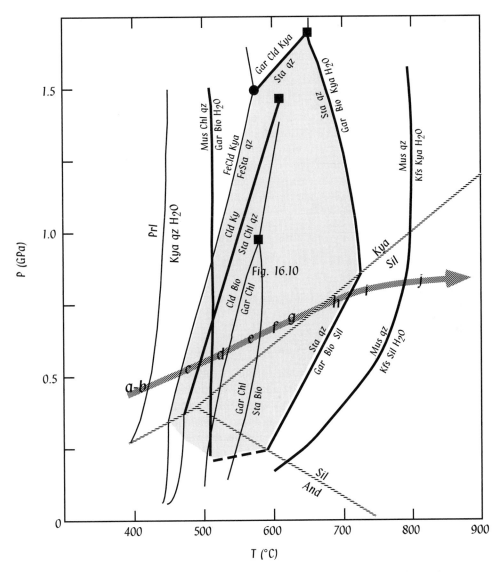

FIGURE 16.10 *Various reaction curves that are appropriate for metamorphism of pelitic rocks are shown.* The P–T path (dark broad arrow) represents the prograde metamorphic sequence shown in Figure 16.9. Figure 16.10a–16.10i correspond to Figure 16.9a–16.9i. The shaded area marks the stability field of staurolite.

Although for the specific bulk composition chosen in Figure 16.9, biotite does not form until at a higher grade, one usually encounters the *biotite isograd* as the first isograd due to the reaction:

Chlorite + K feldspar = Muscovite + Biotite + Quartz + H$_2$O.

This reaction is likely a pseudodivariant reaction, and hence the assemblage chlorite + biotite ± K feldspar ± muscovite may persist in the rock. The Mg/Fe ratio of biotite and chlorite will increase with increasing grade of metamorphism (i.e., the shaded region connecting biotite, chlorite, and K feldspar slide from F to M side in Figure 16.9b).

In a more Al-rich bulk composition, such as the example used in Figure 16.9b, instead of biotite, Fe-rich chloritoid will form (*Fe-chloritoid isograd*) by a reaction between chlorite and pyrophyllite:

$$\text{Fe-chlorite} + \text{Pyrophyllite} = \text{Fe-chloritoid} + \text{Quartz} + H_2O.$$

Chloritoid is an interesting mineral because, at intermediate P–T, it is generally very Fe rich (Mg/(Mg + Fe—generally less than 0.15). However, a very Mg-rich chloritoid (Mg/(Mg + Fe) > 0.5) does form in the high-P–T, eclogite facies conditions.

Again, in Al-rich bulk compositions, the next isograd that would be expected to appear is *kyanite or andalusite isograd* depending on the pressure. For our chosen Al-rich bulk composition, kyanite will form from breakdown of pyrophyllite at about 400°C and 0.5 GPa (AFM diagram not drawn, but should be analogous to Figure 16.9b but replacing Prl with Kya):

$$\text{Pyrophyllite} = \text{Kyanite} + \text{Quartz} + H_2O.$$

One may encounter the *Fe-staurolite isograd* next due to either of the following reactions (Figure 16.9c):

$$\text{Fe-chloritoid} + \text{Kyanite} = \text{Fe-staurolite} + \text{Quartz} + H_2O$$

or

$$\textit{Chloritoid} + \textit{Kyanite} = \textit{Staurolite} + \textit{Chlorite} + \textit{H}_2\textit{O}.$$

Although the Fe-staurolite forming reaction marks the staurolite isograd on this diagram, staurolite may not necessarily appear in the rocks (due to inappropriate bulk composition). The second one is most important because it is usually the reaction that forms the staurolite-isograd in high-Al pelitic rocks. This type of reaction is known as a tie-line flip, where the tie line between chloritoid and kyanite is broken and a new tie line forms between staurolite and chlorite. Tie-line flips affect different types of rocks (whose bulk compositions fall in the A–Sta–Chl or Sta–Ctd–Chl triangles in Figure 16.9b) at once, therefore they make better isograds. Note that any of the following assemblages may form depending on the exact bulk composition: staurolite + chlorite + kyanite, kyanite + chlorite or staurolite + Fe-chloritoid + chlorite. It is obvious from Figure 16.9c that, in Fe-rich pelites (uncommon, but whose bulk composition would fall inside the Chl–Ctd–Bio triangle), the assemblage Fe-chloritoid + biotite + Fe-rich chlorite may form.

Figure 16.9d shows that *garnet isograd* would be expected to come next:

$$\textit{Chloritoid} + \textit{biotite} + \textit{H}_2\textit{O} = \textit{Garnet} + \textit{Chlorite}.$$

This reaction is seen as a tie-line flip from chloritoid-biotite tie line to garnet-chlorite tie line (note Figures 16.9c and 16.9d). The garnet isograd reaction shown previously is an important one because the P–T range over which garnet + chlorite assemblage can occur is very small: T = 520–540°C at ~0.5–0.75 GPa. However, note that the dissolved spessartine (Mn-garnet) component in garnet can expand the stability field over which garnet + chlorite can occur.

Next comes the *Chloritoid-disappearance isograd*:

$$\textit{chloritoid} = \textit{staurolite} + \textit{garnet} + \textit{chlorite}$$

(compare Figures 16.9d and 16.9e).

Figure 16.9f shows that the Gar–Chl tie line breaks down with increasing grade in favor of a Sta–Bio 2-phase field, such that in our chosen bulk composition the assemblage garnet will disappear and staurolite-chlorite-biotite assemblage will form.

However, for more Fe-rich bulk compositions, the stable assemblage is staurolite-garnet-biotite. Thus, we have the *biotite isograd*.

Staurolite-chlorite assemblage breaks down with increasing metamorphic grade, and kyanite + biotite assemblage becomes stable (compare Figures 16.9f and 16.9g):

$$\text{staurolite} + \text{chlorite} = \text{biotite} + \text{kyanite} + H_2O.$$

This usually marks the appearance of the *kyanite isograd* in low-Al pelitic rocks.

In our example, as the grade is increased from $g \rightarrow h$ (Figure 16.10), sillimanite replaces kyanite as the stable Al_2SiO_5 polymorph (i.e., the appearance of the *sillimanite isograd*). Sillimanite typically nucleates as fine fibrous masses in the rocks within muscovite or biotite as kyanite get replaced by muscovite. At a higher grade of metamorphism (granulite facies), sillimanite form clusters of long prismatic crystals.

As the metamorphic intensity is increased from $h \rightarrow i$ (Figure 16.10), staurolite breaks down due to the following reaction (*staurolite-out isograd*):

$$\text{staurolite} = \text{garnet} + \text{biotite} + \text{sillimanite} + H_2O.$$

The shaded region in Figure 16.10 thus represents the overall P–T field within which staurolite may occur.

The next important reaction that would be expected to occur in a pelitic rock (like the one in our example) is the breakdown of muscovite and the formation of the *K feldspar isograd* as the metamorphic intensity is increased from i to j:

$$\text{muscovite} + \text{quartz} = \text{K feldspar} + \text{sillimanite} + H_2O.$$

Note that the phases representing this reaction cannot really be shown in an AFM diagram.

At even higher grades (but usually lower pressure), reactions in the granulite facies conditions may occur that give rise to the formation of Mg-cordierite and pyroxenes in the rock, with the resulting assemblages depicted in the appropriate AFM diagrams (Figures 16.10j and 16.10k).

Diagrams like Figure 16.10, which depict possible reactions in various metamorphic rocks, are generally based on experimental studies of simplified systems with fewer components backed up by laboratory and field studies. For example, most of the reactions shown in Figure 16.10 are based on some of the key reactions in the system KFMASH, which is an abbreviation for $K_2O–FeO–MgO–Al_2O_3–SiO_2–H_2O$. The reactions form a type of grid of pseudounivariant curves, which is commonly referred to as the *petrogenetic grid* because such grids, if applied to rocks of appropriate bulk composition, provide P–T history of the rock.

A Petrogenetic Grid for Mafic Protoliths

The reader has already been briefly exposed to what mineralogical changes occur in a mafic (basaltic) rock during metamorphism in an earlier section (discussion on facies, facies series, and the ACF diagram). The purpose of this section is to highlight some of the important reactions that occur in rocks of mafic composition.

In the beginning stage of metamorphism of mafic (more appropriately, basaltic-to-andesitic rocks), the most important reaction that occurs is one of hydration, in which hydrous minerals begin to replace the anhydrous assemblage plg + cpx. In the lowest grade rocks, relict phases and/or relict texture are often preserved. The water may be supplied in a hydrothermal environment near a midoceanic ridge or by groundwater in a plate-convergent margin. The availability of water is of key importance in the metamorphism of mafic volcanic rocks. The lowest grade assemblages are albite (from albitization of plagioclase) + chlorite (hydration of cpx) + carbonates and zeolites. Laumontite is a zeolite that forms at extremely low-grade, bordering diagenesis (Figure 16.11a), followed by another zeolite—wairakite at a slightly higher temperature. Zeolite → prehnite-pumpellyite facies transition is not particularly clear-cut, but is thought to take place around 300°C, 3–4 kb (Liou, 1971). The breakdown of wairakite to prehnite-bearing assemblages marks the facies transition (Figure 16.11b).

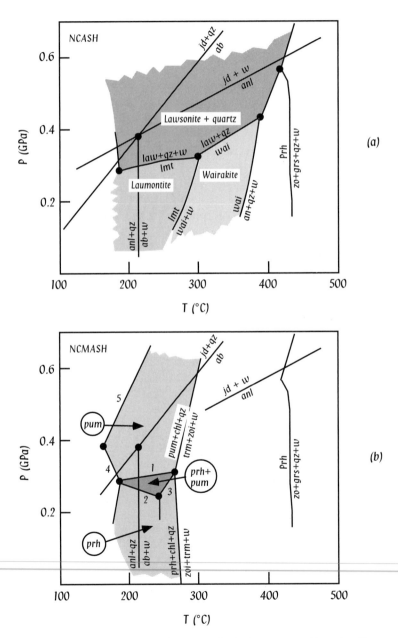

FIGURE 16.11   (a) Petrogenetic grid for zeolite facies metamorphism of mafic rocks in the system Na_2O–CaO–Al_2O_3–SiO_2–H_2O (NCASH). The laumontite and wairakite shaded areas are two subfacies of the zeolite facies. (b) Petrogenetic grid relevant to the formation of prehnite and pumpellyite in the system Na_2O–CaO–MgO–Al_2O_3–SiO_2–H_2O (NCMASH).

At P–T conditions higher than the prehnite-pumpellyite facies, several different reactions occur, resulting in very different P–T-series assemblages. In a high-P–T series, the progression is from zeolite-bearing assemblages → prehnite-pumpellyite bearing rocks → blueschist facies → eclogite facies. A key feature for blueschists is the development of a sodic amphibole (with high proportion of the glaucophane end-member component) that appears bluish under plane-polarized light. The sodic amphibole first begins to form due to the breakdown of chlorite + albite + actinolitic amphibole-bearing assemblages and the formation of riebeckite and glaucophane-rich amphibole ± epidote-bearing assemblages due to the following reactions (Bucher and Frey, 1994):

$$Trm + Alb + Chl = Law + Amp$$

$$Trm + Alb + Chl = Amp + Zoi + Qtz + H_2O$$

$$Alb + Chl + Qtz = Amp + Par + H_2O.$$

Garnet would form within the blueschist facies rocks due to the breakdown of paragonite and lawsonite:

$$Law + Amp = Par + Pyr + Zoi + Qtz + H_2O$$

$$Par + Chl + Qtz = Amp + Pyr + H_2O.$$

Note that, within the blueschist facies, the amphibole composition changes from Fe-rich (riebeckite component) to Na-rich (glaucophane component) with increasing grade. As pressure and temperature are raised further, the glaucophane-rich amphibole + paragonite will eventually break down and typical *eclogite facies* rocks, composed of garnet + sodic pyroxene (omphacite, jadeite), will form due to the reactions:

$$Alb = Jad + Qtz$$

$$Gln + Par = Pyr + Jad + Qtz + H_2O.$$

The approximate locations of these reactions are shown in Figures 16.11 and 16.12.

In an intermediate-P–T facies series, the sequence of reactions occurring within each facies and at facies boundaries are obviously quite different. Prehnite-pumpellyite facies is followed by greenschist facies. The minerals chlorite, actinolite, and epidote, which are all green, are responsible for the green color of greenschists. The transition from prehnite-pumpellyite to greenschist facies is marked by the breakdown of pumpellyite and stabilization of chlorite + albite + actinolite + epidote + quartz assemblages (Figures 16.11b, 16.12):

$$Prh + Chl + Qtz = Epi + Act + H_2O$$

$$Pum + Chl + Qtz = Act + Epi + H_2O$$

$$Law + Pum = Epi + Chl + Qtz + H_2O.$$

Actinolite forms at a lower temperature in Fe-rich bulk protoliths relative to Mg-rich rocks. Greenschist facies is followed by amphibolite facies, which is characterized by hornblende + plagioclase assemblages (e.g., see Reaction 4 in Figure 16.12), due to some broad reaction as:

$$Epidote (zoisite) + Chlorite + Quartz = Amphibole + Plagioclase + H_2O.$$

An important change occurs in the plagioclase composition: It jumps from An_3 (albite) to An_{17-18} (oligoclase) due to an immiscibility gap on the An–Ab solvus. The first appearance of oligoclase (oligoclase isograd) is generally taken as the beginning

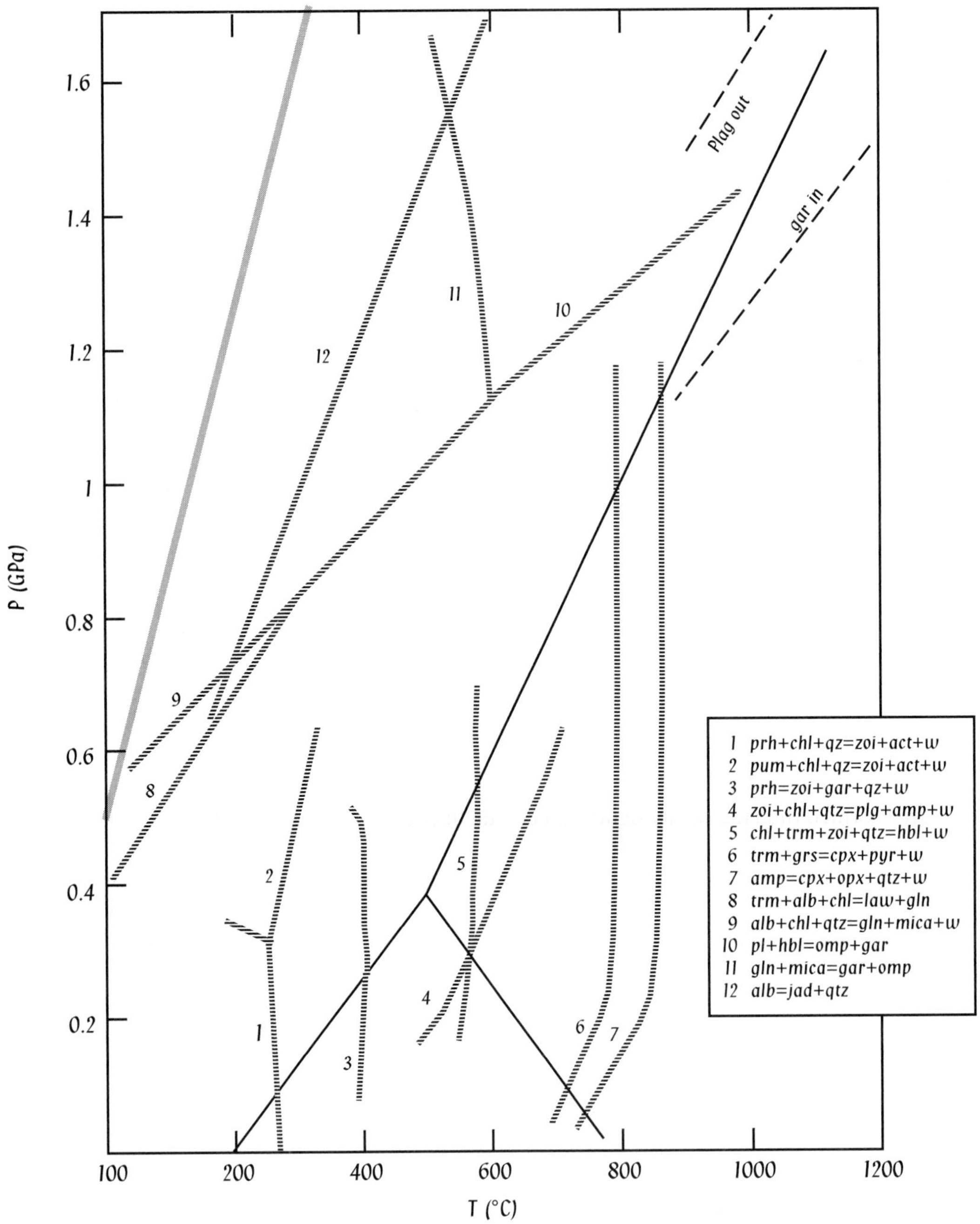
P (GPa)
1.6
1.4
1.2
1
0.8
0.6
0.4
0.2
100
200
400
600
800
1000
1200
T (°C)
Plag out
gar in
1
2
3
4
5
6
7
8
9
10
11
12
1 prh+chl+qz=zoi+act+w
2 pum+chl+qz=zoi+act+w
3 prh=zoi+gar+qz+w
4 zoi+chl+qtz=plg+amp+w
5 chl+trm+zoi+qtz=hbl+w
6 trm+grs=cpx+pyr+w
7 amp=cpx+opx+qtz+w
8 trm+alb+chl=law+gln
9 alb+chl+qtz=gln+mica+w
10 pl+hbl=omp+gar
11 gln+mica=gar+omp
12 alb=jad+qtz

FIGURE 16.12 *A simplified petrogenetic grid for mafic rocks.*

of the amphibolite facies. The other important change is in amphibole composition from actinolite to hornblende, which results from a series of complex reactions. The minerals epidote and chlorite decrease in abundance and eventually disappear between ~550° and 600°C within the amphibolite facies. Chlorite breakdown reactions produce garnet, which becomes modally more abundant with increasing grade. Therefore, the amphibolite facies rocks above ~600°C contain hornblende + plagioclase (An_{30-50}) + garnet ± biotite. At temperatures close to the amphibolite-granulite facies transition, clinopyroxene appears, and a typical assemblage is hornblende + plagioclase + clinopyroxene + garnet (see Reaction 6 in Figure 16.12):

Amphibole + Zoisite + Quartz = Plagioclase + Clinopyroxene + H_2O.

The transition from amphibolite to granulite facies is gradual; it takes place over ~200°C interval and is marked by the first appearance of orthopyroxene. During the transition, the modal abundance of amphibole decreases and is ultimately replaced by pyroxenes. The general reaction that marks the incoming of orthopyroxene is (Reaction 7 in Figure 16.12):

Amphibole = Cpx + Opx + Plagioclase + H_2O.

The transition from granulite facies to eclogite facies for a variety of basaltic rock types have been experimentally investigated by the Ringwood-Green group at Australian National University and Ito and Kennedy at University of California, Los Angeles (reviewed by Ringwood, 1975: Reaction 10 in Figure 16.12). The general reaction for this transition may be written as:

Opx + Plagioclase = Garnet + Cpx + Quartz.

However, this transformation from plagioclase-bearing to plagioclase-free assemblage occurs through a transition from pyroxene granulite (Pl + Cpx + Opx ± Ol ± Sp) to garnet granulite (Gar + Pl + Cpx ± Opx ± Qtz) to eclogite (Gar + Omp). The pressure range over which this transition occurs at a given temperature is strongly dependent on bulk composition (Figure 16.12).

Eclogites form in areas of oceanic plate subduction from blueschist facies assemblages (low-P–T facies series). This type of eclogites register high P but low T, and therefore are referred to as low-T or LT eclogites (Bucher and Frey, 1994). In areas of continental collision, such as the one between the Indian and Eurasian plates that formed the Himalayas, significant crustal thickening occurs. The crust may achieve thickness of >80 km. In such situations, eclogite is very likely to constitute the deep crust, followed above by amphibolites and maybe granulites. This eclogite type may be expected to record intermediate temperatures and is called an MT (medium-T) eclogite. Eclogites also record high temperatures and may form from granulites in a collisional or an extensional environment. The high temperature may be due to their direct crystallization from deep magmas or to heat supplied to granulitic lower crust by magmas. This type of eclogite is called a HT (high-T) eclogites.

ULTRA-HIGH-PRESSURE METAMORPHISM

According to Coleman and Wang (1995) "the term ultra-high pressure metamorphism refers to a metamorphic process that occurs at pressure greater than ~28 kbar (2.8 GPa, the minimum pressure required for the formation of coesite at ~700°C)" (p. 2). Until about 1984, no one could have imagined that continental crustal rocks could be metamorphosed at P > 2.8 GPa and be exhumed to the surface. An outstanding finding by Chopin (1984) of inclusions of coesite, a very high-pressure (>2.8 GPa) polymorph of SiO_2, in garnets of a quartzite from the Dora Maira area of Western Alps of Italy generated a new area of study in metamorphism—ultra-high-pressure-metamorphism (UHPM). Since that time, coesite- and microdiamond (or graphite pseudomorphs after diamond)-bearing UHPM rocks have been discovered in many new sites in Europe and Asia, where continent–continent collision has occurred. These new discoveries are giving new insights into how continental crust may be subducted deep into upper-mantle depths and exhumed back to the surface. It is beyond the scope of this book to include a comprehensive review of all the recent work that has been done on UHPM, and the interested reader is encouraged to read the edited books by Coleman and Wang (1995) and Hacker and Liou (1998).

Significant attention has been given in recent years to the discoveries of coesite- and/or microdiamond-bearing eclogites from Norway, the *Dabie Mountains* of central China, Jiangsu and Shandong (*Su-Lu*) area of eastern China, and diamond-bearing gneiss from the Kokchatev massif of Russia. The peak metamorphic conditions, in terms of depth (pressure) and temperature, attained by the UHPM rocks of these three areas can be seen in Figure 16.13 (largely based on Figure 10.7 of Wang et al., 1995; cited in Coleman and Wang, 1995). This figure shows the location of the polymorphic transformation boundaries quartz = coesite (Bohlen and Boettcher, 1982) and graphite = diamond (Kennedy and Kennedy, 1976). These curves provide minimum pressure estimates (by virtue of the presence of coesite or microdiamond) for the peak metamorphic conditions in Dora-Maira, Su-Lu, and Dabie mountain areas. Whereas the occurrence of coesite and/or microdiamonds in such rocks are clearly indicative of ultra-high-pressure conditions, the temperatures of such peak metamorphism have largely been estimated from the garnet-clinopyroxene thermometer. Garnet-clinopyroxene thermometer is based on the temperature sensitive reaction involving the exchange of Fe^{2+} and Mg between garnet and clinopyroxene.

The rocks in which such UHP minerals have been found are considerably varied:

Region	UHP Minerals	Host Rock
Dora Maira (Italy)	Coesite	Pyrope-quartzite
Dabie Shan (China)	Coesite, Microdiamonds	Eclogite
Su Lu (China)	Coesite	Eclogite
Western Gneiss Region (Norway)	Coesite	Eclogite
Fjorloft Island (Norway)	Microdiamonds	Kya-Bio-Gar Gneiss
Kokchetav (Russia)	Coesite, Microdiamonds	Eclogite (retrograded to Gar-Bio Gneiss)

In most cases, the high-P assemblages have been retrograded to lower P–T assemblages during exhumation, and the UHP minerals are only found as rare relics or inclusions within resistant minerals such as garnet, pyroxene, epidote, or zircon. The protolith is most commonly an eclogite that occurs as isolated boudins[1] within strongly deformed gneisses.

Several different models exist regarding the exhumation of these UHPM bodies to shallow levels. However, it is commonly accepted that:

1. They are associated with cold subduction areas because the peak temperature recorded by the high-P assemblages is generally very low relative to the normal continental geotherm.

2. The UHP minerals were preserved in these rocks because fluids could not access them during exhumation, such that they were not retrograded to some lower-grade assemblages.

---

[1] Boudins are resistant (competent) blocks that break apart and resist ductile deformation. In UHPM occurrences, such boudins appear as a long chain of necked or discontinuous broken blocks surrounded by highly deformed matrix.

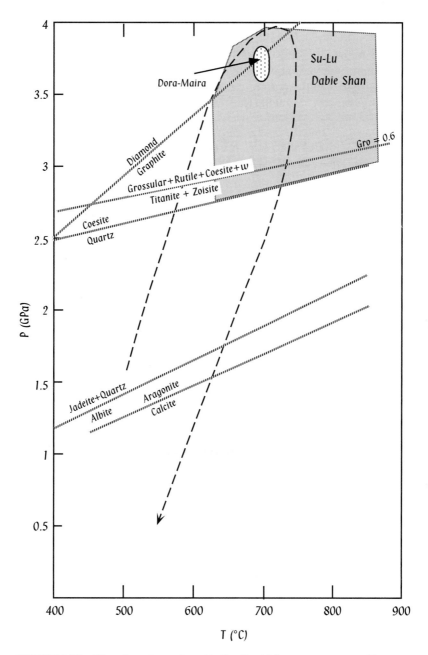

FIGURE 16.13 *Mineral reactions relevant to the ultra-high-pressure metamorphism (UHPM).* The two shaded areas mark the peak metamorphic conditions for the Dora Maira and Su-Lu areas. The dashed curve with arrow represents an inferred P–T path followed in UHPM.

3. The UHP assemblages are cooled during exhumation, with the temperature of the blocks rarely reaching greater than about 800°C (Hacker and Peacock, 1995).

Figure 16.13 shows a likely path (dashed, schematic) that may be taken by one of these UHPM blocks during continental collision and subsequent exhumation. The fact that these rocks consistently record unusually cool temperatures (relative to what would be expected at such great depths) is a clear indication of their development through cold subduction conditions, which is to be expected in a continent–continent collision.

From the foregoing discussion, it is perhaps apparent that an important task of the metamorphic petrologist is to estimate the pressure and temperature conditions at various times in the path of a metamorphic rock as it is buried, metamorphosed, and subsequently exhumed. How this is done is briefly reviewed in the next section.

GEOTHERMOBAROMETRY, REACTION KINETICS, AND PRESSURE–TEMPERATURE–TIME (P–T–T) PATHS

In his attempt to determine P–T conditions of metamorphism, the petrologist must use reactions that are sensitive to pressure and temperature, respectively. Such reactions may be univariant (or pseudounivariant) or divariant (or pseudodivariant). A quick example of a pressure-sensitive univariant reaction is the quartz-coesite reaction (Figure 16.13) because, over a large T range (500–800°C), the pressure range at which quartz may transform to coesite is very narrow (2.7–3 GPa). Relative to the quartz-coesite curve, the graphite-diamond univariant curve is less P sensitive (Figure 16.13). The sensitivity of mineral–mineral reactions to pressure versus temperature is the basis for geothermobarometry, which is the subject that deals with P–T determination of rocks.

Metamorphism is a continuous process in the sense that a rock may undergo progressive metamorphism during burial or subduction, attain peak P–T conditions at some depth, and subsequently be exhumed during which it must undergo retrograde metamorphism. Any reaction that occurs during this process may or may not reach equilibrium depending on kinetic factors that determine the rate at which such a reaction may progress. For example, geological time and fluid activity play an important role in metamorphism because both of these factors determine how much progress individual reactions may make at any particular P–T conditions. It follows that reaction progress determines whether relict (i.e., unreacted) phases would remain in a rock, and preservation of such phases is a key to elucidating the complete P–T–time history of the rock. What follows is a brief discussion of geothermobarometry, reaction kinetics, and P–T–t paths taken by metamorphic rocks.

Geothermobarometry

As indicated previously, the subject of geothermobarometry deals with estimation of temperature and pressure of equilibration of rocks using T- and P-sensitive reactions. Based on our discussion in Chapter 7 on the Clapeyron equation, we know that a P-sensitive reaction should have large ΔV and small ΔH and ΔS, and a T-sensitive reaction should have large ΔH and ΔS and small ΔV. Based on experimentally obtained datasets or from published thermodynamic values for various reactions, one can calculate Clapeyron slopes for such reactions and thus formulate geothermometers and geobarometers.

Figure 16.14a shows a hypothetical example of two univariant Curves 1 and 2, Reactions 1 and 2 being P-sensitive (geobarometer) and T-sensitive (geothermometer) reactions, respectively. A rock containing all of the phases (A, B, C, D, and E) in equilibrium would have formed at the P–T point where the two curves intersect. However, experimental or thermodynamic determination of any such curve is associated with some uncertainty of its precise location, which is shown as the shaded

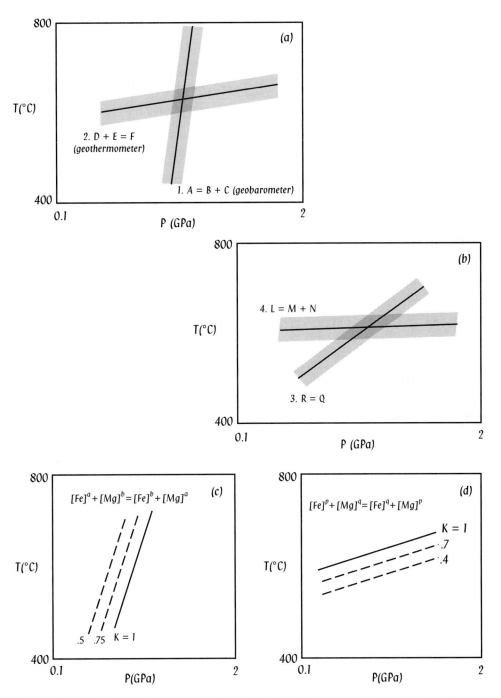

FIGURE 16.14 *Geothermobarometry.* (a) and (b) are for discontinuous reactions, and (c) and (d) are for exchange reactions (see text for details).

error envelope around each of the two curves. As a result, the P–T point will have an error envelope (dark shaded area) around it. Figure 16.14b shows two different sets of reactions (3 and 4), one of which is a somewhat less P-sensitive polymorphic transformation R = Q. Because the angle at which the 3 and 4 intersect is smaller than that between 1 and 2 (compare Figure 16.14a with Figure 16.14b), the error envelope around the point of intersection is significantly larger.

Figure 16.14c and Figure 16.14d show two hypothetical exchange reactions, one of which is sensitive to T and the other to P. As noted earlier, in an exchange reaction between reactant and product minerals, the minerals exchange chemical components as the pressure or temperature (or both) are modified. Such exchange between reactants and products must occur to keep ΔG of the reaction equal to zero (i.e., equilibrium condition). Consider the case of an exchange reaction in which two chemical components Fe^{2+} and Mg are being exchanged between two phases, a and b, and a and b both being (Fe^{2+},Mg) solid solutions:

$$[Fe^{2+}]^a + [Mg]^b = [Fe^{2+}]^b + [Mg]^a$$

The equilibrium constant, K, for this reaction can be written as:

$$K = \left[a_{Fe}^{\ b} \cdot a_{Mg}^{\ a} \right] / \left[a_{Fe}^{\ a} \cdot a_{Mg}^{\ b} \right],$$

where a means activity of some component in a particular phase (e.g., $a_{Fe}^{\ a}$ means activity of component Fe in Phase a, see earlier discussion in Chapter 7). How the activity of a particular component in a multicomponent solid solution may be calculated is somewhat of a complicated subject, and there are several excellent books that the reader may wish to consult (e.g., Saxena, 1973; Wood and Fraser, 1976; G.M. Anderson, 1996).

Briefly, the activity of a component in a multicomponent solid depends on the composition of the solid solution and whether the solid solution in question behaves ideally or nonideally. In a binary system $A - B$, $a_A^{\ p} = X_A^{\ p} \cdot \gamma_A^{\ p}$, where $X_A^{\ p}$ is mole fraction (= moles of A/(moles of A + moles of B) of Component A in Phase p and γ is called the *activity coefficient*. In an ideal solid solution, γ is equal to 1, and therefore activity = mole fraction. For a nonideal solution, γ is not equal to 1, but such cases are not dealt with here, and the reader is referred to the important textbooks listed in the Bibliography. For pure end-member compositions, such as pure forsterite, activity is always equal to 1. Olivine solid solutions, to a first approximation, show ideal behavior. Therefore, in the case of an olivine of composition $Fo_{80}Fa_{20}$, if ideal behavior may be assumed, then mole fraction and activity of pure forsterite may be related: $a_{Fo}^{\ ol} = X_{Fo}^{\ ol} = 0.8$.

Now returning to thermobarometry, let us consider two hypothetical Fe = Mg exchange reactions: Reaction between Phases a and b is pressure sensitive (Figure 16.14c) and that between c and d is temperature sensitive (Figure 16.14d). Figures 16.14c and 16.14d show the variation of equilibrium constant, K, for each of these reactions as a function of P and T. Each dashed curve represents a constant value of K. Although not shown in these figures, one can map the divariant region in each case over which both phases may be stable and K varying between 0 and 1. Determination of an exact value of K, which is possible through analysis of the minerals by an electron microprobe, will pinpoint the P (Figure 16.14c) or T (Figure 16.14d).

Next we use some actual examples to illustrate how P–T estimation may be done on a rock. A detailed review of all geothermobarometers applicable to metamorphic rocks is well beyond the scope of this book. The focus here is on mineral–mineral reactions (such as exchange reactions, solvus reactions, etc.); other methods, such as isotopic and fluid inclusion thermobarometry (which are nonetheless important), are not discussed. The interested reader may wish to consult the text by K. Bucher and M. Frey (1994, Sec. 4.7) and references therein.

Garnet-Clinopyroxene Thermometers

In the case of garnet-clinopyroxene thermometer, which is important for eclogites, garnet pyroxenites and so on, the reaction of concern is the one in which the coexisting garnet and clinopyroxene grains exchange Mg and Fe^{2+} cations:

$$[Mg]^{gt} + [Fe^{2+}]^{cpx} = [Mg]^{cpx} + [Fe^{2+}]^{gt}$$

A relatively simple thermometer by Ellis and Green (1979) based on experimental calibration of K for this reaction is still quite popular:

$$T(°C) = \left[[3030 + 10.86*P(kbar) + 3104*[X_{ca}]^{gt}]/[\ln K_D + 1.9034] \right] - 273$$

$K_D = (X_{Fe^{2+}}/X_{Mg})^{gt}/(X_{Fe^{2+}}/X_{Mg})^{cpx}$ and $X_{Fe^{2+}}$ and X_{Mg} are mole fractions ($X_{Mg} = Mg/(Mg + Fe^{2+}$, etc.). Note that this thermometer assumes ideal exchange of Fe^{2+} and Mg between garnet and clinopyroxene (i.e., $K_D = K$, such that the activities are assumed to be equal to the mole fractions). Figure 16.15 shows $\ln K_D$ versus T plot for a range of garnet/cpx K_D values at two different input values of pressure (3 vs. 4 GPa). From this figure, it is clear that even if one errs in using a pressure value by about 1 GPa, the temperature estimate is affected by no more than 50°. Thus, the garnet-clinopyroxene thermometer is an excellent thermometer. Note that more recent studies indicate that this thermometer overestimates the temperature of granulites and garnet-clinopyroxenites by 50° to 150°C (Sen, 1988; Green and Adams, 1991). It is therefore not surprising that several newer calibrations of this reaction have appeared in the literature. For example, Krogh (1988) presented the expression:

$$T(°C) = \left[[1879 + 10*P(kbar) - 6137*[[X_{ca}]^{gt}]^2 + 6731*[X_{ca}]^{gt}]/[\ln K_D + 1.393] \right] - 273.$$

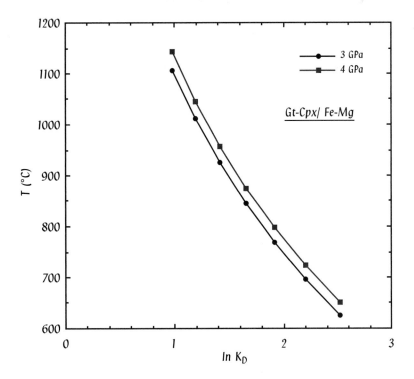

FIGURE 16.15 *Garnet-clinopyroxene thermometry.*

According to Bucher and Frey, Krogh's thermometer apparently gives more reasonable estimates for LT and MT eclogites.

Garnet-Biotite Thermobarometry

In Garnet-Biotite thermobarometry, the exchange reaction is once again Mg, Fe^{2+} exchange:

$$[Fe^{2+}]^{gar} + [Mg]^{bio} = [Mg]^{gar} + [Fe^{2+}]^{bio}.$$

Many thermometers based on garnet-biotite Fe = Mg exchange have been presented, of which a commonly used one is that of Ferry and Spear (1978):

$$T(°C) = [[2089 + 9.56*P(kbar)]/[0.782 - \ln K_d]] - 273,$$

where $K_d = (Fe^{2+}/Mg)^{bio}/(Fe^{2+}/Mg)^{gar}$.

Although the garnet-biotite exchange reaction is largely T dependent, Spear and Selverstone (1983) developed an interesting petrogenetic grid for pelitic bulk compositions that allows determination of both P and T based on compositions of biotite and garnet (Figure 16.16). In this method, ideal mixing is assumed in both phases, and the compositional variation of garnet in P–T space is mapped in terms of $[X_{alm}]^{gar}$-isopleths (i.e., lines of constant mole fraction of almandine component in garnet$_{solid\ solution}$). Biotite compositional variation is expressed in terms of $[X_{ann}]^{bio}$-isopleths (i.e., lines of constant mole fraction of annite component in biotite$_{solid\ solution}$). Figure 16.16 shows that, although both $[X_{alm}]^{gar}$ and $[X_{ann}]^{bio}$ isopleths vary as a function of both P and T, their slopes are sufficiently different such that their intersections can be used as a geothermobarometer.

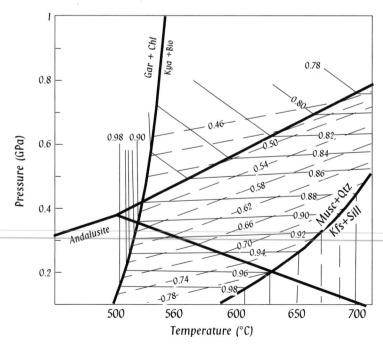

FIGURE 16.16 X_{alm}^{Gar}-isopleths (thin lines) and X_{ann}^{Bio}-isopleths (dashed) in coexisting biotite and garnet in meta-pelitic rocks.

Two-Pyroxene Thermometry

In Chapter 7, the reader was first exposed to the concept of solid–solid immiscibility between two phases. Basically, two minerals may show complete miscibility (i.e., solid solution) at high temperature, but at a lower temperature (during retrograde metamorphism) such phases may exsolve each other due to the presence of an immiscibility gap or solvus. Figure 16.17 illustrates solid–solid immiscibility between diopside and enstatite over a pressure range of 10 to 20 kbar (from Lindsley, 1986). The sensitivity of the diopside limb to temperature and relative insensitivity to pressure make it a

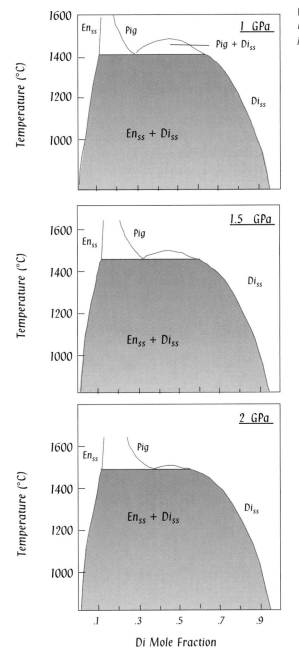

FIGURE 16.17 *Pyroxene immiscibility gap at three different pressures.*

useful thermometer for rocks like lherzolites and two-pyroxene (ortho- and clinopyroxene) granulites. As the two pyroxene phases in a metamorphic rock equilibrate with each other at any given temperature, their compositions are fixed on the two limbs of the miscibility gap (or solvus *sensu lato*) by the particular isotherm in question. Their compositions can only change, according to the temperature-composition diagram shown in Figure 16.17, when the temperature is changed. This compositional exchange of the two coexisting pyroxenes can be expressed as follows:

$$[Mg_2Si_2O_6]^{opx} = [Mg_2Si_2O_6]^{cpx},$$

where $[Mg_2Si_2O_6]^{cpx}$ and $[Mg_2Si_2O_6]^{opx}$ signify enstatite components in coexisting clinopyroxene and orthopyroxene phases, respectively. K for the prior reaction may be written:

$$K = [a_{En}^{cpx}]/[a_{En}^{opx}],$$

where a_{En}^{opx} and a_{En}^{cpx} represent activities of the enstatite component in coexisting ortho- and clinopyroxenes. Because pyroxene solid solutions do not behave ideally, a number of pyroxene activity models exist in the literature, and a good starting point is the reference book entitled "Pyroxenes" (see Bibliography). Since the intention here is only to demonstrate how a thermometer is formulated, ideal solubility is assumed such that K can be related to mole fraction (X) [a parameter that can be determined from microprobe analysis of the coexisting pyroxenes in a rock]:

$$K \approx [X_{En}^{cpx}]/[X_{En}^{opx}].$$

The procedure to calculate these mole fractions is based on cation site occupancies in the atomic structure of pyroxenes as laid out in Wood and Banno (1973):

1. from formula cations (i.e., cations per six oxygens) allocate Ca and Na in M2 octahedral site and all of Si in the T(tetrahedral) site;

2. allocate a portion of Al to the T site to make the T-site total = 2;

3. allocate the rest of Al and all of Cr and Ti to the smaller M1 site;

4. distribute Mg and Fe cations to M1 and M2 sites according to their proportion in the mineral formula. After Stages 1 to 3, M2 site will need 1–Ca–Na cations of Mg + Fe to fill it up. Similarly, M1 site will need 1–Ti–Cr–Al (in M1 site) cations of Mg + Fe to fill that site. Now Mg/(Mg + Fe) of the pyroxene in question needs to be calculated, and then amounts of Mg and Fe in M1 and M2 sites can be individually calculated assuming that the ratio Mg/(Mg + Fe) is the same in both sites—that is,

$$[Mg/(Mg + Fe)]^{mineral} = [Mg/(Mg + Fe)]^{M1} = [Mg/(Mg + Fe)]^{M2}$$

5. Since Mg of the enstatite component only occurs in the M2 and M1 sites of the pyroxene in question, the mole fraction of enstatite in that pyroxene may be written as:

$$X_{en}^{pyroxene} = [X_{Mg}^{M1} \cdot X_{Mg}^{M2}].$$

X_{Mg}^{M1} is mole fraction of Mg in M1 site, which is the same as Mg cations present in the M1 site. The same logic holds for the M2 site. Mole fractions of enstatite in coexisting ortho- and clinopyroxene may thus be calculated:

$$X_{en}^{opx} = [X_{Mg}^{M1} \cdot X_{Mg}^{M2}]^{opx}$$
$$\text{and } X_{en}^{cpx} = [X_{Mg}^{M1} \cdot X_{Mg}^{M2}]^{cpx},$$

and K can in turn be calculated from them $(K \approx [X_{En}^{cpx}]/[X_{En}^{opx}])$.

ln K for this reaction is very sensitive to temperature, but is much less sensitive to pressure variations. Thus, this reaction is a good geothermometer for lherzolites (Sen and Jones, 1989):

$$T(°C) = [4900/[1.807 - \ln K]] - 273.$$

In the spinel lherzolite stability field, Al_2O_3 content of pyroxene varies almost entirely as a function of temperature, and therefore it is an excellent geothermometer for spinel lherzolites. However, in garnet lherzolites and plagioclase lherzolites, the Al_2O_3 content of pyroxenes varies both as a function of P and T (Figure 16.18). Combination of the two-pyroxene thermometer (which is shown as constant mole% En–in–cpx lines) with Al_2O_3 isopleths (i.e., lines of constant weight% Al_2O_3–in–opx) creates a petrogenetic grid in the plagioclase lherzolite and garnet lherzolite fields. Thus, knowledge of these components in coexisting ortho- and clinopyroxene in a plagioclase or garnet lherzolite xenolith can be plotted on the grid shown in Figure 16.18 to obtain an estimate depth (pressure) and temperature of the region from where the xenolith was picked. A number of scientists led by two pioneers—Ian Macgregor (presently at the National Science Foundation) and F. R. (Joe) Boyd (Geophysical Laboratory)— have successfully used this concept to obtain fossil geotherms from garnet lherzolite xenoliths in kimberlite pipes (Figure 16.19).

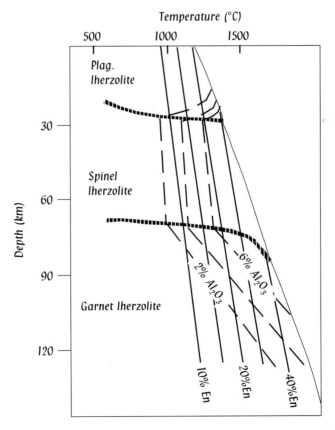

FIGURE 16.18   *Al_2O_3 (wt%)-in-orthopyroxene isopleths and En (mole%) -in-clinopyroxene isopleths for ultramafic rocks.*

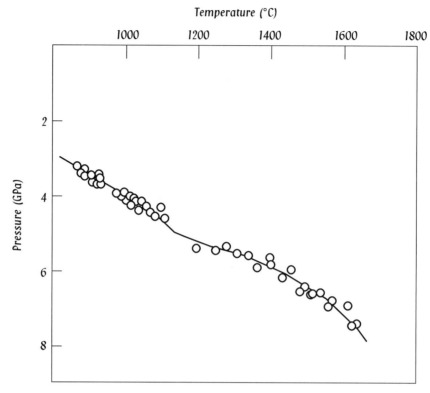

FIGURE 16.19 *P–T estimations on kimberlitic xenoliths from South Africa.*

The GASP Geobarometer

The G (garnet)–A (aluminosilicate)–S (silica)—P (plagioclase) geobarometer is widely used and based on the net transfer reaction:

$$CaAl_2Si_2O_8 = Ca_3Al_2Si_3O_{12} + 2Al_2SiO_5 + 2SiO_2.$$

$$\text{(anorthite)} \quad \text{(grossular)} \quad \text{(ky/and/sill)} \quad \text{(quartz)}$$

The equilibrium constant, K, for this reaction can be written as:

$$K = \left[\left(a_{kya}\right)^{kya}\right]^2 \cdot a_{qz} \cdot \left(a_{gro}\right)^{gar} / \left[\left(a_{an}\right)^{pl}\right]^3.$$

Because quartz and Al_2SiO_5 polymorphs are essentially pure phases, their activities are equal to 1. In contrast, the composition of natural plagioclase and garnet can vary considerably due to solid solutions. Assuming ideal solution, the following simplifications may be made: $\left(a_{gro}\right)^{gar} = (X_{gro}{}^{gar})^3$ and $\left(a_{an}\right)^{pl} = (X_{an}{}^{pl})$ in applying the GASP barometer (follows later) to natural rocks. However, because these minerals do not actually exhibit ideal solution behavior, many alternate barometers have appeared in the literature that take into account departure of these solutions from ideal behavior.

Figure 16.20 shows a plot of K as a function of P–T (from Spear, 1993, Figure 15–8). This geobarometer is well suited for amphibolites and granulites. Using the experimental data of Koziol and Newton (1988), assuming $\Delta C_p = 0$, and using a constant $\Delta V = -6.608$ J/bar, Spear (1993, Equation 15–48) proposed the following equation for GASP:

$$-48,357 + 150.66\, T(K) + (P - 1)(-6.608) + RT \ln K = 0.$$

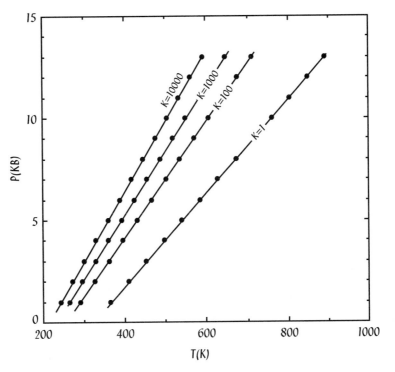

FIGURE 16.20   *The GASP geothermobarometer (see text for details).*

Rearranging, the prior equation becomes:

$$P \text{ (bars)} = 1 + \left[\left[48{,}357 - RT \ln K - 150.66\, T(K) \right] / (-6.608) \right].$$

This is the simplest form of the GASP geobarometer.

Practice of Geothermobarometry

In practice, the most reliable P–T estimates of the peak equilibrium conditions of a metamorphic rock would be the one that is constrained by one or more of the following:

1. Two reactions whose lines of constant Ks intersect at a large angle in P–T space can provide strong constraints on P–T of formation of a rock, although both of the Ks may vary with respect to P and T.

2. An invariant assemblage, such as the kyanite/andalusite/sillimanite triple point, provides an excellent control on P–T of rock formation if such an assemblage is actually found (rare).

3. Intersection of two discontinuous reaction curves, one of which may be sensitive to P and the other to T, may provide valuable P–T information.

4. One may also use several sets of geothermometers and geobarometers for one or more rock types that are spatially and temporally related by the same metamorphic event.

When it comes to applying the mineral geothermobarometers to estimate the P–T of equilibration of metamorphic rocks, many complexities may arise, a large part of which has to do with the difficulty of obtaining equilibrium in laboratory experiments, on which the geobarometer or geothermometer must be based. Often the activity models for individual phases may be inaccurate, which would render significant inaccuracy in the thermometer or barometer. Different reactions may have different closure temperatures (e.g., an exchange reaction between two minerals may stop equilibrating after cooling to some temperature following the peak metamorphic temperature reached by a rock, whereas a second exchange reaction between two other minerals in the same rock or an adjacent rock may continue to much lower temperatures). Thus, the two different reactions may give different final P–T conditions of equilibration of the rock.

Reaction Kinetics

The previous discussion was based on classical thermodynamics in which achievement of chemical equilibrium plays the central role irrespective of the actual processes. We learned that a reaction must proceed if the reactants have a higher free energy (G) than the product(s); at equilibrium, ΔG must be zero. In reality, however, different chemical reactions proceed at different rates, and in a metamorphic rock, reactions between two minerals may not go to full completion. As a result, compositional zoning may persist within individual mineral grains (discussed earlier).

This section focuses on reaction kinetics, which include all rate processes that dictate how fast or slow a reaction may proceed. Although the reactants may have higher free energy than the products, a reaction may not proceed spontaneously because the reactants must cross over an energy hump known as the *activation energy barrier* (Figure 16.21a). Where several different reaction paths are possible, the one that has the minimum activation energy barrier is followed. A stepwise process, in

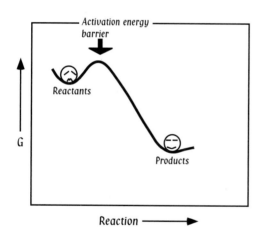

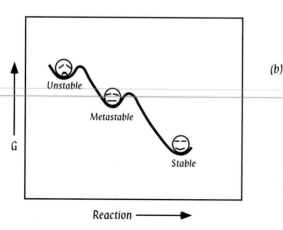

FIGURE 16.21   (a) Schematic diagram showing free energy relations between reactants and products. (b) Free energies of unstable reactants, metastable products, and stable products.

which the reaction may proceed via several steps in which metastable phases are produced, is commonly favored (Figure 16.21b).

Reaction rate (R_r) is expressed in terms of the so-called *Arrhenius relation*:

$$R_r = K \cdot e^{-E/RT},$$

where K is a constant (called *rate constant*), E is activation energy $(J \cdot mol^{-1})$, R is gas constant $(8.3144 \ Jmol^{-1}K^{-1})$, and T is temperature (K). The prior equation can be rewritten as:

$$\log_{10} R_r = \log_{10} K - [(E/RT)].$$

This equation has the form of the equation of a straight line. Plotting experimentally obtained reaction rate versus temperature data for any experiment allows one to calculate the activation energy barrier, E, and the rate constant, K, from the slope and intercept, respectively (Figure 16.22). Figure 16.22 shows such lines for three different reactions—*a*, *b*, and *c*. Note the *a* and *c* are parallel (i.e., their Es are the same), but their intercepts (K) are different. The steeper the slope (E), the more temperature-dependent is the reaction rate. Thus, the reaction rate for *a* is the most temperature dependent, and that for *b* is the least temperature dependent. The reaction rate for *c* is substantially lower than that for *a* because of a lower rate constant. Thus, both rate constant and temperature control how fast or slow a reaction may proceed.

Chemical components (ions) must diffuse in and out of mineral grains in any chemical reaction. This process of diffusion can be of different types, among which diffusion through the volume of a mineral grain (volume diffusion) is of interest in this section. Diffusion of components between reacting phases is controlled by Fick's law:

$$J_z = -D_i[\partial c/\partial z],$$

where J_z is the flux of a component *i* along a direction *z*. $\partial c/\partial z$ is the concentration gradient of Component *i* along direction *z*, and D_i is its diffusion coefficient. Values of D_i range between 10^{-10} and $10^{-14} \ m^2/sec$ in magmas and 10^{-15} and 10^{-21} at 600° to 1,200°C in common silicate minerals. These values indicate that volume diffusion in

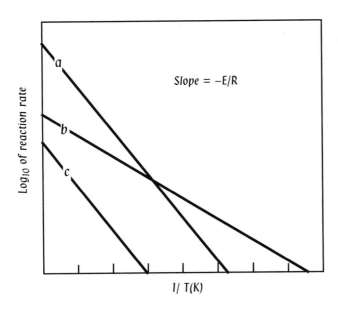

FIGURE 16.22 *Relationship between reaction rate and inverse temperature (in Kelvin) for three hypothetical reactions a, b, and c.* E = activation energy. R = Gas constant.

solids is much slower than through silicate melt or any other fluid. Fick's second law describes diffusion down a concentration gradient along Direction z:

$$dc/dt = D_i(d^2c/dz^2)$$

This equation assumes D_i is constant. However, in most cases, D_i is a function of composition, and this equation is rewritten as:

$$dc/dt = d/dz[D_i(dc/dz)].$$

This equation can be used to calculate concentration gradients as a function of time. In terms of application, a concentration gradient in a zoned crystal can be determined with an electron microprobe. If, by some other means, the time over which such zonation developed can be determined, then the diffusion coefficient may be calculated.

Pressure–Temperature–Time (P–T–t) Paths

The two fundamental controls for metamorphic reactions are free energy and reaction rate. Arrhenius relation (discussed earlier) indicates that reaction rate is exponentially related to temperature: At high temperatures, a reaction will proceed at an exponentially faster rate than at a low temperature. Therefore, when a rock is increasingly metamorphosed (prograde metamorphism), reactions between its constituent minerals will proceed at the fastest speed when maximum (peak) temperature is attained (all other factors being equal, particularly fluid activity). The various stages of prograde metamorphism may be recorded in terms of zoning in minerals. When this rock is brought back up to shallower levels due to uplift and erosion of the overburden (which is related to isostatic adjustment), it should continue to undergo retrograde metamorphism (i.e., reactions should continue, and minerals that are stable at decreasing P–T conditions should form, erasing any evidence of it once being metamorphosed at a much greater P–T). In reality, however, high-grade metamorphic rocks are quite commonly exposed at the earth's surface with little evidence of any retrograde metamorphism except for minor alterations along grain boundaries. Obviously these assemblages are only metastable at the surface. This happens because the retrograde reactions essentially come to a halt at lower temperatures. In addition, fluids, which help in keeping reactions going, do not penetrate the rocks as much.

The P–T conditions during the retrograde evolution of a metamorphic rock may be deciphered using a fluid inclusion thermobarometric method, which is well beyond the scope of discussion here. The complete P–T path followed by such a rock, from the beginning to end of its metamorphic history as a function of time (t), is commonly referred to as a *P–T–t* path. The P–T–t paths of metamorphic rocks can provide important clues about rates of burial, uplift, and erosional processes along plate-convergent boundaries and are discussed in this section.

In an effort to show how actual P–T–t paths may develop in metamorphic rocks, a schematic model is presented in Figure 16.23 (based on Bucher and Frey, 1994). This figure shows how crustal thickening (via continent–continent collision) and subsequent exhumation (owing to isostasy and erosion) may result in three different P–T–t paths for three rocks that are initially located at three different depths within the underthrusting plate. The isotherms in the underlying slab are purely schematic, and their nature would be controlled by several factors, the most important of which are the initial geotherms, the speed of convergence, and radioactive (or magmatic) heating in the base of the crust.

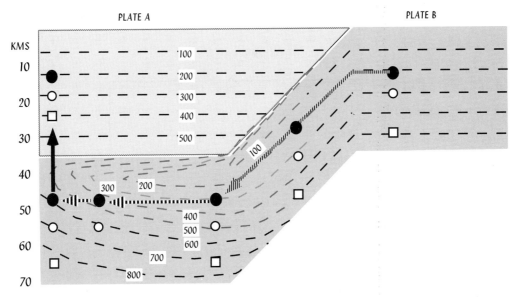

FIGURE 16.23 *A schematic model showing collision between two plates:* dashed lines—isotherms; solid circle, unfilled circle, and square—three reference points (rocks) in the downgoing slab that track the changing P–T conditions as the underthrust slab moves forward and eventually is uplifted; arrows—track the progress of the subducted lithosphere. (After Bucher and Frey, 1994)

Figure 16.24 shows the P–T–t paths for the three rocks while ignoring that these rocks are brought all the way up to the surface but, for brevity, are returned to the same levels where they originally started. It is important to note that all of the paths record a cold–hot–cold sequence (i.e., they are buried cold relative to a steady-state geotherm and are then heated at a maximum temperature and finally uplifted and cooled). These paths do not define a steady-state geothermal gradient. Note that the thick line obtained by connecting the T_{max} (temperature maximum) recorded by each of the three rocks does not coincide with the geotherm. This thick line is called a *metamorphic field gradient (MFG)*. Long ago, Miyashiro (1961) recognized such P–T gradients defined by metamorphic terranes and called them *metamorphic facies series* (discussed in a previous section), which are the equivalents of MFGs.

When P–T–t paths from metamorphic terrains from all over the world are plotted (Figure 16.25; Ernst, 1988), some general observations may be made: (a) Most paths are anticlockwise (not illustrated as such in Figure 16.25) and, as indicated earlier, are probably results of crustal thickening. (b) There are clockwise paths as well, and these occur mostly in the higher T part of the diagram and are likely the result of initial magmatic heating. There are several important aspects of these P–T–t paths: (a) Because T_{max} generally coincides with P_{max} in crustal thickening models, P_{max}, T_{max} conditions probably record the maximum amount of crustal thickening. (b) Time and amount of thickening together give the thickening rate. (c) The uplift rates may be modeled from P–T–t paths as well. Rapid convergence of cold crust leads to the formation of blueschists. If such blueschists are kept at depth for a long time, radioactive heat production would eventually increase their temperature and convert these rocks to a higher P–T rock. Therefore, the only way a blueschist can occur at the surface is through rapid uplift to the surface. P–T–t paths and MFGs are clearly an important direction in metamorphic petrology, and the interested reader is strongly recommended to read the important reference work by Spear (1993).

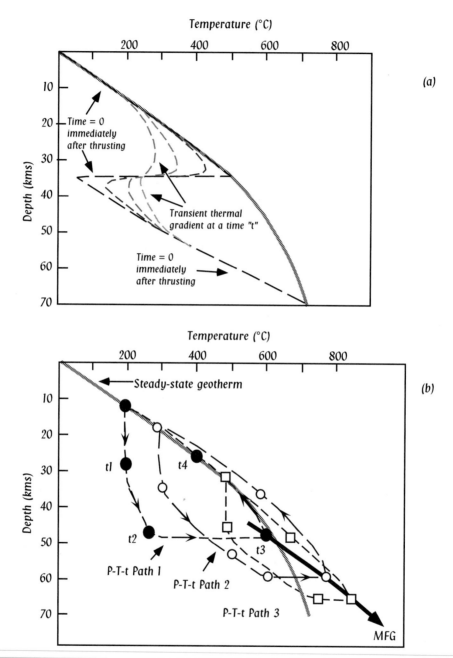

FIGURE 16.24   (a) Changing thermal gradient as a function of time within the upper and lower plates of Figure 16.23. Dark line—normal conductive geothermal gradient in a single lithosphere. (b) Schematic diagram showing P–T–t paths recorded by the rocks at three different depths (circle, unfilled circle, square in Figure 16.23). t1, t2, and so on, mark a time sequence from the oldest to youngest. MFG—metamorphic field gradient (discussed in text).

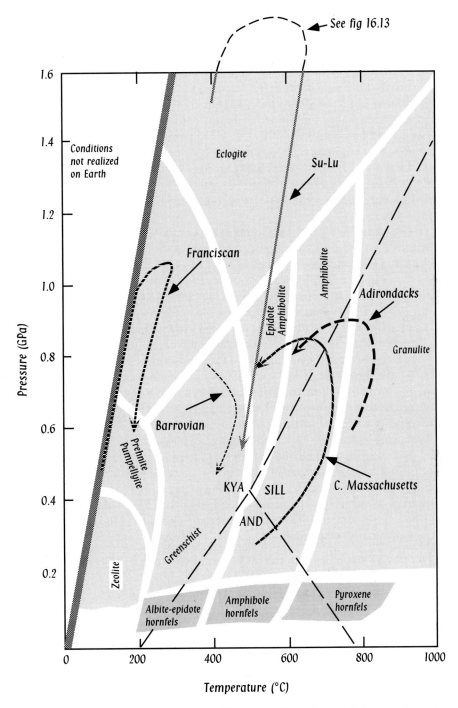

FIGURE 16.25 *Some representative P–T–t paths from the United States and elsewhere* (Sources: Ernst, 1988; Coleman and Wang, 1995).

SEISMOMETAMORPHISM

In most tectonic settings, earthquakes occur at relatively shallow depths ($<$30 km), but in subduction zones, earthquakes occur as deep as 670 km below the earth's surface. Earthquakes are triggered by the sudden rupture of a fault that releases vast amounts of energy. At depths greater than about 40 km, normal brittle failure should be inhibited by the high normal stresses across a fault; in other words, at great depths, the weight of the overlying rocks should create too much friction for fault slips to occur. Earthquakes that occur at depths $>$40 km, therefore, must be triggered by some other mechanism. In recent years, some scientists have focused their research on possible connections between metamorphism and earthquakes in subduction zones (cf. Green and Burnley, 1989; Kirby et al., 1991, 1996a, 1996b; Peacock and Wang, 1999). These studies suggest that both intermediate-depth earthquakes (50–300-km depth) and deep earthquakes (300–670-km depth) might be triggered by metamorphic reactions.

Intermediate-depth earthquakes appear to be triggered by devolatilization reactions within the subducting oceanic crust associated with the transformation of metabasalt and metagabbro to eclogite (Kirby et al., 1996a; Peacock and Wang, 1999). As seen earlier, many devolatilization reactions would be expected to occur in a subducting oceanic crust, beginning with clay- and zeolite-breakdown reactions at shallow depths and culminating in eclogite-forming reactions at deeper levels. Fluids released by these devolatilization reactions increase local pore pressure and decrease effective stresses, thereby promoting slip along preexisting faults (Kirby et al., 1996a).

Deep earthquakes occur within the subducting oceanic lithosphere, the coolest part of the subducting slab. Experimental observations suggest that deep earthquakes are triggered by a shear instability associated with the metastable transformation of olivine to spinel (Green and Burnley, 1989; Kirby et al., 1991, 1996b; see Chapter 17 for olivine-spinel transformation). Because the interior of the subducting lithosphere is relatively cold (500–700°C), the kinetics of metamorphic solid–solid reactions may be quite sluggish, and olivine may persist metastably to depths greater than 400 km.

In SW Japan, arc volcanism is sparse (and highly silicic), and subduction-zone earthquakes do not extend beyond 65-km depth. In NE Japan, arc volcanism is more robust, and intraslab earthquakes extend down to $\sim$200-km depth. Peacock and Wang (1999) constructed detailed thermal-petrologic models of the SW and NE Japan subduction zones, concluding that:

1. the subducting oceanic crust beneath NE Japan is $\sim$300°C cooler than the crust subducting beneath SW Japan;
2. eclogite-forming reactions and dehydration reactions occur much deeper beneath NE Japan than beneath SW Japan, thereby explaining the difference in observed earthquake depths; and
3. the warm subducting oceanic crust beneath SW Japan dehydrates at a relatively shallow level and may partially melt. In contrast, dehydration reactions occur at deeper levels beneath NE Japan, where released fluids infiltrate the hot overlying asthenosphere triggering arc volcanism.

The relationship between metamorphism and earthquakes is an exciting new area of research. Progress has been relatively slow in this area because it requires the collaborative effort of scientists from several different subdisciplines—mineralogy, petrology, rheology, seismology, and thermal modeling. Detailed studies like that of Peacock and Wang (1999) are needed in well-studied areas where vast amount of different types of information (thermal, seismological, etc.) are available to be integrated.

SUMMARY

The basic purpose of the concepts of *zones*, *grades*, and *facies* is to distinguish rocks based on metamorphic intensity. Facies is a particularly powerful concept and uses mineral assemblages to distinguish among assemblages belonging to different facies. The concept of metamorphic facies series is very important in that it is part of a big picture that related plate tectonics to metamorphism. The three fundamental facies series are: high-P–T series, intermediate-P–T series, and low-P–T series. These facies series may be juxtaposed due to plate convergence.

Graphical representation of metamorphic reactions is an important way to depict such reactions. Certain key reactions give us information on P–T conditions of metamorphism. Mineral thermobarometry gives P–T information on metamorphic rocks. Metamorphic rocks generally retain the peak P–T conditions. However, a detailed probing into them can reveal the history of evolution of metamorphic belts (i.e., their P–T–t paths). Seismometamorphism is an interesting new development that relates subduction zone earthquakes to metamorphic reactions.

BIBLIOGRAPHY

Best, M.G. (1982) *Igneous and Metamorphic Petrology*. W.H. Freeman (San Francisco).

Blatt, H. and Tracy, R.J. (1995, 2nd ed.) *Petrology: Igneous, Sedimentary and Metamorphic*. W.H. Freeman (New York).

Bohlen, Jr. and Boettcher, A.L.C. (1981) Experimental investigations and geological application of orthopyroxene barometry. *Am. Mineral.* **19**, 951–964.

Bucher, K. and Frey, M. (1994) *Petrogenesis of Metamorphic Rocks* (6th ed. of H. Winkler's textbook). Springer-Verlag (New York).

Carmichael, D.M. (1969) Intersecting isograds in the Whetstone lake area, Ontario, *J. Petrol.* **11**, 147–181.

Chopin, C. (1984) Coesite and pure pyrope in high grade blueschists of the western Alps. *Contrib. Mineral. Petrol.* **86**, 107–118.

Ellis, D.J. and Green, D.H. (1979) An experimental study of the effect of Ca upon garnet-clinopyroxene Fe–Mg exchange equilibria. *Contrib. Mineral. Petrol.* **71**, 13–22.

Ernst, W.G. (1976) *Petrologic Phase Equilibria*. W.H. Freeman (San Francisco).

Ernst, W.G. (1988) Tectonic history of subduction zones: inferred from retrograde blueschists p–t paths. *Geology* **16**, 1081–1084.

Ferry, J.M. and Spear, F.S. (1978) Experimental calibration of the partitioning of Fe and Mg between biotite and garnet. *Contrib. Mineral. Petrol.* **66**, 113–117.

Green, H.W. II and Burnley, P.C. (1989) A low self-organizing mechanism for deep- focus earthquakes. *Nature* **341**, 737–737.

Hacker, B.R. and Liou, J.G. (1998) When continents collide: geodynamics and geochemistry of ultrahigh-pressure rocks. Kluwer (Boston).

Kirby, S.H., Durham, W.B. and Stern, L.A. (1991) Mantle phase changer and deep earthquake faulting in subducting lithosphere. *Science* **252**, 216–225.

Kirby, S.H. et al. (1996a) Intermediate-depth intraslab earthquakes and arc volcanism as physical expressions of crustal and uppermost mantle metamorphism in subducting slabs. In G.E. Bebout et al. (eds.) Subduction Top to Bottom. Geophysical Monograph 96. *Am. Geophys.* Union (Washington, D.C.), 195–214.

Kirby, S.H. et al. (1996b) Metastable mantle phase transformations and deep earthquakes in subducting oceanic lithospher. *Rev. Geophys.* **34**, 261–306.

Koziol, A.M. & Newton, R.C. (1978) Redetermination of the anorthite breakdown reaction and improvement of plagioclase–garnet–Al_2SiO_5–quartz barometer. *Am. Mineral.* **73**, 216–223.

Liou, J.G. (1971) P–T stabilities of laumontite, wairakite, lawsonite, and related minerals in the system $CaAl_2Si_2O_8$–SiO_2–H_2O. *J. Petrol.* **12**, 379–411.

Pattison, D. and Newton, R.C. (1991) Reversed experimental calibration of the garnet-clinopyroxene Fe-Mg exchange theromometer. *Contrib. Mineral. Petrol.* **101**, 87–103.

Peacock, S.M. and Wang, K. (1999) Seismic consequences of warm versus cool subduction metamorphism: examples from Southwest and Northeast Japan. *Science* **286**, 937–939.

Philpotts, A.R. (1990) *Principles of Igneous and Metamorphic Petrology*. Prentice-Hall (Upper Saddle River, N.J.).

Ringwood, A.E. (1975) *Composition and Petrology of Earth's Mantle*. McGraw-Hill (New York).

Saxena, S.K. (1973) *Thermodynamics of Rock-Forming Crystalline Solutions*. Springer (New York).

Sen, G. (1988) Petrogenesis of spinel lherzolite and pyroxenite suite xenoliths from the Koolau shield, Oahu, Hawaii: Implications for petrology of the post.-eruptive lithosphere beneath Oahu. *Contrib. Mineral. Petrol.* **100**, 61–91.

Sen, G. and Jones, R. (1989) Experimental equilibration of multicomponent pyroxenes in the spinel peridotite field: implication for practical thermometers and a possible barometer. *J. Geophys. Res.* **94**, 17871–17880.

Spear, F.S. and Selverstone, J. (1983) Quantitative P–T paths from zoning in minerals: Theory and some tectonic applications. *Contrib. Mineral. Petrol.* **83**, 348–357.

Thompson, J.B. Jr. (1957) The graphical analysis of mineral assemblages in pelitic schist. *Am. Mineral.* **42**, 842–858.

Thompson, J.B. Jr. (1982) Composition Space: an algebraic and geometric approach. *Rev. Mineral.* **10**, 1–31.

Thompson, J. B. Jr. (1982) Reaction space: an algebraic and geometric approach. *Rev. Mineral.* **10**, 33–51.

Turner, F.J. (1968) *Metamorphic Petrology.* McGraw-Hill (New York).

Wood, B.J. and Banno, S. (1973) Garnet—orthopyroxene and garnet—clinopyroxene relationships in simple and complex systems. *Contrib. Mineral. Petrol.* **42**, 109–124.

Wood, B.J. and Fraser, D.G. (1976) *Elementary Thermodynamics for Geologists.* Oxford University Press. (Oxford, U.K.).

CHAPTER 17

The Deep Earth: Mantle

The following topics are covered in this chapter:

Introduction
Minerals of the deep mantle (pressure above 3 GPa): composition, and phase relations
Some final thoughts
Summary

ABOUT THIS CHAPTER

Previous chapters dealt with minerals and rocks that can be seen on the earth's surface. This chapter is mainly concerned with the unseen minerals from deeper parts of the earth that are never (with a few rare exceptions) brought up to surface. How do we know what they are? Several pieces of evidence, including seismology, experimental petrology, and meteorite studies, tell us what the deep-mantle minerals may be. The principal source of information comes from seismology, which is a study of seismic waves traveling though the earth. Seismic waves traveling through the interior of the earth are routinely recorded by seismographs. Velocities and paths of propagation of these waves are dependent on physical properties of the rocks they must go through. Thus, a close scrutiny of these seismic waves tells us how density and velocity properties change as a function of depth. Laboratory experiments at very high pressures and temperatures tell us what minerals may be stable at different depths. Meteorites also provide some circumstantial evidence for the mineralogy of the deep earth. Experimentally produced minerals, whose density and velocity properties have also been measured in the laboratory, can then be mixed in various proportions to get the best possible match between the seismically defined layers and estimated mineral proportions at various depths.

This chapter provides a general overview of the mineralogy of the deep mantle. It deals with two classes of deep minerals: those believed to be volumetrically significant, and others that may be sources of water.

INTRODUCTION

The upper mantle is almost entirely peridotitic. We know this from xenoliths in kimberlites and alkalic mafic lavas, as well as from ophiolites. Eclogite, which also occurs as xenoliths in kimberlites, must also be present in much smaller amounts in the upper mantle. Geothermobarometry on such xenoliths indicates that the deepest samples from the upper mantle generally come from no deeper than 250 km. What minerals must occur at greater depths? As pointed out previously, our knowledge of deep-mantle mineralogy is based on phase-equilibrium experiments, seismology, and cosmochemistry. Actual physical clues to the constitution of the deep mantle come from rare minerals in meteorites and mineral inclusions in diamonds. A brief summary of some of these exciting developments is presented next following a short discussion on earthquake-based Earth models.

SEISMOLOGY AND EARTH'S INTERIOR

Two types of waves are triggered by an earthquake: body and surface. Body waves travel through the interior of the earth, and surface waves are restricted to the earth's surface. Body waves are of two types: compressional or P (for primary) and shear or S (for secondary). There are two important differences between P and S waves: (a) S waves do not travel through liquid, but P waves do; and (b) P waves travel faster than S waves in any solid medium. The velocity of a P or S wave depends on the physical properties, particularly density, of the material through which it travels. Seismologists (i.e., scientists who study earthquakes) have compiled vast data on velocities and paths of seismic body waves created by thousands of earthquakes over many years and recorded at 3,000 seismic stations. Modern fast computers have allowed seismologists to create reference models that filter the available data in terms of average velocity and density variations as a function of depth within the earth (Figure 17.1). Perhaps the best-known model is the Preliminary Reference Earth Model (PREM; Dziewonski and Anderson, 1981; later modified by Montagner and Anderson, 1989). The density-depth profile (Figure 17.1) put forth in PREM is an important constraint on the types and proportions of minerals that may constitute various parts of the earth's interior.

The existence of seismic discontinuities, which are marked by distinct jumps in P or S wave velocities, has been known for a long time. The presence of such discontinuities has allowed division of the earth's internal layers into crust, upper mantle, transition zone, lower mantle, D″, outer core, and inner core. For example, the Mohorovicic discontinuity (Moho) is one such discontinuity that separates the crust from the mantle. Gutenberg discontinuity separates the mantle from the core. Within the mantle, there are at least two prominent discontinuities at 410- and 660-km depths,

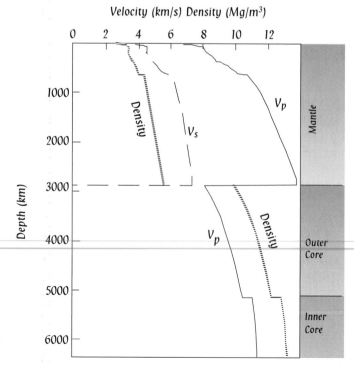

FIGURE 17.1   *Seismic velocity and density variations within the Earth based on modified Preliminary Reference Earth Model (or PREM).*

marking the top and bottom of the transition zone. A third discontinuity, although much less prominent, occurs at 520 km (i.e., within the transition zone). As mentioned in Chapter 1, most recently a highly irregular seismic boundary has been identified at 1,400 to 1,600 km (i.e., within the lower mantle).

The origin of seismic discontinuities is of great interest to geoscientists. Most of these discontinuities in the mantle signify mineralogical phase transformations that may or may not be accompanied by chemical changes. The only exception to these is the newly found 1,400- to 1,700-km discontinuity, which is thought to be chemical in nature (and not a phase transition; Chapter 1). The transition zone is particularly interesting because it is believed to play a key role in a number of global Earth phenomena such as mantle convection and deep-mantle plumes.

Significant technological development over the last decade has allowed mineralogists to conduct high-pressure experiments relevant to the phase equilibria of the deeper mantle. Based on these new experiments, immense progress has been made in topics ranging from deep-mantle mineralogy, storage of water in the mantle, and recycling of the subducted slab. It is well beyond the scope of this chapter to cover all or even most aspects of high-pressure mineralogy. The interested reader is well advised to consult the references listed at the end of this chapter. Only a broad coverage, with some sweeping observations, of deep mantle mineralogy is given here.

DRY MINERALS FROM THE DEEP

In contrast to the earth's crust, which is composed of a tremendous variety of minerals, the earth's mantle appears to be relatively simple and consists of only a few minerals. The term *dry* is meant to include volatile-free minerals, such as pyroxene, olivine, and so on. However, Bell and Rossman (1992) showed that even these dry minerals can contain a small amount of structurally bound H_2O. For example, pyroxenes contain 200 to 500 ppm H_2O, and therefore should be referred to as *nominally anhydrous* minerals. Based largely on high-pressure laboratory experiments, a list of common volatile-free minerals expected to occur in the earth's mantle are given in Table 17.1. It is important to point out that some of these laboratory-produced minerals have indeed been found as disequilibrium assemblages in shock-metamorphosed chondritic meteorites, named *Tenham, Acfer 040,* and *Coorara* (Table 17.1; Stoffler, 1997, and references therein), and as equilibrium inclusions in diamonds from Sao Luis (Brazil and McCammon et al. 1997). Table 17.2 lists the chemical composition of the mineral inclusions in Brazilian diamonds. It is apparent that the variety of observed high-pressure minerals in meteorites and diamonds is rather small relative to the phases that have been produced in experiments. This is to be expected, however, bacause not all of the experimental minerals can be abundant in the earth's lower mantle and transition zone.

What are the most major minerals of the deep mantle, and what are their relative abundances? What types of reactions occur within and between these minerals as a function of depth? Can their abundances and phase-transformation reactions explain the observed seismic discontinuities and density distribution within the earth as required by such models as PREM? To answer such fundamental questions with the help of experimental petrology, the first step is to select an appropriate bulk composition for the unseen portion of the earth's mantle. Herein lies a great difficulty: There is no guarantee that the lower mantle and transition zone are chemically identical to the upper mantle (about which we seem to know more because we have actual samples). Also, we cannot safely assume that the lower mantle is chemically similar to the transition zone. Detailed questions regarding the bulk chemical compositions of the different parts of the mantle are likely to continue to be debated for many years to come (e.g., McDonough and Sun 1995).

TABLE 17.1 *Mantle mineralogy (anhydrous or nominally hydrous minerals).*

Mineral	Composition	Zero-Pressure Density (g cm^{-3})	Occurrence
Olivine	$\alpha-^{VI}(Mg_{0.87-0.9},Fe)_2{}^{IV}SiO_4$	3.37	Ol,opx,cpx,sp, and gt are
Orthopyroxene	$^{VI}(Mg_{0.87-0.9},Fe)^{IV}SiO_3$	3.3	found in upper mantle
Clinopyroxene	$^{VI}Ca(Mg_{0.86-0.9},Fe)^{IV}Si_2O_6$	3.32	xenoliths. Diamond and
Spinel	$^{IV}(Mg,Fe)^{VI}Al(Cr)_2O_4$	3.55	graphite rarely occur in
Garnet (pyropic)	$^{VIII}((Mg,Fe)_{.95}Ca_{.05})^{VI}Al_2{}^{IV}Si_3O_{12}$	3.68	xenoliths
Wadsleyite*	$\beta-^{VI}(Mg,Fe)_2{}^{IV}SiO_4$	3.69	Wadsleyite, ringwoodite,
Ringwoodite*	$\gamma-^{VI}(Mg,Fe)_2{}^{IV}SiO_4$	3.72	perovskite, ilmenite,
$(Mg,Fe)-SiO_3$ Majorite*	$^{VIII}(Mg,Fe)_3{}^{VI}(Mg,Fe)^{VI}Si^{IV}Si_3O_{12}$	3.59	majorite occur
Silicate Perovskite*	$^{VIII}(Mg,Fe,Al)^{VI}SiO_3$	4.15	in shock-metamorphosed
Silicate ilmenite	$^{VI}(Mg,Fe)^{VI}(Si,Al)O_3$		chondritic meteorites
Magesiowustite**	$^{VI}(Mg,Fe)O$	4.10	*Tenham* and *Coorara*.
Stishovite	$^{VI}SiO_2$	4.29	Perovskite, a garnet
TAPP**	$(Mg,Fe,Al)Al_2Si_3O_{12}$		phase (TAPP), and
Ca-perovskite**	$^{VIII}Ca^{VI}SiO_3$		magnesiowustite occur
			as inclusions in diamonds
			from Sao Luiz (Brazil).

Main Sources: Deer, Howie, and Zussman (1995), Anderson (1989), Brown and Mussett (1993, Table 7.4), Thompson (1992, Table 1), Stoffler (1997).

*Phases found in meteorites (see Stoffler, 1997, for references).

Phases found as diamond inclusions (Wilding et al., 1991, Proceedings of the 5th International Kimberlite Conference; Harte and Harris, 1994, Mineralogical Magazine **58A, 384; McCammon et al., 1997, Science 278, 434–436).

*** ^{IV} 4-fold coordination; ^{VI} 6-fold coordination; ^{VIII} 8-fold coordination.

TABLE 17.2 *Chemical composition of diamond inclusions from Sao Luis.*

	Magnesiowustite					Perovskite		TAPP*	
	1	2	3	4	5	6	7	8	9
SiO_2	0.01	0.00	0.02	0.07	0.04	51.41	57.04	41.41	42.24
TiO_2	0.01	0.01	0.02	0.04	0.01	0.02	0.15	0.03	0.04
Al_2O_3	0.13	0.06	0.04	0.12	0.04	10.04	1.33	23.33	24.17
Cr_2O_3	1.06	0.03	0.38	0.66	0.34	1.19	0.4	2.99	2.41
Fe_2O_3	5.72	1.33	0.96	0.88	0.00	4.28	0.84	4.10	3.81
FeO	68.41	58.61	42.49	25.58	23.52	1.28	3.04	1.29	1.76
MgO	23.13	39.75	56.37	71.31	74.77	30.21	36.25	24.95	24.36
NiO	0.10	0.28	0.40	1.40	1.25	0.02	0.00	0.01	0.02
CaO	0.01	0.00	0.00	0.00	0.02	0.65	0.06	0.13	0.11
Na_2O	1.05	0.11	0.06	0.12	0.18	1.05	0.03	0.16	0.09
K_2O	0.00	0.00	0.00	0.00	0.01	0.26	0.02	0.02	0.00

Note. This table is a slightly simplified version of Table 1 in McCammon et al. (1997, *Science* **278**, p. 434).

*TAPP—a type of garnet: tetragonal almandine-pyrope phase.

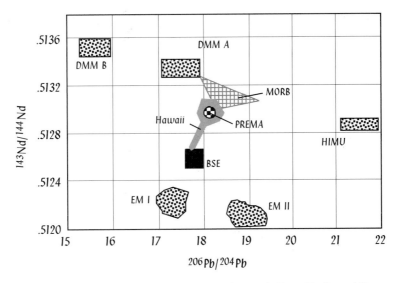

FIGURE 17.2 *Isotopic reservoirs within the earth's mantle (from Zindler and Hart, 1986).* DMM—Depleted MORB Mantle. EM—Enriched Mantle. HIMU—High μ. BSE—Bulk Silicate Earth. PREMA—Prevalent Mantle. The fields for MORB and Hawaiian lavas are only shown here.

The concept of a single bulk chemical composition for the entire mantle may seem preposterous to some petrologists/geochemists in light of strong isotopic and elemental heterogeneities in mantle-derived basalts (Figure 17.2). For example, based on isotopic variability in basalts, Zindler and Hart (1984) proposed the existence of at least five isotopically distinct mantle reservoirs.

Both lateral and vertical heterogeneities in the mantle have also been found in mantle tomographic studies (Figure 17.3), in which large numbers of earthquakes

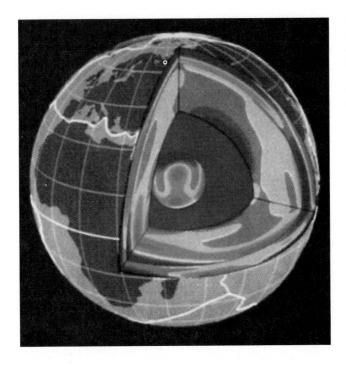

FIGURE 17.3 *Seismic tomography of the earth showing hot materials (light gray) in the mantle beneath oceans and cool mantle materials (deeper gray) beneath continents.* (From Davidson et al., 1997, *Exploring Earth*, Prentice Hall).

are studied to obtain a three-dimensional picture of velocity variations through the earth's interior (much the same way CAT scans image the brain). These observed three-dimensional seismic variations are explained largely in terms of lateral and vertical temperature variations (i.e., as *hot* and *cold* reservoirs). However, it is possible that some of the observed variations are due to seismic anisotropy (i.e., preferred orientation of grains that allows seismic waves to go faster along some preferred crystallographic direction than along other directions).

Although we know much about isotopically inferred reservoirs and seismically modeled hot and cold domains in the mantle, it is unclear whether or how the seismic domains correspond to the isotopic reservoirs.

Notwithstanding the difficulties summarized previously, we do have some constraints from seismic velocities and densities, composition of chondritic meteorites, and upper-mantle xenoliths on the lower mantle and transition zone compositions. Among the different bulk starting compositions used in laboratory experiments by different scientists, two are distinct—one is a pyrolite or peridotite and the other is a chondritic meteorite. Some other scientists follow a slightly different approach: They study phase relationships of end-member minerals such as forsterite and enstatite, two dominant components of the upper mantle, and their solid solutions. Based on such data, these authors then extrapolate their findings to the earth system, which is a multicomponent system. What follows is a brief summary of the various findings, starting with some simple but relevant systems.

Olivine's Behavior at High Pressure

Presnall and Walter (1993) carried out multianvil experiments over the pressure range of 10 to 16.5 GPa (i.e., a depth range of about 310–500 km) on forsterite. By combining other published data, they extrapolated the phase relations to about 25 GPa (Figure 17.4). They noted that forsterite breaks down to β-phase (wadselyite) at ~ 14 to 16 GPa (about 450 km) over a large temperature range. β-phase ultimately dissociates to a mixture of Mg-perovskite and periclase (MgO) somewhere ~ 22 GPa (about 660 km). Another Mg_2SiO_4 polymorph, γ-phase, appears only in the subsolidus region at a lower temperature. The reader may recall that the seismic discontinuities at the top and bottom of the transition zone (TZ) occur at ~ 400 and 660 km, respectively. Thus, based on this phase diagram, one could argue that the 400-km discontinuity is due to the polymorphic transformation of forsterite to wadsleyite and the 660-km discontinuity to perovskite + periclase forming reaction.

Although the forsterite phase diagram is an excellent starting point, it is generally recognized that the mantle olivine must have ~ 10 to 12 mole% Fe_2SiO_4 component in solid solution. Therefore, the Mg_2SiO_4–Fe_2SiO_4 system is more relevant to TZ pressures. Experimental petrologists from Japan have studied the relevant magnesian portion of this system at different pressures (Figure 17.5). Ito and Katsura (1989) showed that at 1,400°C magnesian olivine (α-phase) first inverts to β-phase at a pressure range over 12.5 to 15 GPa, and then β-phase inverts to γ-phase at a pressure of 24 to 25 GPa along their respective inversion loops. Consider, for example, how phase inversions would occur in an olivine of composition M—a typical mantle olivine (Figure 17.5). When M reaches 1, it would begin to invert to a β-phase of composition 1'. With increasing pressure, inversion would continue. All along, compositions of coexisting olivine and β-phase would become more magnesian; however, at any particular pressure, the β-phase would always be richer in Fe than the corresponding olivine. The inversion would be complete at 2', and at higher pressures only β-phase would

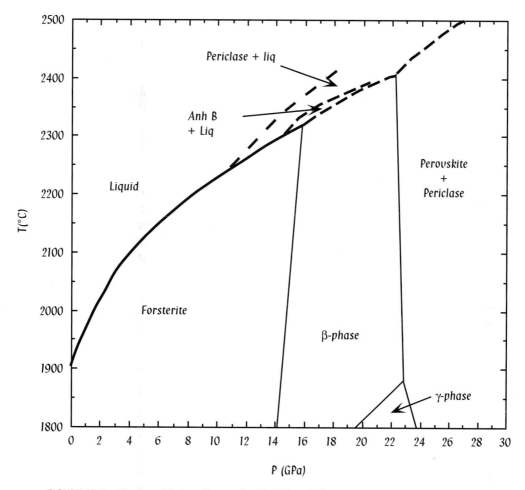

FIGURE 17.4   *Phase equilibrium diagram for Mg_2SiO_4 at high pressures.*

occur until point 3 is reached. At 3, γ-phase would appear. As pressure increases, more β-phase would invert to γ-phase. At point 4, the last β-phase would disappear, and thus, at P > ~21 GPa, only a γ-phase would remain. Around 24 GPa, the temperature in the mantle is more likely to be ~1,700°C (Philpotts, 1990). The 1,700°C isothermal diagram shows that Mg-perovskite and periclase [(Mg,Fe)O] would form:

$$\gamma - [Mg_{.9}Fe_{.1}]_2SiO_4 \quad = \quad [Mg,Fe]SiO_3 \quad + \quad [Mg,Fe]O.$$

γ-phase   Mg-perovskite   Fe-periclase or
                                                magnesiowustite

Although the Fe/Mg ratios of periclase and Mg-perovskite cannot be read out of this diagram (Figure 17.5), Ito and Takahashi (1989) indicated that the periclase/magnesiowustite and Mg-perovskite in their experiment had $Mg_{.83}Fe_{.17}$ and $Mg_{.97}Fe_{.03}$, respectively.

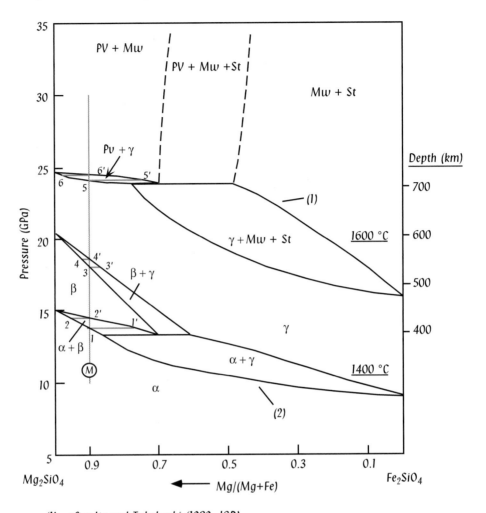

(1) - after Ito and Takahashi (1989, JGR)

(2) - after Katsura and Ito (1989, JGR)

FIGURE 17.5 *Subsolidus equilibrium in the system Mg₂SiO₄–Fe₂SiO₄ at pressures relevant to the earth's transition zone.*

Based on a thermodynamic analysis of the system Mg_2SiO_4–Fe_2SiO_4 at very high pressures, Akaogi et al. (1993), presented a P–T diagram showing the nature of the various phase-transformation boundaries (Figure 17.6). They showed, as others have before them (most notably, A.E. Ringwood and coworkers from Australian National University), that if the earth's mantle is peridotitic in bulk chemical composition, then the 400- and 660-km discontinuities can be explained by $\alpha \rightarrow \beta$ and $\gamma \rightarrow$ Mg-perovskite + periclase reactions. These reactions are virtually independent of pressure, and thus can be used to pin down the temperature at 400 and 660 kms, respectively: The 400- and 660-km discontinuities under normal mantle should be at temperatures of ~1,400 and 1,700°C, respectively. These authors also pointed out that a cool downgoing slab beneath a subduction zone may go through these transitions at somewhat different depths. (Figure 17.6).

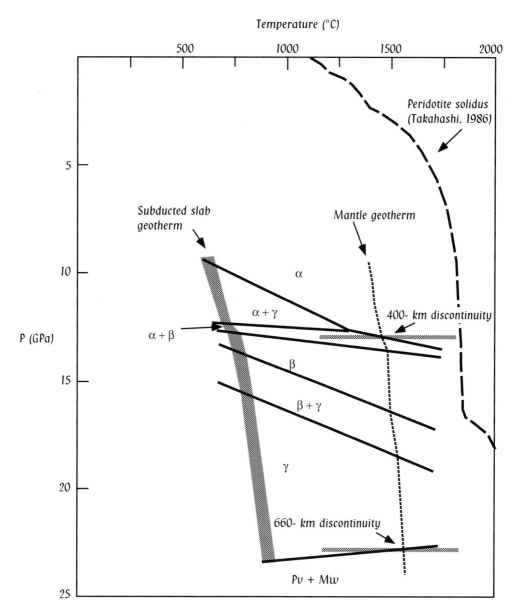

Temperature (°C)

Note: From Akaogi et al. (1989 JGR). The γ --> Pv + Mw curve is based on Presnall and Walter (1993)

FIGURE 17.6 *Phase diagram of the earth's mantle as presented by Akaogi et al. (1993).* Relevant geothermal gradients for subducted slabs and normal mantle are also shown.

Enstatite's Behavior at High Pressures

Enstatite is the second major mineral in mantle peridotite. Presnall and Gasparik (1993) carried out experiments on pure $MgSiO_3$ at pressures up to 17.8 GPa in a multianvil apparatus. Saxena et al. (1996) assembled all available experimental data on the subsolidus phase relations in the system $MgSiO_3$ and performed a rigorous thermodynamic analysis. Figure 17.7 is based primarily on Presnall and Gasparik's

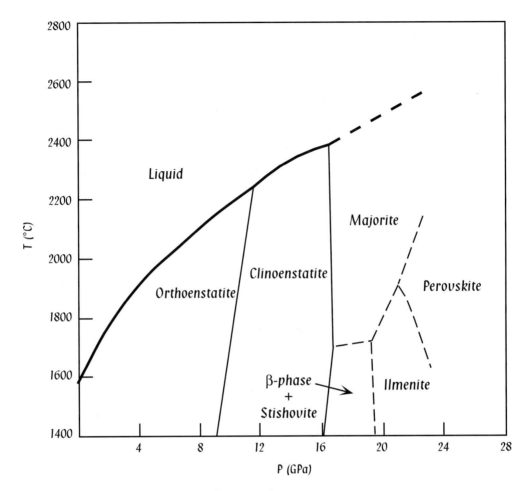

FIGURE 17.7 *Phase diagram for MgSiO₃ at mantle pressures.*

work, and the dashed boundaries are based on Saxena's analysis. Up to the pressure of 24 GPa, five different MgSiO₃ polymorphs appear in this system:

> majorite (with garnet structure),
> Mg-perovskite (with atomic structure similar to real perovskite-CaTiO₃),
> Mg-ilmenite (with atomic structure similar to ilmenite-FeTiO₃),
> orthoenstatite (orthorhombic), and
> clinoenstatite (monoclinic).

Of particular importance are the clinoenstatite/majorite and clinoenstatite/β-phase + stishovite boundaries, which are virtually independent of temperature. Although the location of these boundaries at ~17.8 GPa does not coincide with any seismic discontinuity, it is possible that the presence of ~12 mole% FeO in the natural mantle system drives this boundary to ~14 GPa, where the 400-km discontinuity occurs.

We have already noted in our discussion of the Mg₂SiO₄–Fe₂SiO₄ system that β-phase appears at ~13.5 GPa. Thus, β-phase must be a major constituent of the upper part of the TZ. Stishovite and majorite may also be important. These studies on the simple system MgO–FeO–SiO₂ present an important quantitative understanding on the nature of the TZ. Because the mantle contains at least two other major components—namely, CaO and Al_2O_3—it would be pertinent to ask how these reactions occur in something like a peridotitic or pyrolitic mantle. What follows is a discussion of phase relations in more real mantlelike systems.

Phase Relations in Natural Multicomponent (Real Mantle) Systems

Using a natural upper-mantle peridotite as a starting material in their experiments, Zhang and Herzberg (1994) produced a phase diagram, a portion of which is shown in Figure 17.8. As far as the mantle seismic discontinuities are concerned, most notable features are: (a) the very narrow pressure range over which α/β and β/γ transitions

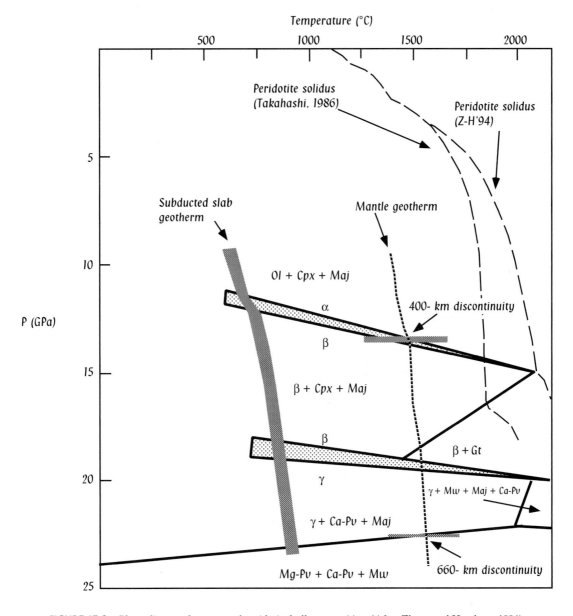

FIGURE 17.8 *Phase diagram for a natural peridotite bulk composition.* (After Zhang and Herzberg, 1994)

occur, (b) the appearance of Mg-perovskite + magnesiowüstite assemblage at a minimum pressure of ∼22 to 24 GPa, and (c) the occurrence of majorite and Ca-perovskite. Phases like majorite and Ca-perovskite are to be expected because of the presence of CaO in the starting material, which was not the case in the previously mentioned phase equilibrium studies of simple systems. Note that the incoming of β-phase and Mg-perovskite + magnesiowüstite correspond really well with the 400- and 660-km discontinuities assuming a mantle adiabat of ∼1,400°C.

Experiments by other authors on chondritic starting materials yielded some differences; however, the major phase-transition curves correspond well with those in a peridotitc/pyrolitic system. Therefore, these experiments substantiate a proposal put forth by A.E. Ringwood and colleagues in the late 1960s that the 400- and 660-km seismic discontinuities result from phase transitions involving α/β and γ/Mg-perovskite + magnesiowüstite. The most significant difference between a peridotitic versus chondritic starting mix is in the proportion of minerals that would be expected along a mantle geotherm at any given pressure (Agee, 1993; Figure 17.9). Peridotite/pyrolite starting material yields much greater amounts of olivine than chondrites, and therefore more β- and γ-phases are formed from a peridotitic starting mix than from a chondritic bulk composition. However, the pyroxene component, which transforms to majorite, is greater in the chondritic bulk composition. Therefore, more majorite and pyroxenes are to be expected from a chondritic starting mix.

Lower Mantle and Perovskite: Some Additional Comments

As shown in Figure 17.8, all experiments point to a lower mantle that is dominantly composed of Mg-Perovskite with a periclase (MgO) or magnesiowüstite (Mg,Fe)O. There has been some debate about the stability of perovskite at very high pressures corresponding to the deep lower mantle. For example, based on data on pure $MgSiO_3$, Saxena et al. (1996) concluded that Mg-perovskite $MgSiO_3$ breaks down to periclase (MgO) and stishovite (SiO_2) at pressures between 58 and 85 GPa within the earth. There is also an added controversy regarding whether a free aluminous phase, such as majorite, can occur deep in the lower mantle.

Irifune (1994) and Kesson et al. (1998) carried out experiments on a pyrolite-starting composition and discovered that Al_2O_3 can be easily accommodated by Mg-perovskite, and a free aluminous phase, corundum (Al_2O_3), cannot be stable in the lower mantle. They also observed that majorite completely breaks down to Mg-perovskite. Noting that Mg-perovskite can dissolve as much as 25% Al_2O_3 at 135 GPa, Kesson et al. (1998) concluded that this dissolved alumina increases the stability of Mg-perovskite to the base of the lower mantle. Table 17.3 lists mineral analyses reported in the experiments conducted by Irifune (1994) and Kesson et al. (1998). (It is interesting to note that Kesson et al. did not actually list such a highly aluminous Mg-perovskite in their experimental runs. The maximum Al_2O_3 in their experimental charges is 5%.) However, the maximum Al_2O_3 of 10% in Mg-perovskite was found in Sao Luis diamond inclusions (Table 17.2), but we do not know what sort of depth it may have come from.

Irifune (1994) calculated that the relative abundances of the minerals in the uppermost part of the lower mantle are: Mg-pv—76% to 77%, Ca-pv—8%, and Mw—15% to 17% (Figure 17.10). This brings to question whether ilmenite-structured silicate (proposed by Saxena and coworkers) occurs at all in the lower mantle. At this point, its presence is doubtful at best.

A number of authors have presented forceful arguments in favor of greater silica and iron in the lower mantle relative to a pyrolite/peridotite model. One of these is based on the difference between PREM densities in the lower mantle versus those

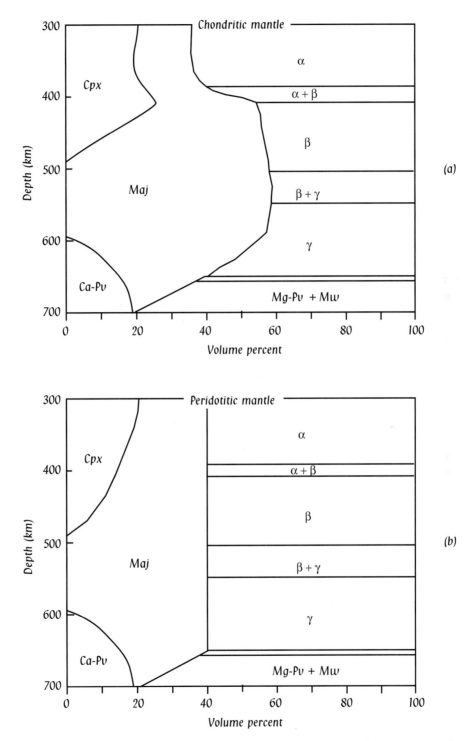

FIGURE 17.9   *Agee's (1993) calculated models showing modal proportions of constituent minerals in two different bulk compositions for the earth's mantle—chondritic versus peridotitic.*

TABLE 17.3   *Chemical composition of minerals in experiments.*

P(GPa)	24	135	28	135	25	135	23	24
	Mg-Pv		Ca-Pv		Mw		Majorite	
	1	2	3	4	5	6	7	8
SiO_2	53.42	55.40	49.06	51.20	0.22	—	51.64	49.79
TiO_2	0.36	0.50	0.14	—	—	—	0.20	0.18
Al_2O_3	3.68	5.00	1.15	1.00	0.15	0.80	10.31	12.34
Cr_2O_3	0.52	0.50	—	—	0.32	0.80	1.32	1.07
FeO^*	6.34	4.50	0.48	2.20	21.69	24.0	3.98	4.00
MgO	34.27	39.75	1.19	6.10	75.59	71.90	27.57	22.15
NiO	—	0.28	0.20	—	0.80	0.40	—	—
CaO	0.20	1.20	43.10	36.60	—	—	4.39	8.59
Na_2O	—	0.11	0.21	0.9	0.28	1.60	0.51	1.21
K_2O	—	—	—	1.80	—	—	—	—

Note. The data at 135 GPa are from Kesson et al. (1998); all others are from Irifune (1994).

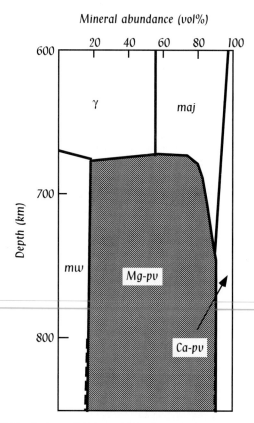

FIGURE 17.10   *Irifune's (1994) model for the modal variations in the mantle.*

calculated from experimentally generated assemblages with a peridotitic-pyrolitic starting composition. Brown and Mussett (1993) pointed out that a $Mg/(Mg + Fe)$ ratio of 0.86 for the lower mantle, relative to 0.88–0.90 for the upper mantle, can explain the density difference. The lower mantle is also believed by Brown and Mussett and other authors to have much higher content of SiO_2 relative to the upper mantle, which would be reflected in a greater content of perovskite than would be allowed by a pyrolite model. Brown and Mussett gave the following abundances of the lower-mantle minerals: Mg-perovskite—81.5 wt%, magnesiowustite—9.1%, Ca-perovskite—5.8%, and corundum—3.6%. However, as indicated earlier, the most recent experiments indicate that corundum cannot be stable in the lower mantle, and its abundance would have to be redistributed among the other three minerals.

Based on what we know today about the phase relationships of these various major minerals of the lower mantle and transition zone, a summary figure may be drawn (Figure 17.11) that shows the occurrence of these minerals as a function of depth. It is unlikely that the entire lower mantle, and particularly the transition zone, is homogeneous. It is possible that considerable lateral and vertical proportions of these minerals exist. Figure 17.11 is intended for only the generic case—away from hot spots and subduction zones. There have been some important experiments on the fate of subducted materials as a function of depth, and these are summarized next. However, precious little is known about the deep structure or constitution of hot spots.

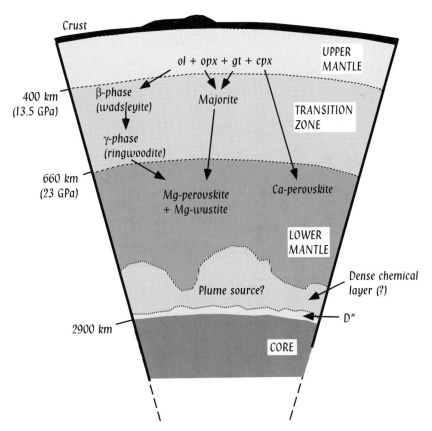

FIGURE 17.11   *A summary diagram showing the major mineral phase changes across the earth's transition zone.*

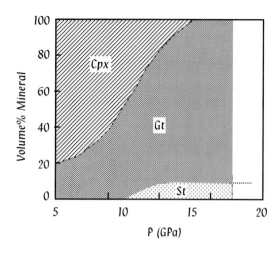

FIGURE 17.12 *The ANU (Australian National University) model for the modal proportions of dominant constituent minerals in a subducted slab as a function of pressure.*

Dry Minerals of Deeply Subducted Slabs

T. Irifune, T. Sekine, A.E. Ringwood, and others at Australian National University have carried out many important experiments on basaltic rocks of the ocean crust at very high pressures to examine what happens to the oceanic crust when it is subducted to deep mantle. Figure 17.12 summarizes some important aspects of their findings:

1. The basaltic rocks convert to eclogite at pressures between 5 and 10 GPa (i.e., well before the slab would reach the top of the transition zone at ~13.5 GPa).

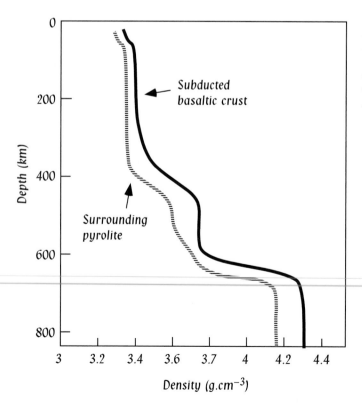

FIGURE 17.13 *Density difference between the (eclogite) crust and peridotite (basal lithosphere) parts of the subducting slab as a function of depth (based on publications from ANU).*

2. Between 10 and 15 GPa, clinopyroxene dissolves entirely into garnet$_{solid\ solution}$. At P > 15 GPa, the eclogite crust becomes an essentially monomineralic rock—garnetite. Stishovite, a high-pressure SiO_2 polymorph, may occur in small amounts at such high pressures.

3. At a pressure of ~23 GPa, garnetite must transform to a Mg-perovskite and Mg-wüstite. Stishovite may persist up to greater pressures.

4. The subducted crust remains denser than the mantle portion of the subducted slab at all depths (Figure 17.13). The first sharp increase between 350 and ~480 km in crustal density is due to eclogite → garnetite transformation. The next sharp increase at ~660 km (i.e., bottom of the transition zone) is due to transformation of garnetite to perovskite + magnesiowustite.

5. Because of the density relationships, the subducted crust may detach from the mantle portion of the slab and continue to descend into the lower mantle. The mantle portion of the slab may reach the 660-km discontinuity and accumulate there as a megalith because its density is less than the lower mantle.

WET MINERALS FROM THE DEEP MANTLE

As indicated in the previous section, small amounts of H_2O in the deep mantle may be stored in the atomic structure of nominally anhydrous minerals, and larger amounts of it may be stored in true hydrous minerals. Table 17.4 gives a list of several hydrous minerals that may occur in the mantle. Among these, phlogopite, amphibole, serpentine, and talc occur in mantle xenoliths and ophiolites. The 10Å-phase, and the so-called *alphabet phases* (including hydrous-A, hydrous-B, etc.) are commonly referred to as *dense hydrous mantle silicates* (DHMS).

H_2O plays an important role in subduction-related volcanism (discussed earlier). Questions regarding how water is transported down into the mantle, how far down it is transported, and how much of it is released back through volcanic eruptions versus how much is left behind in the mantle are all of interest to the earth scientist. This area is being actively researched through high-pressure, multianvil experiments

TABLE 17.4   *Important dense hydrous magnesium silicates (DHMS) of the mantle.*

Mineral	Composition	Zero-Pressure Density (g cm^{-3})	Occurrence
Phlogopite (phl)	$KMg_3AlSi_3O_{10}(OH)_2$	2.784	Phl, amp, ser, and talc
K-richterite (amp)	$K_{1.9}Ca_{1.1}Mg_5Si_{7.9}Al_{0.1}O_{22}(OH)_2$	?	are found in mantle
Serpentine (ser)	$Mg_3Ai_2O_5(OH)_4$	2.55	xenoliths and ophiolites.
Talc (tc)	$Mg_3Si_4O_{10}(OH)_2$	2.78	
10A phase	$Mg_3Si_4O_{14}H_6$	2.65	
Hydrous A (hy-A)	$Mg_7{}^{IV}Si_2O_8(OH)_6$	2.96	
Hydrous B (hy-B)	$Mg_{24}{}^{VI}Si_2{}^{IV}Si_6O_{38}(OH)_4$	3.368	Others have been found
Superhydrous B (su-B)	$Mg_{10}{}^{VI}Si_1{}^{IV}Si_2O_{14}(OH)_4$	3.21	in experiments.
Brucite (bc)	$Mg(OH)_2$	2.37	
Chondrodite (cho)	$Mg_5{}^{IV}Si_2O_8(OH)_2$	3.06	

Source: Thompson (1992, Table 1).

by a number of scientists. What follows is a brief summary of the stability of some of these minerals in pressure-temperature space and their possible roles as water transporters in the mantle.

Bose and Ganguly (1995) performed experiments on the system $MgO–SiO_2–H_2O$ and provided a comprehensive review of other data relevant to the question of H_2O-involved reactions that may occur in the subducted slabs. Figure 17.14 shows the stabilities of some of the important DHMS minerals in the pure system $MgO–SiO_2–H_2O$, along with the thermobarometric conditions that might be expected to persist in young (≤ 5 Ma) versus old (≥ 50 Ma) slabs subducting at rates of 3 to 10 cm/year and at an average angle of 26.6° (from Bose and Ganguly, 1995). The jump in temperature at ~2 GPa in these thermobaric profiles is due to subduction-induced convection in the overlying wedge.

Particularly interesting is the occurrence of both vapor-evolved and vapor-conserved reactions that likely occur in the subducted slab. Following is a summary of the more interesting reactions:

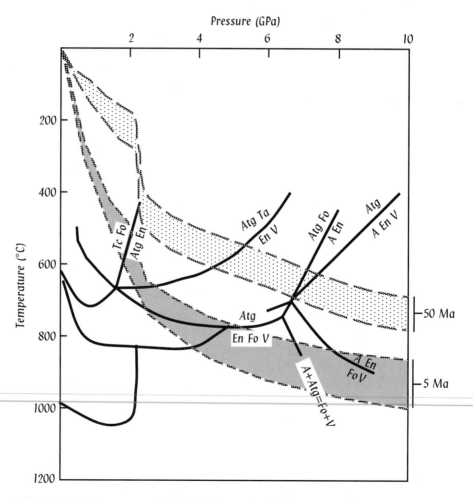

FIGURE 17.14   *The most recently published diagram for the stability of DHMS in the system MgO–SiO₂–H₂O (Bose and Ganguly 1995).* Abbreviations used: Tc–talc, Atg–antigorite, En–enstatite, Fo–forsterite, A–hydrous phase A, V–vapor.

1. In old slabs, talc may be stable to as much as 150 km, whereas in the young slabs, it can only be stable down to ~75 km. Talc breakdown produces antigorite (a hydrous phase) and enstatite, and the reaction conserves vapor:

$$31 \text{ Talc} + 45 \text{ Forsterite} = \text{Antigorite} + 135 \text{ Enstatite.}$$

2. Between ~75- and 200-km depth, antigorite may break down and release a vapor phase:

$$\text{Antigorite} = 20 \text{ En} + 14 \text{ Forsterite} + 31 \text{ H}_2\text{O.}$$

   Bose and Ganguly noted that breakdown of amphibole in the subducted crust may release water into the overlying wedge and stabilize antigorite.

   In colder slabs, antigorite may break down but not release any vapor at ~200 to 225 km:

$$2 \text{ Antigorite} + 113 \text{ Forsterite} = 31 \text{ Phase A (a DHMS)} + 153 \text{ Clinoenstatite.}$$

   Based on relative stabilities of various hydrous phases and the solidus of H_2O-bearing lherzolite solidus (see Chapter 9), Bose and Ganguly concluded that the dehydration breakdown of antigorite may be an important reaction in magma generation beneath island and continental arcs, where the bulk of the magma is generally believed to be produced at ~100 to 150 km in the asthenospheric wedge.

3. Another interesting aspect of Figure 17.14 is that DHMS Phase A is likely to remain stable at much deeper levels as the slab penetrates into deeper parts of the upper mantle and maybe the transition zone. Phase A is therefore a major carrier of water back into the deeper mantle.

SUMMARY

Deep-mantle mineralogy is a fairly new area of experimental research. Our understanding of the mantle is evolving rather rapidly. Keeping this in mind, the main points of this chapter may be summarized as follows:

[1] The upper mantle is peridotitic/pyrolitic. The lower mantle is dominantly composed of silicate-perovskite with a bulk composition of [Mg, Fe]SiO_3.

[2] The upper and lower boundaries of the transition zone, marked at 400- and 660-km seismic discontinuities, correspond to the following reactions:

$$\text{olivine } (\alpha\text{-phase}) \longrightarrow \text{wadsleyite } (\beta\text{-phase})$$

$$\text{ringwoodite } (\gamma\text{-phase}) + \text{(some majorite)} \longrightarrow \text{Mg-perovskite} + \text{wüstite.}$$

[3] In the subducted slab, basaltic crust transforms to eclogite at about 3 to 4 GPa. Eclogite goes through a transition (as clinopyroxene gets progressively dissolved into garnet) and becomes monomineralic garnetite at ~15 Gpa (~450 km).

[4] The subducted crust remains denser than the peridotitic portion of the slab. Therefore, the crust and upper-mantle portions of the slab may get detached from each other while traveling through the transition zone.

[5] Water may be stored in small amounts in nominally hydrous minerals and in larger amounts in amphibole, antigorite (and other serpentine minerals), phlogopite, and hydrous Phase A. Based on existing experiments, it seems doubtful that a hydrous phase would be stable in the lower mantle. However, water may still occur in small amounts in the structure of nominally hydrous minerals in the lower mantle. Arc volcanism may be triggered by the release of H_2O vapor breakdown of such hydrous minerals as amphibole and antigorite.

BIBLIOGRAPHY

Agee, C.B. (1993) Petrology of the mantle transition zone. *Ann. Rev. Earth Planet. Sci.* **21**, 19–41.

Akaogi, M., Ito, E., and Navrotsky, A. (1989) Olivine-modified spinel-spinel transitions in the system Mg_2SiO_4–Fe_2SiO_4: calorimetric measurements, thermochemical calculations, and geophysical applications. *J. Geophys. Res.* **94**, 15671–15685.

Anderson, D.L. (1989) *Theory of the Earth.* (Blackwell, Boston), 366 pp.

Bell, D.R. and Rossman, G.R. (1992) Water in earth's mantle: The role of nominally anhydrous minerals. *Science* **255**, 1391–1396.

Bose, K. and Ganguly, J. (1995) Experimental and theoretical studies of stability of talc, antigorite and phase A at high pressures with applications to subduction processes. *Earth Planet. Sci. Lett.* **136**, 109–121.

Brown, G.E. and Mussett, A.E. (1993) *The Inaccessible Earth: An Integrated View of its Structure and Composition* (2nd ed.). Chapman and Hall (New York).

Davidson, J. P., Reed, W.E., and Davis, P. M. (1997) *Exploring Earth.* Prentice Hall (Upper Saddle River, N.J.).

Deer, W.A., Howie, R.A. and Zussman, J. (1995) *An Introduction to the Rock-Forming Minerals* (2nd ed.), Longman group (Essex, U.K.).

Dziewonski, A.M. and Anderson, D.L. (1981) Peliminary Reference Earth Model. *Phys. Earth Planet. Inter.* **25**, 297–356.

Harte, B. and Harris, J.W. (1994) Lower mantle mineral associations preserved in diamonds. *Mineral. Mag.* **58**, 384–385.

Irifune, T. (1994) Absence of an aluminous phase in the upper part of the earth's lower mantle. *Nature* **270**, 131–133.

Irifune, T., Sekine, T., Ringwood, A.E. and Hibberson, W.O. (1986) The eclogite-garnetite transformation at high pressure and some geophysical implications. *Earth Planet. Sci. Lett.* **77**, 245–256.

Ito, E. and Katsura, T. (1989) A temperature profile of the mantle Transition Zone. *Geophys. Res. Lett.* **16**, 425–428.

Ito, E. and Takahashi, E. (1989) Postspinel transformations in the system Mg_2SiO_4–Fe_2SiO_4 and some geophysical implications. *J. Geophys. Res.* **94**, 10637–10646.

Kesson, S.E. FitzGerald, and Shelley, J.M. (1998) Mineralogy and dynamics of a pyrolite lower mantle. *Nature* **393**, 252–255.

McCammon, C.A., Hutchison, M.T. and Harte, B. (1997) Ferric iron content of mineral inclusions in diamonds from São Luiz: A view into the lower mantle. *Science* **278**, 434–436.

McDonough, W.F. and Sun, S.-s. (1995) The composition of the earth. *Chem. Geol.* **120**, 223–253.

Philpotts, A.R. (1990) *Principles of Igneous and Metamorphic Petrology* (Prentice-Hall, N.J.).

Presnall, D.C. and Gasparik, T. (1990) Melting of enstatite ($MgSiO_3$) from 10 to 16.5 GPa and the forsterite (Mg_2SiO_4)-majorite ($MgSiO_3$) eutectic at 16.5 GPa: Implications for the origin of the mantle. *J. Geophys. Res.* **95**, 15771–15777.

Presnall, D.C. and Walter, M.J. (1993) Melting of forsterite, Mg_2SiO_4, from 9.7 to 16.5 GPa. *J. Geophys. Res.* **98**, 19777–19783.

Ringwood, A.E. (1979) *Composition and Petrology of the Earth's Mantle* (McGraw-Hill).

Saxena, S.K. and 6 others (1996) Stability of perovskite ($MgSiO_3$) in the Earth's Mantle. *Science* **274**, 1357–1359.

Stoffler, D. (1997) Minerals in the Deep Earth: A message from the Asteroid Belt. *Science* **278**, 1576–1577.

Thompson, A.B. (1992) Water in the earth's mantle, *Nature* **358**, 295–302.

Ulmer, P. and Trommsdorff, V. (1999) Phase relations of hydrous mantle subducting to 300 km. In Mantle Petrology: Field observations and high pressure experiments: A tribute to Francis R. (Joe) Boyd (Fei, Y., Bertka, C.M. and Mysen, B.O., eds.), *The Geochemical Society, Special Pub. No.* **6**, 259–281.

Wilding, M.C., Harte, B. and Harris, J.W. (1991) Evidence for a deep origin for São Luiz diamonds, *Extended Abstracts, 5th Int. Kimberlite Conf.* 456–458.

Zhang, J. and Herzberg, C. (1994) Melting experiments on anhydrous peridotite KLB-1 from 5.0 to 22.5 GPa. *J. Geophys. Res.* **99**, 17729–17742.

Zindler, A. and Hart, S.R. (1986) Chemical geodynamics, *Ann. Rev. Earth Planet. Sci.* **14**, 493–571.

APPENDIX I

TABLE AI.1 *Microprobe analyses of some silicate minerals.*

	Olivine	Opx	Cpx	Amphibole	Garnet	Spinel	Plagioclase	Phlogopite (mica)
SiO_2	40.52	54.72	54.06	44.57	42.70	0.85	49.87	39.14
TiO_2		0.18	0.16	4.21	0.01	0.13	0.02	5.29
Al_2O_3		4.59	5.61	12.91	21.77	55.23	30.74	16.00
Cr_2O_3		0.18	1.13	0.48		9.77		0.30
FeO*	10.37	7.11	3.47	4.73	19.01	12.84	0.53	7.48
MnO		0.02	0.09		0.30			
MgO	48.71	32.53	14.61	16.69	19.01	20.70	0.11	18.12
CaO	0.10	0.68	19.34	9.82	19.01	0.13	15.30	0.16
Na_2O		0.11	2.13	3.52			3.26	0.59
K_2O				1.35			0.25	8.72
Total	99.76	100.12	100.60	98.28	99.30	99.65	100.08	95.80

Structural Formulae.

Oxyens =	4	6	6	23	24	4	8	22
Si	0.997	1.894	1.940	6.309	6.149		2.288	5.575
Ti		0.005	0.004	0.448	0.001		0.001	0.567
Al		0.188	0.238	2.155	3.697	1.691	1.663	2.687
Cr		0.005	0.032	0.054	0.000	0.201		0.034
Fe	0.213	0.206	0.104	0.560	1.295	***	0.020	0.891
Mn		0.001	0.003			0.007		
Mg	1.787	1.679	0.782	3.474	4.081	0.822	0.008	3.848
Ca	0.003	0.025	0.744	1.489	0.781		0.752	0.024
Na		0.007	0.148	0.503			0.290	0.163
K				0.244			0.007	1.584
Cation Sum	3.002	4.009	3.995	15.747	16.004	3.00†	5.029	15.373

Calculated FeO   0.220
Calculated Fe_2O_3 0.059

FeO* = all Fe counted as FeO.

*** FeO and Fe_2O_3 are calculated by a separate program, in which FeO is incrementally converted to Fe_2O_3 until the cation sum becomes 3.00 and all charges are balanced.

† This total is assumed.

Source: Sen (1988).

A P P E N D I X   I I

Simplified CIPW Norm Calculation

The guidelines given next are simplified from the original CIPW procedures in that not all the normative minerals are calculated. Instead, only those that are of importance in igneous rock classification are calculated. Minerals that are extremely rare in most rock norms (e.g., acmite, chromite, etc.) are ignored. An Excel® spreadsheet that allows the student to calculate norm will be given in my web site currently under construction.

Normative Mineral	Composition	Molecular Weight
Apatite (*ap*)	$3(3CaO.P_2O_5).CaF_2$	336.2
Ilmenite (*il*)	$FeO.TiO_2$	151.7
Magnetite (mt)	$FeO.Fe_2O_3$	231.5
Orthoclase (*or*)	$K_2O.Al_2O_3.6SiO_2$	556.7
Albite (*ab*)	$Na_2O.Al_2O_3.6SiO_2$	524.5
Anorthite (*an*)	$CaO.Al_2O_3.2SiO_2$	278.2
Diopside (*di*) components		
Wollastonite (*wo*)	$CaO.SiO_2$	116.2
Enstatite (*en*)	$MgO.SiO_2$	100.4
Ferrosilite (*fs*)	$FeO.SiO_2$	132
Hypersthene (*hy*) components		
Enstatite (*en*)	$MgO.SiO_2$	100.4
Ferrosilite (*fs*)	$FeO.SiO_2$	132
Olivine (*ol*) components		
Forsterite (fo)	$2MgO.SiO_2$	140
Fayalite (fa)	$2FeO.SiO_2$	204
Quartz (*q*)	SiO_2	60.1
Nepheline (*ne*)	$Na_2O.Al_2O_3.2SiO_2$	284.1

Calculation Procedure

Step 1 (Calculate molecular numbers). Divide wt% abundance of various oxides (Column A in Table AII.1) by their respective molecular weights (Column B). These numbers (i.e., molecular numbers) are listed in Column C. Note that after each of the steps you need to keep a running tab of how much of each of the oxide remains.

Step 2   (Calculate apatite). If P_2O_5 is reported in the rock analysis, then allocate all of P_2O_5 in Column C to apatite. Allocate $3.33 \times P_2O_5$ under the apatite column (Column F in Table AII.1). (*Running tab: remaining CaO = CaO in column C—CaO used to make apatite. There is no residual P_2O_5 after this step.*)

Step 3   (Calculate ilmenite). Allocate all of TiO_2 in Column C and an equal amount of FeO to ilmenite (Column D in Table AII.1). (*Running tab: TiO_2 available after this step—TiO_2 used = 0. FeO remaining = FeO (column C)—FeO used to make ilmenite.*)

Step 4   (Calculate magnetite). Allocate all of Fe_2O_3 and an equal amount of FeO to form magnetite (Column D). (*Running tab: available Fe_2O_3 = 0, FeO = FeO (remaining after Step 3)—FeO used to make magnetite.*)

Step 5   (Calculate provisional orthoclase: or'). Allocate all of K_2O, an equal amount of Al_2O_3, and six times as much SiO_2 to make provisional orthoclase. (*Are you keeping a running tab?*)

Step 6   (Calculate provisional albite: ab'). Allocate all of Na_2O, an equal amount of Al_2O_3, and six times as much SiO_2 to form provisional albite.

Step 7   (Calculate anorthite: an'). If there is any excess of Al_2O_3 after Step 6, then it is used to make anorthite. Allocate available Al_2O_3, an equal amount of CaO, and twice as much SiO_2 to make anorthite.

The steps that follow involve (Mg,Fe) solid solutions of pyroxenes and olivine.

Step 8   (Add FeO and MgO). In calculating proportion of various components in pyroxenes and olivine, one must first calculate the following ratios (see discussion in Chapter 8):

$$MR = MgO/[MgO + FeO], FR = FeO/[MgO + FeO]$$

It is assumed that

$$MgO/[MgO + FeO]^{mineral} = MgO/[MgO + FeO]^{rock}$$

Step 9   (Calculate diopside). Allocate CaO remaining after Step 7 and an equal amount of (MgO + FeO) and twice as much SiO_2 to make diopside.

Step 10   (Calculate hypersthene and olivine). If available (after Step 9) $SiO_2 >$ remaining (MgO + FeO), then allocate all of (MgO + FeO) and an equal amount of SiO_2 to form hypersthene. Any remaining SiO_2 may be calculated as quartz (q).

However, if (MgO + FeO) $> SiO_2$, then the rock is olivine-normative, and therefore a different set of steps needs to be followed to calculate olivine and hypersthene. As discussed in Chapter 8, assume x = no. of hypersthene moles and y = no. olivine moles. S = available SiO_2, and M = available MgO + FeO. Calculate x and y as follows:

$$x = 2S - M \quad \text{and} \quad y = S - x$$

If at this step all of SiO_2 is used up then break down diopside, hypersthene, and olivine mole fractions into their components as follows:

Diopside
   wo = CaO (step 8)
   en = $(MgO+FeO)^{\text{allocated to making diopside (step 8)}} \cdot MR$
   fs = $(MgO + FeO)^{\text{allocated to making diopside (step 8)}} - en$

Hypersthene

en = $x \cdot$ MR

fs = $x -$ en

Olivine

fo = $y \cdot$ MR

fa = $y -$ fo

Step 11 (Calculation of nepheline and recalculation of albite). From provisional albite calculated in Step 6, add the Na_2O, Al_2O_3, and SiO_2 back to their respective columns. See whether available $SiO_2 \geq 2Na_2O$; if so, divide the available Na_2O, Al_2O_3, and SiO_2 into nepheline and albite as follows:

Let p = nepheline moles and q = albite moles. N = available Na_2O. S = available SiO_2.

$p = (S - 2N)/4$

$q = N - p$

Allocate p amount of Na_2O, an equal amount of Al_2O_3, and twice as much SiO_2 to make nepheline. Allocate q amount of Na_2O, same amount of Al_2O_3, and six times as much SiO_2 to make albite.

[Note that similar steps may be added to calculate normative leucite (lc, $K_2O.Al_2O_3.4SiO_2$) and orthoclase from provisional orthoclase. However, such calculation is generally not necessary for most common rock types.]

Step 12 Convert mineral mole fractions so calculated into weight percentages by multiplying them by their respective molecular weights (see bottom half of Table AII.1).

TABLE AII.1 *CIPW norm calculation of an alkali basalt (SiO$_2$ undersaturated rock).*

	A	B	C	D	E	F	G	H	I	J	K	L	M	N	O	P	Q	R
											Diopside			Hypersthene		Olivine		
Oxide	wt%	Mol.wt	Mol no.	mt	il	ap	or	ab'(prov)	ab(real)	an	wo	en	fs	en	fs	fo	fa	ne
SiO$_2$	49.69	60.084	0.83				0.06	0.252	0.21	0.144	0.084	0.06	0.028	0.12	0.058	0.036	0.016	0.014
TiO$_2$	2.05	79.9	0.03		0.026													
Al$_2$O$_3$	12.68	101.96	0.12				0.01	0.042	0.035	0.072								
Fe$_2$O$_3$	1.38	159.69	0.01	0.009														
FeO	11.19	71.846	0.16	0.009	0.026								0.028		0.058		0.031	
MnO	0.18		0.00															
MgO	9.97	40.304	0.25									0.06		0.12		0.071		
CaO	9.09	56.08	0.16			0.006				0.072	0.084							
Na$_2$O	2.59	62	0.04					0.042	0.035									0.007
K$_2$O	0.90	94	0.01				0.01											
P$_2$O$_5$	0.27	142	0.00			0.002												
	99.99																	

		Mols.	NORM. Mol.%.	Mol Wt	CIPW NORMWT%
mt		0.009	**1.61**	232	**2.09**
il		0.026	**4.65**	152	**3.95**
ap		0.002	**0.36**	310	**0.62**
or	PLAG	0.01	**1.79**	556	**5.56**
ab	PLAG	0.035	**6.26**	524	**18.3**
an	PLAG	0.072	**12.88**	278	**20**
wo	DIOP	0.084	**15.03**	116	**9.74**
en	DIOP	0.056	**10.02**	100	**5.6**
fs	DIOP	0.028	**5.01**	132	**3.7**
en	HYP	0.12	**21.47**	100	**12**
fs	HYP	0.058	**10.38**	132	**7.66**
fo	OL	0.036	**6.44**	140	**5.04**
fa	OL	0.016	**2.86**	204	**3.26**
ne	OL	0.007	**1.25**	284	**1.99**
			100.00		**99.6**

Notes. Calculations pertaining to the above norm problem. After calculating provisional ab, an, or, mt, il, ap..., all of CaO must go into making diopside. MgO, FeO in diopside must be calculated using the assumption:

$$[MgO/(MgO + FeO)]^{Di} = [MgO/(MgO + FeO)]^{Remaining} \quad [Eq.II.1]$$

Available:

SiO$_2$ .371

FeO .121 $[FeO + MgO]^{Remaining} = 0.247 + 0.121 = 0.368.$

MgO .247

CaO .084 $[MgO/(FeO + MgO)]^{Remaining} = 0.247/0.368 = 0.672.$ [Eq.II.2]

Now, according to diopside formula (CaO.(MgO + FeO).2SiO$_2$), MgO+FeO should be equal to CaO

That is, $[MgO + FeO]^{Di} = 0.084$

Therefore, plugging the above into Equations 1 and 2:

$$[MgO]^{Di}/0.84 = 0.672; MgO \text{ of Diopside} = 0.056$$
$$FeO \text{ of Diopside} = 0.084 - 0.056 = 0.028$$

		wo	en	fs	Remaining
SiO$_2$	.371	.084	.056	.028	0.231
FeO	.121		.028		0.093
MgO	.247		.056		0.191
CaO	.084	.084			0

FeO + MgO = 0.284, which is more than available SiO$_2$ needed to make hypersthene. Therefore, these have to be recast into a combination of olivine and hypersthene molecules.

Let x = hy moles, y = ol moles, S = SiO$_2$ M = MgO + FeO

hy = $x = 2S - M = 2(0.231) - 0.284 = 0.178$ ol = y = M − x
= 0.284 − 0.173 = 0.106

Calculation of en and fs in hypersthene:
hy = 0.178. i.e., $[MgO + FeO]^{hy} = 0.178$, and $[SiO_2]^{hy} = 0.178$
$[MgO/(MgO + FeO)]^{hy} = [MgO/(MgO + FeO)]^{balance} = 0.672.$

$MgO^{hy} = 0.672*0.178 = 0.12$
$FeO^{hy} = 0.178 - 0.12 = 0.058.$

	en	fs
SiO$_2$	0.12	0.058
FeO		0.058
MgO	0.12	

Calculation of fo and fa in Olivine:
Using a similar method, except SiO_2 is 1/2 the amount of FeO,MgO

$$[MgO]^{ol} = 0.071, [FeO]^{ol} = 0.031$$

At this point, a check of available SiO_2 shows that we have used too much SiO_2 (0.854) while we have a total SiO_2 of only 0.827 (i.e., we have a silica-deficient situation). Therefore, we need to redistribute provisional albite molecules into albite and nepheline to ensure that there is no deficiency of SiO_2.

x = ab moles, y = ne moles, S = available SiO_2, N = Na_2O, N = 0.042 S = 0.225, x = (S − 2N)/4 = 0.035, y = N − x = 0.007, ne moles = 0.007, ab moles = 0.035.

TABLE AII.2   *Norm calculation of a rhyolite (SiO_2 oversaturated rock).*

	A	B	C	D	E	F	G	H	I	J	K
	wt%	mol. wt.	mole prop.	il	mt	or	ab	an	hy	co	qz
SiO_2	71.30	60.08	1.188			.258	.354	.066	.029		.481
TiO_2	0.31	79.9	0.004	.004							
Al_2O_3	14.32	102	0.14			.043	.059	.033		.005	
Fe_2O_3	1.21	160	0.008		.008						
FeO	1.64	71.85	0.023	.004	.008				.011		
MgO	0.71	40.3	0.018						.018		
CaO	1.84	56.08	0.033					.033			
Na_2O	3.68	62	0.059				.059				
K_2O	4.07	94.2	0.043			.043					

Norm

		Moles		Mol. wt.		wt%
	il	.004	×	152	=	0.61
	mt	.008	×	232	=	1.86
	or	.043	×	556	=	23.91
	ab	.059	×	524	=	30.92
	an	.033	×	278	=	9.17
hy	en	.018	×	100	=	1.80
	fs	.011	×	132	=	1.45
	co	.005	×	102	=	0.51
	q	.481	×	60	=	28.86

Note. In the above calculation there is an excess of Al_2O_3, which is calculated as normative corundum (c, Al_2O_3).

APPENDIX III

A Brief Introduction to Isotope Geochemistry

The isotopes of an element must have the same atomic number (Z) but different masses (M). One can distinguish between radiogenic and nonradiogenic isotopes: The former is produced as a daughter isotope from a parent radioactive isotope via breakdown reactions (commonly referred to as radioactive *decay*); the latter does not have an origin through radioactivity. Examples of radiogenic isotopes are: ^{87}Sr (i.e., Sr-isotope with an atomic mass of 87), ^{143}Nd, ^{207}Pb, and ^{206}Pb. Nonradiogenic isotopes of these same elements are: ^{86}Sr, ^{144}Nd, and ^{204}Pb. Isotope geochemistry has two fundamental applications, one of which is in dating rocks, a subject known as *geochronology*, and the other in deciphering petrogenetic processes. It is beyond the scope of this book to cover isotope geochemistry in any detail, and only some general concepts concerning the isotope systematics of the rubidium (Rb)—strontium (Sr), samarium (Sm)—neodymium (Nd), and uranium (U)—lead (Pb) isotopic systems are presented. The interested reader may consult a number of textbooks available on the topic (e.g., DePaolo, 1988; Faure, 1986).

In any system that involves the breakdown of radioactive parent atoms into daughter atoms (which may or may not be radioactive), the number of atoms of the parent isotope (N_o) at time (t) = 0 and the number of atoms of the daughter isotope (N) produced through time t are related by the following relation:

$$N = N_o e^{-\lambda t} \qquad \text{[Eq. AIII.1]}$$

λ is a constant for any particular isotopic system and is known as the *decay constant*. The decay constant is known experimentally for each individual isotopic system.

Consider a mineral that contains N atoms of the parent isotope and D atoms of the daughter isotope at the present time (t). Then $N_o = N + D$. Substituting this relation into Equation (AIII.1), one obtains

$$N = (N + D)e^{-\lambda t}. \qquad \text{[Eq. AIII.2]}$$

Rearranging, we obtain:

$$N/(N + D) = e^{-\lambda t}$$

or

$$D = N(e^{\lambda t} - 1) \qquad \text{[Eq. AIII.3]}$$

or

$$e^{\lambda t} = (N + D)/N$$

$$\lambda t = \ln\left(1 + (D/N)\right),$$

i.e.,

$$t = (1/\lambda).\ln\left(1 + (D/N)\right) \qquad \text{[Eq. AIII.4]}$$

Equation AIII.4 is the basis for geochronology, which is the science of determining the ages of minerals and rocks.

Rb–Sr system

In this system, atoms of the radioactive parent isotope ^{87}Rb break down into the atoms of the daughter isotope ^{87}Sr and release a β-particle in the process. Based on Equation AIII.3, one may write:

$$^{87}\text{Sr} = {}^{87}\text{Rb}(e^{\lambda t} - 1).$$

The rock may have originally contained some initial number of atoms of ^{87}Sr, which we may call $^{87}\text{Sr}_{\text{initial}}$. Over time t, an additional amount of ^{87}Sr atoms have been produced via decay of ^{87}Rb. Thus,

$$^{87}\text{Sr}_{\text{measured at time } t} = {}^{87}\text{Sr}_{\text{initial}} + {}^{87}\text{Rb}(e^{\lambda t} - 1). \qquad [\text{Eq. AIII.5}]$$

Scientists are able to determine ratios of isotopes with much greater precision than absolute values of individual isotopes. Therefore, Equation AIII.5 is converted to the following equation by converting the previous into ratios by dividing each isotope by a nonradiogenic isotope of Sr, which is ^{86}Sr:

$$\left[\frac{^{87}\text{Sr}}{^{86}\text{Sr}}\right]_{\text{measured at time } t} = \left[\frac{^{87}\text{Sr}}{^{86}\text{Sr}}\right]_{\text{initial}} + \left[\frac{^{87}\text{Rb}}{^{86}\text{Sr}}\right](e^{\lambda t-1}). \qquad [\text{Eq. AIII.6}]$$

An important factor in the appropriateness of a given isotopic system to date rocks is its *half-life*, which is the time taken for half of the atoms of the parent isotope to decay into atoms of daughter isotope. The Rb–Sr system has a half-life of 48.8×10^9 years. Thus, its use in geochronology is restricted to dating very old rocks—from the age of the earth to 10^7 years (Brownlow, 1996). There are other isotopic systems that have a much shorter half-life and may only be used to date younger rocks. For example, the ^{14}C–^{14}N system has a half-life of 5,730 years, and therefore this system can be used to date rocks that are no older than 70,000 years (Brownlow, 1996).

We can use the Rb–Sr system to show how it can be used to date rocks with the use of an isochron diagram (Figure AIII.1). Consider a rock b with its component minerals a, c, and d. a, b, c, and d all originally had very different $^{87}\text{Rb}/^{86}\text{Sr}$ ratios ($a = 0.15$, $b = 0.25$, $c = 0.3$, $d = 0.4$), although they all had the identical $^{87}\text{Sr}/^{86}\text{Sr}_{\text{initial ratio}}$—0.70300. With time, more and more atoms of the daughter isotope ^{87}Sr will be produced from the parent isotope ^{87}Rb. Thus, $^{87}\text{Sr}/^{86}\text{Sr}$ ratio will increase and $^{87}\text{Rb}/^{86}\text{Sr}$ ratio will decrease in a, b, c, and d. a, with its lowest $^{87}\text{Rb}/^{86}\text{Sr}$ ratio, will see the least increase in $^{87}\text{Sr}/^{86}\text{Sr}$ ratio, whereas d will experience the maximum change. As shown in Figure AIII.1, at any given time (three examples shown—0.1, 1, and 2 billion years), however, a straight line, called the *isochron*, will pass through the four points corresponding to the $^{87}\text{Sr}/^{86}\text{Sr}$, $^{87}\text{Rb}/^{86}\text{Sr}$ values of a, b, c, and d at that time. No matter what age, all isochrons for a given rock (provided that the system was closed at all times—i.e., it was not subsequently affected by metamorphism and other processes) must go through the $^{87}\text{Sr}/^{86}\text{Sr}_{\text{initial ratio}}$. Note that, although the example focuses on a single rock and its constituent minerals, one can use the same technique for rocks that are petrogenetically related (e.g., members of an igneous differentiation series, each of which started out with very different $^{87}\text{Rb}/^{86}\text{Sr}$ ratios). For example, igneous rocks from a layered intrusion may be dated this way.

Another important application of isotope geochemistry is in understanding how the different isotopic reservoirs within the earth, such as continental crust, oceanic crust, parts of the suboceanic mantle, and so on, have evolved over time. Analysis of oceanic basalts suggests that the oceanic mantle has a low present-day $^{87}\text{Rb}/^{86}\text{Sr}$ ratio

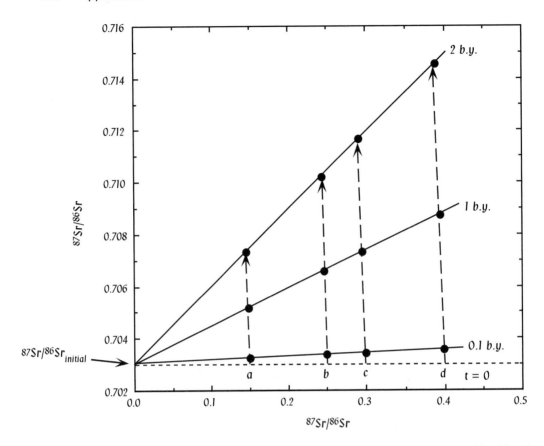

FIGURE AIII.1 *Rb–Sr isochron diagram. a,c, and d are the three constituent minerals of rock b.* Although the minerals and the bulk rock all have the same $^{87}Sr/^{86}Sr$ initial ratio, their $^{87}Rb/^{86}Sr$ ratios are different at the time of the formation of this rock (i.e., time $t = 0$). Their $^{87}Sr/^{86}Sr$ and $^{87}Rb/^{86}Sr$ change with time, as illustrated by the three isochrons at 0.1, 1, and 2 billion years since the formation of this rock.

of about 0.027 relative to that ($\sim$0.24) of the mean continental crust (i.e., averaging continental crust of all ages; Cox et al., 1979). This fundamental difference in the Rb/Sr ratio between the continental and oceanic crusts existed throughout the history of the earth and is the reason for a 2-b.y.-old continental crust to have a present-day $^{87}Sr/^{86}Sr$ ratio as high as 0.724, whereas most ocean floor basalts have a $^{87}Sr/^{86}Sr$ ratio of 0.7030 to 0.7037. It follows that granitic magmas produced by melting of old continental crust should have very high $^{87}Sr/^{86}Sr$ values relative to the oceanic basalts.

Figure AIII.2 illustrates a hypothetical example of the evolution of three isotopic reservoirs; each of the evolutionary paths is calculated assuming an initial Rb/Sr ratio and a $[^{87}Sr/^{86}Sr]_{initial\ ratio}$. The first one is the path that would have been taken by the undifferentiated bulk earth (i.e., if the earth did not differentiate into its component layers, particularly the continental crust). The starting point of this path is the location of the Basaltic Achondrite Best Initial (BABI) ratio of $^{87}Sr/^{86}Sr$ 0.69897 (Papanastassiou and Wasserburg, 1969), which is generally accepted to be the ratio with which the earth formed some 4.5 billion years ago. For a constant Rb/Sr ratio of 0.027, the present-day $^{87}Sr/^{86}Sr$ ratio would be 0.704 (Wilson, 1989). If a continental crust (Rb/Sr = 0.18) had segregated some 2.5 billion years ago from the bulk earth reservoir, then its $^{87}Sr/^{86}Sr$ would have rapidly evolved to a present-day ratio of 0.718. The depleted (i.e., depleted of continental crustal components) oceanic upper man-

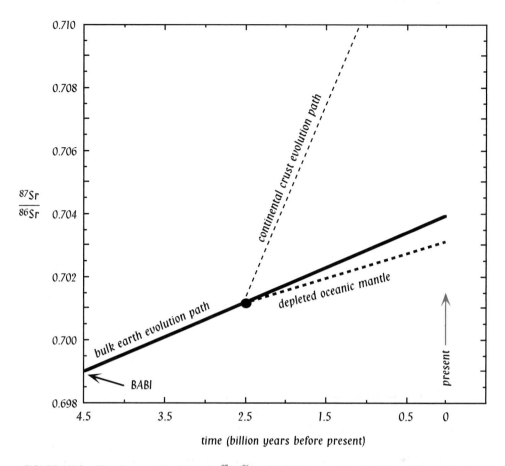

FIGURE AIII.2   *This diagram shows how the $^{87}Sr/^{86}Sr$ of BABI (representing the bulk earth) would evolve with time since the earth's formation 4.5 billion years ago.* Also shown are two lines, one of which traces the evolution of $^{87}Sr/^{86}Sr$ of the continental crust (with a higher $^{87}Rb/^{86}Sr$ than the bulk earth) separated from the bulk earth some 2.5 billion years ago. The mantle reservoir that lost such continental crust would evolve along a very different path, which is perhaps recorded by the depleted oceanic mantle rocks.

tle, having a lower Rb/Sr (0.024), would then have a lower present-day $^{87}Sr/^{86}Sr$ ratio than the calculated present-day bulk earth ratio. To easily explore the differences between various isotopic reservoirs within the earth, reference is often made to ε_{Sr}, which is calculated with the help of the following equation:

$$\varepsilon_{Sr} = \left[\frac{(^{87}Sr/^{86}Sr)^{t}_{sample}}{(^{87}Sr/^{86}Sr)^{t}_{CHUR}} - 1 \right] \times 10^{4}, \qquad \text{[Eq. AIII.7]}$$

where $(^{87}Sr/^{86}Sr)^{t}_{sample}$ refers to the $^{87}Sr/^{86}Sr$ ratio of the sample at time t, and the denominator reflects the same ratio at time t for the bulk earth or Chondritic Uniform Reservoir (CHUR; DePaolo and Wasserburg, 1979). Bulk earth always has a ε_{Sr} value of zero. Positive ε_{Sr} values are commonly called *enriched values:* Continental crust has very high positive values. Negative values, a characteristic of oceanic basalts, are referred to as *depleted* ε_{Sr} values (Figure AIII.3; modified after Wilson, 1989). A similar expression for ε_{Nd} was presented by DePaolo and Wasserburg (1979) for the Nd-isotopic system (discussed later). Combining ε_{Sr} and ε_{Nd}, as seen in Figure AIII.3, has proved to be a powerful tool in petrological synthesis of the earth and its processes (also discussed in Chapter 11). Because closed-system magma-differentiation processes do not alter the isotopic composition of magmas, isotopic compositions of intrusions

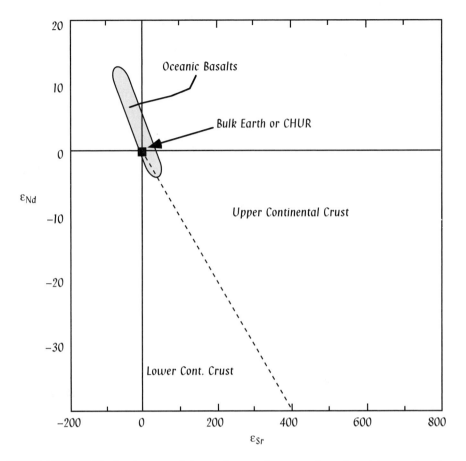

FIGURE AIII.3 *Fields of various oceanic basalts, lower and upper continental crustal reservoir rocks are shown in an ε_{Nd}–ε_{Sr} diagram.*

and lavas can be used to trace contributions made by various sources (i.e., lithosphere, asthenosphere, plume, etc.) to the magma. In case of magma contamination (i.e., open system), the nature of the contaminants and the extent of contamination may also be deciphered with the help of these isotopic ratios. Isotopic compositions of rocks can be used to resolve problems of as large a magnitude as global tectonics and mantle convection to much smaller scale of magma contamination. Perhaps a unique advantage of isotopes is that they record ancient events in the earth's history that otherwise would not be known.

Sm–Nd System

Samarium (Sm) and neodymium (Nd) are light, rare earth elements and generally behave as incompatible elements (i.e., they prefer magma over crystalline phases during crystal–liquid separation, be it partial melting or crystallization). Nd is slightly more incompatible than Sm. ^{143}Nd is a daughter isotope produced by α-decay of the radioactive isotope ^{147}Sm. The Sm–Nd system has a half-life of 106×10^9 years, and the decay constant is 0.654×10^{-11} yr^{-1}. Like the Rb–Sr system, the Sm–Nd radioactive reaction may be expressed as:

$$\left[\frac{^{143}\text{Nd}}{^{144}\text{Nd}}\right] = \left[\frac{^{143}\text{Nd}}{^{144}\text{Nd}}\right]_{\text{initial}} + \left[\frac{^{147}\text{Sm}}{^{144}\text{Nd}}\right](e^{\lambda t} - 1),$$

where ^{144}Nd is a nonradiogenic isotope of Nd.

The measured ^{143}Nd/^{144}Nd value for Bulk Earth or CHUR is corrected to a ^{146}Nd/^{142}Nd (0.636151) or ^{146}Nd/^{144}Nd (0.7219). This value as corrected to the latter is 0.512638 (see Wilson, 1989).

As stated earlier, partial melting results in a greater Sm/Nd ratio of the residuum because Nd is more incompatible than Sm (i.e., concentrates in magma more than Sm). It follows that as time progresses rocks crystallizing from such partial melts will evolve to lower ^{143}Nd/^{144}Nd than the residue because the residue will have a greater amount of ^{147}Sm/^{144}Nd ratio acquired during the partial melting event (Figure AIII.4).

Partial melting leading to the formation of a depleted mantle is a single-stage process. The portion of the mantle so depleted of magmatic material may subsequently undergo enrichment events (as well as multiple depletion events) in which small amounts of melt percolating through it might add back the components that were lost in the earlier melting event. There have been many well-documented cases of isotopic enrichment of the depleted mantle (e.g., Nixon, 1987). In fact, all portions

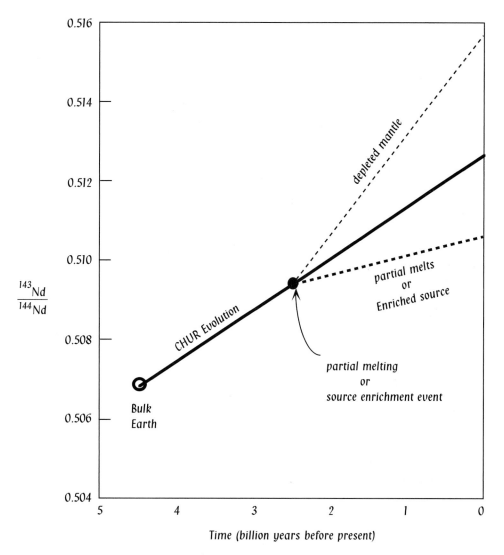

FIGURE AIII.4 *Nd-isotopic evolution of the Bulk Earth, depleted mantle, and enriched sources (or partial melts).* (See text for further discussion).

of the mantle probably go through many stages of enrichment and depletion. Enriched mantle will have a higher Nd/Sm ratio (since a partial melt will have a higher Nd/Sm ratio than its source) than the depleted mantle, and therefore the enriched mantle will evolve toward an increasingly lower $^{143}Nd/^{144}Nd$ ratio with time, much the way rocks formed from partial melts would be expected to behave.

As with the Rb/Sr system, DePaolo and Wasserburg (1976) introduced a parameter, ε_{Nd}, that compares rock materials with the evolutionary path of Bulk Earth or CHUR:

$$\varepsilon_{Nd} = \left[\frac{(^{143}Nd/^{144}Nd)_{initial}}{CHUR_{initial}} - 1 \right] \times 10^4,$$

where $^{143}Nd/^{144}Nd_{initial}$ is the initial ratio for the rock/mineral and $CHUR_{initial}$ is the corresponding initial ratio of CHUR at the time of the formation of the rock/mineral.

By definition, Bulk Earth or CHUR has an $\varepsilon_{Nd} = 0$. Rocks/magmas with positive ε_{Nd} values originate from a depleted source (i.e., a source that has previously undergone melt extraction), and those with negative ε_{Nd} are said to be enriched because they are derived from a magma-added source. Whereas the bulk of oceanic rocks have positive ε_{Nd}, old continental crust is characterized by strongly negative ε_{Nd} (Figure AIII.3).

The Sm–Nd isotopic system has a particular advantage over other isotopic systems in deciphering the evolution of continents through studies of detrital sediments. It turns out that fine-grained clastic sediments record the Sm/Nd ratios of their source rocks, and these ratios are virtually unmodified by sedimentary processes, which is not true of Rb/Sr or U/Pb systems (O'Nions, 1992). Continental crust with some reasonable Sm/Nd ratio extracted from the depleted mantle some 3 billion years ago will evolve to strongly negative ε_{Nd} values. Sediments derived from them will have the same characteristic ε_{Nd} as the source rocks. Using an Sm/Nd and $^{143}Nd/^{144}Nd$ ratios, a model age of such sediments may be calculated, which would essentially indicate the separation age of the continent from the depleted mantle. This model age is referred to as the *crustal residence age* because it represents the time spent by such sediments in the continental environment. Remarkably, clastic particles from the entire globe seem to give a rather tight range of crustal residence ages with a mean of 1.7 billion years, which may be taken as the mean age of the continental crust (i.e., a weighted mean of crustal materials of all ages; O'Nions, 1992).

U–Th–Pb Systems

Unlike the Rb/Sr and Sm/Nd isotopic systems, the U–Th–Pb systems go through a chain of radioactive decay reactions in which several intermediate radioactive isotopes are produced, which in turn decay into other ones and so on. This topic is rather complex and cannot be covered in any detail. Uranium (U) has three naturally occurring radioactive isotopes: ^{238}U, ^{235}U, and ^{234}U. Th occurs mainly as a radioactive isotope— ^{232}Th, and as a few other short-lived isotopes that are produced as U-series decay products. Lead (Pb) has several naturally occurring isotopes: ^{204}Pb, ^{206}Pb, ^{207}Pb, and ^{208}Pb, of which ^{204}Pb is nonradiogenic. The endproducts of three important chain reactions involving the formation of lead isotopes are as follows:

$$^{238}U \rightarrow {}^{234}U \rightarrow {}^{206}Pb$$
$$^{235}U \rightarrow {}^{207}Pb$$
$$^{232}Th \rightarrow {}^{208}Pb$$

Correspondingly, the following equations may be written relating the appropriate daughter isotope to its parent:

$$\frac{^{206}\text{Pb}}{^{204}\text{Pb}} = \left[\frac{^{206}\text{Pb}}{^{204}\text{Pb}}\right]_{\text{initial}} + \frac{^{238}\text{U}}{^{204}\text{Pb}}\left(e^{\lambda t} - 1\right)$$

$$\frac{^{206}\text{Pb}}{^{204}\text{Pb}} = \left[\frac{^{207}\text{Pb}}{^{204}\text{Pb}}\right]_{\text{initial}} + \frac{^{235}\text{U}}{^{204}\text{Pb}}\left(e^{\lambda t} - 1\right)$$

$$\frac{^{208}\text{Pb}}{^{204}\text{Pb}} = \left[\frac{^{208}\text{Pb}}{^{204}\text{Pb}}\right]_{\text{initial}} + \frac{^{232}\text{Th}}{^{204}\text{Pb}}\left(e^{\lambda t} - 1\right)$$

Common lead (i.e., lead found in the earth) is a mixture of (a) lead that is present from the beginning of earth history (primeval lead), and (b) radiogenic lead that is produced via radioactive decay of radioactive isotopes of U and Th (cf. Hall, 1987). The measured isotopic compositions of primeval lead, as estimated from a mineral called troilite (Fe_2S) in Cañon Diablo meteorite, are as follows (Chen and Wasserburg, 1983):

$^{206}\text{Pb}/^{204}\text{Pb} = 9.3066$

$^{207}\text{Pb}/^{204}\text{Pb} = 10.293$

$^{208}\text{Pb}/^{204}\text{Pb} = 29.475$

Age determination using U/Pb isotopes is not as straightforward as in the Rb/Sr and Sm/Nd systems because uranium and lead are easily removed during weathering and metamorphic processes, and smooth, straight-line isochrons are rarely obtained. It is somewhat less of a problem in $^{208}\text{Pb}/^{204}\text{Pb}$ dating because Th is less mobile than U. Normally the same rock is dated with the three isotopic systems listed before, and generally one obtains discordant ages determined by the three methods due to variable loss of U and Pb. Typically, a plot of $^{206}\text{Pb}/^{238}\text{U}$ versus $^{207}\text{Pb}/^{235}\text{U}$, known as a *concordia diagram*, is made (Figure AIII.5). ^{238}U and ^{235}U have half-lives of 4.47 and 0.7 billion years, respectively (i.e., the former decays at a much slower rate than the latter; and this differential decay rates result in the curvature of the concordia diagram). Rock and mineral samples from a petrogenetically related suite generally plot on a straight line, such as the one shown extending from 2.5 Ga (Giga years = billion years) to 1 Ga. One interpretation may be that the suite was formed 2.5 billion years ago but was affected by metamorphic processes some 1 billion years ago. However, it is also possible that Pb was lost continuously during the metamorphic evolution of these rocks, in which case the lower age has no particular significance. The strength of age interpretation is dependent on how closely the rock samples plot toward the concordia curve. In passing, it is worthwhile to note that core to rims of individual-zoned zircon crystals in clastic sediments and granitoid rocks can be often be dated with a U–Pb dating method to reveal their evolutionary history.

U–Pb–Th system can also be used to reveal petrogenic processes operating within the crust and mantle. A plot of $^{207}\text{Pb}/^{204}\text{Pb}$ versus $^{206}\text{Pb}/^{204}\text{Pb}$ has been found to be particularly useful in this context (Figure AIII.6). The ratio $^{238}\text{U}/^{204}\text{Pb}$ is known as μ. In Figure AIII.6, the two curves show the closed-system evolution of Pb-isotopic ratios for two cases: one with a high μ and the other with a low μ. Isochrons on this diagram are straight lines radiating from the primeval lead composition. Stony meteorites containing variable amounts of U and Th plot along such an isochron of 4.52 to 4.57 billion years (Hall, 1987).

The isochron for present day (time $t = 0$) is called the *geochron*. Note that the two closed-system cases shown in Figure AIII.6 are examples of what is often called

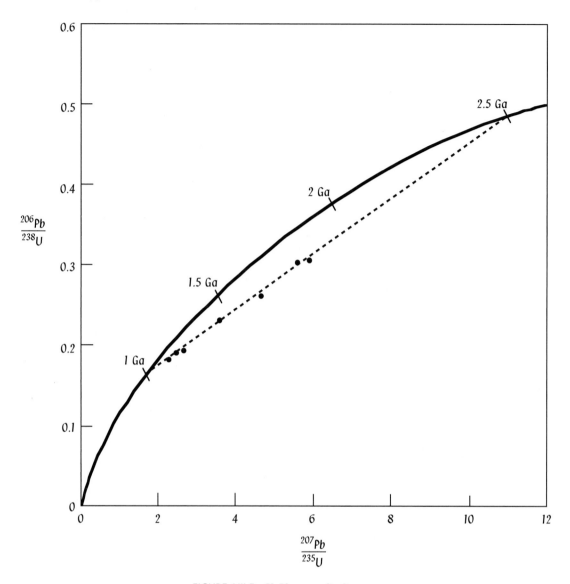

FIGURE AIII.5   *U–Pb concordia diagram.*

single-stage evolution. However, lead may not often evolve in a closed system, and the U/Pb ratio may get reset by metamorphic or other processes, which may result in multistage lead evolution (Figure AIII.6 shows an example of a two-stage lead evolution). Interpretation of stages of lead evolution is often not straightforward. Rocks can fall both to the left and right of the geochron: Those plotting to the left are said to give positive model ages, whereas those to the right give negative model ages. A rock can have a negative model age when it separates from a single-stage evolution curve, evolves in a very high μ environment, ends up with excess ^{206}Pb, and plots to the right side of the geochron. Most oceanic Pb falls along linear arrays on the right

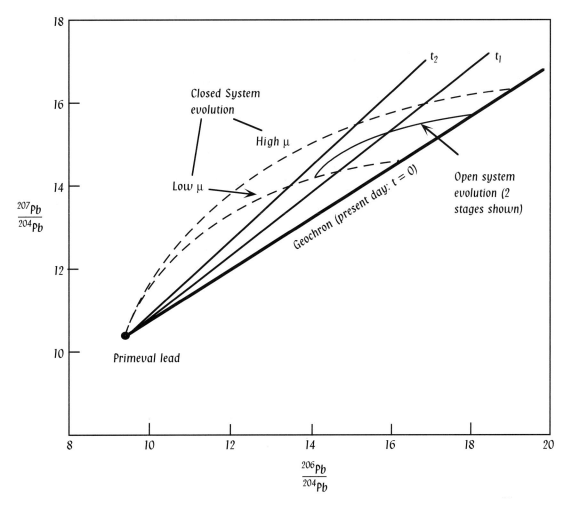

FIGURE AIII.6   *Evolution of lead isotopes.*

side of the geochron, and therefore gives negative model ages. The oceanic lead characteristics have been interpreted in a number of ways, including one of mixing between magmas tapping distinct sources (Tatsumoto, 1978).

Isotopes of Pb, Sr, and Nd have been particularly helpful in identifying the end-member sources or reservoirs that may be tapped by magmas (e.g., Zindler and Hart, 1986; Wilson, 1989). The principal end members are: depleted mantle (DM), enriched mantle I (EMI), enriched mantle II (EMII), and St. Helena component (SHC). Their distinct isotopic characteristics are listed in Wilson (1989). MORBs are principally derived from a DM source, whereas Hawaiian magmas appear to tap a mixture of DM and EMI (enriched plume) sources. Aside from Pb, Sr, and Nd isotopes, new insight into the earth processes is being sought through studies of isotopic ratios of He, Os, and Hf. A discussion of these new advancements is well beyond the scope of this book. The reader is encouraged to read about them in isotope geochemistry textbooks for such purpose.

BIBLIOGRAPHY

Brownlow, A.H. (1996, 2nd ed.) *Geochemistry*. Prentice Hall (Upper Saddle River, New Jersey).

Chen, J. and Wasserburg, G. J. (1983) The least radiogenic Pb in iron meteorites. In *14th Lunar and Planetary Science Conference, Abstracts*, **Pt. 1**, 103–104.

Cox, K.G., Bell, J.D., and Pankhurst, R.J. (1979) *The Interpretation of Igneous Rocks*. Allen and Unwin (Boston).

DePaolo, D.J. (1988) *Neodymium Isotope Geochemistry*. Springer-Verlag (New York).

DePaolo, D.J. and Wasserburg, G.J. (1976) Nd isotopic variations and petrogenetic models, *Geophys. Res. Lett.* **3**, 249–252.

DePaolo, D.J. and Wasserburg, G.J. (1979) Petrogenetic mixing models and Nd-Sr isotopic patterns, *Geochim. Cosmochim. Acta* **43**, 615–627.

Faure, G. (1986 2nd ed.) *Principles of Isotope Geology*. Wiley (New York).

Hall, A. (1987) *Igneous Petrology*. Wiley (New York).

Nixon, P.H. (1987, ed.) *Mantle xenoltihs*. Wiley (New York).

O'Nions, R.K. (1992) *The Continents*. In Understanding the Earth: a new synthesis (Brown, G.C., Hawkesworth, C.J. and Wilson, E.C.L., eds.), 145–163.

Papanastassiou, G. and Wasserburg, G.J. (1969) Initial strontium isotopic abundances and the resolution of small time differences in the formation of planetary objects, *Earth Planet. Sci. Lett.* **5**, 361–376.

Tatsumoto, M. (1978) Isotopic composition of lead in oceanic basalt and its implication in mantle evolution, *Earth Planet. Sci. Lett.* **38**, 63–87.

Wilson, M. (1989) *Igneous Petrogenesis: a Global Tectonic Approach*. Unwin Hyman (Boston).

Zindler, A. and Hart, S.R. (1986) Chemical geodynamics, *Ann. Rev. Earth Planet. Sci.* **14**, 493–571.

Index